KEITH COLLIER
British Columbia Institute of Technology

Fundamentals of CONSTRUCTION ESTIMATING and COST ACCOUNTING

PRENTICE-HALL, INC., Englewood Cliffs, N.J.

Library of Congress Cataloging in Publication Data

Collier, Keith.
Fundamentals of construction estimating and cost accounting.

Bibliography: p.
1. Building–Estimates. 2. Construction industry–Accounting. I. Title.
TH435.C72 1974 692′.5 73–10378
ISBN 0–13–335604–3

This book is dedicated to

Pauline Collier

10 9 8 7 6 5 4 3 2

Printed in the United States of America

Prentice-Hall International, Inc., *London*
Prentice-Hall of Australia, Pty. Ltd., *Sydney*
Prentice-Hall of Canada, Ltd., *Toronto*
Prentice-Hall of India Private Ltd., *New Delhi*
Prentice-Hall of Japan, Inc., *Tokyo*

Contents

Construction Drawings

* Drawings marked with an asterisk are included in the standard full-page size, and also in double full-page size for clarity.

It is suggested that when the drawings have been removed from the book, that they be kept flat in a large envelope pasted to the inside of the back cover.

Preface

I have assumed that whoever studies construction estimating has already learned something about building construction because that is the proper order; and I have not written about the science of building construction except to illustrate examples of estimating practice. Besides, there is more than enough to say about estimating, and for this reason it has been necessary to make a distinction between fundamentals and variables. Consequently, I have left it to the periodicals to more ably provide current cost data for estimators in different areas; and I have left it to the readers to obtain their own reference material on contracts and construction materials. Much of this reference material can be obtained free, or for a nominal charge, from original sources, and its reproduction here would not be worthwhile. Some useful material is listed in the Bibliography, and students should start to collect a reference library as soon as they can.

Construction estimating and building economics have a universal application, and I have tried to show the fundamental things that underlie them and how they apply to the building construction process, particularly in conjunction with that most neglected aspect of construction management—cost accounting. Good construction management is not possible without estimating and cost accounting; and although building economics has yet to be formalized and made into a discipline, cost accounting will be one of the means by which this is accomplished.

In places, I have used abbreviations of construction terms because this is commonly done in construction estimates; and those abbreviations used are listed before the first chapter. The text also includes certain terms in *italics,* and they are listed in the Glossary, at the end. The terms in *italics* are generally common words; but they are also *key words* in the text with special meanings that are explained in the Glossary. Some are terms specially coined for this text to convey certain concepts, and this usage is elaborated in a preamble to the Glossary.

The immediate acknowledgements I should make follow; others are not diminished because they are not included, nor are these by their brevity; G. Berkenpas, assistant master, BCIT: for all the figures and drawings; G. M. Hardie, FRICS, MCIQS, chartered quantity surveyor and associate master, BCIT: for reading the manuscript, for his assistance in checking some of the examples, and for his useful suggestions; R. Ochotorena, for typing the manuscript; G. E. Parsons, FRICS, MCIQS, chartered quantity surveyor, for reading the manuscript, for his suggestions, and for his aid and counsel over many years; and F. Wools, FRICS, chartered quantity surveyor, for his good instruction.

Special acknowledgement is made to the following persons for their permission to reproduce the building drawings for this text: The Residence, H. E. Kuckein, architect, Vancouver, B.C.; The Warehouse, Dominion Construction Co., Ltd., Vancouver, B.C.; and The Apartment Block, W. Ralph Brownlee, architect, Vancouver, B.C.

Finally, I should acknowledge my debt to the Royal Institution of Chartered Surveyors, and to the Canadian Institute of Quantity Surveyors, for permission to quote from the Methods of Measurement and other documents that they publish, and for the general debt that all members owe to their professional institutes.

K. F. C.

British Columbia
Institute of Technology

List of Abbreviations

Abbreviations are widely used in describing *work*[1] in construction estimates and on drawings. With some abbreviations there is a customary format, whereas others are written according to personal inclination. As a step toward standardization, I have used those formats that I believe to be the most common and the most easily understood, and I have been guided by general quantity surveying practice, and by other writers, particularly by Wass,[2] who rightly advocates the use of open punctuation with abbreviations. However, a few abbreviations make other words, as in the case of inch—in. To avoid confusion in such cases open punctuation is not used.

Notes on Abbreviations

1. Add the letter "g" for "ing" (i.e., ct—coat; ctg—coating).
2. In the absence of a standard abbreviation, the best method of abbreviation is to eliminate all vowels other than those required to avoid ambiguity.
3. The well-known, standard abbreviations that are in general use, including titles of societies, associations, and the like, and those abbreviations found in trade manuals and dictionaries, are generally not included here.

[1] Terms in *italics* are listed in the Glossary.
[2] Alonzo Wass, *Manual of Structural Details for Building Construction* (Englewood Cliffs, N.J.: Prentice-Hall, Inc., 1968). List of Abbreviations, pp. XIII–XIV.

A	area
a.b	as before
AG	asphalt and gravel
agg	aggregate
avg	average
B	breadth
b.f	both faces
BF	board feet
bgdg	bridging
bk	brick
bkwk	brickwork
bldg	building
blk	block
blkwk	blockwork
bm	beam
brd	board
brr	bearer
b/s	both sides
btm	bottom
b/up	built up
b/w	both ways
CA	cost accounting
CF (cf)	cubic feet
c/flshg	counter-flashing
chy	chimney
CI	cast iron
CIQS	Canadian Institute of Quantity Surveyors
circ	circle, circular
clg	ceiling
cmt	cement
col	column
conc	concrete
constr	construct(ion)
cont	continuous
cors	course
ct	coat
C & W	cutting & waste
CY (cy)	cubic yards
D	depth, deep
dbl	double
Ddt	deduct
desc	describe, description
dia	diameter
dim	dimension
do	ditto
dpf(g)	damproof(ing)
dtl	detail
dwg	drawing
ea	each
e.f	each face
elev	elevation
E.O	extra (cost) over
e.w	elsewhere
e/w	each way
EDP	electronic data processing
excav	excavation
extg	existing
extl	external
extr	exterior
FF	first floor
fbm	feet board measure
fin	finish(ed)
flr	floor
flrg	flooring
flshg	flashing
fmwk	formwork
fnd	foundation
frmg	framing
ftg	footing
Ga	gauge
galv	galvanized
GF	ground floor
gnd	ground
grd	grade
H (h)	height (high)
Hdwd	hardwood
hlw	hollow
incl	include, including
insul	insulation
intl	internal
intr	interior
inv	invert
jst	joist
jt	joint
L	length, angle (steel)
lab	labor
lam	laminated
LF (lf)	linear feet
l & m	labor & material
LS	lump sum

LY (ly)	linear yards
M	thousand
mat	material
max	maximum
MBF	thousand board feet
meas	measure
m.g	make good
min	minimum
misc	miscellaneous
mldg	molding
MM	method of measurement
MP	mean perimeter
m.s	mild steel
m/s	measured separately
mtr	mortar
n.e	not exceeding
No	number (enumerated)
n.w	narrow width
o/a	overall
o.c	over centers, on center
opng	opening
o/s	one side
P.C	prime cost
pcs	pieces
perim	perimeter
P cmt	portland cement
pl	plate
pr	pair
proj	projection, projecting
ptn	partition (wall)
pvg	paving
rad	radius
RC (& W)	raking cutting (and waste)
R conc	reinforced concrete
rebar	steel reinforcing bars
reinf	reinforced, reinforcement
RICS	Royal Institution of Chartered Surveyors
rnd	round
SF (sf)	square feet
SFCA	square feet contact area
sht	sheet
shthg	sheathing
sof	soffite
specs	specifications
Sq	square (100 square feet)
s.q	small quantities
std	standard
str	straight
surf	surface
susp	suspended
SY (sy)	square yards
t & g	tongued and grooved
temp	temporary
th	thick
trwld	troweled
UF	upper floor
u/s	underside
V	volume
VB	vapor barrier
W	waste
wd	wood
WF	wide flange (section)
wdw	window
wi	with
wpf(g)	waterproof(ing)
wt	weight
X bgdg	cross-bridging

1

Why Estimate Construction Costs?

To answer this question it is necessary to look at the ways in which buildings are designed and built, at the persons involved, and at the relationships among them. The construction industry is unique for reasons that have existed since man began to build. Buildings are part of the land on which they stand because of the nature of buildings and because of their physical connection with the land beneath them.

The early laws of property recognized the difference between "movables" and "immovables."

> The only natural classification of the objects of enjoyment, the only classification which corresponds with an essential difference in the subject matter, is that which divides them into Movables and Immovables.[1]

[1] Sir Henry Maine, *Ancient Law* (Everyman's Library; London: J. M. Dent & Sons Ltd.; New York: E. P. Dutton & Co. Inc.).

Land and buildings are immovables, and they are called "real property" as distinct from "chattels," which are movables or "personal property." Buildings are legally part of the land because they are fixed to it; and, in another more fundamental sense, because of how they are constructed. This relationship was more apparent in the past when buildings were built from simple materials—when stones were built into walls; clay was burned into bricks; lime was made into mortar; and trees were cut and shaped into posts and beams. Cottages, castles, and churches all contained the same basic materials—earth, stone, and wood—and the differences among them involved size and shape and the work and skill used in building and decorating them. Their common origin was the land, and this special relationship between buildings and land is ancient and natural and is still reflected in the laws of property.

Today, most building materials are manufactured or reconstituted by complex processes. Many building components are preassembled, and some buildings are completely prefabricated. These developments in building technology have obscured or diminished the original relationship between buildings and the land so that today it has become less significant. Nevertheless, this relationship remains the primary reason why the construction industry is unique in comparison to other industries.[2]

Each piece of land is unique and different from every other in many ways—in soil and water conditions, substrata, topography, altitude, and aspect. Or the differences may be few, and sometimes may involve only a difference in location. But this principle of difference is invariable and only the degree of difference varies. Because of the close relationship between land and buildings, the principle of difference also applies to the buildings fixed to the land. And to the extent that buildings are fixtures on the land, so are they unique. The antithesis of this is the mass-produced mobile home with only a tenuous connection to the land. But until all buildings are mobile, and until the historical relationship between buildings and land disappears, the construction industry will remain unique and each building will be unique in at least one respect—its fixed location on the land.

Each site requires a building to be specially designed to suit the site's unique characteristics. In addition, most buildings are different because of the *owner's* requirements, because of his desire for individual expression or for commercial reasons. Suburban estates of mass-produced houses are no exception to this principle of difference if the estate, and not the house, is regarded as the unit of construction. Usually this principle is not too difficult to detect because the houses are often more or less identical, and the only difference is between one housing estate and another, often only because of a different location and a different arrangement of the houses on the site.

[2] Other reasons include (1) the history and nature of the traditional building trades; (2) construction is essential to all other industries, and largely dependent on them; (3) governments use the construction industry as an economic control; (4) the construction industry is made up of many small firms and a few large firms; although this characteristic is changing as the larger firms increase and diversify.

The Persons Involved in Construction

The *owner* is the initiator of construction. He may also be known as a *developer.* He may want to develop the land by building a residence for himself, or he may want to build an office building or a warehouse for business purposes. He may want to make an investment in a building development to produce an income by renting building space to tenants. Or he may want to develop the land with the intention of selling the land and buildings (improvements) for a profit. The *owner* may be an individual person, a company, or a public body such as a school board. Anyone with rights to land who exercises those rights by having *construction work* done on the land is an *owner* in the contractual sense.

If the *owner* is a company specializing in property development, the company may be equipped to carry out all stages of the development from land purchase to rental or sale of the developed property. In most cases, however, the *owner* is not equipped with either the staff or the experience for such an undertaking, and most *owners* require the services of a *designer.* The *designer* then makes a contract with the *owner* to provide certain services in return for a fee, and these services might include:

1. schematic building designs and plans of land utilization
2. developing the design drawings and site plans
3. preparing the documents (drawings and specifications) required for the *construction work*
4. negotiating with or obtaining bids from construction companies for the *construction work*
5. inspecting the *construction work* in progress
6. settling the final account for the *work*

The *designer's* services to the *owner* may include some or all of these items, together with other services related to building and real property development. The role of the *designer* is changing, and the subject of professional services is extensive and complex. Further reading can be selected from the Bibliography.

The *designer's consultants* are specialists in various

aspects of the design and construction industry. Few designers can be expert in so many fields, and consultants are hired by the *designer* to assist him in providing all the services required by the *owner*. As the need for larger and more complex buildings and environmental controls within them has increased so has specialization in design and construction. Such fields of specialization include:

Engineering	**Planning and Design**
structures	hotels and theaters
soils	kitchens and restaurants
acoustics	educational buildings
mechanical services	laboratories
electrical services	industrial plants
	parking and traffic
Management	parks and landscaping
financing	town planning and
feasibility studies	almost any class of
costs and economics	building and property
scheduling	development

The number and variety of consultants retained by the *designer* is determined by the *owner's* requirements for the building and the site and the *designer's* need for specialized assistance.

The *contractor* is the builder and the construction expert who does *construction work* for payment and who enters into a contract with an *owner* for this purpose. He is sometimes called a *general contractor* to distinguish him from the various specialist *subcontractors* described below. In the past the *general contractor* employed a variety of tradesmen to do everything from concrete and masonry to glazing and painting. The *mechanical work*, such as heating and plumbing and gas and electrical services, was often done by *subcontractors*, or very often by tradesmen employed by the *general contractor*. Later, more was done by specialist firms and less by the *general contractor*, who become a manager and a co-ordinator of the specialist *subcontractors* although still doing a portion of the *work* himself, usually the foundations and the structure.

The *subcontractor* is a specialist who has a contract with a *contractor* to do *work* and who is responsible to the *contractor* for its proper execution. The *contractor* is in turn responsible for the *subcontractor's work* to the *owner*. The contract between the *subcontractor* and the *contractor* arises out of and is subsidary to the primary contract; hence the name, subcontract. *Subcontractors* often have contracts with others who may be called *sub-subcontractors*, because they contract for part of the *work* of a subcontract. The *subcontractor* is responsible to the *contractor* for the work of a *sub-subcontractor* in the same way that the *contractor* is responsible to the *owner* for the *work* of a *subcontractor*.

The persons involved in design and construction are shown diagramatically in Fig. 1, and the parties to

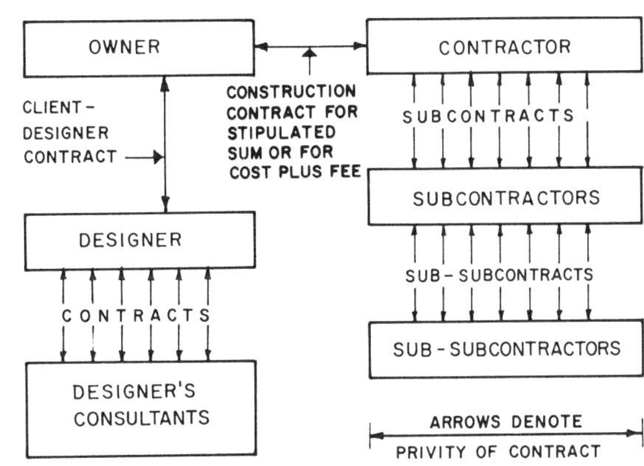

FIG. 1. The Contractual Relationships among persons involved in the design and construction of a project, in which the *owner* has one contract for design services and one contract for construction.

the several contracts are shown connected by double-headed arrows to indicate their contractual relationships. These are the usual relationships among an *owner*, his agent the *designer*, and a *contractor*. Other contractual arrangements are possible. The *designer* may be an employee of the *owner* if the *owner* is a public body or a government department, and in some cases the *designer* may be employed by the *contractor*.

If the *designer* is employed by the *contractor*, the *owner* may have what is commonly called a "package deal," and instead of being party to two contracts (one with the *designer* and one with the *contractor*), the *owner* is party to only one contract for the provision of both design and construction services. The *owner* may also purchase the land together with these services from one company and enter into a "turn key" project through which the *owner* has only to state his requirements and pay to be able to turn the key and enter the completed building.

Construction Contracts

This text cannot contain a complete explanation of the nature of contracts, but some reference to them is necessary because construction contracts are one of the primary reasons for construction estimates.

Civil law in the United States of America, Canada, Britain, and many countries in the British Commonwealth has a common origin in British common law. Consequently, contract law in these countries is basically the same, although specific laws may vary from place to place and civil law in Quebec and Louisiana is based on French law. The student of construction estimating should have a practical working knowledge of the law of the area in which his company does *work*. The nature of a contract is such that it binds the contracting parties together in an exclusive relationship, refered to as "privity of contract" (see Fig. 1.) Contracts are essential and commonplace in modern society,. and a good contract that benefits all the parties also benefits society at large. All technical specialization depends on contracts as the means through which the specialist can apply his knowledge and practice his skill in return for payment.

After the *designer* has approval of the design from the *owner* and local authorities and has obtained the *owner's* instructions to proceed, he prepares the *contract documents,* which will be known as the *bidding documents* until a contract is made. The *designer* then arranges for *bids* to be submitted to the *owner* by construction companies, usually from selected construction companies (through selective bidding) or from any qualified company (through open bidding) in the case of public works. These bids are offers to do the *work* according to the requirements of the *bidding documents.* Alternatively, and depending on the type of construction contract required, the *designer* may enter into negotiations with one or more construction companies on the *owner's* behalf so as to arrange a construction contract between the *owner* and a *contractor* for executing the *work.*[3]

A contract must have certain ingredients to be valid. These may be described as:

1. mutual agreement
2. consideration
3. lawful object
4. capacity
5. genuine intention

Some texts use different terminology, but they all refer to the same things. Not all construction contracts have to be in writing to be valid, but an oral contract is frequently a source of disputes and may be difficult to enforce. Therefore, all construction contracts should be in writing and certain construction contracts are usually required by statute to be written, including contracts for public works and contracts of more than one year's duration.

Mutual agreement is expressed in the offer (*bid*) made by a construction company and in the acceptance of the offer by the *owner.* The offer and its acceptance must be unqualified; that is, without terms and conditions. The basis of mutual agreement is the *bidding documents,* which become the *contract documents* when an offer to do the *work* as described in the documents is accepted.

Consideration in a construction contract, on the one hand, consists of the *work* performed by the *contractor* for the *owner,* and, on the other hand, of the payment made by the *owner* to the *contractor* for the *work* done. The law is not concerned with the amount of the consideration; only that it have some value. If a *contractor* makes an offer that is accepted, he cannot abandon the contract and not complete the *work* because he finds that his offer was too low and the payment insufficient.

Lawful object is not usually a problem in construction contracts because of their nature and because of the preliminary procedures involving development approvals and building permits. A contract to construct a building for an illicit purpose would presumably be invalid, if a certain nineteenth-century legal case is a precedent. In that case, a fashionable lady refused to pay for a new carriage and was taken to court by the carriage-maker. Her defense was that the carriage-maker knew her trade and reputation and he knew that the carriage would be used for illicit purposes; therefore, she claimed, there was no valid contract. The judge found in her favor.

Capacity refers to the capacity of the parties to make a legal and valid contract. To do so, they must not be sentenced criminals, lunatics, enemy aliens (during war), or infants (minors). However, minors can make valid contracts to obtain certain "necessaries" such as food and education.

Genuine intention means that a real and genuine offer and acceptance have been made and that both parties truly and willingly intend to enter into a contract. If one party's intention is not genuine, the contract may be invalid. Invalidation could be the result of one of the following:

1. a mistake, such as in regard to subject matter or identity
2. misrepresentation of facts by one of the parties (without intent)
3. fraud by one of the parties (with intent)
4. duress, such as a threat of violence
5. undue influence, such as is exerted by an educated person over an illiterate person

The legal aspects of construction contracts are many and varied, and only the briefest outline has been given so that construction estimating can be seen in proper context. A contract is based on mutual agreement, which is expressed by offer and acceptance; and

[3] Further reading on the nature of construction contracts and bidding procedures can be selected from the Bibliography.

an offer to construct a building for a certain payment cannot be made without first making an estimate of the costs.

Types of Construction Contracts

There is always a risk in classifying things, because most things do not neatly fit into a few classifications. At the same time, classifications are useful in helping to identify and to compare things, so long as we remember that classifications are made for convenience and that they are not part of reality.

The design and construction industry uses three basic types of construction contracts, and others that are modifications. The types of construction contracts to be examined here are:

Basic types of contracts	**Other types of contracts**
1. stipulated sum contracts	
2. cost plus fee contracts	3. target figure contracts
4. unit price contracts	
	5. management contracts

Types (1) and (2) are examined first, followed by an examination of type (3), which is a modification of and a development from the first two. The unit price contract is examined next, and finally management contracts, which are really not a type of contract but rather an arrangement of several different contracts.

The Stipulated Sum Contract

Few people will make a purchase before they know the price. Hence, the seller is required to establish and indicate the price at which he is prepared to sell; and such is usually the case when the purchaser is an *owner* and the seller is a *contractor.* In the majority of construction contracts, there is a "contract sum," or "contract amount," which is the amount of the payment to be made by the *owner* to the *contractor* for the *work* to be done. This amount is first stipulated by the *contractor* in his offer to the *owner,* somewhat in this form: "We offer to perform all the *work* shown on the drawings and described in the specifications for the sum of $ . . .". The sum is stipulated in the *bid.*

By receiving several such *bids,* or offers, the *owner* expects to get a competitive price; and if he accepts an offer (usually the lowest), he expects to get the *construction work* done for that price. He may endeavor to further secure proper performance of the *work* by requiring a performance bond from a third party who will guarantee that the *contractor* will perform the contract.

It will be apparent that to be able to stipulate a firm price for the *work* the *contractor* must know exactly what the *work* includes. This kind of contract, therefore, requires the *designer* to prepare detailed *bidding documents* before invitations to bid for the *work* are extended. From the drawings and specifications that make up the *bidding documents,* the bidders must then make detailed estimates of the *costs of the work* as a basis for their *bids* to the *owner.*

In order for the bidder to make an accurate estimate of the *costs of the work,* adequate information about the *work* must be given in the *bidding documents.* He also must have knowledge of the conditions at the site and experience in construction so as to be able to judge the probable costs. His knowledge can never be absolutely complete because site conditions are unique and always changing. His experience can never be totally adequate because no two construction projects are ever identical. And in this inevitable inability to foresee all costs lies much of the risk the *contractor* always takes, for there is always some risk of financial loss. But as long as the risk is not unreasonable, the *contractor* is prepared to take it in return for the anticipated *profit* that he includes in his estimate of the costs and in his *bid.* Of course, he cannot overensure against the risk of a loss by including an excessive *profit,* because then his *bid* may no longer be competitive.

The bidders may not be able to make a reasonable estimate of the probable *costs of the work* and thus may not be able to make firm and competitive *bids* to the *owner* if the bidders' knowledge of the site conditions is inadequate, if the site is such as to make it impossible to ascertain these conditions in advance (because of location, substrata, or soil conditions, for example), if the bidders' experience does not include such *work* as that required by the *owner,* or if the information about the *work* is incomplete because of the *owner*'s or the *designer*'s inability to make an immediate decision about the *work,* for any reason. If all or most of the bidders find themselves in that position, the *owner* may not be able to obtain competitive *bids* and a stipulated sum contract. Or, if any or all of the above conditions exist, any *bids* made for a stipulated sum contract may be inordinately high; because all the bidders may have included a high mark-up for *profit* and allowances for contingencies to cover the risk that they are unable to estimate.

The *designer* should foresee if the nature of the *work* or the conditions of the site are such as to make it impossible for the *owner* to get competitive stipulated sum *bids,* and he should then advise the *owner* that another type of contract is necessary.

Cost Plus Fee Contracts

In its simplest form, this type of contract may be made by accepting an offer worded essentially as follows: "We offer to perform all the *work* for all costs (as defined in the agreement) and a fixed fee of $. . ." (or, . . . a fee of *X* percent of all costs). Only the amount of the fee is stipulated.

This kind of contract need not eliminate competitive bidding. If "costs" are fully defined in the documents, the "fee" portion of the contract can be the basis of competitive *bids.* But this kind of contract is "open ended," and the *owner* does not know what his total costs will be until the *work* is finished. Therefore, he carries most of the risk.

Costs are usually defined as all construction costs except certain *job overhead costs,* all *operating overhead costs,* and the *profit,* which together are all covered by the fee. It is imperative that the costs are explicitly defined and that there is proper mutual agreement about them. One method is for the contract to explicitly state which costs are to be paid by the *owner* and to define the fee as including all other costs. Some *standard forms of contracts* define costs as "reimbursable costs." The "nonreimbursable costs" are also defined, and they include any costs not included under "reimbursable costs" and they are covered by the fee.

The fee may be a fixed percentage of the costs as defined in the contract, or it may be fixed within certain limits of cost beyond which the percentage fee varies. Or the fee may be a fixed amount, or may be fixed within certain limits of cost beyond which the fee varies pro rata, or according to a fee scale included in the contract. Many variations are possible, but the theme is always the same. Certain defined "costs" are paid periodically by the *owner* as they are incurred, whereas other *costs of the work* (primarily the *operating overhead costs* and the *profit*) are paid for as a "fee" to the *contractor,* usually in installments, pro rata to and at the same time as the periodic payments of "costs." Sometimes the fee is paid in installments related to specified stages of completion of the *work.*

This all becomes more easily understood by noting that the "costs" paid by the *owner* are costs directly chargeable to the *work* being done, for such tangible and visible things as construction materials, workmen's payroll, the use of construction equipment, and for temporary services and offices on the site. The "fee" paid by the *owner* to the *contractor,* on the other hand, is for less tangible things and for things over which the *owner* and the *designer* have no control; namely, the *contractor's operating overhead costs and profit.* Therefore, the "fee" is determined in the first instance by the *contractor* according to his business needs and policies, even though he may subsequently negotiate for it with the *owner* or bid for it in competition.

Target Figure Contracts

If we imagine the two kinds of contracts just examined to be at the two poles of a "scale of risk," with the **stipulated sum contract** at one end, with most of

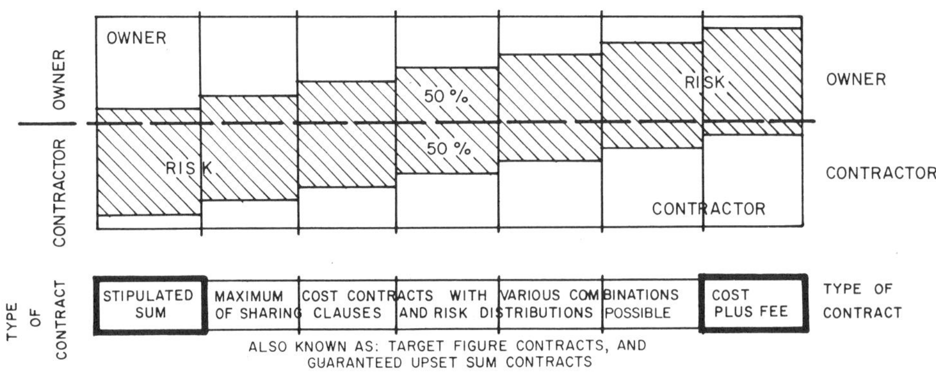

FIG. 2. A Scale of Risk
in construction contracts indicating the possibilities in the distribution of risk between an *owner* and a *contractor.*

the risk with the *contractor*, and with the **cost plus fee contract** at the other end, with most of the risk with the *owner* (as shown in Fig. 2), we can place between these two poles any number of different contracts composed of elements from both kinds of contracts at the poles, and each with a different distribution of risk between *owner* and *contractor*. Chromatically speaking, with black at one end and white at the other, there are any number of shades of gray in between.

Standard forms for stipulated sum and cost plus fee construction contracts are published and sold, but almost all contracts based on a *standard form of contract* require some supplementary conditions and some changes to be made in the *standard forms* to suit some particular requirements. As developments and projects become larger and more complex, it often becomes necessary to make contractual arrangements that are different from those described, and a better contract and a better arrangement for both parties may sometimes be achieved by a more even distribution of risk and by a more flexible form of contract.

The **stipulated sum contract** is essentially inflexible. The detailed requirements of the *work* must be determined and set down in the documents before contracting companies can be invited to make firm offers. Subsequent changes in the *work*, although usually permitted by the conditions of the contract, are contrary to the fundamental nature of a stipulated sum contract. They are usually a nuisance to the *contractor*, or an unreasonable cost to the *owner*. At the same time, the *contractor* has little opportunity to use his full expertise and ingenuity, because the *work* has already of necessity been designed to the last detail, which often helps to perpetuate the unnecessary redesigning of building components that might otherwise be standardized. The main advantage of this kind of contract is that the *owner* obtains a fixed price at the outset through competitive bidding, an advantage that is dissipated by any changes made in the *work* by the *owner* (or the *designer*) during the contract's execution because the costs of such changes are, for the most part, determined by the *contractor*. Attempts by *designers* to draw up stipulated sum contracts to obtain better control of the costs of changes usually succeed only insofar as they are able to introduce elements of other kinds of contracts into the stipulated sum contract.

The **cost plus fee contract** is above all flexible, and the *owner* can make decisions and changes during the contract's execution, although this often creates extra costs. There is little or no incentive for the *contractor* to be economical or efficient, and the general result of this kind of contract is that the *owner* usually ends up spending more than he originally intended. The main advantages are that the *work* can be started before complete documents have been prepared by the *designer*, the *work* can be designed as it proceeds, and the *contractor's* knowledge can be utilized by the *designer*.

The **target figure contract** is made like a stipulated sum contract and administered like a cost plus fee contract. The *bids* to the *owner* are worded essentially as follows: "We offer to perform all the *work* shown on the drawings and described in the specifications for a maximum cost (as defined in the agreement) of \$... and a fixed fee of \$...". The wording is similar to the *bids* for stipulated sum contracts, except that instead of one stipulated sum there are two sums in the *bid*—the maximum cost and the fixed fee.

In target figure contracts, the bidders have to be provided with documents describing the *work* required so that they can estimate the probable maximum cost and the fee. With sufficient information, but lacking many of the details, the bidder can in effect say: "It can be done for not more than \$..., possibly for less," and he makes his offer on that basis. This means, of course, that the *contractor* in this kind of contract must have some part in making subsequent decisions about the *work*.

Up to the point of making a **target figure contract**, the *owner* has obtained the primary advantage of the **stipulated sum** contractual method; namely, competitive bidding and a fixed price. In the contract's performance, the *owner* obtains certain advantages found in the **cost plus fee** contractual method; namely flexibility in making decisions and changes and the opportunity to utilize the *contractor's* expertise when decisions are made. As changes are made by the *owner* and the *designer* in a target figure contract, the *contractor* is asked to estimate the amount by which the change should alter the "target figure" stated in the contract. When this amount is approved by the *designer* and the order to make the change is issued, the "target figure" in the contract is changed accordingly.

The target figure contract usually contains what is called a "sharing clause" under which any savings made in the total *costs of the work* are shared between the *owner* and the *contractor* according to prescribed proportions. Savings occur when the actual total costs paid by the *owner* (excluding the *contractor's* fee) are less than the stipulated "maximum cost." The usual proportions appear to be 75 percent of the total savings for the *owner* and 25 percent for the *contractor*, more or less. On the other hand, if the total costs exceed the "maximum cost" (adjusted for any changes as explained), under the terms of the contract the *contractor usually* has to complete the *work* at his own expense. The *owner* usually pays no more than the "maximum cost" plus the fee.[4]

The "sharing clause" creates an incentive for the *contractor* to be efficient and economical, in addition to the incentive created by the "maximum cost" in the contract for which the *work* must be completed. It appears that a 25 percent share of any savings for the *contractor* is reasonable and effective in contracts performed under conditions that would not normally deter the *contractor* from bidding on a stipulated sum contract. In other words, under conditions that might be classed as "normal," and under which the *contractor* would reasonably be able to estimate probable costs given all the necessary information about the *work*, a sharing clause that provides the *contractor* with a 25 percent share of any savings made in the contract appears to be acceptable to most bidders.

Under less favorable conditions, such as would make an accurate cost estimate impossible and a stipulated sum contract undesirable to a *contractor* (even with the provision of complete and detailed *bidding documents*), it is unlikely that the 75/25 percent sharing of any savings would be very attractive to a construction company. Under such conditions, contracts are more likely to be made by negotiation rather than by usual bidding procedures. After all, a *designer* cannot prescribe the terms of a contract and call for *bids* if the conditions affecting the *work* are such that he cannot anticipate what might be generally acceptable terms to bidding construction companies. Although selective bidding for contracts with prescribed terms might be possible following preliminary discussions with selected companies to determine acceptable terms.

Looking at Fig. 2., a model of a target figure contract could be constructed exactly halfway between the two poles of the contractual scale shown. In such a hypothetical contract, the target figure and the fee would be arrived at by a fairly detailed estimate to make it realistic, but not with the same detail that would be required in a stipulated sum contract. This would mean that reasonable design solutions would have been found for the primary components and major elements of the building, but that other solutions would be possible and might be more economical. The "sharing clause" would prescribe any savings to be shared 50/50 percent between *owner* and *contractor,* whereas any losses (excess of actual costs over the stipulated "target figure") would also be shared 50/50 percent. The model is theoretical, and such a contract might never be suitable for practical use. Nevertheless, it does indicate how this type of contract can be used to stipulate terms to suit any conditions.

The **target figure contract** may be viewed as a hybrid of the two basic types of contracts previously examined. It has many variations and names, including "maximum cost plus fee" and "guaranteed upset sum." The essential features are a stipulated "target figure," or "maximum cost," a "fixed fee," and usually a "sharing clause." The target figure contract may be awarded on the basis of competitive bids made up of the "target figure," or "maximum cost," and the "fixed fee," and therefore is similar to a stipulated sum contract in this respect. But once the contract has been made, it has to be operated like a cost plus fee contract, because the actual costs must be controlled and known so that the difference between actual costs and the stipulated maximum cost in the contract (which is increased or decreased by mutual agreement as changes are made in the *work*) can be determined and the "sharing clause" applied.

Unit Price Contracts

This basic type of contract is more familiar in heavy construction and highway construction than in building construction in North America. In Britain, and in other Commonwealth countries such as Australia, New Zealand, and in many countries in Africa and Asia where the quantity surveying profession is established and where the "quantities method" of bidding and contracting is used, this type of contract and the methods related to it are widely used in building construction.[5]

A *contractor's* estimate prepared prior to making a *bid* for a stipulated sum contract consists of two primary parts:

1. **the measured quantities of work**
2. **the prices of work.**

[4] For purposes of general discussion the terms "target figure" and "maximum cost" are taken here to mean the same. However, if the terms of a contract establish an absolute maximum cost to the *owner*, the term "maximum cost" (and "maximum cost plus fee contract") is more descriptive. Whereas, if the terms of a contract require the *owner* to pay part of any costs over the amount stated in the contract, the term "target figure" (and "target figure contract") is better. The difference lies mainly in the sharing clause.

[5] The quantity surveying profession is discussed in Chapter 13.

For example, such an estimate might include this item:

(a)	*(b)*	*(c)*	*(d)*	*(e)*	*(f)*
033101	Concrete (3000 psi) in continuous footings	200	CY	$21.00	$4,200.00

(a) code number (for identification)
(b) description (abbreviated from specifications)
(c) quantity (measured from drawings)
(d) unit of measurement (cubic yards)
(e) *unit price* (per cubic yard)
(f) cost of *item of work* (cost = quantity × *unit price*)

and the estimator measures all the *work* in this way and prices it, item by item, with general items for temporary services, site offices, and supervision also included in the estimate.

In the *bidding documents for a* **unit price contract** there is a *"schedule of unit prices"* prepared by the *designer* (or by a *designer's consultant*) that contains the measured quantities for all the *items of work* in the contract, as in columns (a), (b), (c), and (d) of the above example. The bidder does not have to measure the *work*: he only has to estimate the *unit prices,* apply them to the quantities provided, and compute the total costs. *Overhead costs and profit* are usually allowed for in each *unit price,* but some overhead items and their costs may be entered separately in the *schedule of unit prices.*

The primary reason for using unit price contracts in heavy construction and highway construction is the difficulty in determining in advance the exact *quantities of work* required to be done in these types of construction, which depend so much on ground conditions. Some of the decisions that determine such things as the amount of excavation necessary, or the amount of fill required, can only be made once the *work* is underway. Consequently, it is impossible for construction companies to make accurate estimates of costs and to submit firm *bids.* It would be possible for the *owner* to enter into a cost plus fee contract for this kind of *work,* and this is sometimes done. But most *owners* insist on a contract that gives them a tighter control of the costs, and this may be obtained by a unit price contract. Changes in the *quantity of work* are easily made in this type of contract. But changes in the actual *items of work* required are no easier to make than in a stipulated sum contract, although such changes are less frequently necessary in heavy construction, which involves relatively few *items of work* compared with building construction.

As the *contractor* completes sections of the *work* in a unit price contract, the quantities of *work* are measured at the site by both the *contractor* and the *designer,* or their representatives, so that mutual agreement about the amount of *work* done is reached and the agreed amounts are recorded. The *contractor* is then paid for the actual quantities of work done at the *unit prices* in the contract. If quantities of *work* vary beyond specified limits, the contract may require the parties to agree to adjustments in the *unit prices* or the contract may provide for such adjustments.

In Britain, and in other countries where quantity surveyors are employed, "bills of quantities" are used for the majority of major building projects. Because the *items of work* in the superstructure of buildings can usually be accurately measured from the drawings by the quantity surveyor, and because the measured quantities will accurately represent the actual quantities of *work* done by the *contractor,* it is not usually necessary to remeasure the *work* in the superstructure of a building at completion. In these building contracts, only the *work* in the substructure (below ground) is usually remeasured, because this *work* is so often subject to variations owing to subsurface conditions. Nevertheless, any changes ordered in the *work,* and any discovered or suspected errors in the "bills of quantities" are cause for measuring at the site any or all *items of work* as required by either the *contractor* or the quantity surveyor. Payment for the *work* is made on the basis of the *unit prices* in the contract, and if changes involve new *items of work* for which there are no *unit prices* in the contract, the method of payment will be as set down in the contract, usually on the basis of negotiated *unit prices* or on a "cost plus fee" basis.

Any type of contract can provide for *items of work* to be done on a different contractual basis from the others. This is frequently the case with *"extra work"* ordered during a contract, but it can also apply to *items of work* prescribed in the original contract. For example, a "stipulated sum contract" may contain some *work* described and measured in a "schedule of quantities" to be done at *unit prices.* This provision gives the *designer* some latitude and flexibility in deciding on the quantity, location, and arrangement of such items as fixtures and movable partitions, or other *items of work* included in a schedule for this purpose.

Management Contracts

The term "management contract" has become popular in the last few years along with other terms from the new management sciences. So often, new terms for old, commonsense methods are taken up and become fashionable for a time; but often these terms are not properly understood, and they mean many things to many people. A "management contract" may have many features, but common to most of them are the following:

1. There are two agents of the *owner,* instead of one as before; namely:
   a. the *designer,* primarily responsible for the design; and
   b. the *contract manager,* primarily responsible for translating the design into reality.
2. There is not one *contractor* responsible for all the *work,* but instead several *specialist contractors* all of whom have separate contracts with the *owner* for parts of the *work* required.

The *designer* and the *contract manager* may work together in both designing the project and in seeing that it is built according to the *owner's* requirements. Ideally, they are both appointed by the *owner* at the outset so that they can complement each other's skills and experience. In this respect, this arrangement is not unlike the "package deal" that has been available to *owners* for some time. But whereas the so-called "package deal" has been offered for the most part by designer-builders for commercial development, the management contract is used today in both commercial developments and in institutional developments such as universities. It is often initiated by an *owner* who sees the advantage in having two agents with complementary skills but with separate responsibilities and interests.

The increasing complexity of technology and building construction processes has brought us to the point where the *designer* and the *contractor* cannot be responsible to the *owner* for all the specialized *work* of many of the specialist *subcontractors.* For example, how can a *contractor* (in a stipulated sum contract) effectively take responsibility for the *work* of an *elevator subcontractor* when the *contractor* (and the *designer*) has only a superficial knowledge of elevator installations and their operation? Obviously, the proper thing is for the elevator specialist to have a contract with, and a direct contractual responsibility to, the *owner.* Hence, the practice in management contracts is for all persons doing parts of the *work* to have a contract directly with the *owner.* These contractual relationships are shown in Fig. 3(A).

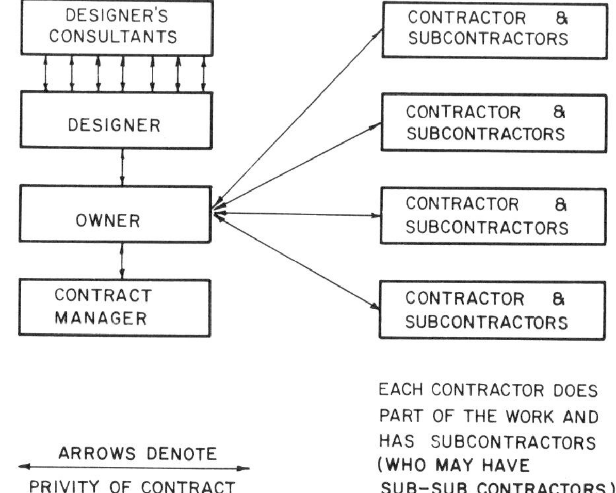

FIG. 3(A). The Contractual Relationships among persons involved in the design and construction of a project in which the *owner* has one contract for design services, one contract for construction management services, and several other contracts for construction services covering all the *work* in the project.

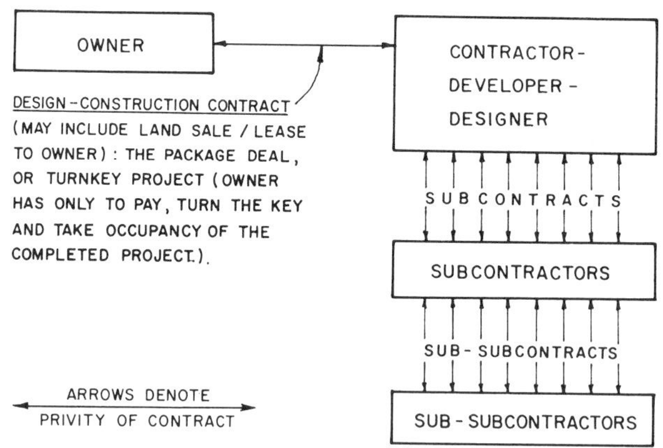

FIG. 3(B). The Contractual Relationships among persons involved in the design and construction of a project in which the *owner* has only one contract for both design and construction services (which may also include the sale of land for the site).

The first function of a *contract manager* is to provide economic and technical knowledge during the designing stages of the project, although the design is the primary responsibility of the *designer.* The next function of the *contract manager,* and his primary responsibility, is to organize and manage the *work* at the site in the best interests of the *owner.* The *con-*

tract manager usually prepares cost estimates at various stages of the design and of the *work* in progress, and it is his responsibility to see that the *work* is completed within the budget established at the outset.

In some contracts, the *contract manager* may be required or subsequently requested to provide certain site facilities, or to undertake certain parts of the *work*. In other words, he then fulfills the function (at least in part) of a traditional *general contractor*. It is questionable that this is either necessary or desirable if the *contract manager* is to properly fulfill the function of an agent of the *owner*, as described above; and it is obvious that it is not in the *owner's* interests to lose any advantage by awarding *work* to someone in a preferred position. From the *owner's* position, it is far better for the *contract manager* to have no investment or other similar financial interest in the *work* so that his responsibilities are clearly defined. Alternatively, the *owner* can get a "package deal" from a designer-builder, as indicated in Fig. 3(B).

The "package deal" has the great advantage of simplicity in that the *owner* enters into only one contract for what can include everything from the sale of land for the building site to the building complete and ready for occupation. Hence, the other name for this kind of arrangement, the "turn key project." One has only to turn the key and walk into a completed building ready for use. But, of course, the key has to be paid for first, and some criticize these contracts because an *owner*, in dealing with one person only, may not have the advice he needs to guide him in matters of quality and costs. Others say that many of the firms offering "design and build" packages employ architects and engineers, and that these firms can be as reliable as anybody. Some *owners* have found that they can retain the services of an independent cost consultant to advise them in making this kind of contract.

Estimating for Construction Contracts

All kinds of contracts made as a result of competitive offers require the bidders to make estimates of costs before they can make an offer. Even the cost plus fee contract requires a proper understanding and consideration of costs as defined in the *bidding documents*, and a computation of the fee accordingly. If a contract is the result of negotiations rather than of competitive bidding, the negotiations must be based on knowledge of the *work*, and this again requires an estimate. As we shall see, there are other reasons for making estimates, but the primary and fundamental purpose is to make an offer to do *construction work*.

Some estimates are very analytical and detailed, such as an estimate made as the basis of a competitive *bid* for work to be done for a stipulated sum. Other estimates are less detailed, such as an estimate made to determine the approximate cost of a project as the basis for determining the amount of the fixed fee required to do *work* within a cost plus fee contract. Nevertheless, the first purpose is always the same: to obtain a knowledge of the *costs of work* so as to be able to enter into an agreement for its execution. The various kinds of estimates are examined in the next chapter.

Questions and Topics for Discussion

1. What is the essential difference between real and personal property?

2. Why are all buildings unique?

3. What similarity exists between *subcontractors* and *designer's consultants*, and what are the reasons for their presence and importance in the design and construction industry?

4. What are the five ingredients of a valid construction contract?

5. What information does a bidder need to be able to submit a satisfactory *bid* for *work* to be done in a *stipulated sum contract*, and why?

6. How can competitive bids be made for *work* to be done in a *cost plus fee contract?*

7. Explain how a *target figure contract* can contain major features and advantages of both a *stipulated sum contract* and a *cost plus fee contract*.

8. Explain the meaning of the term *unit price*.

9. Explain the meaning of the term *item of work*.

10. Name all the persons who have contracts with the *owner* in the arrangement commonly called a "management contract."

2

What is an Estimate?

An estimate involves calculating the *costs of work* on the basis of probabilities. The calculations are of two kinds, relative to the two primary parts of an estimate: 1. **measurement** (by mensuration), 2. **pricing** (by arithmetic). These two kinds of calculation are simple, and because they are simple they are often not given the attention they deserve. All measurement is approximate, and this fact should be remembered because it is invariable. Only the degree of approximation is variable, and the acceptable degree of accuracy depends on the purpose and methods of the measurement. Since all estimates start with the measurement of *work*, it follows that all estimates are approximate to a variable degree in their first part, the measurement. In the second part of an estimate, the pricing, the degree of approximation is even greater because of the difficulty in predicting all the probabilities of such things as labor productivity and site conditions. The ability to predict probabilities largely depends on the data available from past experience and the estimator's intuition, which is, perhaps, the mental use of data without mechanical means.

All cost data is acquired from experience. If an estimator does not have cost data from his own experience, he must use cost data from another's experience. For this reason, estimators' price books and handbooks are helpful to the novice, and are sometimes a useful guide to an experienced estimator. These handbooks contain data from the writers' experiences and from the collected experiences of others. They are better than no data, but they are much less valuable than an estimator's own experience. These books sometimes make the wrong impression on the novice; the impression that prices and productivity rates are facts that are applicable to all estimates at all times. But all prices are subject to the market and to economic conditions, and they fluctuate continuously. Labor productivity is even more variable, because it is subject to so many conditions, some of which are reviewed in Chapter 5. The most valid and useful data for estimating is that data acquired by the estimator from his own experience and from the contracting experience of the company for which he works. Not only is this data the most relevant to the future *work* done by that company, but also such data can best be analysed and understood by those who have been involved in the *work* that produced the data.

The more valid data the estimator has available, the better he will be able to estimate the probabilities of costs arising from such things as:

site, location and accessibility
subsurface and soil conditions
time and season
climatic conditions
wage agreements
strikes and lockouts
market prices of basic materials
availability of money
the demand for construction
political and economic climates

and the innumerable other things that are continually changing and affecting the economics of construction.

Engineering science is governed by more or less immutable laws, such as the law of gravity, which give the *designer* a reliable basis on which to design a structure. The force of gravity can always be relied on; hence, it is a law. Construction costs are governed by economics, politics, and social conditions, and they are like the sea–fluid, moving, and never still, never the same from one moment until the next. Therefore, estimating cannot be a pure science, but scientific methods can be employed in estimating.

Estimating should be a part of the cyclical management process illustrated in Fig. 4., in which estimating is shown as one of several parts of construction management, each one interdependent with the others. Each part, or function, is discussed in detail in Chapter 3, and in other chapters, but for the moment,

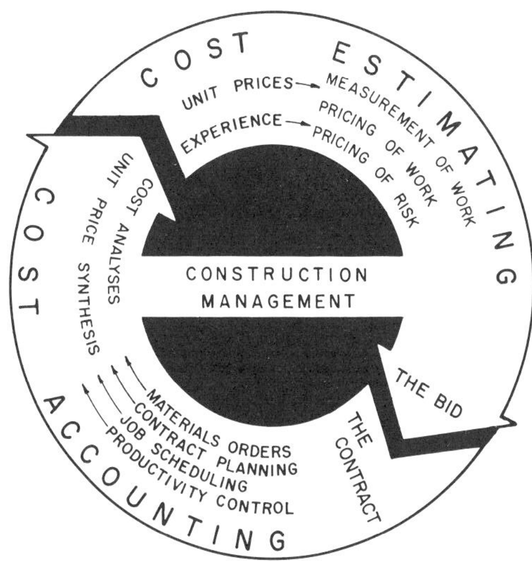

FIG. 4. The Cyclical Process of Cost Estimating and Cost Accounting.

it can be seen from the cyclical diagram that estimating relies on *cost accounting* to provide cost data, because it is only through *cost accounting* that data from past jobs can be acquired and recorded for use in future estimates.

Costs of construction work are classified and examined in Chapter 5 under the following headings:

1. *materials costs*
2. *labor costs*
3. *plant and equipment costs*
4. *overhead costs and profit*

Data on all of these costs are required to make an estimate and to calculate costs in an estimate: **cost = quantity × unit price.**

EXAMPLE (1)[1]

Description of item: Ready-mixed portland cement concrete (3000 p.s.i. × 1½" max size aggregate) placed in continuous footings:

Material	Quantity	*Unit Price*	
Concrete, as specified	150 CY @	$18.00 per CY	$2,700.00
Labor			
Place concrete in footings	150 CY @	$ 3.00 per CY	$ 450.00
Material and Labor Costs	150 CY @	$21.00 per CY	$3,150.00

[1] These examples are given to show the function of *unit prices* in an estimate. The *unit prices* themselves are meaningless unless the location, date, conditions, and specifications of the *work* are known.

EXAMPLE (2)

Description of item: Ready-mixed portland cement concrete (3000 p.s.i. × ¾" max size aggregate) placed by crane in roof slabs:

Material	Quantity	*Unit Price*	
Concrete, as specified	200 CY @	$18.00 per CY	$3,600.00
Labor			
Place concrete	200 CY @	$ 5.00 per CY	$1,000.00
Equipment (crane)	30 hrs @	$12.00 per hr	$ 360.00
Material, Labor and Equipment Costs	200 CY @	$24.80 per CY	$4,960.00

The two parts of the estimate, measured quantities and prices, cannot be viewed separately because each is a function of the other. In **Example (1)**, *labor costs* are $450.00 and the quantity of *work* is 150 CY of concrete in footings.

$$\textit{unit price} \text{ of labor} = \frac{\text{total costs of labor}}{\text{number of units}} = \frac{\$450.00}{150 \text{ CY}} = \$3.00 \text{ per CY}$$

The *unit price* for "labor placing concrete" in this job will not be known to the *contractor* until he has placed all the concrete and has accounted for all the *labor costs.* The first few yards of concrete probably will cost more than $3.00 per CY to place, and the last yards of concrete will probably cost less than $3.00 per CY to place. The *unit price* is, therefore, **an average price per unit of measurement.**

The *unit price* for an *item of work* will vary with each job. The odds against an absolutely identical *unit price* occuring for the same item on two jobs are high because of the large number of variables involved, and this is sometimes presented as an argument against a scientific approach to estimating. But because accuracy is relative, and because the units of measurement and pricing are relatively large, practical accuracy in estimating and *cost accounting* can be achieved. Nevertheless, it is proper for estimators to keep in mind that every *item of work* in every construction job is unique, that it is done under unique conditions, and that accuracy in estimating is relative.

The question might be asked, Why is it necessary to calculate *unit prices,* and why cannot the estimator simply estimate the total costs required to carry out each part of the *work*; why does he need to divide costs by quantity to obtain the *unit price* when all he needs is the total costs of each item? Students of estimating often ask this question, and the answer lies in what has been said before about the estimator's need for cost data. The ability to predict probabilities largely depends on data from past experience.

To estimate the *costs of an item of work,* the estimator may rely entirely on recorded cost data from past jobs; and the most convenient way of recording cost data is in the form of *unit prices* and *unit rates.* To say that the *labor costs* for placing 150 CY of concrete in footings were $450.00 is (for the estimator) to immediately divide cost by quantity and to reflect on the validity of a *unit price* of $3.00 per CY for this item. Or if the current cost of labor is $5.00 per hour, to reflect on the validity of a *unit rate* of 0.6 manhours per CY. To estimate the costs of major and important items in a job, the estimator may determine the required crew size and the total time the crew will probably require to complete an *item of work,* and from this he can calculate the *labor costs.* But only by taking another step, by calculating the *unit price* (the average cost of a unit of *work*), can the estimator compare the price he has just estimated with the costs of similar items in past jobs; and by calculating the *unit rate* he can compare the productivity achieved.

The Contractor's Estimate

To submit a competitive *bid* for a stipulated sum contract, a bidder must make as accurate an estimate of costs as possible and, if the offer is accepted, the sum stipulated in the *bid* becomes the contract sum for which the *work* must be done. Similarly, in a "maximum cost plus fee contract," the bidder largely determines the *costs of the work* to the *owner,* even though the "maximum cost" may be subject to modification by changes in the *work* and by a "sharing clause" in the contract. The **contractor's estimate** is the basic estimate in the design and construction process, and all other estimates are either subsidiary to it or try to anticipate it. The *contractor* more or less establishes the price for the *work* through his estimate, depending on the type of contract within which the *work* is done.

The **contractor's estimate** may be viewed in four parts:

1. estimate of the *costs of work* to be done by the *contractor* with his own forces
2. estimates of *costs of work* to be done by *subcontractors* (and prepared by them for *bids to the contractor*)
3. estimate of costs of *general requirements* for all the *work* provided by the *contractor*

4. estimate of *overhead and profit* required by the *contractor* for doing *work* with his own forces, for supervising and taking responsibility for the *work* of his *subcontractors,* and for providing the *general requirements*

The meaning of the term *work* should be recalled. It includes labor, materials, equipment, and all other services required by the contract and provided by the *contractor.*

Bidding *contractors* and bidding *subcontractors* use essentially the same estimating methods in respect to the *work* they propose to do with their own forces. The *contractor* receives *sub-bids* from *subcontractors* and combines the most favorable *sub-bids*[2] for each part of the *work* to be done by a *subcontractor* with his own estimate of (1) and (3) above. To this he adds estimated allowances for *overhead and profit* (4) to make up the total estimated costs and the amount of the *bid.* The integration of the *subcontractors' sub-bids* and the estimate of the *work* to be done with the *contractor's* own forces is often one of the most difficult parts of preparing a *bid.* Part of this difficulty arises from the variations in scope and content of *subcontractors' bids*; and the *contractor* has to ascertain that each of the *sub-bids* he includes in his estimate and *bid* is for a specific part of the total *work,* no more and no less, so that all parts of the *work* are included, but only once.

Because "mutual agreement" (expressed in the "offer and acceptance") is essential to a good contract, it follows that it is essential that the *contractor's* estimate is based on precisely the same facts and requirements as are contained in the contract, which comes about as a result of the *owner's* acceptance of the *contractor's* offer. In other words, the *bidding documents* must be identical to the *contract documents.* Addenda to the *bidding documents* issued to bidders during the bidding period must be prepared and issued in such ways that there is nothing that can detract from the "mutual agreement" necessary to the contract. Anything the *contractor* is required to do as part of the contract must have been reasonably foreseeable by the *contractor* while making his estimate.[3]

For a "cost plus fee contract," in which there is no maximum cost stipulated, the *contractor's* estimate will usually be an estimate of the "fee" portion only; and this will primarily include *overhead and profit,* without the need for any measurement and pricing of *work.* Nevertheless, "mutual agreement" is still essential to the contract, and the same precepts apply. For complete mutual agreement in a cost plus fee contract, and in a maximum cost plus fee or target figure contract, it is essential that "cost" and "fee" be clearly defined. Lack of distinction between "cost" and "fee" is a common cause of incomplete "mutual agreement" and of misunderstandings that lead to disputes between the *owner* and the *contractor. Standard forms of contracts* provide only general guidelines for properly defining these two important terms. *Contract documents,* particularly the articles of the agreement and the conditions of the contract, must be specially prepared for each contract.

Those who say that *standard forms of contract* should always be used and should never be amended do not understand construction contracts. But *standard forms* are sometimes abused, and this may be the reason for some of the criticism of their use in amended forms. For the sake of clarity, *standard forms of contract* should be incorporated in total by a specific reference to the *standard form* in the project's specifications. Any amendments, omissions, or additions should then be made in the specifications immediately after the reference. Made in this way, changes in *standard forms* are quite clear and easily understood. The practice of copying the whole or parts of a *standard form* into specifications together with amendments is to be deplored, because made in this way the amendments are not apparent and may be overlooked.

An estimator should always be aware of the fact that complete "mutual agreement" is essential to a good contract. If *bidding documents* are ambiguous, or if they are not complementary one to another, or if they are inequitable in their requirements, the estimator should try to get them changed by a written addendum from the *designer.* Failing that, he should decline to make an offer, or calculate and allow for and take the risk. Silent acceptance of deficient documents by bidders leads only to disputes and more deficient documents.

One maxim for the specification writer is: be fair. Do not specify requirements that are inequitable or unreasonable. Do not require bidders to accept an incalculable risk by writing so-called "weasel clauses." Despite the efforts of the Construction Specification Institute and the Specification Writers Association (of Canada), such clauses still appear in specifications and continue to hamper the workings of the construction industry because they cause bad contracts.

An estimator should be aware of the fact that an estimate is more than simply a means of making a *bid.* If a bid is accepted, the result is a contract; and proper contract management requires information for

[2] Usually, the lowest and most competitive bid. Sometimes, the lowest bid is not used because it is incomplete or because it is qualified by certain conditions, or because it is from a firm with whom the *contractor* does not wish to do business.

[3] This is not to say that the *contractor* may not overlook something and omit it from his estimate, but simply that all terms and requirements of the contract should be reasonably capable of being seen by the *contractor* while making his estimate.

purchasing, planning, controlling, and for *cost accounting.* It is essential that an estimate be made with all these objectives in mind, and this aspect of the estimator's work and the use of his estimate is explained in the next chapter.

The Subcontractor's Estimate

Although a **subcontractor's estimate** is for only part of the *work,* the *subcontractor* is responsible to the *contractor* for that part in the same way that the *contractor* is responsible to the *owner* for the entire *work.* This also applies to the *sub-subcontractor* who bids to a major *subcontractor* such as a mechanical contracting company that may be responsible to the *contractor* for up to 50 percent of the *total work,*[4] and often has to incorporate *sub-sub-bids* for parts of the *mechanical work* into its own *sub-bid* for the total *mechanical work.* The contractual relationship between *subcontractor* and *contractor* (and *sub-subcontractor* and *subcontractor*) should reflect the contractual relationship between *contractor* and *owner.* If it does not, it is probably because the *contractor* (or *subcontractor*) has failed to make it so, and he may carry unnecessary and additional risks if he has not delegated risks and responsibilities to his *subcontractors* (or *sub-subcontractors*) through *contract documents* that contain the same requirements as are contained in the *contract documents* between *owner* and *contractor.*

That which already has been written about a *contractor's* estimate can generally also apply to *subcontractors'* estimates. However, in many subtrades a large part of the *work* is done in a factory or a shop and not at the site, which means that the estimating technique for some *subcontractors* is similar to those estimating techniques used in manufacturing industries wherein *work* is done in factories and mass production and controlled conditions prevail.

Another aspect of *subcontractors' estimates and bids* is the practice called "bid peddling" or "bid shopping" in which some *general contractors* may play off one sub-bidder against another by offering a subcontract to one firm for a lower price than that submitted by the firm's competitor. Consequently, bid depositories have been established in many places in an effort to prevent this bid peddling and to regulate bidding procedures.

The idea of a bid depository involves an independent place where all *sub-bids* so designated by the bidding authority (the *designer*) must be deposited and recorded by a certain time. Bidders can then collect the *sub-bids* made to them from the bid depository, say, twenty-four hours after the closing time for *sub-bids.* The *bidding documents* prepared by the *designer* usually have to stipulate the requirements for depositing *bids* and, in particular, require that all bidders must use deposited *sub-bids* for those trades designated for bid depository, because only the *designer* calling for *bids* has the power (as agent of the *owner*) to require *sub-bids* to be deposited. The bid depository system reduces but does not eliminate bid peddling.

Contractors find bid depositories an advantage in that they are ensured of receiving deposited *sub-bids* by some specified time before their *bids* must be submitted to the *owner.* This helps to reduce the frenetic rush that so often precedes the completion of an estimate and the submission of a *bid.*

It should be pointed out that bid depositories cannot change contractual principles. A bidding *subcontractor* is not obliged to deposit *sub-bids* for all bidding *contractors,* and a *subcontractor* may make different *sub-bids* to different *contractors.* The fundamentals of "genuine intention" and "mutual agreement" must exist for a legal and valid contract, and these cannot be induced by any external agent. And bidders are always free to bid or not to bid however and to whomever they please.

Some bid depositories have been scrutinized in the past for possible infractions of the law and, in some cases, their methods of operation have been questioned because they appeared to restrict competition. It can be argued in some cases that they do reduce competition, in fact, if not by intent. If only one *subtrade bid* is deposited for a particular *section of work,* and bidders can use only deposited *bids,* there is not much incentive for a bidder to seek out a more competitive *sub-bid,* because he knows his competitors must also use the one and only *sub-bid* submitted. So each bidder can use that *sub-bid* without fear of competition, knowing that if he is awarded the contract there is a good chance of realizing greater profits by shopping later for a more competitive *sub-bid.*

Some hypocrisy often accompanies bidding practices, and highflown terms such as "the sanctity of bidding" are used by some in explaining the purpose of bid depositories. Yet it always takes two to peddle bids; and if all *subcontractors* refused to peddle, no

[4] The *mechanical work* (including plumbing, heating ventilating, air conditioning, gas supply, and other mechanical systems) may include as much as from 40 to 50 percent of the total *work* in some kinds of construction.

contractor seeking to increase his profits at the expense of others could force them into subcontracts for skimmed payments. The *subcontractor's* greatest strength is his right to withhold his offer from those with whom he does not want to contract.

The majority of contracting firms seem to approve of bid depositories, because they regulate the timing and flow of *sub-bids* and at least reduce the amount of bid peddling. But some *owners* will not permit the use of bid depositories on their jobs, because they believe that bid depositories do restrict competition.

The term *subcontractor* comes from the need to distinguish between a contractor who has a contract with an *owner* and those specialist contracting firms that indirectly do *work* for an *owner* through a subcontract with a *contractor.* In other contractual arrangements, such as "management contracts," the specialist firms have contracts directly with the *owner* and are *contractors* in their own right; and their estimating techniques and procedures are essentially the same as those of the *general contractor.* So although the methods of contracting *work* may be changing, and the traditional stipulated sum contracts with a *general contractor* (often) now are replaced by management contracts, the essential relationship among an *owner* and those who do *construction work* for him remains the same as it has always been, and estimates of construction costs are as essential as ever to construction contracts.

The Designer's Estimate

A **designer's estimate,** which may be made at any time before the *contractor's estimate* is made, is an effort by the *designer* to anticipate the contract amount for which the *work* will be done by the *contractor.* This preliminary estimating is frequently done (in part or in whole) by one or more of the *designer's consultants,* either by the engineering consultants, who usually estimate only the costs of those parts of the *work* that they design, or by a quantity surveyor or other cost consultant who specializes in providing such services to *designers* and *owners.*

These preliminary estimates are made without the full information required for a *contractor's estimate,* and the earliest estimates are frequently made from rough design sketches, without dimensions or details, and from an outline specification and a schedule of space requirements. This conceptual estimating requires not only a knowledge and understanding of estimating as it is done by *contractors,* but also requires a broad knowledge of construction materials, methods, and trade practices; an understanding of building design; a knowledge of comparative costs and economics; and the ability to perceive construction details within the broad lines of design sketches—ideally from past experience with a particular *designer*—so that the estimator knows how the *designer's* sketches will be translated into working drawings and specifications. Preliminary estimates by a skilled practitioner can usually be made to within 5 percent of the arithmetical mean of the *bids* subsequently made for the *work* by *contractors,* and consequently they make it possible for the costs of the building to be planned as well as its functional parts.

Cost planning should be part of the design process for all building developments, because the *designer's estimate* provides the *designer* with information about the *work* that he will need to administer and supervise the contract properly. It is, as well, often an important part of the services the *designer* provides to his client, the *owner.*[5] It enables the *designer* to design the building within the financial limits set by the *owner,* and sometimes saves the *designer* from expensive and distasteful redesigning that might otherwise be necessary if *bids* submitted for the *work* are much higher than the amount of the budget.

The techniques used in making a *designer's estimate* cannot be fully explained here, but a brief outline of some of the methods used is given under the following headings:

1. **cost per place** (per room; per bed; per seat)
2. **cost per cubic foot** (of building volume)
3. **cost per square foot** (of building area on ground floor or of building area on all floors)
4. **cost per story-enclosure**[6] (areas of all floor and roof decks and of all enclosing walls, with applied factors for building height, in stories)
5. **cost per element** (cost of *element* expressed in terms of the building area and cost of *element* in terms of its *unit prices;* i.e., total cost of *element* divided by its quantity)
6. **cost per approximate quantities** (cost of *items of work* obtained from quantity × *unit price,* as in a *contractor's estimate,* but with quantities and prices more or less approximate, depending on available information

[5] See William Dudley Hunt, Jr. (ed.), *Comprehensive Architectural Services—General Principles and Practices,* prepared by the American Institute of Architects (New York: McGraw-Hill Book Company, Inc., 1965).

[6] See Douglas J. Ferry, ARICS, *Cost Planning of Buildings* (London: Crosby Lockwood & Son, Ltd., 1964) for explanations of the Story-Enclosure Method and Elemental Estimating, both of which were developed in Britain by chartered quantity surveyors about 1954.

These methods are listed in the order of efficacy for most estimates, and they are briefly described here.

Cost per place is useful only for *cost planning* of a general nature and on a broad base such as in a school district on the basis of average cost per pupil. It is suitable for budgeting finances, but not for estimating the costs of individual projects.

Cost per cubic foot can be very unreliable unless virtually identical buildings are compared. There is not much relationship between the volume of a building and its costs. Costs come primarily from the walls and decks that enclose the building's volume, and there is no constant relationship between the areas of walls and decks and the enclosed volume.

Cost per square foot is used to indicate the approximate costs of an *owner's* requirements in the early stages, when often only the area is known. More costs are related to a building's superficial area than to a building's volume, and properly applied this method is more reliable than the last.

Cost per story-enclosure realistically takes account of those things that create costs: the areas of the horizontal and vertical planes of a building—the decks and the walls. Also, factors are used to take account of a building's height. It has been largely superceded by the next method.

Cost per element analyzes buildings into basic parts, such as exterior walls and roof decks, and prices each element separately on the basis of the *unit prices* of the *elements*. It approaches the next method in detail and efficacy.

Cost per approximate quantities is the most reliable and the most time consuming, as it comes closest to the estimating methods used by *contractors* and *subcontractors*. By measuring approximate quantities, and by ignoring the *divisions of work* between subtrades, a quantity surveyor can prepare accurate estimates for *designers* in an acceptable period of time and for a reasonable fee.

The methods used in making a *designer's* estimate may include several of the above methods for different parts of the *work*, depending on the information available and on the person doing the estimate. A preliminary estimate might be made by a quantity surveyor on the basis of "cost per square foot," with a conceptual estimate made later (before working drawings and specifications are prepared) on the basis of "cost per *element*" or "cost per approximate quantities." This gives the *designer* an opportunity to design within the budget and to tell the *owner* the costs of his specific requirements.

As the working drawings and specifications are being prepared, the quantity surveyor should periodically review his estimate; and ideally he should be retained to prepare a more detailed estimate from the working drawings and specifications before they are completed, so that the *costs of the work* can be planned as between the various parts of the *work* and within the budget.

A quantity surveyor's fees will depend on the nature of the project and the extent to which his services are required. A single preliminary estimate might cost a few hundred dollars, and comprehensive cost planning services for a complex project might cost 1/2 of 1 percent of the *cost of the work*. Scales of fees are published by institutes of quantity surveyors such as the Canadian Institute of Quantity Surveyors.

Since the purpose of a *designer's* estimate is to anticipate and predict the amounts of the *contractors' bids*, it follows that a cost consultant (or a *designer*) must be familiar with the techniques and procedures of estimating in construction firms to prepare realistic preliminary and conceptual estimates. Consequently, the precepts and techniques explained later are fundamental to making estimates of construction costs of all kinds.

The Owner-Developer's Estimate

A distinction is made between an *owner-developer's estimate* and a *designer's estimate* because they may differ in both form and function, although they may be one and the same thing. The *owner-developer* may have an estimate made—often for a feasibility study—before he retains the services of a *designer*. A feasibility study is made to examine, among other things, the economic feasibility of a project by estimating the *costs of the work*, and the probable *costs in use* of the building development. The probable gross income from the development is also estimated, and from all of this data the net income and the net return on the investment can be estimated.

It has been shown that in some cases the *costs of the work* can have a very significant effect upon the return on investment yielded by a commercial development.[7] On the other hand, it has been shown elsewhere that the costs of constructing institutional buildings are not so significant when they are considered with the total *costs in use* over the life of the building.

[7] Larry Smith, "*Principles of Feasibility for Revenue-Producing Real Estate*," in Hunt (ed.), *Comprehensive Architectural Services—General Principles and Practice.*

Feasibility studies often make it necessary to estimate the *costs of work* without preliminary design sketches of any kind, and with only a list of space requirements. In making such estimates, cost and design data from previous jobs are used; but, inevitably, some assumptions have to be made about the form and nature of the building in a preliminary estimate. And once these assumptions become part of the study, they may impose restrictions on the *designer* and the building's design. Ideally, therefore, the *designer* should participate in a feasibility study from the outset.

Estimates Generally

The object of all estimates is to supply more information about the *work*, not only the probable costs but also the relative costs of the various parts and how long it will take to construct them. The *owner* wants to know how much he should spend on the building, and what return he can expect. The *designer* wants to know that his design can be built within the *owner's* budget. The *contractor* wants to know what the *work* will probably cost so that he can make a *bid* to do the *work* for a stipulated sum or a fee. If proper estimates are not made, the *contractor* may find himself contractually bound to do *work* at a financial loss. Or the *designer* may be obliged to redesign the *work* at no extra fee. Or the *owner* may find that he has raised a development that is not financially sound and therefore is a poor investment. Estimates and the information they provide are, therefore, essential to the construction industry, particularly in times of a universal shortage of capital for development and rising costs.

The Estimator

Above all, the estimator must have an extensive knowledge of construction, a knowledge of construction materials and methods, and a knowledge of construction practices and contracts. An estimator should be able to read and write *bidding documents,* and he should be able to sketch construction details. He must be able to communicate graphically and verbally. He should have a facility with measuring techniques and mathematics, and he requires a good general knowledge of economics and business.

The estimator must be able to visualize what is not apparent, and for this he requires imagination. He must be able to think in abstract terms. He must be analytical and critical. The estimator should be able to see the fundamental form through prolific detail. His work consists of translating graphic and verbal information in the *bidding documents* into probable costs in the light of experience. This is not a direct and mechanical process; it is creative, and it demands a high level of concentration and mental effort.

The estimator must be able to work alone and to make unilateral decisions on matters that are often nebulous. This requires an ability to make decisions in the light of whatever information is available, and later to be able to say that this decision was made because these were the known facts, and that this was the best decision in the light of those facts. It can be a lonely job, and it requires an integrity that makes for honesty with self and with others, and a desire to seek and know the truth of things. It requires common sense and an ability to work for a goal that is not always achieved.

Some estimators are engineers and architects who have come to specialize in construction costs and contracts. Some are immigrants who have been trained elsewhere as quantity surveyors, particularly from Britain, Australia, and other Commonwealth countries.[8] Others have become proficient estimators by starting work in a construction trade, and then entering the estimating office as a trainee. In the past, such trade experience was not unusual and it was invaluable to an estimator. Today, the traditional trades are changing and disappearing, whereas new materials and methods of construction are proliferating. Today, perhaps, the full knowledge of one trade is less preferable than a general knowledge of many, except possibly in the case of a subtrade estimator. Formal training in construction estimating is not available everywhere, but it is offered in some universities, technical colleges, and institutes, and new programs are continually being developed. In some countries, construction economics and quantity surveying are offered as degree-level courses in universities and institutes of technology; and construction cost research programs have been initiated by professional bodies and governments, so that the body of knowledge and techniques is growing and is being disseminated. But the research is too little and the dissemination is too sparse. Construction economics should be part of the education of every architect, engineer, and of every superintendent and foreman in the construction industry.

[8] See Chapter 13, Quantity Surveying.

Questions and Topics for Discussion

1. Define an "estimate" and identify the two primary parts of estimating.

2. Explain the purpose of calculating the *unit prices* of completed *work,* and explain by a formula how it is done.

3. Explain how *standard forms of contract* should be used and modified for a particular construction project, and explain why.

4. What is the primary purpose of a bid depository, and explain briefly how one operates.

5. Why is "cost per cubic foot" not a good method of making a preliminary estimate?

6. Why is a *designer's* estimate usually required?

7. What is the most valid and useful data for estimating, and why?

8. To what other uses can the information contained in an estimate be put by a *contractor*?

9. What are the several distinct parts of a *contractor's* estimate, and which parts of the *work* do they usually include?

10. If a *contractor* intends to build an office building on his own property, for his own use and occupancy, would it be necessary for him to make any kind of estimate of the *costs of the work*? Explain your answer.

3

What is Cost Accounting?

A study on manpower utilization in the Canadian construction industry concluded:

> There is a lack of detailed cost information available to most contractors. . . . There was a lack of general organizational planning at the job site.[1]

This statement is true, not only in Canada but in other countries as well. Yet if the success of a construction company depends on its ability to obtain *work* in a competitive market and to make a *profit,* information and good management are essential, and *cost accounting* (CA) is one of the means. It was explained in Chapter 2 that CA is necessary to estimating in providing cost data from previous jobs. In addition, CA is essential to good construction management in that it provides other information needed for planning, scheduling, and controlling the *work.*

[1] David Aird, *Manpower Utilization in the Canadian Construction Industry,* National Research Council, Division of Building Research, Technical Paper No. 156 (Ottawa, Canada, 1963), p. 38.

These were the first two of several conclusions reached after work studies were made on several construction jobs in two major Canadian cities. This pilot study is recommended for its content and as an example of the research needed by the construction industry.

Cost accounting has one purpose directed toward three primary ends, one of which is immediate, and the other two, in the future. The purpose of CA is to obtain knowledge and information of construction economics. The three primary ends of CA are:

1. to plan, manage, and control the original job (immediate);
2. to plan, manage, and control other jobs (future);
3. to make estimates of the costs of other projects (future).

The last of the three objectives of CA probably is best understood, because it is the logical answer to the inevitable question, From where does an estimator get the cost data to prepare an estimate? There is only one source of valid data for estimating, and that is past jobs, preferably past jobs estimated by the estimator himself or by his colleagues and performed by the construction company for which he is estimating. Failing that, an estimator may seek guidance from the past experience of others published in handbooks. But he would probably do better by talking to his company's foremen and superintendents and using their experience.

Future Estimates Through Cost Accounting

A simple, hypothetical example will explain the application of CA to one *item of work* by a construction company; and similarly, CA can be applied by a construction company to most *items of work* in a job to obtain cost data for future estimates. This may be done by the company's estimator, or it may be done by a cost accountant in the company who will pass on the results of his work to the estimator. It is obvious that if it is done by a cost accountant he must understand estimating and work closely with the estimator.

Looking at the example in Fig. 5, and comparing **actual** quantities and costs with **estimated** quantities and costs, the amount of concrete used and placed is more than the estimated amount. This should lead to a check of the estimated amount taken from the contract drawings. If the 600 CY is found to be correct, the next step is to look for a reason at the site. The foreman's explanation may be that such an increase is not unusual, because some spillage usually occurs and because forms sometimes bulge and expand. If

DESCRIPTION	QUANTITY	UNIT	UNIT PRICE	TOTAL ESTIMATED MATERIAL COST	UNIT PRICE	TOTAL ESTIMATED LABOR COST	UNIT PRICE	TOTAL
ESTIMATED COSTS (in Estimate)								
Conc Fnd Walls 8" thick	600	cy	18.50	11,100	2.50	1500		12,600
ACTUAL COSTS (through Cost Accounting)								
Conc Fnd Walls 8" thick	618	cy	18.50	11,433	3.00	1854		13,287
RECORDED UNIT PRICES (for future estimates)								
Conc Fnd Walls 8" thick (allow 3% Waste)	600	cy	19.06	11,433	3.09	1854		13,287
Conc P.C. $18.50 cy								
Lab Rate $5.00 hr								
RECORDED UNIT RATE (for future estimates)								
Conc Fnd Walls 8" thick (allow 3% Waste)	600	cy			($3.09 / $5.00 per hr)			= 0.62 manhrs/CY

FIG. 5. Estimated and Actual Quantities and Costs of Concrete Foundation Walls: An Example.

similar results continue to appear on other jobs done by the company, the estimator must make some allowance for this increase, or *waste,* in his future estimates. This can be done by adding a percentage allowance to the future measured concrete quantities (e.g., 600 CY + 3% = 618 CY). But the better way is to keep the measured quantity as a "net quantity" of 600 CY (the amount required by the contract) and to add a 3 percent allowance for *waste* to the *unit price.* The "prime cost" of the concrete of $18.50 per CY is also recorded. In this way, all the facts are retained, and the quantities required by the contract are not obscured by adding estimated allowances for *waste.* Instead, *waste* is allowed for in the *unit price* as a variable to be determined for each job, based on experience.

In the case of actual *labor costs,* a total of $1,854.00 has been charged for this *item of work,* which would give a *unit price* of $3.00 per CY ($1,854.00 ÷ 618 CY), based on a gross quantity of 618 CY. But based on the net quantity of 600 CY (the amount required by the contract), the *unit price* is $3.09 per cubic yard, and this is the *unit price* to be recorded for future estimates.

If no conditions or circumstances can be found that might have caused an increase in the *labor costs,* it must be assumed that the *labor costs* were underestimated, and the actual costs should be recorded. If similar results are found on other jobs, then it will be necessary to increase *unit prices* for this item in future estimates.

If the *unit prices* used in the estimate (in the example) were originally established by CA on several previous jobs, it is unlikely that the *unit prices* would be changed in future estimates because of higher costs on only one job. But a warning signal has appeared that should be heeded, and special attention should be given to this item on other jobs so that if a change in a *unit price* is necessary, it is not made too late and after serious losses have occurred.

This example is simply to illustrate the method. Actual *unit prices* of all major items should be recorded by construction firms, and with sufficient data a mean price can be established together with upper and lower limits (established mathematically by the use of standard deviations) to provide a practical guide for the estimator in pricing. *Unit prices* and *unit rates* for *labor costs* are most important, because labor productivity is the most difficult factor of construction costs to estimate. *Unit rates* (calculated from the *unit prices*) are the best form in which to record this data in order to avoid the changing effect of increases in *wage rates,* as shown in the last example.

Management Through Cost Accounting

Referring again to the example in Fig. 5., a more immediate use of CA data can be seen by comparing estimated and actual costs. In the example, the comparison shows a loss on this *item of work* at completion. This is useful but expensive knowledge, which should be applied in the future to avoid a repetition. If a loss is incurred in placing concrete in foundation walls, it would be better to know before all the concrete is placed so that, if possible, losses might be reduced by reorganizing the *work.* Costs of major *items of work* of several days duration can often be controlled in this way so that probable losses are avoided or actual losses reduced, if CA is done on a daily basis. This is only possible with major items because minor *items of work* are usually completed before any warnings can be heeded. But the major items are the very items most in need of control and attention, because an estimating mistake in those items may cause a serious financial loss.

In the above example of concrete foundation walls, if all the concrete is placed over a period of, say, three days, 200 CY may have been placed after the first day. The amount is easily ascertained from delivery tickets. The *unit rate* for labor placing the 200 CY of concrete can be calculated by dividing the day's *labor costs* chargeable to this item by the total quantity placed. A resultant *unit rate* of more than that estimated will indicate that something is wrong, either with the *work* or the estimate. The work methods should be reviewed by the superintendent immediately and before the next concrete is placed. The reason for the higher *labor costs* might be that the labor crew is too large, or that there is a delay between the arrival and departure of ready-mixed concrete delivery trucks, or some other such cause. If no improvement can be made in work methods, the discrepancy must be assumed to be in the estimate. At least, the question has been raised and answered and, if work methods can be improved before more concrete walls are placed, losses might be reduced on this job and eliminated on other jobs and more competitive *bids* or larger profits may be possible in the future.

This example illustrates a simple comparison made by a construction firm between the estimated and actual costs of one *item of work.* But before such a comparison can be made, actual cost data must be obtained; and if the *item of work* is not complete (and an interim comparison is desired before completion), the amount of *work* actually done to date must be measured. This is the substance of *construction cost accounting*: actual costs properly allocated to the right *items of work*, and known quantities of *work* completed for those costs.

So far, only cost data for estimating and cost control have been discussed. Before *work* starts, a job should be planned and scheduled to achieve the highest possible degree of efficiency; and whichever planning method is used, information is necessary. Each part of the *work* must be assessed for its duration; its materials, labor, and equipment requirements; and for its time relationship to other parts of the *work*. If any part of the *work* or any activity on the job is critical, this should be known at the outset; and it should be given special attention in planning and scheduling, because the start of future activities depends on the completion of certain critical, earlier activities. Hence the development of the "critical path method" (CPM) of scheduling the activities that make up a job. Information required for CPM and for any other planning and scheduling methods can be obtained from the estimate.

For example, in Fig. 5., the estimate of the item "conc fnd walls" contains a labor cost of $1,500.00. With a crew of three laborers (placing concrete directly from a ready-mix truck), each man costing $5.00 per hour, the above amount represents a crew time of 100 hours. Obviously, the total time allocated for this *item of work* in any schedule should be the same as the total time allowed in the estimate. In other words, if the *work* can be planned so that three such crews (of three laborers) can place concrete foundation walls at the same time, four days should be allowed for this activity in the schedule.

An Example of Estimating and Cost Accounting by Computer

The following description and quotations from a technical brochure, entitled **"Att Databehandla ett Byggnadsprojekt"** (**Using Data Processing on a Building Project**), are given with the permission of the developers of the Quantity Calculator QC–95, the company of Samdata AB, Huvudkontor: Vretenvägen 8, Fack, 171 20 Solna 1, Sweden. This company was formerly known as Bygg-ADB, which means in English Building-ADP (automatic data processing).

Samdata AB claims that many leading contractors in Sweden use their services and systems and that they have serviced hundreds of construction jobs. The company has a computer center with divisions for key punching, data processing, programming, systems development, and training. It designs special systems for its clients, and also trains their staffs in the systems. It also provides a key-punching service. The computer center contains an IBM System 360, of the third generation.

Samdata AB explains in its brochure, which is available in English, that:

> Everything is changing. The building industry of the future will look different from that of our time. Computer based systems for building projects (are) today a reality: tomorrow a necessity.
>
> The building industry of tomorrow will require an organization devoted to marketing, prepared to give proposals and offers very quickly. The competition will increase, the bidding periods will be shorter, fast decisions and estimating will be a necessity. The solution is data processing.
>
> Future requirements will be faster deliveries for construction projects, more detailed planning, and more accurate production rates. At the same time, building projects will be more complicated with a greater number of subcontractors, and therefore coordination and more efficient planning will be a necessity.
>
> The lowest price will not always get the contract. In many cases, the construction period will decide the competition. To meet target dates [will be] a necessity. The solution is data processing.
>
> To survive in the building business of tomorrow estimating must be carried out so that it is possible to control the estimates, and so that the project engineers will get faster reports on the progress of the projects; so that any necessary action can be taken. The projects will be larger and more expensive. The risks will increase, and therefore a more advanced technique for cost control will be a necessity. The solution is data processing.
>
> Samdata AB's integrated system is similar to a circle where output data from one routine can be used immediately as input data in the next routine, and where experience from one project is automatically stored and used on future projects.
>
> In the circle [in Fig. 6], there are three arrows. If we start at the lower left of the picture, the first arrow represents the estimating phase. The next arrow represents the planning phase, and the third arrow represents the control phase.
>
> Starting with the estimating arrow, the first thing that happens is the quantity survey, which is indicated by the circle lying outside the large circle. All the smaller circles represent programs which Samdata AB uses in its system. The outer circle stands for quantity surveying.[2] The next circle [2] repre-

[2] It is significant that the quantity survey is shown outside the "cycle of automatic data processing" and at the beginning. The work of the estimator and the quantity surveyor is interpretive, analytical, and fundamental, and cannot be done by a computer. But the estimator and the quantity surveyor can benefit from the use of tools such as EDP and the quantity calculator to speed up the process of measurement and to record descriptions and measurements of work. (author's note)

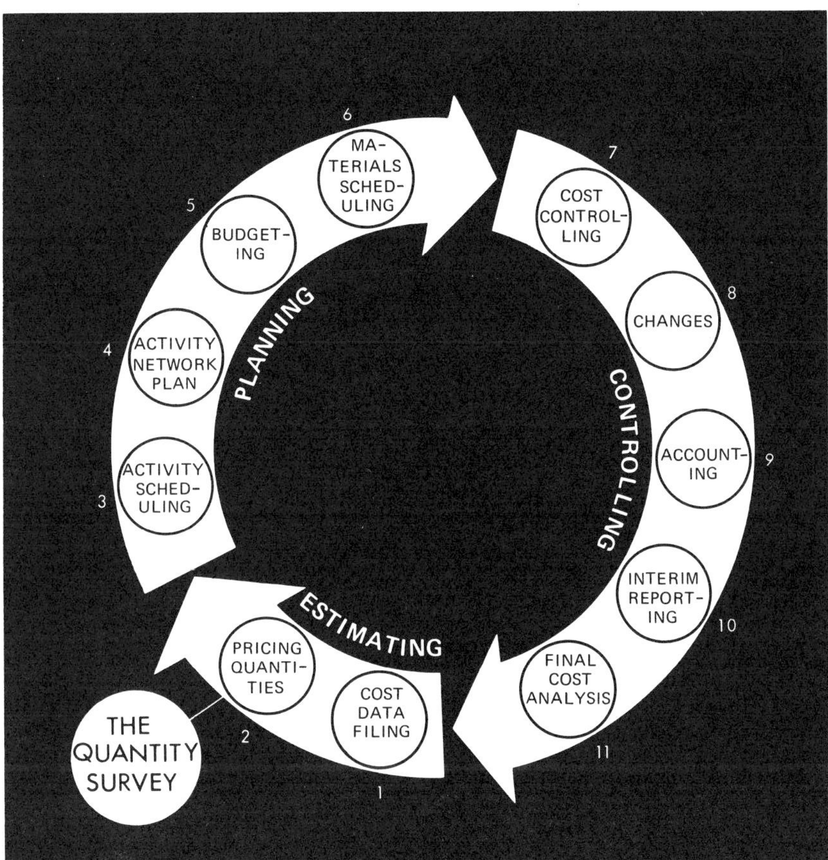

PROGRAMS: (1) COST DATA FILING (Kalkylregister); (2) PRICING QUANTITIES (Kostnads-beräkning); (3) ACTIVITY SCHEDULING (Planerings-underlag); (4) ACTIVITY NETWORK PLAN (Nätverksplan); (5) BUDGETING (Time and (Costs) (Tids-ochkostnads-budget); (6) MATERIALS SCHEDULING (Leveransplan); (7) COST CONTROLLING (Litterer-adförkalkyl); (8) CHANGES IN WORK (Kalkyiandringar); (9) ROUTINE ACCOUNTING (Kostnadsspecification); (10) INTERIM REPORTING (Avstämningsrapport); (11) FINAL COST ANALYSIS (Efterkalkyl).

NOTE: The above titles do not appear in the original and have been introduced into this diagram and the text by the author-translator.

FIG. 6. The Estimating-Planning-Controlling Cycle of Construction.

sents the program for pricing the quantities. This program gives, as its output, the final estimate. For cost estimating it is necessary to use stored data on specifications and costs. These data come from the program [1] preceding the circle for cost estimating. This circle is called "kalkylregister" (cost data file) and represents the programs for updating the standard specifications and standard costs.

In the planning arrow there are four different programs. The first [3], called "Planeringsunderlag" (activity schedule) represents a technique for getting production rates and costs for activities in the network diagram which is prepared for the project. The next circle [4] called "nätverksplan" (network plan) represents the program for network calculations which among many other things, has an advanced technique for resource planning. The two other circles [5 and 6] represent programs for budgeting and preparation of bill of materials.

Finally, the control arrow consists of five different programs. The first [7] called "littererad förkalkyl" is a program for a reorganization of the estimate so that it will fit into the system used for cost control. The second program [8] is for changes in the budget. The third program [9] is used in cooperation with the routine accounting in a contracting company. The fourth circle [10] represents the program for preparing periodic reports on the project. And, finally, the last program [11] is for preparing a cost analysis. On the basis of this last report the standard costs and standard production rates are updated and entered into the stored register [1] with these data. And so the circle is closed.

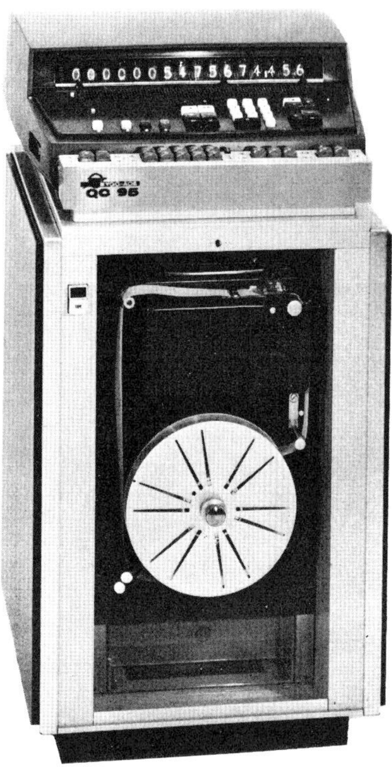

FIG. 7(A) The QC-95 Quantity Calculator.

FIG. 7(B) The Quantity Calculator in Use.

Estimating

With the Samdata AB Quantity Calculator QC-95–a unique Swedish product–it is possible to count and measure items directly on the blue-prints and enter them into a paper tape which then can be fed into a computer for further data processing [see Fig. 7A and B].

The Quantity Calculator has many advantages; since the take-off can be speeded up considerably, and most of the handwriting can be eliminated.

Calculations and typing are done automatically by the computer and no extra punching is needed, as this is done by the Quantity Calculator.[3]

With the Quantity Calculator the quantities of a project can be specified into a large number of so-called planning units. This very detailed specification, which can be done in a considerably shorter time than is needed today, makes it possible to use the original estimate for the future planning and control: an important part of the integration idea behind Samdata AB's system. When using the Quantity Calculator for take-off, two pencils are used for counting and measuring the different items on the blue-print. The results of the estimating phase are, first of all, printed lists for immediate use; and, second, stored data on magnetic tape for later use in connection with planning and control [See Fig. 8A and B].

"Mätprotokollet" is a list showing exactly the measurements and quantities registered on the Quantity Calculator.

"Mängd- och kostnadsberäkningen" [quantity and cost estimate] is the final estimate for the project, or, if required, a schedule of quantities with the cost figures omitted.

"Kalkylregistret" [cost data file] is a document showing all standard specifications and standard costs which are stored on the contractor's tape. It is this document which is used for reference when using the Quantity Calculator. "Kalkylregistret" [cost data file] makes it possible to check and change specifications and prices at any time.

The combination Quantity Calculator and Data Processing gives faster take-off and calculations, gives increased accuracy with more reliable prices and more correct calculations, and it gives an integration with planning and control through the data stored on the magnetic tape.

[3] "The quantity calculator has three main functions: **measuring, calculating,** and **punching.** These functions can be used together or one at a time.

For the **measuring** there are two special pencils attached directly to the quantity calculator. The first pencil is a sort of tracker with a serrated wheel at the tip and is used to take-off linear distances. The other pencil is used for counting single items and also works as a ballpoint pen so that the items can be marked on the blue-print.

The **calculations** are performed in an electronic desk calculator which can be used either just for calculations, or as a part of the quantity calculator, and, in the latter case, together with the pencils and the punching equipment.

The **punching** is performed by a paper tape punch which also is part of the QC-95. Data to be punched are entered on a special keyboard and the output is an 8-channel paper tape which can then be fed into a computer for further data processing.

To the QC-95 can also be attached a hard copy device so that it is possible to see what is printed on the paper tape."

BYGG-ADB — MÄTPROTOKOLL

Firma	Projekt	Etapp	Utskriftsdatum	Sida
100	1000	1	27-06-66	3

Firma: AB BYGGFÖRETAG — Projekt: KV. SMULTRONET — Etapp: HUS 1 OCH 2

Kalkylnummer (x anger standardkalkyl)	Planeringsdel	Mätdata	Mängd
14012x	11	0.30 x (8.00 + 41.50 + 7.50 + 9.40 + 7.00)	22.00
	12	0.30 x (8.00 + 30.60 + 4.40 + 7.00)	15.00
	13		15.00
	22		15.00
	23		15.00
	21	0.30 x (8.00 + 7.00 + 41.50 + 7.50 + 10.00 + 6.00)	24.00
14011x	11		113.00
	12		
	13		
	21		
	22		
	23		
14013x	11		
	12		
	13		
	21		
	22		
	23		
14912	11		
	21		
	12		
	13		
	22		
	23		

BYGG-ADB — MÄNGD- OCH KOSTNADSBERÄKNING

Firma	Projekt	Etapp	Utskriftsdatum	Sida
100	1000	1	27-06-66	7

Firma: AB BYGGFÖRETAG — Projekt: KV. SMULTRONET — Etapp: HUS 1 OCH 2

Kapitel: 14 — Kapiteltext: BETONGARBETEN

Kalkylnummer (x anger standardk.)	Beskrivning (Planeringsdel / Delmängd)	Mängd	Enhet	1.MTRL + UENTR À-pris	1.MTRL + UENTR Kronor	2. TRÄ À-pris	2. TRÄ Kronor	3. GROV À-pris	3. GROV Kronor	4.MUR À-pris	4.MUR Kronor
14011x	VÄGGFORM TOM 0.25 M KANTSÅGADE BRÄDER										
	11 113.0										
	12 119.0										
	13 119.0										
	21 121.0										
	22 119.0										
	23 119.0										
	TOTALT	710.0	LM	2.00	1420	2.40	1704	0.60	426		
14012x	VÄGGFORM TOM 0.40 M KANTSÅGADE BRÄDER										
	11 22.0										
	12 15.0										
	13 15.0										
	21 24.0										
	22 15.0										
	23 15.0										
	TOTALT	106.0	M2	5.70	604	7.45	790	2.15	228		
14013x	VÄGGFORM TOM 2.0 M KANTSÅGADE BRÄDER										
	11 35.0										
	12 35.0										
	13 35.0										
	21 35.0										
	22 35.0										
	23 35.0										
	TOTALT	210.0	M2	3.75	788	5.40	1134	1.60	336		

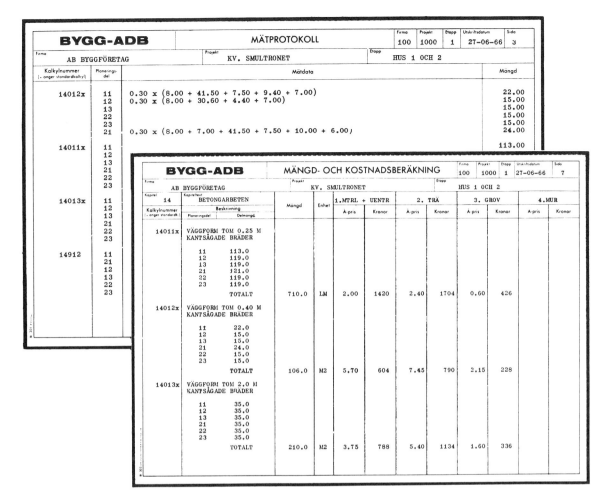

FIG. 8(A) **Estimator's Quantity Sheet and Schedule of Quantities Sheet.**

FIG. 8(B) **Cost Data File Sheet.**
"The three lists shown and described contain entries for wall forms (vägform) and column forms (pelarform) of wood boards, and round column forms. Entries include: the dimensions and quantities; unit prices, and costs of materials; carpenters' time, and laborers' time. The units are linear meters (LM) and square meters (M2)."

BYGG-ADB — KALKYLREGISTER FÖR SPECIFIKATIONER OCH SNABBKALKYLER

Firma	Utskriftsdatum	Sida
100	27-06-66	8

Firma: AB BYGGFÖRETAG — À-priser

Kalkylnummer	Produktionsmetod	Beskrivning	Enhet	1.MTRL	2.TRÄ	3.GROV	4.MUR	5.UENTR	Nr	Nr	Nr
14011		VÄGGFORM TOM 0.25 M KANTSÅGADE BRÄDER	LM	2.00	2.40	0.60					
14012		VÄGGFORM TOM 0.40 M KANTSÅGADE BRÄDER	M2	5.70	7.45	2.15					
14013		VÄGGFORM TOM 2.0 M KANTSÅGADE BRÄDER	M2	3.75	5.40	1.60					
14014		VÄGGFORM ÖVER 2.0 M KANTSÅGADE BRÄDER	M2	3.60	5.20	1.60					
14015		PELARFORM H TOM 0.40 M KANTSÅGADE BRÄDER	M2	10.00	8.80	2.10					
14016		PELARFORM H TOM 0.70 M KANTSÅGADE BRÄDER	M2	10.00	8.80	2.10					
14017		PELARFORM H TOM 2.00 M KANTSÅGADE BRÄDER	M2	10.00	8.80	2.10					
14018		3/4x TREKANTLIST	LM	0.45	0.40	0.05					
14019		RUND PELARFORM DIAM 25	LM	11.50	12.00	2.50					
14020		RUND PELARFORM DIAM 35	LM	12.00	12.00	2.50					
14021		RUND PELARFORM DIAM 50	LM	12.50	12.00	2.50					

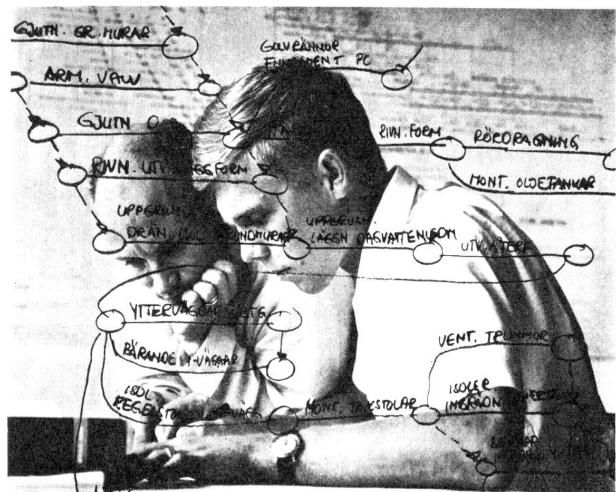

FIG. 9(A) Preliminary Network Diagram.
"The preliminary network diagram, shown in part, indicates how activities are planned manually before the final schedule is produced by computer."

FIG. 9(B) Network Activity Schedule.
"The activity schedule (planeringsunderlag) shown contains entries for those same items shown in Figs. 8(A) and 8(B), together with other related items of work."

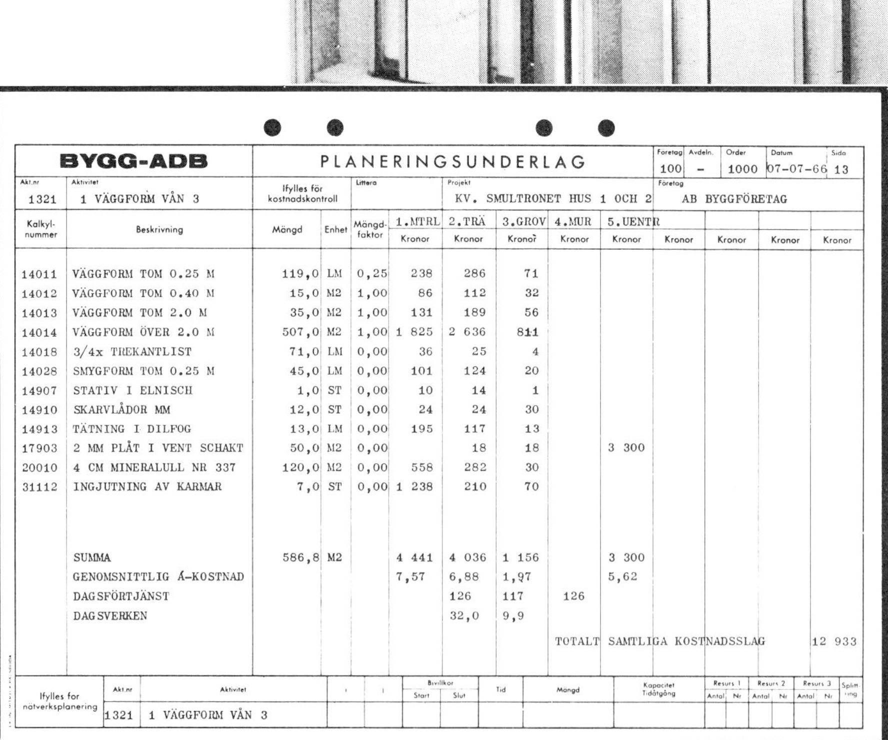

BYGG-ADB — PLANERINGSUNDERLAG

Företag	Avdeln.	Order	Datum	Sida
100	-	1000	07-07-66	13

Akt.nr	Aktivitet	Ifylles för kostnadskontroll	Littera	Projekt	Företag
1321	1 VÄGGFORM VÅN 3			KV. SMULTRONET HUS 1 OCH 2	AB BYGGFÖRETAG

Kalkyl-nummer	Beskrivning	Mängd	Enhet	Mängd-faktor	1.MTRL Kronor	2.TRÄ Kronor	3.GROV Kronor	4.MUR Kronor	5.UENTR Kronor	Kronor	Kronor	Kronor	Kronor
14011	VÄGGFORM TOM 0.25 M	119,0	LM	0,25	238	286	71						
14012	VÄGGFORM TOM 0.40 M	15,0	M2	1,00	86	112	32						
14013	VÄGGFORM TOM 2.0 M	35,0	M2	1,00	131	189	56						
14014	VÄGGFORM ÖVER 2.0 M	507,0	M2	1,00	1 825	2 636	811						
14018	3/4x TREKANTLIST	71,0	LM	0,00	36	25	4						
14028	SMYGFORM TOM 0.25 M	45,0	LM	0,00	101	124	20						
14907	STATIV I ELNISCH	1,0	ST	0,00	10	14	1						
14910	SKARVLÅDOR MM	12,0	ST	0,00	24	24	30						
14913	TÄTNING I DILFOG	13,0	LM	0,00	195	117	13						
17903	2 MM PLÅT I VENT SCHAKT	50,0	M2	0,00		18	18		3 300				
20010	4 CM MINERALULL NR 337	120,0	M2	0,00	558	282	30						
31112	INGJUTNING AV KARMAR	7,0	ST	0,00	1 238	210	70						
	SUMMA	586,8	M2		4 441	4 036	1 156		3 300				
	GENOMSNITTLIG Á-KOSTNAD				7,57	6,88	1,97		5,62				
	DAGSFÖRTJÄNST					126	117	126					
	DAGSVERKEN					32,0	9,9						
								TOTALT	SAMTLIGA KOSTNADSSLAG				12 933

Ifylles för nätverksplanering	Akt.nr	Aktivitet			Bivillkor Start	Bivillkor Slut	Tid	Mängd	Kapacitet Tidåtgång	Resurs 1 Antal	Resurs 1 Nr	Resurs 2 Antal	Resurs 2 Nr	Resurs 3 Antal	Resurs 3 Nr	Splittring
	1321	1 VÄGGFORM VÅN 3														

Planning

The planning phase is started by drawing up a Network Diagram prepared in teamwork by a planning engineer and the foreman on the building site. In this first phase there are no time-allowances whatsoever in the diagram. When it comes to the use of network technique in building industry Samdata AB has, for a long time, taken the lead. The experience gained from a large number of projects under different circumstances and for different clients has served as the basis when developing the Samdata AB system.

The estimate and network diagram are linked together so that the quantities and costs of the estimate are distributed over the activities in the network diagram. When doing this, the data earlier stored on the magnetic tape are used. The output is the so-called "Planeringsunderlag" which is used first, for deciding on the durations of the different activities, and, second, as a tool for the foreman on the building site. ["Planeringsunderlag" could be called an "Activity Specification," showing in detail the amount of work that has to be carried out under the different activities.]

The network diagram is run through a computer; which means that the starting and finishing dates of the various activities are calculated under different conditions so that an efficient use of the available resources is achieved.

The advanced technique for resource planning which has been developed by Samdata AB makes it possible to have the computer determine the number in the crews on the building site, so that an even manpower level is obtained.

The final network diagram can, in certain cases, be drawn automatically by a lineplotter and gives a very detailed picture of how the planning for the project is carried out.

Control

The collection of data, required for the control of the building project, must be coordinated with the accounting procedures that always exist within a contracting company. A complete integration means that all notations about different data have to be done once only and then are available on the magnetic tapes both for the accounting routines and for the production routines. In the Samdata AB system there are also programs for the accounting routines; but,

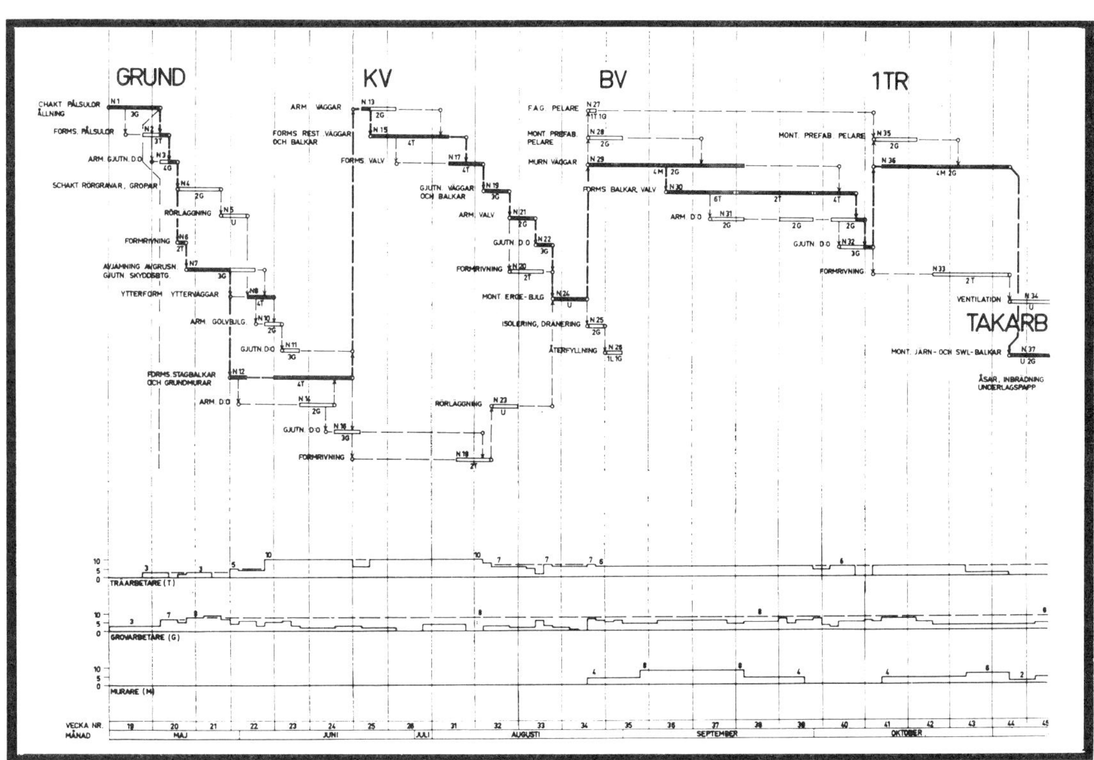

FIG. 9(C) Final Network Diagram.
"The final network diagram, shown in part, schedules the various work activities, such as forming and pouring concrete walls & beams, and erecting prefabricated columns. It also shows, at the bottom, the scheduled requirements for workmen (carpenters, laborers, and masons), each week, according to the activities."

a description of these is not within the scope of this brochure.
However, . . . Samdata AB has developed . . . programs for invoice handling, machine costs, payrolls and general accounting.
The control results in the so-called "Avstämningsrapport" [periodic report] which usually is run through the computer once every month. This "Avstämningsrapport" gives, for each account, a comparison between estimated and actual costs and also shows completed quantities and actual unit costs and, furthermore, gives a forecast of the expected final costs.
The last "Avstämningsrapport" [periodic report] for the project is also the final cost analysis. From this report it is possible to get the data about costs and production rates which the contractor wants to enter into his register with standardized specifications and costs.
The circle is thereby closed, and the goal is achieved. This means that, from the existing production, all the necessary information for estimating and planning of future projects has been collected."

This description is reproduced because it simply and clearly indicates the essential relationships among the several phases of construction management and the importance to construction firms of information and data that can be obtained from estimates and past jobs. But it should not be assumed that this cyclical process of estimating–planning–controlling is only possible with the aid of a computer. CPM scheduling, for example, can be done without computers, and much of the facility with which planning and controlling can be carried out depends on the way the estimate has been prepared. Many estimators are aware of this when they measure *work* in stages (such as floor by floor, for a high-rise building), and an estimate should always be made with other uses beyond the *bid* in mind.

Questions and Topics for Discussion

1. What are the primary purposes of *cost accounting*?

2. What is the best form in which to record data of labor productivity for the various *items of work* in a construction job?

3. Explain briefly how *cost accounting* can sometimes help a *contractor* to avoid or reduce a loss on a construction job.

4. How does an estimate of the *costs of work* directly relate to a schedule for the *work*?

5. Draw a diagram to show the *estimating and cost accounting* cycle, indicating the main construction management functions within the three primary phases of the cycle.

6. What is the initial function that precedes data processing in construction estimating and management, and why cannot it be done by a computer?

7. What is the primary advantage to an estimator in having EDP, and why? Is there a disadvantage?

8. If *cost accounting* shows, at the end of a work day, that the first fifty cubic yards of a concrete retaining wall (containing a total of 200 yards) have cost more to place than was allowed in the estimate, what possible conclusions might be drawn?

9. Explain in detail how data from an estimate can be utilized in any one of the essential parts of the management of a construction project.

10. If *cost accounting* accounts for the *costs of work* done, what other information about the *items of work* is essential to make the costs meaningful?

4

Construction Economics: an Introduction

At present, estimating is not a science, if by a science we mean "a branch of knowledge or study dealing with a body of facts or truths systematically arranged and showing the operation of general laws.[1] Building construction is based on science, and measurement is scientific; but in construction estimating there is no "body of facts or truths systematically arranged and showing the operation of general laws." Among all the estimators in all the construction companies there is a great reservoir of facts and an abundance of knowledge and practical experience. But this information is not systematically arranged and until it is we shall see few general laws in operation. There is a great need for research and dissemination of facts about construction economics.

In her book entitled *Economic Philosophy*, Joan Robinson refers to

1 *The Random House Dictionary of the English Language* (1966).

economics as: "a vehicle for the ruling ideology of each period as well as partly a method of scientific investigation," and goes on to say: "All along it [economics] has been striving to escape from sentiment and to win for itself the status of a science."[2] If this can be said of economics, with its impressive array of thinkers, writers, experts, and schools, and with its established place in society and government, what can be said for construction economics and estimating?

Since the Second World War, statistics and statistical methods have been increasingly used in many fields. This same period also has witnessed the great development in electronic data processing (EDP) and computers, which has stimulated the use of statistical methods in matters of probability because of the ability of computers to deal with huge quantities of data.

It appears that construction estimating has not yet been widely affected by the use of statistical methods and EDP; although some important developments have occurred, such as the one in Sweden described in the last chapter.[3] Yet construction is a basic industry, producing between one seventh and one fifth of the gross national product,[4] and the complexities of the estimating process appear to make it an ideal subject for scientific methods and computers. In the same way that an international crisis produces new sciences, scientific methods, and techniques, so, perhaps, another crisis of a different kind will eventually bring about a change and new developments in the construction industry.

We do not really know what buildings cost, and why they cost what they do. Precisely what affect do building regulations, the season, the climate, the design, the *designer*, and the present methods of bidding and arranging contracts have on construction costs? No one really knows. Some fundamentals appear obvious, but have they been scientifically tested? Some research is being done, but it appears to be insignificant considering the general lack of knowledge of economics in the design and construction industry.

One fact, however, has become clear. And that is the interrelation between Time and Cost in construction. Time is the essence of a contract, and time is also money. Interim financing is a major construction expense; and the risks arising out of having to predict costs over extended periods often cause higher costs for the *owner* or more losses for construction firms. It has become obvious to many that traditional methods of designing and constructing buildings are no longer effective.

An Efficient Industry?

The design of each major building is more or less original because of the requirements of the *owner* and the site. But, many buildings could be constructed almost entirely of standard components except for certain plastic materials such as concrete and waterproofing in the substructure. Nevertheless, in many *designers'* offices, draftsmen are employed to draw and redraw building components and details that are essentially the same in function and appearance from one project to the next. Often the details in the drawings and the descriptions in the specifications have been copied from manufacturers' catalogues or from previous jobs, thus creating the illusion of originality. This practice is still taught in many architectural and engineering schools, and in many technical institutes; as a result, modular construction is not common, the use of standard components is not always widespread, standard drawing practices are hardly used, and specifications are too often ponderous conglomerates of misused standards and jargon.

Considered objectively, there appears to be no reason why design and construction should be separated, as they so often are in English-speaking countries. There may have been advantages to this separation in the past, but do they still exist? And if they do, can we say that they are paramount and are not outweighed by other advantages that could be obtained by integrating design and construction?

What were the advantages to society of a profession specializing in design with no commercial interest in construction? Has the last century and a half clearly demonstrated the success of this concept? Could not professionalism have been nurtured in the construction fraternity if professionalism was to be the means of attaining better buildings and a better urban environment? And has professionalism brought us to this goal? Or, has it led us away from the ideal of a profession of builders with a responsibility to society to both design and build that which is sound, commodious, and delightful? We do not underrate the importance of the environment to man's well-being, and the special responsibility to society of those given the privilege to shape it. Rather, we would ask if a better built environment has been achieved by separating the designer from the builder. This is not to say

[2] Joan Robinson, *Economic Philosophy* (Pelican Books, 1964), pp. 7, 25.

[3] For some other developments see Appendix A.

[4] The lower figure applied to the United States of America and the higher figure to Canada during the last decade.

that all the skills required could be found in the members of some kind of super profession of designer-builders. But then we do not expect any of our professionals to be universal men, and specialization is now necessary in all professions.

In the balance, it does seem that separating the functions of designing and constructing has caused problems that are still with us. By isolating and exalting one, the other has not been raised up. However, looking back we must respect the accomplishments of our designers and builders. Looking forward we may also see the need for change.

The present method of estimating and bidding for stipulated sum contracts has not changed radically in the last 100 years. But in that time construction materials and methods have changed. In the nineteenth century, cast-iron columns and beams were first used in a building frame for a seven-story mill designed by two engineers, Boulton and Watt; the last named the same Watt who invented the steam engine.[5] At that time, new buildings appeared that were products of the industrial revolution and owed little to the past. Since then, building construction has been mildly stimulated by war and space technology, but design and contracting methods have not kept pace. Drafting techniques are often still essentially the same as they were generations ago. Specification writing is just beginning to appear as a learned skill, but it is still not taught in many architectural and engineering schools. University courses in construction management and estimating are almost as rare as courses in specifications and contracts.

Many large, modern buildings are being built under new contractual arrangements and by new systems; but most are still built by a contracting system which relies on inadequate methods of planning and communication and a wasteful method of making contracts. Some approximate figures (which cannot be called statistics) may indicate one area for research.

Several estimators have estimated that the cost of bidding for stipulated sum contracts is about one- to two-tenths of 1 percent of the total job cost for each general bidder. The sub-bidders estimating costs are additional. The *contractor* usually does about 25 percent of the total job with his own forces. The other 75 percent is done by *subcontractors*. By direct proportion, the *subcontractors'* total bidding costs would be about three times those of the *contractor*. But the *contractor* has the additional responsibility for assembling the total estimate and bid, including the sub-trades. To be conservative, we will assume all the *subcontractors'* bidding costs are 1½ times as much as the *contractor's* bidding costs. There are usually several bidders, so let us assume that there are six general bidders and an average of three bidders for each subtrade on a project estimated to cost $1 million.

COST OF BIDDING FOR A $1 MILLION BUILDING PROJECT

General bidders:	$1 million × 0.002 × 6 bidders =	$12,000.00
Subtrade bidders:	$1 million × 0.003 × 3 bidders =	9,000.00
	Estimated Total Costs of Bidding	$21,000.00

Although based on several assumptions, this figure indicates total bidding costs of from 1 to 2 percent of the total *costs of the work*.[6] Of the total costs above, only $5,000.00 spent by the successful bidders is productive. The other $16,000.00 spent by the unsuccessful bidders is wasted. Nevertheless, it has to be paid, and it must be included in the total *overhead costs* added by all bidders to their estimates. It is, therefore, paid for by the customers of the construction industry.

Another major disadvantage of the present system of contracting for stipulated sums is the military-like chain of communication from *owner/designer* down to the *contractor*, and from *contractor* down to the *subcontractors*. This arrangement may have been adequate when *designers* were generally knowledgeable about all the traditional trades, and most of the construction methods and details were commonplace and well known. But today, construction has become too complex for the *designer*—even with a team of specialist consultants to help him—to make all the decisions and to prescribe all the materials and methods before the *contractor* is selected and can be consulted.

Designers and developers, generally, do not appear to have changed their attitudes about buildings if present building designs and developments are evidence. Most buildings are still designed and built for one limited use, such as an office block, a school, a church, and so on. And each of these uses is exclusive of other

[5] See S. Giedion, *Space, Time and Architecture* (Cambridge, Massachusetts: Harvard University Press; London: Oxford University Press, third edition, 1954), Part III, *"The Evolution of New Potentialities,"* for an account and illustrations of this and other construction developments in the nineteenth century.

[6] Some other estimates of bidding costs are much higher, but comparisons are difficult without precise definitions. A *general contractor's* estimator can often put together a bid for $1 million job in less than four weeks, and several estimators have each consistently bid from $12 million to $20 million worth of *work* a year, including *subtrade work*. But, the required time and bidding costs for *subcontractors* and for smaller projects is often much greater.

uses and is limited to eight, six, or even two hours a day, depending on the building's function and the denomination of the users. Of all the plant and equipment and fabricated resources of modern society, none is more expensive to produce and cheaper to operate than buildings, and none is used less efficiently.

Buildings contain few moving parts—doors, window vents, and some fans and pumps. They do not normally suffer from vibration, overheating, and worn bearings like equipment and machinery, and depreciation is not a major expense. The greatest expense and the economic failure of most buildings and their sites appears to stem from the fact that they are not used enough before they become obsolete. To many *owners* and *designers*, a building has only one special and limited use; and our urban areas are still organized according to rural timetables and planned for the nineteenth century.

The construction industry is conservative not so much because it is opposed to change, but because it is bound by obsolete laws and regulations and operated by *designers* who cannot experiment, by trade unions that are divisive, and by *contractors* who are conservative because they are caught between the others. Research and experimentation is essential to the scientific method, yet it is practically non-existent in the construction industry, which is second only to agriculture in its importance to people. There are a few building research institutes but presumably with small budgets and less influence to judge from their effect on the average construction company.

Almost anyone can become a *contractor*. The primary requirement is to stay in business for the first five years or so and to learn by experience and survival. Many do not survive the first five years. About 50 percent of all construction business failures occur in the first five years of operation, and most are due to lack of experience and plain incompetence. There is relatively little suitable training and education available for the young man who decides that he wants to go into the construction business, although this situation has started to change in the last few years. Perhaps more than from anything else, the industry has suffered from the absence of a professional approach to practical building construction. Much of the knowledge necessary for operating a successful construction business, particularly the knowledge of construction economics, has not been formalized, accumulated, and passed on as in other professions. The professionals have, for the most part, concerned themselves with design and have left its execution to the *contractors*, as though the idea and the act could be separated without some loss. Design is essentially an individual act, and it seems that *designers* generally are not of a kind to pool and disseminate their knowledge in the manner of other professionals. Finally, the design and construction industry suffers—or rather, everyone suffers—from the practice of awarding public contracts to the lowest bidder, regardless of whether or not the bidders have been selected or approved by the *owner* and the *designer* and irrespective of their qualifications and experience.

The common reaction to criticism of the industry is to point to the many fine construction works that have been designed and built and to say that such criticism is unwarranted in the light of the industry's achievements in the last 100 years. But it is not to detract from past achievements to say that the next 100 years will be much different, and that we must change and adopt and use new methods of designing, contracting, and constructing buildings now and again, and again, in the future.

Much of what was learned in the past is of value today, and will be of value tomorrow. In the past, builders and estimators were not unaware of the many probabilities involved in estimating and planning a project. It was simply that there were so many probabilities that there was neither the time nor the techniques to deal with more than the most immediate and obvious. Human intuition has always played a major part in estimating and planning. Its value cannot be overstated—and it will never be entirely replaced. But now it can be raised up and given full scope by scientific methods and electronic hardware (EDP) that can handle immense quantities of data in microseconds and faster:

> The advent of transistorized computers (second generation, starting in 1958), increased processing speeds to microseconds (millionth of a second), and the present-day, solid-state, printed-circuit computers (third generation, starting approximately in 1963) are operating in nanoseconds (billionth of a second).[7]

Computers, EDP, new information and communication systems, and new construction techniques such as "systems building" and mass production of building components, are slowly changing the design and construction industry, particularly its contracting and estimating methods. Now we should be able to collect and systematically arrange a vast body of facts about the economics of construction and to observe the operation of economic laws that at present are still theories and assumptions at best, and that for the most part are probably still unrecognized.

The greatest advances appear to have been made

[7] James A. Saxon and Wesley W. Steyer, *Basic Principles of Data Processing* (Englewood Cliffs, N. J.: Prentice-Hall, Inc., 1967), pp. 181, 182.

in industrial and heavy construction wherein engineers have used their traditional and practical tool, mathematics, to rationalize construction economics; and at least one large corporation has developed a statistical approach to the probabilities of construction costs.[8]

The Economics and Costs of Construction

The economics of construction cannot be examined in isolation because they are a small part of general economics. But certain simple fundamentals of construction economics and costs stand out, and these are examined here and in the next chapter.

Construction costs have two origins:

1. the *owner/designer,* through the *owner's* requirements and the design (examined in this chapter)
2. the *contractor/subcontractors,* through the competitive market and their own organizations (examined in the next chapter)

These origins cannot always be separated because one is dependent in part on the other, and the *contractor's* determination of costs always depends in large part on the design of the *owner/designer.* At the same time, there are certain aspects of construction costs that are beyond the control of the *owner/designer* and are determined in the construction market. Therefore, although realizing the limitations, it is expedient to examine construction costs from these two aspects.

Costs Through the Design

It is primarily the *owner* who originates the costs of construction through his construction requirements and his ability to pay for them. Although at first he may not always know exactly what his requirements are, and may be able to express them only in general terms, he will probably have a clear idea of his financial limitations. Therefore, it is important that the *designer* (or one of his consultants) be competent to advise the *owner* on construction costs as well as design.

The ways in which the design can affect costs of a building are practically innumerable. Some of the principal ways include shape and size, materials and methods of construction, and design style.

[8] See Appendix A, Data and Probabilities of Construction Costs.

Shape and Size. These factors must be considered together because they are two aspects of the same thing—the three-dimensional nature of the building, its form and its mass. Buildings have volume and they have surface area. The volume is the enclosed space, which is the building's reason for existence. The enclosure—the exterior walls, roof, and floor—has surface area. Under present circumstances, spherical buildings are generally less practical than buildings made up of horizontal and vertical planes, although Buckminster Fuller may have started a movement toward hemispherical buildings with his geodesic dome construction style.[9] Therefore, the cubic building is usually still the most economical shape, if volume is the primary consideration (see Fig. 10). A square has the lowest ratio of perimeter to area for rectangles, and the surface area of exterior walls is a function of the perimeter. Therefore, the closer a plan approaches a square, and the closer a building's shape approaches a cube, the more economical the enclosed building space will be, all other things being equal. Unfortunately, few buildings work well as a cube, but the general principle can be applied to obtain the most economical shape and size commensurate with other needs.

Straight lines and planes are usually cheaper than curved and irregular lines and planes in buildings. Right angles are cheaper than irregular angles, and rectangles are cheaper than circles and irregular shapes. These facts are not invariable and building technology may change them. For example, a thin shell reinforced concrete barrel-vault roof can sometimes be more economical than a flat roof; and geodesic domes are an economical shape to enclose spaces for certain purposes.

Higher buildings cost more to build than lower buildings of equal capacity because of the extra costs of a structure required to resist winds and earthquakes as well as to support its own weight. There are also extra costs involved in hoisting men and materials, in safety requirements, and in limited working spaces in high buildings. High buildings also require additional services, such as elevators and pumps. The exterior walls of high buildings must be designed to withstand higher wind pressures, which distort the components and cause dust, air, and water to infiltrate. Heat losses at higher elevations are usually greater because of greater temperature differences, and the exposure to which a high building is subjected requires special design considerations for other

[9] Buckminster Fuller, "Preview of Building," Chapter 11 in Robert W. Marks (ed.), *Ideas and Integrities* (New York: Collier Books, 1969).

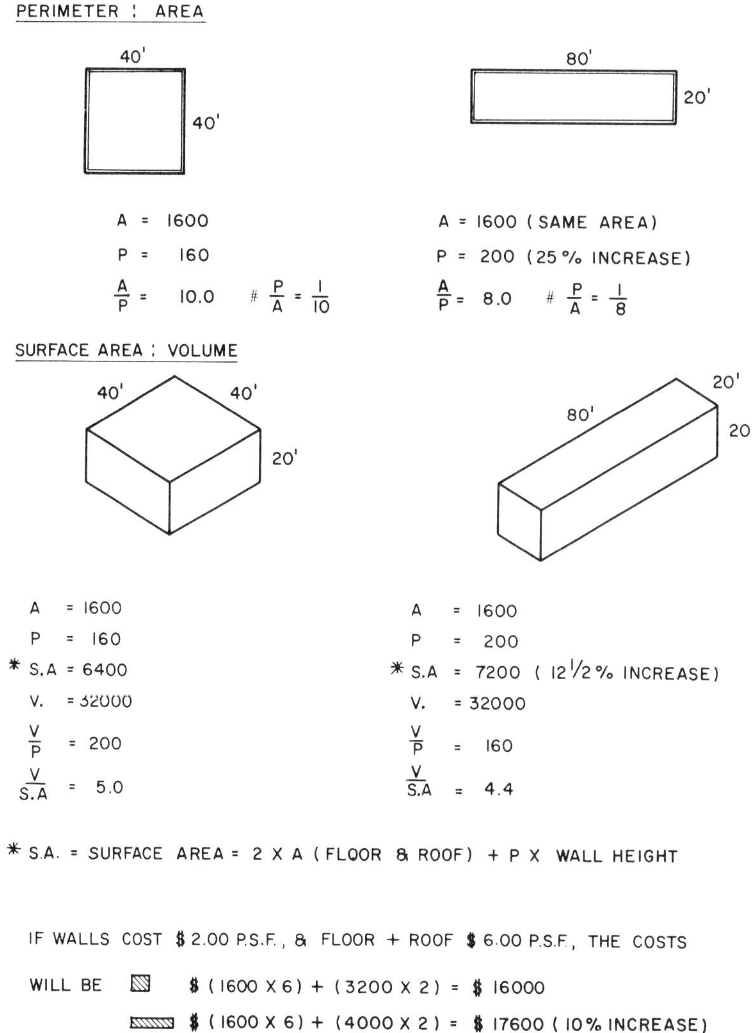

PERIMETER : AREA
40'
40'
80'
20'
A = 1600
P = 160
A/P = 10.0 # P/A = 1/10
A = 1600 (SAME AREA)
P = 200 (25% INCREASE)
A/P = 8.0 # P/A = 1/8
SURFACE AREA : VOLUME
40'
40'
20'
80'
20'
20'
A = 1600
P = 160
* S.A = 6400
V. = 32000
V/P = 200
V/S.A = 5.0
A = 1600
P = 200
* S.A = 7200 (12 1/2 % INCREASE)
V. = 32000
V/P = 160
V/S.A = 4.4
* S.A. = SURFACE AREA = 2 X A (FLOOR & ROOF) + P X WALL HEIGHT
IF WALLS COST $ 2.00 P.S.F., & FLOOR + ROOF $ 6.00 P.S.F, THE COSTS
WILL BE $ (1600 X 6) + (3200 X 2) = $ 16000
$ (1600 X 6) + (4000 X 2) = $ 17600 (10% INCREASE)

FIG. 10. Shape and Size: An Example of the Effect of Shape on the Costs of a Building.

conditions, including thermal movement, anticorrosion, lightning protection, and cleaning and maintenance. All of these features make high buildings more costly.

Both shape and size affect costs, and smaller buildings are more expensive per unit area. It is not the enclosed volume of space in a building that costs money, but rather the surface area and quality of the enclosure (the horizontal and vertical planes that are the walls, roofs, and floors), and the environmental services, such as plumbing, heating, ventilating, air conditioning and lighting.

Materials and methods of construction. These costs are partially determined by the *owner/designer* and are partially determined by external factors, such as:

1. climate
2. availability of material and labor
3. laws and regulations
4. attitudes of people toward buildings

These factors affect construction costs through their influence on the building's design and on the materials and methods of construction.

Climate, particularly rainfall and temperature, should be a primary consideration of the *designer* in selecting materials and construction methods. Because of advances in building technology, climate is often ignored or defied—sometimes to the *owner's* discomfort and loss. Fashion may dictate the design and supercede local customs based on climate. Because of climate, houses in Florida, for example, generally cost less per square foot than similar houses in northern states, because of the difference in building design and heating requirements. Climate also influences costs through its affect on the efficiency of labor at the construction site.

Availability of materials and labor affects local construction costs. Structural steel made in North America may be more expensive in the State of Washington and in the Province of British Columbia than it is in the State of New York and in the Province of Quebec because of the high freight costs for shipping steel from eastern steel mills. Consequently, reinforced concrete structures are more common than steel structures in the West. But imported steel may cause a change.

During periods of intense construction activity the costs of certain trade work may increase, because of a shortage of skilled tradesmen and the necessity of employing whoever is available; including less skilled and less productive tradesmen. Also, supply and demand affects the costs of construction, the same as every other product.

Laws and regulations affect building design and construction costs through local building codes, bylaws, fire regulations, and other ordinances. Unfortunately, some codes and regulations are obsolescent and many are not entirely applicable to modern construction. Building codes often vary from place to place without reason, and consequently North America is covered with a patchwork of different building jurisdictions. Although Canada has made good progress toward the widespread use of a national building code, there is still some parochial resistance. One solution is to give *designers* much more responsibility to ensure that the buildings they design fulfill certain performance specifications. The *designer* is usually better qualified than an official enforcing a code, and after all it is not usual to have an official from city hall watch a surgeon operate on his patient.

Attitudes of people, such as customers and neighbors, toward a building's design and appearance may greatly influence the design and the choice of materials and methods of construction, and thereby affect the building's cost. This is particularly true in the case of residences and retail and commercial buildings in which appearance is a means of attracting customers. *Contractors'* and *subcontractors'* attitudes toward *owners* and *designers* also affect construction costs. There appears to be evidence that in certain instances the *owner* and the sources of his finances affect the price level of construction.

Some *owners* and some *designers* have reputations as hard customers. One *designer* had this kind of reputation with local painting contractors, and it was claimed that this caused an increase of 10 percent on all painting contracts done under that *designer's* supervision. But it was never quite clear whether this meant that the *owner* simply had to pay for what he received or whether the 10 percent was truly a premium. Some *designers* apply the contract (which they have written) to the letter in supervising the *work* and judging its performance, whereas others are more reasonable and practical. *Construction work* is performed under the terms of a contract, contracts are made among persons, and personal attitudes and relationships always affect costs.

In selecting materials and methods of construction for a building, the *owner* will be more or less guided by his *designer,* and since this selection involves the opinions of two persons and the finances of one of them, there are many considerations involved. Some considerations primarily involve economy, and these considerations include two aspects of costs:

1. *Initial costs of the work*
2. *Subsequent costs in use*

Initial costs of the work are those costs incurred by an *owner* in having the *work* done in the first instance. These include the *costs of materials, labor, equipment, overheads, and profit,* which are examined in more detail in the next chapter, and the cost of design services.

Subsequent costs in use are those costs incurred by an *owner* that arise out of his ownership and use of the building. For example, the *costs in use* of a metal and glass entrance during the life of a building might include:

1. washing the glass three times each week
2. cleaning the metal four times a year
3. overhauling and oiling the hinges and closers once a month
4. repairing hinges, closers, and handles once a year
5. replacing handles every three years
6. replacing closers every five years
7. replacing doors every ten years
8. replacing complete metal and glass entrance after thirty years.

In many cases, the *costs in use* are greater than the *initial costs of the work;* and in some cases, higher initial costs would result in lesser *costs in use* and an overall saving. If the *owner* intends to retain ownership, both the *costs of work* and the *costs in use* should be estimated when the building is designed, because only in this way can the true economics of the building be determined and the most prudent decisions be made. But if the *owner* is speculating and intends to sell as soon as possible, his attitude may be different.

Design Style. This element includes the other features and characteristics of a building in addition to size and shape, the less obvious and more subtle details that are not usually apparent but taken together have an immediate effect on the costs of construction, and sometimes also a continuing effect on the *costs in use.*

Formwork and finishing for exposed concrete might be used as an example. The *designer* might specify that a certain type of form shall be used, that it will be erected with no joints closer together than 4 feet, that all joints will be laid out symmetrically about the center lines of walls, that all construction joints will be made on formwork joint lines, and that on stripping the forms the concrete surfaces will be treated to produce a rough textured finish. These requirements, if properly fulfilled, may greatly improve the appearance of the concrete walls, but they will increase the *cost of the work.* On the other hand, the wall surfaces thus treated may not require paint, and the *costs in use* may thereby be reduced.

Details such as finishes cut to a line or held back from abutting on another surface and the absence of trim to cover joints that must be made clean and true when left exposed are some of the other innumerable minor details that make a difference in the initial costs. Anything unusual, original, and untried will probably cost more, because innovations often reduce productivity at first, and the ultimate benefits, if any, are gained later by others.

There are countless *items of work* that may sometimes increase the costs of a job unnecessarily, such as items that may be included without any real consideration of their importance in a job, and items of a quality far above that which is necessary. Gold door knobs and faucets are not commonplace, but some other items are—fine surface treatments on virtually unexposed surfaces, expensive metals and metallic coatings used where cheaper ones would do; high-grade lumber used in concealed places; imported hardwoods used instead of cheaper domestic woods; polished plate glass installed where absolutely undistorted vision is not essential, instead of heavy sheet which may be stronger; unsuitable finishes that require continuous maintenance or early replacement; heating and cooling systems designed for extremes of temperature likely to occur only once or twice a year; metal access doors and cover plates for mechanical valves and controls that are gems of nonferrous metal workmanship. The list is a long one.

To say that certain details and items are costly is not to say that they should not be used, but only that everyone concerned should know what and where the *costs of the work* are, and then decide. Either the *designer* must be educated to an awareness of construction costs and economics, or a cost consultant should be employed to assist the *designer* in planning the *costs of the work.*

The costs of a construction material or a building component are often as important as its physical characteristics, because costs are often one of the decisive factors in choosing a particular material or component. And if "cost" is important to the *owner,* it should embrace both the *initial costs of the work* and the *costs in use.* The function and purpose of many buildings is directly related to investment, costs, and income; and in any case, in one way or another, all buildings have their economic considerations—in their design and function, in their construction, and in their relationship to land space. The words "economics" and "ecology" are both derived from the same Greek word *oikus* (house), and through this fundamental relationship we can see that construction economics is another one of the many problems that beset us and have recently become a matter of popular concern.

To be able to measure a thing is to know it and to begin to understand it and to be able to describe it; and it is not an exaggeration to say that the techniques of analysis and synthesis and the organization and the use of data that are fundamental to estimating have a much wider application in judging and evaluating the probabilities of other things.

Questions and Topics for Discussion

1. If a *contractor* bids for twelve jobs before he gets one, who pays for the costs of bidding for the eleven that were not obtained, and how?

2. What general developments can make estimating more of a science, and why?

3. Who is usually the first to determine the costs of a construction project, and how?

4. Explain briefly how the shape and size of a building can affect its costs.

5. Explain briefly why higher buildings usually cost more to build than lower buildings of equal capacity.

6. What are the primary external factors that help to determine the materials and methods of construction in a building, beyond the persons directly involved?

7. Explain the significance of *costs in use* to an *owner*.

8. Compare the costs of painting a concrete wall every 5 years over a building's life of, say, 50 years, to the costs of covering the wall with ceramic tiling when it is first built. What are the advantages and disadvantages of both approaches?

9. Why do some landlords abandon the apartment buildings that they own?

10. Is it good business to have a service garage rotate the tires on your car, if the charge is $2.00 per wheel? Explain your answer by calculations.

5

Construction Costs

In the preceding chapter, some aspects of construction economics and costs related to a building's design were examined. In this chapter, those more tangible aspects of costs that are primarily related to the competitive market and to the *contractor* and his *subcontractors* are covered.

Costs Through the Construction Market

After the *owner's* requirements have been translated by the *designer* into the drawings and specifications, which will be part of the *contract documents,* the *costs of work* have to be determined by the bidders, who will made offers (*bids*) to do the *work* or otherwise negotiate and arrange to do the *work* for the *owner.*

The *costs of work* always have to be established for a particular job in a particular place as of a particular period of time, and the costs are subject to all those economic conditions that affect the prices of all things at all times and in all places. These conditions are countless, but there are certain factors that help to reduce the size and complexity of estimating their effects on the thousands of *items of work* involved.

The risk in making an offer to do *work* and doing the *work* for a fixed price under certain conditions is not usually carried by one person. Not only is the risk shared by the *contractor* and his *subcontractors,* but each *subcontractor* will usually pass on part of his share of the risk (and potential *profit*) to others, including the *sub-subcontractors* and *suppliers.*

Each person sharing in the performance of a job has also shared in determining the total *costs of the work* to the *owner* by making some of the decisions, estimates, offers, and agreements that constitute a primary construction contract and its subcontracts and sub-subcontracts. To see how these decisions are reached and how the estimates and contracts are made, it is necessary to analyze the *costs of work* and to classify the parts in an attempt to comprehend the whole.

Costs of work may be classified as follows:

1. *Material costs*
2. *Labor costs*
3. *Equipment costs*
4. *Overhead costs*
5. *Profit*

Specific *costs of work* may sometimes have a place in more than one of these classifications, and other costs may not always clearly belong under any one of them. The primary value of these classifications is that they help us examine and analyze construction costs. If a cost does not fit into one of the classifications, the classifications are not necessarily wrong; there simply might be more of them. But fewer classifications are more manageable, and too many classifications defeat their purpose.

Material Costs

At times, unusual conditions vividly illustrate that prices can be extremely variable. In the Cariboo region of central British Columbia, vast grazing lands are tracked over each autumn by hundreds of amateur hunters from the cities to the south. Driving conditions become bad when the winter snow falls; and many hunters are ill-prepared to face the severe weather. Gasoline is not readily available once the blacktop road is left behind, and supplies must be carried. Lack of experience leads many hunters to underestimate their needs, and the ground conditions increase gasoline consumption. Consequently, the few isolated ranches are often visited late by many hunters seeking out precious gasoline. Even traditional hospitality is sometimes overtaxed. One old pump bore a rancher's rather shaky sign: **"After 7:00 p.m.—first gallon costs $5.00."** If the usual price were $1.00 per gallon at that place, a night's purchase of 2 gallons would average $3.00 per gallon; and a purchase of 6 gallons would average $1.67 per gallon.

Price[1] is subject to supply and demand and is affected by many other things, including:

1. quality
2. quantity
3. time
4. place
5. buyer
6. seller

Most estimators have been asked questions such as "What is the price of a yard of concrete?" To answer this kind of question, it is necessary to ask several questions in return, such as:

"What type and quality concrete is required?"
"How much concrete do you want to buy?"
"When do you want it delivered?"
"Where do you want it delivered?"
"From whom do you want to purchase the concrete?"
"What is your relationship with the *supplier*?"

All of these questions must be answered before the original question can be answered with any accuracy.

Quality affects price, because a higher quality usually means a higher price. Is 3000 p.s.i. concrete or 4000 p.s.i. concrete required? Higher strength is usually achieved by adding more portland cement to the mix, and cement costs ten times more than stone and sand. So, a higher price is usually required for higher strength concrete.

Quantity affects price, because usually the larger the quantity purchased, the lower the price. Seven cubic yards of ready-mixed concrete may cost $20.00 per cubic yard, delivered. Seventy yards may cost $19.50 per cubic yard, whereas 3 cubic yards may cost as much as $35.00 per cubic yard. A larger quantity of material may mean that certain fixed marketing costs are spread thinner, and freight costs may be reduced by using a larger and more economical carrier. Thus, the *supplier* can pass on some of the savings to the purchaser of large quantities through a lower selling price.

Time often affects price in several ways, because it affects the production and delivery of a product. If

[1] **Price** and **cost** are commonly interchanged. But "price" is used mainly with objects offered for sale, whereas "cost" is a much more general term. **Price** and **value** are often not interchangeable, for the most valuable things are priceless.

concrete is required at a certain time outside of normal business hours, the *supplier* may have to pay overtime rates to his employees and increase the price accordingly. Some products are cheaper or more expensive at different times of the year, particularly if there is a seasonal demand, or if weather affects production or delivery. Time affects costs, because of changing markets, fluctuating supply and demand, inflation, and other economic factors. Price is, therefore, invariably tied to a specific date or at least to a specific period of time.

Place and location usually affect price through the means and distance of delivery and the accessibility of the site. Handling goods is costly, and changes in methods of delivery from factory to site affect the final cost to the buyer. Freight costs are also determined by the type of goods and by the quantity delivered. Unforeseen costs of handling, unloading, and getting materials into store or to the job site are not uncommon. Contracts for supplying materials and building components are not always explicit about these things, and the estimator must anticipate and allow for all such costs.

The buyer affects the price at which he buys through his creditability as a customer, through the quantity he purchases, and by the marketing level at which he is able to purchase. **The supplier** affects the price because he, too, is a buyer as well as a seller of materials and services. Similarly, the relationship between a lessor and a lessee affect the rates at which plant and equipment are rented.

There is no such thing as a "fixed price" for anything. Products bear labels saying "Suggested Retail Price . . . ," which are usually disregarded; and the international price of gold goes up and down. Some basic construction materials are notoriously variable in price, and copper prices fluctuate frequently. In 1969, in British Columbia, one of the world's largest producers of softwood lumber, the local price of construction lumber rose from about $100.00 to almost $200.00 per thousand board feet, and then fell back again to the lower price within a period of about a year. It has since risen again, in 1971.

List prices are usually not the prices at which goods are sold, but are simply a datum, or price level, established as a basis for selling. Selling prices are frequently quoted by percentage adjustments to list prices. This facilitates price quotation and adjustment. One customer may be quoted a price of "List less 10 percent." Another customer may be quoted a price of "List less 10 percent, less 5 percent" for the same goods. The second quotation is not the same as "List less 15 percent." "List less 10 percent" may indicate a current and general price adjustment. "List less 10, less 5" may indicate an additional special discount of 5 percent below the general price level for a large order or for a special customer. These discounts are generally known as "trade discounts." If the "List less 10 percent" price level remains valid for some time, a new price list may be issued containing all the prices listed at 10 percent below the former list prices.

In some cases, the list price may be a retail price, and the prices to another class of purchaser, such as *contractors*, may be list price less a discount. At the same time, a third class of purchasers, say, large contracting firms, may get a larger discount. Quotations to supply goods in large quantities for a particular project may be based on yet another price specially discounted even further below list price. Generally, list prices do not mean very much; it is the discounts that are much more significant.

Discounts from list prices can be confusing, and maybe they are meant to be. But they are essential as a means of setting different price levels. A manufacturer may sell large quantities of his product to distributors who have extensive storage facilities and maintain large stocks. They in turn may sell to suppliers with smaller facilities and stocks, who in turn may sell to retailers. At each marketing level the same functions exist—"selling, buying, transporting, storing, risk-bearing, standardizing and grading, financing, and providing market information."[2] All of these marketing functions create costs, and thereby create different prices at different marketing levels. This explains why the home handyman pays for lumber from the retail store at twice the price or more paid by the construction company for larger quantities delivered to a job site.

An estimator may keep current price lists of materials for reference, and he obtains revisions as they are issued. But for basic materials or products that represent a significant part of the total costs of a project, the estimator usually obtains specific quotations from manufacturers or *suppliers* for a particular project. If the quantity is large, a quotation may be obtained from the manufacturer. In other cases, it may be necessary to deal with a *supplier*, depending on the marketing policies of the manufacturer. With some products, manufacturers may bypass *suppliers* and sell directly to contracting firms, particularly when the order is large enough to eliminate the need for handling and storage at an intermediate level.

Other types of discounts include cash discounts for early payment before the due date and discounts for volume of business. Most prices are quoted on the

[2] Vernon A. Musselman and Eugene H. Hughes, *Introduction to Modern Business*, 4th ed. (Englewood Cliffs, N.J.: Prentice-Hall, Inc., 1964), p. 423.

basis of a certain credit period allowed between the time the goods are shipped and the time when payment for the goods is to be made. This period varies, and it may be for thirty days (or less), sixty days, ninety days, or whatever period is agreed. If payment is made earlier, the cost of loaning and using money will decrease, and the seller can share some of the resultant savings in the cost of credit with the purchaser in the form of a cash discount.

A further type of discount is that given annually (or more frequently) by a *supplier* to a customer on the basis of the total volume of business done between them over a certain period. This discount, or rebate, is similar to a trade discount, but it does not relate to any particular order or job and it does not usually show on any invoices and statements. This makes it difficult for the *owner* and the *designer* to ascertain actual *material costs* in "cost plus fee" contracts, in which (according to the *standard forms of contract*) the *owner* undertakes to pay the *cost of materials* after all trade discounts and rebates have been deducted. It might be argued that the described discount above is not a trade discount, but that argument is inconclusive.

The costs of moving materials from place to place are a part of the costs of materials, and an estimator must ascertain what handling and transportation will be necessary so that provision can be made for these costs, either in quotations from suppliers, or in *subcontractors'* bids, or in his own primary estimate of costs. These costs may include not only freightage and cartage from factory or supply house to the site, but, also handling costs of loading and unloading and any hoisting to levels above ground, which includes not only man-hours in handling, but also the costs of using hoists or elevators. In the case of materials like sand and gravel, cartage is often the major part of the cost and the distance from source to site is a primary consideration. Sometimes, quotations and *bids* do not specify what is included in the way of handling and hoisting, and this omission sometimes leads to misunderstandings, disputes, and unexpected costs. The estimator should try to avoid these problems by carefully describing his requirements and closely scrutinizing all quotations and *bids.*

Suppliers' and subcontractors' bids should make it clear if the *bid* is made on the assumption that another's elevator, hoist, or other equipment and men will be available for use at the site for unloading and handling, and there should be proper mutual agreement about any charges to be made.

Storage is another part of the cost of materials. This element may also involve risk of damage and loss during storage or the costs of insurance, protection, and security. Some of these costs, such as temporary storage buildings at the site and general security provisions, are better classified and estimated under *overhead costs,* as described later in this chapter.

Taxes, particularly sales taxes, are another part of the cost of materials. An estimator must ascertain whether or not quoted prices include all appropriate taxes, because not all price lists and quotations make this point clear. The usual practice is for local sales taxes not to be included in listed prices, because the price list is often used in several states and provinces, each with different sales tax rates or with none at all.

All building materials and components are the products of labor. The question in *construction work* is "Where does the labor occur?" For example, portland cement is produced by labor in a plant, and it is shipped to a construction site. At the site, the cement may be mixed with sand and applied to walls as cement plaster. Part of the total labor required for cement-plastered walls occurs in the cement plant, and part at the site. The same division of labor applies in principle to all *construction* work, and changes gradually are occurring in the division of labor. Generally, the trend is for more labor to occur in plants and shops and less at the site, and as this change occurs the nature of construction *work* changes, and *suppliers* replace *subcontractors.*

When we speak of *material costs* we refer to the costs at the site to the *contractor,* and to the *owner;* and it appears that these costs will become an increasingly larger part of the total costs as fewer materials requiring site labor and more factory-made, standard components are used in construction. Other changes will consequently occur. Purchasing will become more important, and the skilled tradesman less so.[3] Detail design will be done more and more by the manufacturers of building components, and entire buildings will be designed by their manufacturers. Manufacturers of components and materials may diversify by incorporating construction companies to create markets for their own products; and in some cases the designing, manufacturing, erecting and servicing of buildings will be done by one company, which will manufacture, lease, erect, dismantle, scrap, and reconstitute its products.

Labor Costs

These costs may be examined from two aspects:

3 The fast rising costs of tradesmen at job sites and the attitudes of some unions are two of the forces that will revolutionize the construction industry by diminishing the amount of *work* done on job sites, and by increasing the use of factory-made, standard components in buildings.

1. **Labor rates:** the hourly rates of employing workmen, based on total *labor costs* divided by the total number of hours worked.

2. **Productivity:** the rates of production by workmen employed; the amounts of *work* done in the specific periods of time paid for.

Labor rates include all *labor costs,* both the direct and indirect costs. *Direct labor costs* are wages and other payments made to workmen. *Indirect labor costs* are other payments made by the *contractor* on the employee's behalf, and these include fringe benefits (according to current wage agreements) and statutory payments (according to law). The *wage rate* is the direct cost per hour. The *labor rate* is the total of direct and indirect costs per hour.

Time worked over the standard day (of usually 7, 7½, or 8 hours, depending on location and wage agreement) is paid for as overtime at rates usually one and one-half or two times the basic *wage rate.* In addition, "travel time" may also be paid for, in addition to a guaranteed minimum number of hours per day, or per week, specially on out-of-town jobs in remote locations. For example, on an out-of-town job:

Guaranteed minimum week, say,	= 60 hours
Standard week, (7½ hours × 5 days)	= 37½ hours
Overtime hours	= 22½ hours

If the first 5 hours of overtime are paid at one one-half times the basic rate, and the balance at two times the basic rate of, say, $8.00 per hour, the *effective wage rate* will be $10.67 per hour.[4]

Standard week: 37½ hrs	@ $8.00	=	$300.00
Overtime: first 5 hrs	@ $12.00	=	$60.00
balance 17½ hrs	@ $16.00	=	$280.00
Total 60 hrs	@ $10.67	=	$640.00

This is in effect an increase of one-third in the *wage rate;* and additional costs on out-of-town jobs may include subsistence allowances for board and lodging and traveling costs to and from the job site at start and finish, and sometimes during the job.

A guaranteed minimum day, or "show up time," is paid in some cases to workmen who show up for work at the site but cannot work because of weather, or another reason. This kind of payment is sometimes also required for workmen hired from a union hall who, if found to be unsuitable, are discharged soon after arrival at the site. Premium rates are usually payable for unusual conditions—for work at high levels, dirty work, underground work, work in water, work under water, and work in compressed air (such as in caissons).

Fringe benefits, which are part of *indirect labor costs,* may include paid vacations (holiday with pay), pension fund payments, group insurance premiums, payments to funds for the health and welfare of employees, and payments to funds for apprentices, training and trade promotion. These payments may be based on time worked or on a proportion of wages or they may be levied as a lump sum.

Wage rates and fringe benefits are generally established through negotiations (sometimes preceded by strikes or lockouts) and agreements among employers and the various and many trade unions in the construction industry. Despite the interdependency of the construction trades, and the fact that their cooperation and coordination are essential to construction jobs, there is often not much cooperation and coordination among construction trade unions off the site. Each local union usually negotiates a separate and individual wage agreement with employers, often at a different time and often for a different period from other local unions. There are a few national agreements. *Wage rates* and fringe benefits consequently vary greatly among trades and localities.[5]

Statutory payments, which are *indirect labor costs,* are paid by employers to government on their employees' behalf, according to the appropriate statutes. In the United States these payments include contributions to social security and Medicare, unemployment insurance, and workmen's compensation insurance. In Canada the payments are similar, except that payments are made to the Canada Pension Plan instead of to social security.

To summarize, *wage rates* are total payments paid to employees (direct costs) divided by the total hours worked, whereas *labor rates* are all *labor costs* (direct and indirect) divided by the total hours worked. *Wage rates* can be established from payrolls, but *labor rates* are usually best calculated on a weekly or two-week period, or on the basis of some other pay period.

Some estimators use *wage rates* in calculating *unit prices* and in pricing *labor costs* in estimates, in which case *indirect labor costs* are added to total *direct labor costs* as a percentage, usually between 20 and 25 percent, according to the trades involved. Other estimators calculate and use *labor rates* to price *labor costs* in estimates. The choice of method depends on the various trades and rates involved, the type of estimate

[4] Because *wage rates* vary so much according to time and place, different rates are used in some examples.

[5] For labor law and labor relations in construction see Richard H. Clough, *Construction Contracting,* 2nd ed. (New York: Wiley-Interscience, 1969), Chapters 13 and 14. For comparative labor rates see *Building Construction Cost Data* (published annually) (Duxbury, Massachusetts: Robert Snow Means Company, Inc.,) and several construction periodicals.

and contract, the company's accounting methods, and individual preferences.

Productivity is the one aspect of all costs that is the estimator's greatest challenge. How many man-hours are required to complete each *item of work*? The variables of productivity are innumerable, and it is one of the major factors in all construction costs. Yet not much research has been done in this subject considering its importance. The main obstacles to research appear to be the size and diversity of the industry, the large number of small construction companies and the small number of large companies, and the reluctance of most companies to pool data for their mutual benefit.

Construction companies must know the *costs of the work* that they do. This requirement appears obvious —yet many companies do not always know what their costs are. Even if they practice some kind of *cost accounting* they do not always know, because the results are often inadequate on their own to provide the company with all the information it needs. Therefore, most construction companies need other sources of information, which is why so many estimators buy published cost data.

Pooling and sharing cost data sounds to some like socialism and a denial of the free enterprise system; but this is because they misunderstand the procedure. It is not necessary for construction companies to disclose their discounts, their *overhead and profit,* and their *unit prices.* It is not necessary that they give away anything that would reduce proper competition. All that is involved is that they share certain factual data and information about such things as the affects of season, weather, design, productivity, and other economic factors on the costs of construction. This sharing could be done through local construction associations in such ways that no individual source of information could be identified, and all participating members would benefit.

Now we are in a time when information collection and handling and data processing is a basic industry. But information about the design and construction industry is, for the most part, still dispersed and unprocessed, and the time has come when we need to systematically collect and analyze and use this information. An example of the kind of research needed is the study on manpower utilization referred to in Chapter 3, which indicated one of the major deficiencies of the construction industry when it concluded that: "There is a lack of detailed cost information available to most contractors."[6]

[6] David Aird, *Manpower Utilization in the Canadian Construction Industry,* National Research Council, Division of Building Research, Technical Paper No. 156 (Ottawa, Canada, 1963) p. 38.

The first reaction to the Study is said to have been vigorous, possibly because it said, in its summarization of the work studies and observations made of laborers and carpenters on several jobs, that it found the workmen were "idle for no apparent reason" for about 17½ percent of their total work day. The average total "idle time," including time spent waiting for men, materials, and equipment, was observed to be about 25 percent of the carpenters' work day, on the sites where the study was carried out. The Study's conclusions, however, placed almost the entire blame for this inefficiency with construction management. Generally, the conclusions were that there was a lack of good management and that this resulted in inefficiency, poor utilization of labor, and "involuntary idleness" on the part of workers. However, the Study should be read in its entirety to understand all of its implications and limitations. One portion of the Study does not make an irrefutable case; but its findings and conclusions are supported in part by statistics from other sources, such as Dun and Bradstreet Ltd., whose business information systems, services, and sciences provide "publications and services for management needs." These items include annual "Key Business Ratios" and "Failure Records" for various industries, which show each year that most business failures can be traced to incompetent management.

Productivity is fundamental to the economics of construction. The estimator, as part of management, has to relate past productivity and past costs to future jobs through *cost accounting.* But, the estimator should not simply be a surveyor and user of productivity data. The estimator can assist in achieving higher productivity by pointing out those situations wherein cost data indicate that productivity has fallen below the median, or norm; by making economic comparisons of the use of different materials and methods of construction; by comparing productivity in one branch, or company, with that in other branches, or other companies; and by providing data for use in scheduling and controlling *construction work.* To manage well, it is necessary to know exactly what the company is doing. The greatest obstacle to good management is ignorance and a lack of information.

Productivity is essentially variable, and the variables exist in (1) the workers, and (2) the worker's surroundings, equipment, and tools. The variables of the worker include:

> anatomy, brawn, contentment, creed, earning power, experience, fatigue, habits, size of man, skill, temperament, and training. The other variables include: appliances, clothes, colors, entertainment, ventilation, lighting, quality of material, reward and punishment, size of unit moved, special fatigue-eliminating de-

vices, surroundings, tools, union rules, and weight of unit moved.[7]

These variables indicate the problem of productivity. Each case consists of a different set of variables, and those construction estimating texts that use the term "labor constants" to identify productivity rates are guilty of creating a false impression of constancy. Similarly, those handbooks and texts that publish construction productivity rates in terms of man-hours expressed to two and three decimal places are also misleading. The data may have been analyzed and presented in good faith, and the published results may be mathematically correct averages, or medians; but, to say that framing and erecting stud walls requires 29.1 carpenter man-hours per thousand board feet of 2 × 4 lumber may create false impressions of accuracy. We should ask how the figure of 29.1 man-hours was obtained. If, as is probable, it is an average figure from several framing jobs, we should ask how many jobs were analyzed. For example, suppose the figures obtained were as follows:

Production Rates	*(man-hours)*
Job no. 1	35.33
Job no. 2	34.82
Job no. 3	31.01
Job no. 4	29.76
Job no. 5	29.30
Job no. 6	28.99
Job no. 7	28.34
Job no. 8	27.30
Job no. 9	23.81
Job no. 10	22.34
10)	291.00
Average (for 10 jobs)	29.10 man-hours.

Looking at these production rates for ten different jobs, it is apparent that if the average were based on more or less jobs, the average would be different. There are not enough data to plot a "distribution curve"; but if there were data from 100 jobs available, it might be possible to draw a "distribution curve," and we might get a better picture from the statistics. A mean, or average, can be distorted by one unusual figure, and the median, or the mode, may be a better guide. Better yet would be the ability to mathematically predict the probabilities of variations from medians for major items of an estimate, and to indicate their probable effects on the total estimated *costs of work*.[8]

[7] Frank G. Gilbreth, *Motion Study* (New York: D. Van Nostrand Company, Inc., 1911).

[8] See Appendix A for more on predicting probabilities.

Equipment Costs

A more complete title for this section should be *"plant and equipment costs,"* to identify the two kinds:

1. **Plant,** which refers to those things that are "planted" in position, such as a concrete-mixing plant
2. **Equipment,** which refers to those things that are mobile, such as a bulldozer.

This distinction in terms is not always made, but in the interest of greater precision such distinctions should be encouraged.

Once again, a reminder is necessary that classifications are only a convenience that do not change the reality of things one bit. Things are what they are, no matter what we call them. We may give names to groups of similar things for our own convenience and thereby may deceive ourselves if we come to think of the name as the reality.

The economics of plant and equipment are widely discussed and thoroughly examined in many publications, and one of the primary questions involved is whether to own or to rent equipment. In estimating, it makes no essential difference whether plant and equipment is owned or rented, because in either case the estimator should allow for all the *plant and equipment costs* in his estimate at realistic rental rates.

If plant or equipment is rented, the rental rate will be established by offer and acceptance as in any other contract. If the plant or equipment is self-owned, an economically realistic rental rate should be estabished for use within the company, and that rate should take into account all the economic facts of owning and using plant and equipment.

If such an accounting is not made, one of two conditions will result. Either the plant or equipment will be charged for in the estimate at an inflated rate, the total estimated *cost of the work* will thereby be inflated, and to that extent the estimate will not be competitive. Or the plant or equipment will be charged for in the estimate at a deflated rate, and, to that extent, the estimate will be too low and the *owner* will receive something for nothing. In such a case, the *contractor* will suffer a loss. Owned plant and equipment must be charged out at economical rates just as if it were rented at competitive rates.

Money invested in plant and equipment should produce an income, the same as any other investment. The amount of income that should be produced varies in the market; but obviously, to show a *profit*, the amount should be greater than the premium the company has to pay to borrow money from the bank. The income is the net income produced by the plant and equipment after all costs are deducted. These costs are of two classes:

1. **Owning costs:** the costs of owning plant and equipment.

2. **Operating costs:** the costs of using the plant and equipment over and above the *owning costs*.

All *plant and equipment costs* can be put under one of these two headings. But in some cases the distinction is not always clear, because some costs are not always obviously in one class or the other. This condition exists because we have only two cost classifications for the sake of simplicity. We could have a third heading for those costs that relate to both owning and operating, but then some of the advantages in classifying the costs would be lost.[9]

Owning Costs

These may be identified as:

1. **Depreciation** (loss in value from any cause)

2. **Maintenance** (major repairs and replacement of parts)

3. **Investment** (costs arising from investment and ownership).

Depreciation is usually the biggest single cost. It arises out of wear and tear—that is, physical depreciation and obsolescence. As the plant or equipment is superceded by better designed, more efficient, and more desirable plant or equipment, it becomes obsolescent and it loses value.

There are several classical methods of calculating depreciation, and each method will give a slightly different result. This does not necessarily mean that one method and one result is right and the others are wrong. The several different methods of calculating depreciation enable accountants to record and keep accounts of different assets that decline in value because of depreciation, and different methods are better suited to different situations. Of course, the method used should, if possible, more or less reflect the value of the plant and equipment at different times of life. Depreciation rates for deductible allowances in calculating income tax are laid down by the tax authorities, but these rates may sometimes be determined by temporary economic and political strategies rather than by observable facts.

The most common and the most easily understood method of calculating depreciation is known as the "straight line method." For example:

1. Original cost of equipment delivered, including taxes $12,000.00
2. Less salvage value after 5 years working life .... $ 2,000.00
3. Total depreciation over 5 years working life $10,000.00
4. Average annual depreciation over 5 years working life .... $ 2,000.00

"Original cost" must include all attributable costs, including taxes and freight. "Salvage value" may be only "scrap value" or nothing at all if the equipment is used at an isolated location. An "average annual depreciation" of $2,000.00 does not mean that at age 2½ years the equipment will be worth precisely $7,000.00. However, its value at that age should be approximately $7,000.00, depending on the amount of maintenance required and done, the use and abuse of the equipment, and any new developments in this type of equipment that might make the older model more difficult to sell. You can probably relate this situation to your own automobile, which also depreciates.

Two other classical methods of calculating depreciation that are explained in many books on business and accounting are the (1) declining balance method, and (2) sum of digits method. These are less straightforward than straight line depreciation. The first method depreciates the item by the same percentage amount applied to the undepreciated balance each year, and so, in theory, the item never fully depreciates but does eventually reach a practical salvage value. The "sum of digits" method is not unlike the "declining balance" method; but instead of a constant percentage deduction, a declining fraction of the total costs is deducted each year.[10]

Maintenance costs vary with the type of plant and equipment and with the type of *work* done. Mobile equipment doing heavy *work*, such as a crawler-

[9] An explanatory example of the method of estimating the *owning costs* and the *operating costs* of plant and equipment is given in Chapter 10, Pricing Work: General.

[10] With the **declining balance method,** up to twice the percentage annual depreciation found by the straight line method is deducted from the declining balance each year until the estimated salvage value is reached. Thus, for an estimated 5-year life (and 20 percent straight line depreciation) deduct up to 40 percent of the declining balance each year.

With the **sum of digits method,** the total depreciation is found by first deducting any estimated salvage value from the original cost (total investment); and depreciation is deducted from the declining balance each year as a fraction calculated from the sum of the digits obtained from the estimated life in years. Thus, for a 5-year life the sum is $15 (= 1 + 2 + 3 + 4 + 5)$, and in the first year, 5/15ths of the depreciation is deducted; in the second year, 4/15ths of the balance is deducted; and in the fifth year, 1/15th of the balance is deducted, leaving only the salvage value.

tractor, might have total maintenance costs equal to the total depreciation costs over the equipment's working life. That is to say, if the total depreciation of the equipment is $10,000.00, the equipment owner can expect to pay $10,000.00 in maintenance costs over its working life. At the other end of the scale, the maintenance costs for equipment such as cranes might be from 12½ percent down to as low as 5 percent of the total depreciation. Maintenance costs are usually indicated as a proportion of the depreciation because there is a logical relationship between these two costs and the working life. Consider how much it costs to maintain your own automobile compared with the annual depreciation.

Investment costs include:

1. interest on investment
2. insurances and taxes on plant and equipment
3. storage costs arising from land and buildings used to store equipment.

You can relate all of these expenses to your own car. Do not forget the property taxes on your garage, or carport, if you have one. Interest on investment may be either interest paid by purchaser to a finance company or to the seller if the equipment is bought on credit. Or, if the equipment is bought for cash, it is the interest that would have been received if the cash had been invested. This cost of money should also include an allowance for inflation of, say, 4 to 5 percent per year, since this is apparently the approximate annual rate at present. If inflation is not accounted for, apparent *profit* may not actually exist.

Insurances for the plant and equipment are necessary to protect its owner against loss, and the premiums will depend on the type of plant and equipment. Plant will usually cost less to insure than mobile equipment. Tax levies vary according to location and regulations, and in some places, only plant affixed to land is taxed.

Storage costs include the use of land and any buildings used to store and protect the equipment when not in use. In some cases, these may be classed as *overhead costs.* The costs of a storage and maintenance building for equipment would also include costs similar to those arising from owning plant and equipment; namely, depreciation, maintenance, and investment costs.

Operating Costs

These costs may be identified as:

1. **Fuel** (including lubricants and additives)
2. **Running repairs** (including minor repairs and replacement of small parts)
3. **Transportation** (including transporting to and from site; setting up and dismantling)
4. **Operator** (including wages and fringe benefits).

Once again, relating these expenses to the operating of your own automobile will help you to understand them better. Fuel oil (diesel) and gasoline are used at hourly consumption rates related to the "brake horsepower" of fully loaded equipment. Running repairs depend on the plant and equipment and its use, and there is some connection between this item and the cost of maintenance. Consumable items, the use of which is directly proportional to the amount and weight of *work* done, are better included under running repairs. These items include such things as fan belts, cables, hoses, cutting edges, and the like. Rubber tires are a large expense, and on rented equipment they may be classed as a maintenance cost. Transportation to and from the site may involve the use of other equipment to carry equipment with tracks that is not allowed on highways, and it also includes the costs of any scout cars required to lead and follow large equipment traveling on highways. Setting up and dismantling plant or equipment, such as a tower crane, can be a major expense and may require the use of a large mobile crane.

The costs of the equipment's operator are similar to those *labor costs* previously examined in this chapter. The estimator should also consider the cost of additional labor working in conjunction with plant and equipment. For example, excavation equipment and cranes frequently require a man working on the ground in front of the equipment to assist the operator by signaling to him or to do minor handwork with a shovel, and the *labor costs* should be attributed to the use of the equipment.

Overhead Costs Generally

These are *construction costs* of any kind that cannot be attributed to any specific *item of work.* If costs can be attributed to an *item of work,* they should be, and these costs will then come under one of the classifications already discussed; namely, *material costs, labor costs,* or *equipment costs.* Otherwise, they are *overhead costs.*

Overhead costs can be classified by applying an extension of the same reasoning by which we define *overhead costs.* That is, if costs can be attributed to a specific job site because the costs arise only out of that particular job, those costs are *job overhead costs.*

If costs cannot be attributed to any particular job, they are *operating overhead costs.* The scope and content of these two classifications vary with the conditions and requirements of the contract. That is to say, certain *job overhead costs* on one job (or in one company) may be classified as *operating overhead costs* on another job (or in another company). Once again, the purpose of classifications should be remembered.

Job Overhead Costs

These costs can often be estimated in the same way as other *costs of the work,* because they include *material costs, labor costs,* or *plant and equipment costs,* even though they are of a general nature. Many *job overhead costs* are specifically referred to in the contract, whereas others are not and must be anticipated by the estimator. There are two primary contractual sources for these costs:

1. the **articles of the contract** (the general conditions and the agreement)
2. the **general requirements**[11] of the contract

Some of the general requirements in a contract are of a general nature only insofar as they are required for several *specific items of work,* and therefore they may actually relate to the costs of those specific items rather than to the job as a whole as *job overhead costs.*

Articles of the contract may give rise to the following *job overhead costs*:

1. Liquidated damages (a risk to be priced by the bidder)
2. Taxes and duties (when not for specific *items of work*)
3. Legal fees and costs (related to the *bid* and the contract)
4. Consultants' fees and costs (where required to be paid for surveying, building layout, testing, etc.)
5. Contract documents (extra copies to be copied or purchased)
6. Site staff (salaries and allowances; offices and accommodation)
7. Personnel expenses (related to staff)
8. Fees and premiums (for permits, bonds, and insurances)
9. Protection of life, *work,* and property (*temporary work* required)
10. Contingencies (for delays, damage, emergencies, inefficiency, and the like)
11. Financing (short term, as required by the *work*)

[11] General requirements are Division 1 of the "Uniform Construction Index" (see Bibliography).

Liquidated damages are predetermined money damages payable to the *owner* should the *contractor* cause him to suffer damages (loss) through failure to substantially complete the *work* of a contract by a completion date stated in the contract. These damages are "liquidated" (i.e., settled in advance) and are made part of the terms of the construction contract to create an incentive for the *contractor* to perform the contract and to complete the *work* on time, and to make immediate and direct restitution to the *owner* for damages should the *contractor* default by not completing the job on time. The *owner* could sue the *contractor* for breach of contract and for actual damages incurred through late completion of the *work.* But in order to avoid a court action and its attendant costs in time and money, liquidated damages are calculated and agreed in the contract, in advance, so that, should it be necessary, the *owner* can obtain restitution through the contract instead of through the courts. Liquidated damages are, in fact, related to the actual damages suffered by the *owner.* That is to say, liquidated damages are not a penalty for late completion. There is a legal distinction between liquidated damages and a penalty. A penalty is not particularly related to actual damages and may not be valid unless equitably balanced in the contract by a bonus for early completion. But a bonus is not necessary in a contract with liquidated damages. If liquidated damages are prescribed in a contract, a bidder must consider:

1. the desirability of bidding for a contract requiring liquidated damages for late completion, rather than bidding for other *work* that may be available
2. the probabilities that the *work* of the contract can, or cannot, be completed within the period of time stated in the agreement
3. the amount (if any) to be included in the estimate and *bid* to offset the risk of having to pay liquidated damages

Taxes and duties are usually attributable to *specific items of work,* in which case, they are part of those items' costs. Otherwise, they are a *job overhead cost.*

Legal fees and costs may arise from legal services in respect of a specific contract—for example, fees for a legal opinion on an unusual term of a contract. Otherwise, legal fees are probably an *operating overhead cost,* such as a retainer paid by a construction company to its legal adviser.

Consultants' fees and costs might arise from design and testing services required to be provided by the *contractor.* In that case, the fees and costs should be ascertained and included in the estimate. If they are attributable to specific *items of work,* they should be part of the costs of those items. Fees may be paid to a professional surveyor for surveying a site or for laying out a building, and in such cases they would be part of *job overhead costs.*

Contract documents may have to be copied or purchased by the *contractor,* particularly if he requires more copies than the number specified in the contract to be provided by the *designer.*

Site staff may include the following:

1. project managers and engineers
2. superintendents
3. quantity surveyors
4. cost accountants
5. timekeepers and first aid men
6. purchasing agents and expediters
7. storemen, watchmen, etc.

The question to be answered is, Can all or part of a staff member's salary, or wages, or expenses be properly charged to a particular job? If a staff member, such as a purchasing agent, is at the company's head office and his salary cannot be charged to specific jobs, then it is an *operating overhead cost.* The size and location of the job usually determines the size and scope of the site staff, and the *job overhead costs.* Larger jobs and jobs that are not close to the head office usually entail larger *job overhead costs* because they require more site staff.

Personnel expenses might include such things as transportation, supplies, and equipment, and might also include camp costs and catering costs for staff and workmen on out-of-town jobs. Overseas jobs often require personnel expenses for medical treatment, air fares, and special payments of income taxes and bonuses.

Fees and premiums include fees for building permits, bid bonds, performance bonds, payments bonds, and the like, as well as premiums for builder's risk insurance and public liability insurance. All of these costs are directly related to the job. Some insurance premiums may be part of *operating overhead costs,* because a construction company often has general insurance coverage for the usual risks, and the general premiums do not relate to any one job.

Protection of life, work, and property includes such items as the costs of safety measures; the risk and costs of having to protect, or repair and make good the *owner's* property at the site; or the property of others, because of disturbance or damage resulting from the *work.* Such costs cannot always be charged to specific *items of work* and are, therefore, a *job overhead cost,* unlike the costs of protecting specific *items of work,* such as newly laid masonry or concrete, which can be charged as part of the costs of those items.

Contingencies may be allowed in an estimate for any number of probabilities and risks. For an out-of-town job on which local labor will be employed, the estimator may price the *items of work* as usual, and then make an estimated allowance of, say, 10 percent on the *labor costs* to allow for the lower productivity of local labor. An estimated allowance may be made for probable delays due to weather conditions, or late deliveries. If *work* is halted at the site, certain *job overhead costs* will continue. In *alteration work,* there may be a risk of damage to *existing work* that should be allowed for in the *bid.* Blasting, for example, may weaken or damage nearby structures. The extent of contingencies required in an estimate for any such items depends on the type of contract and the job conditions.

Financing is an important cost because the *contractor* is paid in arrears for "work done," usually at the end of each month during the course of the contract. This means that in most contracts, the *contractor* must have some money available to perform the *work.* Most contracts also provide for the *owner* to hold back a percentage of the costs of the completed *work* each month to provide for the requirements of lien statutes, or to provide security for the *owner,* or to create an incentive for the *contractor* to perform the contract. In times of scarce money, the cost of financing a job often becomes critical, especially for smaller construction companies. In all contracts in which the *contractor* has an investment there is a cost of financing that is a *job overhead cost.*

Financing a contract sometimes gives rise to a practice known as "front end loading," or an "unbalanced bid." This practice gives a *contractor* the opportunity to try to get overpayments from the *owner* in the first months of the contract, so that the *owner* is effectively financing his own job and the *contractor* has to put in less of his own money. This feature is incorporated in a **stipulated sum contract** by submitting a schedule of values[12] to the *designer* before the first interim payment for *work* is made, with the earlier *items of work* over-valued and the later items under-valued. In a **unit price contract,** the *unit prices* for the earlier *items of work* are inflated, and *unit prices* for later items are deflated, so that the total estimated cost is still competitive. Many *designers* are

[12] The "schedule of values" is an analysis of the contract sum showing the value (costs) of the various parts of the *work* and is submitted by the *contractor* to the *designer* for his approval before the first contractual payment for *work* is made.

aware of these practices, and try to circumvent them by not approving schedules of values that contain *items of work* that are over-valued. This means that a *designer* must have reliable cost information about the *costs of the work* in a contract, such as an accurate estimate prepared by himself or by a cost consultant.

The *contractor* has to allow for the cost of some financing on most jobs and so do the *subcontractors* and *sub-subcontractors*. In fact, the *contractor* does not carry all the financing costs, only those costs relative and proportionate to the cost of those parts of the job done by the *contractor's* own forces. The rest of the financing costs are incurred by the *subcontractors* and their *sub-subcontractors*, each according to the value of his subcontract. And in some contracts the *subcontractors* may have proportionately higher financing costs than the *contractor*, simply because they receive their payments through the *contractor*, who may find reasons and excuses to hold onto their money and use it without right or good cause. Likewise, *sub-subcontractors* may suffer at the hands of some *subcontractors*. Some prime contracts (between *owner* and *contractor*) contain provisions to ensure proper and timely payments to *subcontractors*, because it is usually in the *owner's* interests that the *subcontractors* should be fairly treated.

Finally, it should be pointed out that a large proportion of the interim financing required during most construction jobs comes from the *suppliers* who extend credit to the general and specialist construction companies who are their customers. The cost of this credit is included in the prices of the materials supplied, and it is, therefore, passed on to the *owner*.

General Requirements listed in the Uniform Construction Index include:

1. schedules and reports
2. samples
3. shop drawings
4. temporary facilities
5. cleaning up
6. closing out

Sometimes these requirements duplicate the General Conditions. Some of them generally refer to costs that are attributable to specific *items of work* rather than to *job overhead costs*.

Schedules and reports may include critical path schedules, sometimes to be prepared by specialist consultants; site photographs taken periodically to record progress and *work* to be covered and concealed; and inspection reports of completed *work* to be obtained and paid for by the *contractor*.

Samples and Shop drawings are usually required for specific *items of work* and are, therefore, part of their costs. They are usually for *work* done by *subcontractors*, who therefore provide them.

Temporary facilities may be any number of different things to be provided by the *contractor* for performing the *work*, and might include:

1. temporary offices, stores, and facilities for staff and workmen
2. temporary means of communication, such as, telephone, telex, and electronic secretaries
3. temporary services, including water, electric power, gas, and drainage; also, temporary heating and ventilation for the building while *work* is in progress
4. temporary roads and site drainage
5. temporary building enclosures for winter *work*
6. temporary screens, fences, walkways, and signs
7. temporary security and safety precautions, such as barricades, lights, and security services
8. temporary parking and access facilities on the adjoining property of others

Temporary offices, sheds, and other structures may be either owned or rented by the *contractor*. Either way, they should be charged to the job in much the same way as any other item of plant. Similarly, the means of communication, and such things as temporary fences, screens, barricades, and heaters, may be either rented or owned. They are usually charged to the job on a "time" basis and include the costs of transporting them to and from the site, of erecting them, and of dismantling them. Costs for these items can be established in the same way as *plant and equipment costs*.

Temporary services, such as water and electric power, are charged according to current consumption rates, plus costs of permits, installing, and removing meters, pipelines, wiring, and the like. Heating and lighting a building during construction, either by temporary heaters and lights or by using the building's permanent services once they are installed, can constitute a large *job overhead cost* that is not always easy to estimate because it depends on climatic conditions. Experience from previous jobs done under winter conditions, preferably in the same locality, is invaluable to the estimator in these cases.

Cleaning up the site and the building and disposing of rubbish is a continuing expense. Costs may include labor; rental of bins from a disposal company; and rental of rubbish chutes, incinerators, and other plant and equipment, such as hoists and trucks. It is not uncommon for these costs to be underestimated or overlooked completely in an estimate.

The costs of cutting and patching to make it possible to integrate the *work* of the many *subcontractors* might be included in this classification. To a large extent, this *work* is made the responsibility of the various *subcontractors,* but it is not always possible for the *contractor* to delegate it all.

Closing out costs might include drawings and diagrams of services and concealed *work* as executed by the *contractor,* permanent bench marks, boundary marks, survey monuments, and similar items. Also, the costs of repairs, correcting defects, and any *maintenance work* required by the contract might be included here. Final cleaning of the building, preparatory to occupation by the *owner,* is included in the "general requirements" in some specifications. A better practice is to make this *work,* which is often done by janitorial and maintenance firms, the subject of a separate section of the trade specifications, and a subcontract.

Because *job overhead costs* are usually referred to in the articles of the agreement and the general conditions of the contract, and in the general requirements of the specifications, it is important that the estimator should carefully read and understand these parts of the *bidding documents,* as well as the trade sections of the specifications and the drawings. A common fault is to assume that these contractual requirements are always "typical" and the same as were required for previous jobs, when, in fact, this may not be so. Also, it should be remembered that the *contract documents* do not indicate all general requirements for the *work.* Usually, only those of direct interest to the *owner* and to the *designer* are specified, and the *contractor* must decide for himself what other things may be necessary to perform the contract. For example, general requirements for winter *work* may not be explicit in the contract; but, they may be implicit, in that the performance and completion of the *work* according to the contract's requirements may make temporary enclosures and heaters essential.

Operating Overhead Costs

Operating overhead costs are those costs that cannot be attributed to any particular job. It is characteristic of these costs that they are incurred by a construction company whether or not it is actually doing *construction work.* They are often known as "head office overheads"; but this term may be misleading, and the term *"operating overhead costs"* is more descriptive of the costs of operating a construction business, as distinct from the costs of running specific construction jobs.

The detailed analysis of *operating overhead costs* is not usually part of an estimator's duties. It is usually the duty of the construction firm's accountant to provide the estimator with the necessary information so that a proper and adequate provision for these costs can be included in each estimate. The accountant has to predict the annual *operating overhead costs* and the (dollar) volume of *work* that the company will do in each coming year. These two amounts will then enable him to calculate the theoretical percentage allowance to be made in each estimate to provide for the *operating overhead costs.* If the company does not do the predicted amount of *work,* then these costs may have to be taken out of an already diminished *profit.* For example, if the predicted costs for the next year are:

Operating overhead costs	\$ 600,000.00
Volume of work	\$20,000,000.00

the percentage allowance for these costs in estimates for projects to be done in the next year would have to be:

$$\frac{\$600{,}000 \times 100}{\$20{,}000{,}000} = 3 \text{ percent.}$$

If at the end of the year the company has done only a \$15,000,000 volume of *work,* with all estimates containing a 3 percent allowance for *operating overhead costs,* only \$450,000 may have been obtained toward these costs, and the remaining \$150,000 will have to come out of *profits.* Some firms may apply different percentages to *labor costs* and *material costs* to allow for *overhead costs* in their estimates, because jobs with higher *labor costs* often require higher *overhead costs.* Other methods of allowing for *operating overhead costs* are used, but probably the most common is a percentage addition to the total of all *costs of the work.*

Operating overhead costs may include:

1. **Managements and staff,** including:
   a. Salaries, fringe benefits, and expenses
   b. Transportation (to job sites)
   c. Other expenses attributable to staff and not chargeable to a specific job
2. **Business offices,** including:
   a. Rent (or depreciation and operating costs of owned premises)
   b. Office furniture and equipment (leasing costs or depreciation allowances)
   c. Office supplies and consumable stores
   d. Other such expenses attributable to the operation of

the company but not attributable to a specific construction job

3. **Communications,** including:
   a. Telephone, telex, and the like
   b. Promotion and advertising

These *overhead costs* are not very different from those of any other company in business, and more information about them can be found in many general texts on modern business.

In summary, we have seen that *overhead costs* are of two basic types: those that are attributable to a specific job, and those that are not; and that the distinction between these two types of *overhead costs* is not always clear and fixed.

In a stipulated sum contract, the *owner* is not concerned with the details of these costs, whereas a *contractor* can usually identify a *job overhead cost* as a cost arising out of a particular job, as distinct from an *operating overhead cost.* In a **cost plus fee contract,** the distinction between these two basic types of *overhead costs* becomes very important to both the *owner* and the *contractor,* and a subject for proper definition in the contract. This importance issues from the fact that the job overhead costs are paid directly by the *owner* as part of "cost"; that is, they are reimburseable costs as defined in the contract. The *operating overhead costs,* on the other hand, are not reimburseable costs and come out of the "fee" paid to the *contractor.* If "cost" is loosely defined in the contract, the *contractor* may be able to charge to "cost" (for direct payment by the *owner*) certain costs that normally should be *operating overhead costs* payable by the *contractor* out of his "fee."

"Cost" is generally defined in the *standard forms* of cost plus fee contracts. But, as we have seen, construction jobs produce unique, not standard, products. Therefore, *standard forms of contracts* are not usually suitable for specific projects without amendments and supplements. Nevertheless, the *standard forms* do provide a sound basis for most construction contracts, and their use is to be encouraged in the interest of mutual agreement. By proper use of *standard forms,* the standard and commonplace can be indicated by specific references, and thus the particular and unusual is made more apparent by spelling it out in the *bidding documents* unencumbered by repetition of the usual.

Profit

Profit is one of the primary motivations of all business, but it is not always the only one. Many small construction companies exist because the principal prefers to work at construction, and as long as he earns the living he needs and expects he counts his business as a success.

An important aspect of *profit* is that it is one measure of the efficiency and success of a company, but it is not always easy to apply. A "break-even point" is reached when total income reaches total expenditure, and after that point further income is a *profit.* But how is *profit* to be measured and expressed?

A distinction must be made between *profit* and *profitability. Profit* is the difference between total income and total expenditure, and is stated in dollars, or in any other currency. *Profitability* is a better measure of business efficiency and success, and is, perhaps, best stated as a return on investment. It must be emphasized that net *profit* and *profitability* are determined by deducting all expenditure from all income. This means, among other things, deducting from income proper salaries for the principals of the company, even if the company is a small one-man operation. Otherwise, the excess of income over expenditure will be false, and a true picture of *profit* and *profitability* will not emerge.

Profitability as a return on investment indicates the measure of business success, if we accept the premise that business is undertaken for the purpose of *profit.* If no investment were made in a business, the available capital could be invested in stocks or bonds to produce a *profit.* If, instead, the capital is invested in a construction business, then a *profit* should be produced commensurate with the amount of capital invested and the risk of doing business. How much *profit,* or what degree of *profitability,* is necessary? Several ratios are used by Dun and Bradstreet, Ltd., business consultants, to examine the economic health of many lines of business, including construction. These ratios include: (1) current assets to current debts; (2) current year profit on tangible net worth; (3) sales to tangible net worth; (4) sales to inventory; (5) fixed assets to tangible net worth; (6) current debt to tangible net worth; and (7) total debt to tangible net worth. The ratio of "current year profit on tangible net worth" is described by Dun and Bradstreet, thus:

> Tangible net worth is the equity of stockholders in the business, as obtained by adding preferred and common stock plus surplus (less deficits) and then deducting intangibles. The ratio is obtained by dividing Profits by Tangible Net Worth. The tendency is to look increasingly to this ratio as a final criterion

of profitability. Generally, a relationship of at least 10 percent is regarded as a desirable objective for providing dividends plus funds for future growth.[13]

Market Conditions

It would be wrong to give the impression that all estimates and *bids* for *construction work* are made up of *costs of labor, materials, equipment, job overhead costs,* and *operating overhead costs,* all estimated from determined facts and calculated probabilities, and with a *profit* margin precisely computed according to current economic indicators. In fact, there are other things that at times may have a greater influence on the amount of a *bid* than any variations of those costs just discussed, and the most important of these is the demand for *construction work.*

There are times when a construction company, having bid for several projects in succession, finds that it has all the *work* that it needs for the next year, or longer. There are other times when a construction company with an inescapable *operating overhead* is without *work* to pay for those costs of being in business. At such times as these, *bids* are often submitted in ways that appear at the time to be the most expedient under the circumstances.

When work is scarce, bidders may completely omit *profit* from their estimates, hoping to obtain a contract that will at least help to pay for some of the *operating overhead costs* and enable the company to avoid the loss of key staff. At such times, owned plant and equipment may not be fully charged for in the estimate in an effort to be competitive, and other proper charges may not be made and greater risks than usual may be taken.

When *work* is abundant, at least adequate *profits* are included in *bids,* and sometimes jobs are obtained despite an unusually high profit margin allowed, because there was little competition. Then, all costs are included in the estimate, and allowances for contingencies that may have been included are sometimes found to be unnecessary, thus producing additional *profits.*

Perhaps the biggest problem for the construction industry is the erratic flow of *work*—a matter of feast or famine—in part because of the unique nature of the industry and its custom-designed products. Part of the solution, therefore, lies in the development of industrialized building systems and in the prefabrication of standard building components and complete buildings so as to enable more of the *work* to be done away from the site.

Questions and Topics for Discussion

1. Who shares with the *contractor* in determining the *costs of the work* to the *owner,* and who also shares the risk in doing the *work* for a stipulated sum? Explain precisely how and why.

2. Name and briefly explain the primary factors that affect the costs of construction materials.

3. Describe a type of discount or rebate, in addition to trade discounts and cash discounts.

4. Explain the meaning of the term *labor rate* and the difference between *labor rate* and *wage rate.*

5. Describe the fundamental difference between the *owning costs* and *operating costs* of construction equipment.

6. What is usually the biggest single cost of equipment? Explain briefly how it can be calculated for estimating purposes.

7. Define *overhead costs,* generally.

8. Explain how and why a *job overhead cost* on one job might be classed as an *operating overhead cost* on another job and give an example to illustrate your answer.

9. Explain the statement "Financing is a *job overhead cost.*"

10. What is the difference between *profit* and *profitability,* and how are they measured?

[13] *Key Business Ratios,* Dun and Bradstreet, Ltd.

6

Precepts of Estimating and Cost Accounting

The systematic measurement and pricing of *building construction work* has a history that, among English-speaking people, goes back at least 300 years, and probably twice that far. During the last 100 years the industry has been stimulated by society's growth and industrialization and by the work and interests of certain professional bodies. A few principles and precepts have gained general but not universal recognition, and although the term "principle" has been used by some of the professional bodies involved, for the present it may be prudent to call them all precepts.

Precepts of Estimating

The five precepts have been entitled:

1. Precept of Purpose
2. Precept of Veracity
3. Precept of Verification
4. Precept of Measurement
5. Precept of Accuracy

Each of these precepts is explained below, and its practical applications are examined. In subsequent chapters, these precepts are followed in the examples and enlarged on in the explanations.

The Precept of Purpose in Estimating

1. The primary purpose is to make a *profit* by obtaining and doing *construction work* by the most efficient means available, and this requires that costs be estimated.[1]

2. The specific purposes of estimating are to calculate the *costs of construction work* on the basis of probabilities and to obtain and to record information about *work* for management purposes.

The Precept of Veracity in Estimating

1. All *costs of the work* should be properly accounted for in the estimate insofar as is practical by allocating all foreseeable costs to the proper *items of work*, or to overhead items.

The Precept of Verification in Estimating

1. All estimated costs and other data in an estimate should be compared with the actual costs and data obtained by *cost accounting*, and insofar as is practical, the estimated costs and other data should be either verified or corrected as soon as possible.

The Precept of Measurement in Estimating

1. All *work* should be measured "net in place" (installed) as required by the contract, and "*waste*", and the like, should be allowed for in the *unit prices* of the estimate.[2]

2. No *item of work* should be measured in place of another *item of work*.

3. No *item of work* should be measured in combination with another *item of work* so as to obscure the quantities or costs of either.

4. All *work* should be measured in an estimate so as to make its pricing and subsequent verification as practical and accurate as possible.

The Precept of Accuracy in Estimating

1. All *work* should be measured and priced as accurately as is practically possible, commensurate with the type of *work* and the units of measurement used.

2. Delusions of accuracy should be avoided by an appropriate degree of accuracy in estimating and in *cost accounting*.

These five precepts are explained in greater detail in the sections that follow, and some illustrations are presented. The precepts also are demonstrated later in the estimating examples in Chapter 9.

Purpose in Estimating. *Profit* cannot be more important than producing the things that society needs; but the two purposes, *profit* and production, are interdependent and each is necessary to the other. *Profit* is a desirable and necessary motivation.[3] It creates a challenge, and it is a measure of efficiency. *Profit* is excess of income over expenditure and in order to know the amount of *profit* all costs must be known and accounted for and deducted from income.

The given purposes of estimating are valid for all kinds of estimates, whether they are made by *designers* or by *contractors*, even though the methods of estimating may vary. The purposes of planning and controlling the *work* by using data from the estimate make a construction estimate necessary for all *work*, whether or not it is the subject of competitive bidding. Estimates are not made solely for the purpose of bidding, but also as a means of efficient execution.

Veracity in Estimating. Unless each *item of work* bears all the costs attributable to it, its true cost and the true cost of other items cannot be determined and false costs may be applied to similar *items of work* in other estimates. An estimate should represent the best

[1] The fundamental purpose is, of course, to provide something that society requires, but this consideration enters into questions that are beyond the scope of a text on estimating.

[2] Some estimators disagree with this precept and the arguments for it presented here. However, despite some particular exceptions, many others accept it as a general precept conducive to greater accuracy and better cost control.

[3] State-planned economic systems in some socialist countries have been stimulated by profit motivation after other motivations were found inadequate. At the same time, our society tends to produce many things for which profit appears to be the primary purpose and only justification.

possible effort to calculate all the probable costs and to come as close as possible to the actual costs, which will not be known until the completion of the *work*. An estimate should also provide as much information about the project as is possible and necessary for proper management, including information for planning the *work* and controlling the costs while the job is in progress. To reflect such things, an estimate must be properly organized and laid out, with adequate supporting calculations and descriptive notes to make it clear how the *items of work* have been measured; what assumptions (if any) have been made; and what factors have been allowed for such things as *swell and shrinkage, waste, laps,* and the like, so that the logic of the estimate can be followed and applied.

The precept of veracity in estimating applies to all the different kinds of estimates described in Chapter 2. However, since preliminary estimates are usually required and prepared before all the information about the design of the *work* is available, such estimates are invariably made from incomplete information and a number of assumptions. These assumptions should be stated in the written estimate.

Preliminary estimates made by a *designer,* or by a *designer's consultant,* or by any other person, must make clear the assumptions on which they are based. As the assumptions become facts, or are changed, the estimate must be revised accordingly. Preliminary estimates should develop into intermediate estimates, and from these into final estimates, as design decisions are made and as the *bidding documents* are prepared.

An estimate should be true to the facts as they are known, at all times. Any assumptions that are substituted for facts should be made clear in the estimate and pointed out to the *owner* to avoid friction later. This procedure will also draw attention to those parts of the *work* that still require decisions. Questioning an assumption will sometimes produce a decision and, from the decision, a fact. However, if this precept is not applied, a preliminary estimate will probably become a false estimate as the design proceedes, to the possible embarrassment of the *designer.*

Verification in Estimating. It is imperative that costs and other data in an estimate are compared with the actual costs and data obtained during the course of the *work*. This enables the estimate and data to be verified, and corrected if necessary, as the *work* proceeds. In this way, valid data is available for immediate use on the job for planning and scheduling the *work* and controlling the costs, and for later use in making other estimates. This is the **cyclical process of estimating and cost accounting,** shown diagramatically and explained in Chapters 2 and 3.

Measurement in Estimating. *Work* should be measured net in place and as required by the contract, because this is the only unchanging basis for the *quantities of work*. The alternative is to measure *work*, including extra material, for *waste*. But the amount of *waste* is always a variable, and it varies among all *items of work,* among all jobs, and among *contractors.* *Waste* is an aspect of *material costs* related to the material and its installation that must be estimated and included. But it is better to allow for it in the *unit price* (usually as a percentage), and not in the measured quantities of the estimate, so as to distinguish it as a variable to be estimated or at least considered each time and for each job.

Waste is material that is not required by the contract to be installed as part of the *work*, even though it is often a necessary part of doing the *work*. The *owner* in a stipulated sum contract is not concerned with *waste*. It concerns only the *contractor* who has to do the *work* for the stipulated sum and who has some control over the amount of *waste*. The *contractor* should be concerned with the amounts of *waste* in all contracts in order to control it, and to be able to estimate the *waste* in future jobs. The amounts of wasted materials should be checked while the *work* is in progress so that unnecessary *waste* can be avoided and the estimated allowances for *waste* can be verified or modified for use in future estimates. If the *work* has not been measured net, it is necessary to deduct the *estimated waste* from the measured quantities (which include *waste*) to find the "net quantity" so that a relationship between net quantity and *actual waste* can be established. It is impossible to verify the amounts of *waste* without knowing the net quantities. But there are times when the gross quantities of materials should also be measured, particularly if expensive materials are being used. The amount of *waste* in certain types of *work* may be very small—in some cases, virtually a constant factor that has been verified and established by long experience. In these cases it may be sufficiently accurate to include the *waste* in the measured quantities of *work*. But the amount of *waste* can only be verified by measuring the net quantity installed, and by comparing this quantity with the gross quantity supplied to the job. And this procedure should be followed from time to time with all established allowances for *waste*.

Waste should not be confused with *laps,* which are material parts of the *work* required by the contract and by proper construction practices. *Laps* should always be measured and included in the quantities of an estimate as part of the *work* whenever the *laps* are not an invariable and integral part of the *work* being measured. However, when the *laps* are an invariable and integral part of the *work,* such as the *laps* in a roofing system, the *laps* are not measured, only the net area covered by the *work.* The measurement of *laps* and the treatment of *laps* and *waste* are shown in several of the examples in Chapter 9.

The practice of measuring one *item of work* in place of another is not uncommon, but it should not be done except in preliminary estimates made by a *designer.* In such estimates, it is not usually necessary to know the costs of particular *items of the work,* or the *costs of work* performed by particular trades. The object is to find the *total costs of the work,* or the costs of the major parts, such as the structure, finishes, and the mechanical and electrical systems. Therefore, substituting one item for another may be acceptable in some preliminary estimates of cost, simply as an expedience. However, in an estimate that is to be the basis of a *bid,* the practice of measuring one item in place of another may cause confusion; and it may make *cost accounting* impossible. The different *items of work* should each be measured separately, even though they may exist only in combination. For example, plaster is usually applied to a base, such as concrete or masonry; but unless the base is one constructed by the plasterer, such as lathing, it should be measured and priced separately.

In preliminary estimates it is quite usual to measure certain *items of work* in combination. For example, reinforced concrete may be measured in cubic yards, including the steel reinforcing bars, and the formwork. The quantities of steel can be estimated quite accurately in this way if previous jobs have been analyzed. There is a logical relationship between the quantities of steel reinforcing bars and the quantities of reinforced concrete in a structure, through the structural design. There is much less and often no relationship between the contact area of formwork and the quantity of concrete, because there is no constant relationship between the area and the perimeter of the cross-sections of concrete components.

Measuring *items of work* in combination in preliminary estimates may be carried still further, and the *work* may be measured and priced as *elements* of a building. For example, an *external cladding element* might consist of a concrete wall with rigid, plastic foam insulation adhered to the inner face and covered with painted hardwall plaster. The outer face of the concrete wall might be covered with a clay brick veneer. All of these different items are combined to make the *external cladding element,* which can be priced by a combination of the appropriate *unit prices,* as shown below.

EXTERNAL CLADDING ELEMENT

	Price per SF
8″ thick reinf conc wall, incl forms and steel	$3.30
rigid insulation, inside	0.33
hardwall plaster, inside	0.38
two coats paint, inside	0.17
4″ thick clay brick veneer, outside	1.92
Unit Price of Element	$6.10

If properly based on valid, historical cost data, estimating by *elements* can be used by construction firms to produce "maximum cost" figures for maximum cost plus fee contracts, which were explained in Chapter 1. It is generally the best method for making preliminary estimates.

Measurement and pricing are the two complementary parts of estimating, and since measurement precedes pricing, the measurement must be done so as to make accurate pricing possible. An estimator must consider the *item of work* being measured, its costs, and how it will be priced; and he must measure the *work* accordingly. This approach will help an estimator to avoid measuring in ways and in units that obscure the costs of an item and make accurate estimating and *cost accounting* impossible. This procedure is further explained in Chapter 8, Measuring Work: Particular.

In addition to measuring for the purpose of pricing an estimate, an estimator should measure for subsequent *cost accounting,* because it is through *cost accounting* that he will obtain data for construction management and for future estimates. *Cost accounting* usually requires the use of a "code" to identify the *items of work* so that costs can be segregated and charged to the appropriate items. An estimator should, therefore, measure *items of work* in an estimate according to the cost code. Cost codes are discussed further in Chapter 12, Cost Accounting Practices.

Generally, there is no need for an estimator to measure *work* in greater detail than is needed for *cost accounting* purposes, because the measured details of the *estimated work* can never be verified, or, if necessary, corrected. There are, however, some important exceptions to this general rule. For example, in measuring formwork to a concrete wall in which there are formed offsets and recesses, the extra labor and material used in forming the offsets and recesses should be measured and priced. However, building the wall forms and forming the recesses is one operation and therefore, it may not be possible to isolate the extra costs of forming the offsets and recesses (see Fig. 11). Obviously, the offset and recessed wall shown in the figure will cost more to build than the

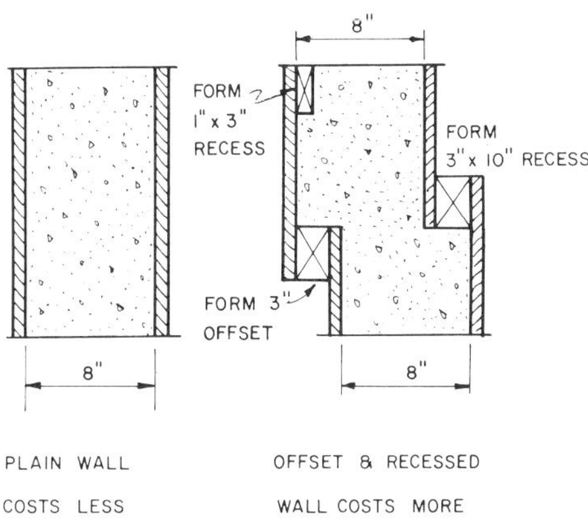

FIG. 11. Forming Concrete Walls

plain wall. But how much more? The estimator may never really know. Nevertheless, the extra cost cannot be ignored, and an estimator should price the *work* of "forming offsets and recesses" by theoretical analysis and experience. The estimator's judgment of the probabilities of costs may never be fully tested in such cases, because the usual *cost accounting* methods cannot segregate the costs of forming offsets and recesses from the total costs of forming walls because of the integrated nature of the *work*. If costs of such items must be accurately determined, it would be necessary to use "work study" methods employing techniques that are beyond the scope of *cost accounting* and that may not always be expedient or desirable.

With the accumulation of cost information on forming walls it may be possible to more or less establish the extra cost of forming walls with offsets and recesses, as compared with forming plain walls; but, this will never be entirely conclusive. Consequently, some estimators wrongly make this a justification for ignoring the costs of such items, while others may only guess at an allowance for such extra costs. A practical approach to this kind of problem is discussed later.

Accuracy in Estimating. This is the passion of the estimator, and it is both his goal and his nemesis. He tries to estimate the *costs of work*, and he knows he cannot completely achieve it. Accuracy is often illusory, such as when the total estimated costs equal the total actual costs and a detailed comparison shows that they are equal only because of several compensating differences. At other times accuracy may be a delusion because the estimator does not really understand accuracy in estimating.

An estimator must strive for accuracy; but accuracy is relative, and the degree of accuracy must be modified according to the *items of work*, the units of measurement, the *quantities of work*, the *unit prices*, and the type of estimate. A candy bar may be casually divided into halves, and only children might quarrel over their shares. A gold bar must be precisely measured and weighed to divide it equally. It is useless to measure *excavation work* to the nearest inch, or to measure concrete to a fraction of a cubic foot, and to then "round off" the total quantities to the nearest cubic yard. Similarly, it is a delusion of accuracy to price each *item of work* in an estimate to the nearest cent, or to make a bid of $341,935.69 (three hundred and forty-one thousand, nine hundred and thirty-five dollars, and sixty-nine cents). In both cases, the estimator has not stopped to consider how accurate his measurements can be, and need to be. He believes in accuracy, and sometimes pursues it blindly without thought. One of the difficulties for the beginner in estimating is to decide how accurate his estimate should be. Pedantry in estimating shows that accuracy is not properly understood. No estimator can make an estimate more accurate than the information from which he works. The degree of accuracy is determined by the quantity and the kind of *work*, and the estimator must decide on the necessary degree of accuracy for each *item of work* in an estimate.

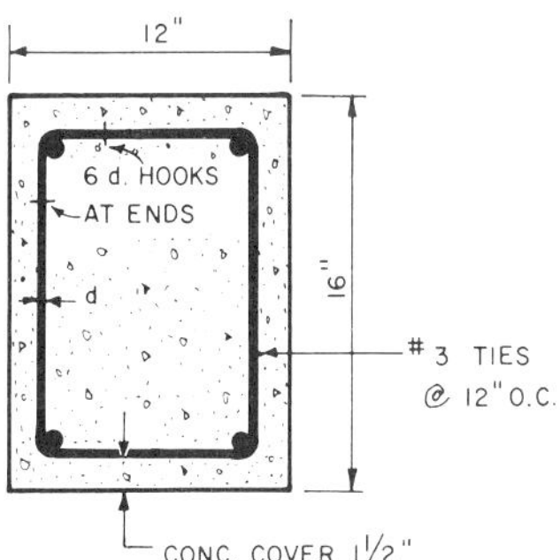

NOTES:

1, OUTSIDE PERIMETER OF CONCRETE COLUMN 2 (16 + 12) = 56 IN.

2, LENGTH OF # 3 TIES MIN. 2(13 + 9) + (2 X 2¼) = 48½ IN. (MORE IF HOOKS ARE LONGER THAN MIN. REQUIRED LENGTH)

FIG. 12. Reinforcing Bars in a Concrete Column

For example, an estimate may include #3 steel bars as ties around vertical reinf bars in conc columns, as in Fig. 12. These ties may cost $.06 per linear foot, (about $.18 per pound), for the costs of material and labor in supplying, bending, and installing the ties. The beginner is often unsure about how to measure the lengths of ties from the structural drawings, which show the sizes of the columns, but which often only indicate the ties diagramatically, or simply by a note on the drawings. In his difficulty, he may not stop to consider the significance of the problem. He should ask himself, How many columns and approximately how many ties are required? If there are only a few columns, he may measure the length of the ties directly from the column sizes, because the perimeter of a column's cross-section will approximately equal the length of the ties in the column.

If there are many ties, this method of measurement may not be sufficiently accurate. But more precise measurement of the tie length will be of little value if the estimator underestimates the number of ties by overlooking the fact that "#3 ties @ 12″ o.c." means that the ties shall not be spaced at more than 12 inches apart, and that in some places ties will have to be spaced closer than 12 inches, such as in a column 9½ feet high, which requires eleven ties. In addition, all the column ties often amount to less than 1 percent of the total steel in a reinforced concrete building.

Precepts of Cost Accounting

Each of the five precepts of estimating has an equivalent precept in the other part of the system—*cost accounting* (CA). Generally, these precepts apply to CA in much the same way as they apply to estimating, but in addition there are some special applications.

The Precept of Purpose in CA

1. The primary purpose is to make a *profit* by obtaining and doing *construction work* by the most efficient means available, and this requires an accounting of actual costs.

2. The specific purposes of CA are to verify and to correct, if necessary, the costs and other information in an estimate, to provide verified data for planning and controlling the *work* in the immediate job (and in future jobs), to provide information for other estimates, and to obtain the total of the *actual costs of work* in order to determine the amount of *profit*.

The Precept of Veracity in CA

1. Every *item of work* and overhead item should carry all the actual costs attributable to that item.

2. All *actual costs of work* should be accounted for by CA, insofar as is practical, and any costs that cannot be attributed to an *item of work* should be attributed to an appropriate overhead item.

The Precept of Verification in CA

1. All actual costs and other data in a *cost accounting* should be compared with the costs and assumptions made in the estimate, so that, insofar as is practical, the estimate can be verified or corrected.

The Precept of Measurement in CA

1. All *work* done should be measured net in place as required by the contract, and any difference between the material quantity of an *item of work* required by the contract and the material quantity of that *item of work* actually used should be classed as *waste*.[4]

2. No *item of work* should be measured in place of another *item of work* (unless a change has been required by the contract and is so identified).

3. No *item of work* done should be measured in combination with another *item of work*, except where so measured in the estimate.

4. All *work* should be segregated and measured in CA as it is segregated and measured in the estimate, in order to verify and correct the estimate and to provide information for other estimates.

The Precept of Accuracy in CA

1. All *work* should be measured and costed as accurately as possible, commensurate with the type of *work* and the units of measurement used.

[4] This does not refer to changes in quantity resulting from a contractual change in the *work* ordered by the *owner* or the *designer*. *Extra work* ordered in accordance with the general conditions of a contract becomes part of the *work* of the contract.

Swell and shrinkage of excavated and fill materials are dealt with here in the same way as *waste*.

2. Delusions of accuracy in CA should be avoided by using an appropriate degree of accuracy in CA and in estimating.

These five precepts are explained below, insofar as their application to CA differs from their application to estimating.

Purpose in CA. The purposes of CA have been explained before, in the first three chapters. CA cannot be examined in isolation, but only as part of the total construction management process, along with estimating. CA is a means of acquiring knowledge about *construction work* as it is actually done, particularly knowledge about the *costs of work* as it is actually done. Without a CA system, a *contractor* may be able to obtain some limited knowledge of completed *work* and its total costs. At least, a *contractor* usually wants to know whether a job made a *profit* or a loss. But some companies do not even know whether or not they have made or lost money, because their records are incomplete or do not exist. Without CA, a *contractor* or his superintendent may have some ideas about the job and about its successes and failures—where money was made and where it was lost. But without CA, it is unlikely that any facts will be recorded that would be useful to an estimator in the future. Without CA, a *contractor* may not know if an *item of work* or a complete job is losing money until it is completed. The purpose of CA is to provide facts. But it may not be enough to justify CA by saying that knowledge is always better than ignorance. More specific reasons may be needed. They are (1) to have information in order to manage the job in hand, (2) to make other estimates, and (3) to manage other jobs in the future.

Veracity in CA. *Cost accounting* is useless if it does not deal with facts. The first argument against CA is usually that superintendents and foremen cannot always be relied on to provide the true facts, that often they only provide information which will benefit themselves, and that they sometimes manipulate cost figures for their own ends. Of course, that happens. But that is not enough to reject the use of CA if we are convinced of its value.

Construction staff who cannot do what is reasonably required of them are in the same position as any other incompetent staff. Many construction foremen and superintendents are convinced of the value of CA and cooperate in making it work. But some do not cooperate, and often they do not stay long with one firm. Having been hired to supervise a job, they may charge costs to the wrong *items of work* or to those items that are actually costing less than was estimated in order to doctor those items that are costing more than was estimated. They may even charge the costs of one job to another in order to make a losing job look better. Finally, when exposed, their argument may be that they are not really dishonest, because in any event all the costs are chargeable to that job, or to that company, and that there is no theft involved. But in a sense there is theft involved, apart from negligence and dishonesty. There is theft involved in the sense that the company's estimators may be misled by false information, and as a result the company may suffer a loss on a future contract.

It is an essential part of the work duties of foremen and superintendents to report on the progress, productivity, and the costs of the jobs on which they are employed; and if a foreman or a superintendent cannot report properly, he should not be employed in that position. Many superintendents and foremen have not received adequate training in CA and in other aspects of supervisory work, and this may explain why some of them do not perform as they should. In such cases, education about the need for CA and the importance of accurate reporting by site staff may be worthwhile.[5]

Verification in CA. This is one of the primary purposes of CA—to verify the costs and other data in an estimate, or otherwise to correct them in the light of actual data.

Measurement in CA. *Work* done at the site and measured for CA—progress reports, interim payment applications, and the like—should be measured in the same way that it has been measured in the estimate; otherwise, valid comparisons cannot be made between the estimate and the CA. As already stated, the primary rules of measurement are to measure *work* net in place and to measure each *item of work* separately. *Work* at the site can be measured more easily from the drawings, and by checking its actual dimensions. If an *item of work*, such as masonry walls, is still largely incomplete, it may be necessary to actually measure

[5] Some observers have pointed out that the construction industry is unusual in that it recruits certain management staff from workmen who are trade union members, and that these superintendents must sometimes face a dilemma in properly carrying out the supervisory duties of management, considering their relationship with their union and its members. Some construction firms now recruit superintendents who have had formal training as engineers or technologists and preferably some business training rather than promoting tradesmen in the field. Nevertheless, practical trade experience is invaluable to a superintendent, and obviously the ideal would be a tradesman who had returned to college for technical and management training.

each portion of completed *work* in place. If, on the other hand, the same item is largely completed, with only a small part of the *work* yet to be done, it will be easier to check the dimensions and to measure the *work* that is not yet done and to deduct this amount from the total amount of the item in the estimate. Of course, if there is any doubt about the accuracy of the amount of *work* in an estimate, the estimated amount should be checked, and the amount of *work* actually done should be measured. By measuring *work* at the site from the drawings, rather than measuring every dimension from the *actual work,* the risk of confusion arising out of differences between dimensions on the *contract drawings* and actual but incorrect dimensions of *work* is avoided.

For example, the drawings may show concrete footings 24 inches wide by 12 inches deep, whereas the actual width of the footings may be 24½ inches in places. The half-inch of extra width should be ignored in measurements, because it is not required by the contract; the footings should be measured as 24 inches wide as required by the contract. To put it another way, only that *work* that is required by and paid for in the contract should be measured in CA. This does not mean that any extra material used in the *work* is not accounted for; the actual quantities of materials delivered to the site are recorded and eventually compared with the amounts required by the contract and measured in the estimate. If there have been no changes required in the contract, any increase in the amount of material used over the amount required in the contract should be classed as *waste* and recorded accordingly.[6] If there is persistent *waste* associated with the use of a particular material, the estimator should know of it and so should the superintendent, because he may be able to reduce or eliminate the *waste* in future jobs. The general rule of measurement for CA is to measure the *work* by the same methods and units as are used in the estimate.

Accuracy in CA. What has been said about accuracy in estimating applies equally to CA because one is the reflection of the other. Attempts to be very accurate and highly analytical in CA may eventually result in a failure to achieve anything, an eventuality that is discussed further in Chapter 12.

To be able to measure something is to know something about it, and these **precepts of estimating and cost accounting** will help estimators and cost accountants to know something more about the *construction work* they have to deal with. That is why the purposes and objectives of their work must be properly understood, and that is why accuracy, veracity, and verification are so important. Otherwise, what is first learned may be diminished; or worse, mistakes and errors may be perpetuated.

Questions and Topics for Discussion

1. Briefly explain the purposes of estimating.
2. Why is it important that each *item of work* should bear all the costs attributable to it?
3. Why is it important that all the *costs of the work* in a construction job should be properly accounted for?
4. Why measure *work* net in place when it is apparent that more material must be used than has to be installed?
5. Explain the difference between *waste* and *laps* in construction materials.
6. Discuss the statement: "Generally, there is no need for an estimator to measure *work* in greater detail than is needed for *cost accounting* purposes." Is this always so, and why?
7. What are the most important qualifications of accuracy in estimating, and why?
8. Briefly explain the purposes of *cost accounting* relative to estimating.
9. Who is the most important person in proper *cost accounting*, and why?
10. State briefly the primary rules for measuring *work* done on a construction site for purposes of *cost accounting*.

[6] No matter the type of contract, oversized concrete footings are essentially *waste* because of the method of construction. The *owner* should not have to pay for any unrequired *work* unless, of course, it is done deliberately to effect a tangible saving, such as eliminating formwork to the footings.

7

Measuring Work: General

The object of *measuring work* is to price the quantities in order to estimate the *costs of the work.* Measurement leads to pricing, and all measurement should be done so as to make accurate pricing possible. Measurement is not an end in itself; and matters of measurement, such as accuracy and tolerances, can only be determined in the light of pricing and costs.

Measuring work in an estimate consists of three parts:

1. **Descriptions** (entering the description of each *item of work*)
2. **Dimensions** (entering the dimensions of each *item of work*)
3. **Quantities** (calculating the quantities of each *item of work*).

These parts will be examined in the following sections.

Descriptions of Work

When both measurement and pricing in an estimate are done by the same estimator, it is possible, but not necessarily desirable, for the estimator to enter very brief descriptions of items in the estimate. The estimator then relies on his memory for a fuller description of each item when he prices the estimate. This practice may be acceptable for *common items of work*, but it is not suitable for uncommon items and complex estimates. Considering the purposes of *estimating and cost accounting* (explained in the preceding chapters), the estimate should contain as much useful information as is practical. The descriptions should be explicit and definitive, and the precise meaning and scope of each *item of work* should be clearly indicated in the estimate so that each item can be accurately priced. This requirement does not demand verbose descriptions, and it is possible to write brief and effective descriptions of *items of work*.

The descriptions of the *work* originate in the *bidding documents*, and it is essential that the estimator study and refer to these documents while estimating. Some estimators do not, and the specifications are sometimes almost completely ignored after an initial perusal. But many *designers'* specifications are too long and badly written, despite the standard formats and guide specifications produced by the Construction Specifications Institute (CSI) in the U.S.A., and by the Specification Writers' Association of Canada (SWAC), with which all estimators should be familiar.

Estimators take a chance in not paying enough attention to specifications and in not using them as they should. If they did, *designers* would produce better specifications. The producers of poor specifications will continue in their ways until the users insist that they must be better.

Some estimators do not recognize or accept the full responsibilities of an estimator's position and have to look to their principals and others for guidance and instructions about all contractual matters. A qualified and competent estimator should have a good working knowledge of contracts and *contract documents* and be able to use that knowledge effectively.

An estimator's first reading of the specifications should be a quick but complete survey of the entire document. He should use the basic techniques of "rapid reading," and a course in this subject would be an advantage to many.[1] In the initial survey, the estimator should reduce the scope and content of the entire project to its essentials, and study the "table of contents" at the beginning of the document. After the initial survey, the estimator should then go through the appropriate *sections* of the specifications for the *work* he is estimating, either marking the document or making separate notes to pick out all those requirements that are neither standard, obvious, nor inevitable. Most specification clauses, however, are standard, obvious, and inevitable because most buildings are for the most part built with commonplace materials and systems. The use of unusual materials and systems is the exception, not the rule.

For example, an estimator sees concrete footings shown on the drawings. He reads the *concrete section* of the specifications and sees that it includes a general reference to an ASTM specification for concrete with which he is familiar. He also sees that concrete in footings must have a compressive strength at 28 days of 3500 pounds per square inch, that the maximum size for the aggregate in the concrete is 1½ inches, and that sulphate resistant portland cement is to be used in the footings. This information gives him the required description of the item "conc ftgs" in the estimate, and it might be written:

> *"conc ftgs 3500 psi (max 1½" agg) with SR p. cmt."*
> ("Concrete footings; compressive strength at 28 days 3500 pounds per square-inch (maximum size of aggregate, 1½"), with sulphate-resistant portland cement".)

Certain *sections* of specifications always require more detailed study and more notes and references than others, because they usually deal with special or unique conditions and requirements. Concrete footings are used under most buildings, but a normal type portland cement concrete is generally used. On the other hand, concrete piles are not common, and any specification *section* on piling will require careful reading and interpretation. Obviously, an estimator must have an extensive knowledge of construction materials and methods, as well as the ability to read and understand drawings and specifications.

The estimator should also examine and read other specification *sections* related to the *work* that he is estimating in order to be sure that he understands the scope of the *work* for which he will bid. If it is *drainage work*, he should read the *sections* on excavation and plumbing; but it is unlikely that he would have to read the *section* on painting.

Some of the information needed to described *items of work* in an estimate is not in the *bidding documents*, and it must come from the estimator through his understanding of *construction work*. For example,

[1] The rapid reading techniques taught by one international school do not require any mechanical devices, and have proven to be an asset to many in the reading of documents. Different writings and different documents require different reading techniques.

formwork for cast-in-place concrete is not usually shown on drawings, and often it is specified in the most general way. An estimator measures from the drawings those surface areas of concrete that will be in contact with formwork (the "contact area"), and he measures any special features formed in the concrete surfaces. The rest of the information required for estimating the costs of formwork must come from the estimator himself. This illustrates again that an estimator must know and understand the *construction work* that he measures and prices, and he must be able to describe the *items of work* from his own knowledge and in such ways as will enable the items to be accurately priced. Specific examples are given in Chapter 9.

Generally, *bidding documents* do not indicate the procedure and the sequence of *work*, nor do most construction contracts. The *designer* illustrates the arrangement and specifies the quality of the *work*, and also may prescribe how some or all of it will be done. The organization and the sequence of the *work*, however, is generally the responsibility of the *contractor*, although there are sometimes contractual exceptions to this rule. Organization of the *work* affects its costs and, therefore, an estimator may have to consider job organization, procedures, and *work* sequences in making an estimate.

In the case of a standard construction project, consisting of more or less common *items of work*, the question of procedure and sequence of *work* may not be a problem for an estimator because he should know the usual procedures and sequences involved. When dealing with a project requiring unusual construction materials and methods, however, an estimator may need advice and assistance. It may be necessary for the estimator to work with other members of the *contractor's* staff, such as engineers and superintendents, to decide on how the *work* will be done before the estimate can be made. Similarly, a standard construction project, but one involving unusual conditions and circumstances that may greatly affect the *costs of the work*, may require that an economical method of construction be found before an estimate and a competitive bid can be made. Even in the most straightforward projects there are many costs that are not indicated or referred to in the *bidding documents* (particularly job overhead items) that an estimator must know about and include in his estimate.

For example, the costs of the West Coast Transmission Building, shown in the photograph on this page, could not have been estimated until certain techniques and procedures had been resolved by the *contractor*, because of the unusual design that utilizes heavy, vertical cables to support the suspended floors. The reinforced concrete central core was slip-formed, and the last concrete to be placed in the core was pumped to a height of over 200 feet. The steel cables were hung over the beams at the top, down the sides of the central core, and were attached to the steel floor decks as the floors were erected. The cables were later encased and protected against corrosion, and finally the building was enclosed by the reflective curtain walls.

The Westcoast Transmission Building
Vancouver, B.C., Canada. Architects, Rhone and Iredale, Vancouver, B.C.; Structural Engineer, Bogue B. Babicki, Vancouver, B.C. (Reproduced by permission of the architects.)

One of the most common errors in estimating often occurs at the outset, when an estimator spends insufficient time in examining the drawings and making notes from the specifications, as described above. He often begins to measure the *work* prematurely, because the temptation to "get going" is hard to resist. Yet time spent at the outset in thinking about the *work* and in studying the documents is always well spent. Junior estimators are often conscious of their confusion when they first look at the drawings of a new project, and they think that starting the estimate will somehow help to dispel their confusion. This may be so for experienced estimators, but for the less experienced a deliberate effort must be made at the beginning to analyze and understand the *work* shown on the drawings and described in the specifications before actually starting the estimate.

This preliminary effort by an estimator is always worthwhile, and sometimes it can be very rewarding. Sometimes the estimator discovers that he can reduce his work by judicious measurement, because the building has features that are repetitive or symmetrical. Often what appears to be a complex set of drawings can be simplified by ignoring the details and minor features and by reducing the building to an arrangement of vertical and horizontal planes laid out about one or more axes.

After the estimator has examined the *bidding documents* and made notes in the appropriate *sections*, the next step is to make an outline of the estimate, consisting of a list of the *items of work* to be measured and priced, grouped under section headings. For example:

CAST-IN-PLACE CONCRETE

Ftgs (SR p. cmt) (3000 psi)	Extr walls, 8″ (4000 psi)
Fnd walls, 8″ (3500 psi)	Sill projs to extr walls
Cols and Bms (3500 psi)	Intr walls, 8″, 6″ (3500 psi)
Flat slabs (3500 psi)	Stairs and lndgs (4000 psi)
Slab on grd (3500 psi)	Mach bases (3500 psi)

Such an outline may not be necessary for an experienced estimator, but it is essential to the beginner if only as a means of giving him pause and thought before starting the estimate. It helps the estimator to organize the estimate and to avoid omissions, and it helps him make concise but adequate descriptions by clarifying how headings should best be used in the estimate. Alternatively, the estimator may choose to use a prepared checklist or a cost code, which achieves the same purpose, as explained in Chapter 12.

It should be remembered that descriptions describe *work*, not only materials. All *items of work* contain *labor costs,* and some items contain only *labor costs.* There is often a tendency to emphasize the "materials" aspects of an item, and to minimize or even to forget the "labor" aspects. *Plant and equipment costs* are also important, and their occurrence should be made clear in the descriptions of items. Measurement precedes and is a preparation for pricing, and all *costs of work* must be priced. For this reason, all items should include all *material costs, labor costs,* and *plant and equipment costs* unless otherwise described, such as "labor only" items.

All descriptions in an estimate are assumed to describe a "positive" item; but often it is necessary to enter a "negative" item—that is, an omission—better described as a "deduction." For example, in measuring the exterior walls of a building, it may be necessary to deduct the areas of doors, windows, and other openings. Such openings are called *voids* when within the boundaries of the measured areas, as in Fig. 13. The term *want* is usually used to identify a deduction made to adjust a deliberate over-measurement, as in Fig. 14, in which the area is first measured by the overall dimensions and the over-measurement is then adjusted by deducting the *want.* This method of measuring areas is recommended in estimating because it simplifies the measurement of complex areas and, in the case of an error, it helps to ensure an over-measurement rather than an under-measurement.

The term *void* is sometimes used to identify a deduction that is made within the boundaries of a measured area because one *item of work* (in the major area) is displaced by another *item of work* (in the minor area, or areas), as in Fig. 15, in which the major area is, say, one type of flooring, and the minor area is another type of flooring in a particular room or area. The practice of writing "negative" items (deducts, omissions) in a different color is obsolete now that photocopies of estimates are frequently made that do not always show the different color.

One way of shortening the descriptions of items in estimates is to use abbreviations, such as those to be found in this book. At present, there are no standard abbreviations for construction, but some abbreviations are widely used and readily recognized. Omitting vowels and dropping the last syllable (or syllables) are the most usual ways of abbreviating words in descriptions of *work.* Standardization of terminology becomes more necessary as computers are used and as computer systems are developed for construction. In some countries dictionaries of standard construction terminology are published for this purpose.

Another means of abbreviation is the use of a code, usually numerical or alphabetical or a mixture of both, as described in Chapter 12. Such a code is used primarily to identify and segregate costs in *cost accounting.* But the same code can be used in estimates, thus eliminating the need for verbal descriptions of items included in the code. There is some reluctance to use only a code for describing items in estimates because of the risk of error and misinterpretation. Such an error in an estimate could be serious. A similar error in a *cost account* does not have the same immediacy and effect, although it might cause a future loss.

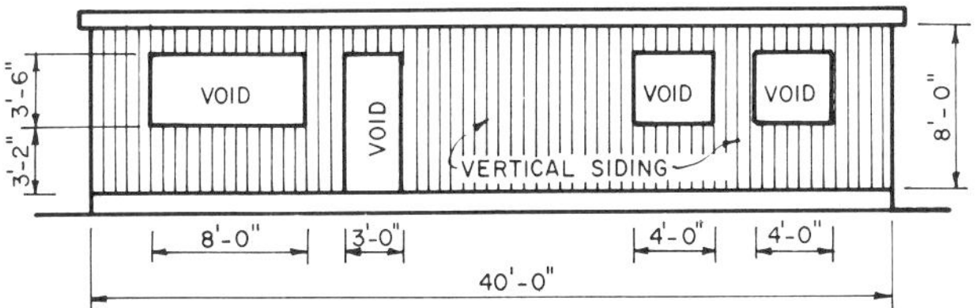

FRONT ELEVATION

AREA OF VERTICAL SIDING (FRONT ELEVATION) — 320 SQ. FT.

DEDUCT (VOIDS / OPENINGS)

		DEDUCT
		20 SQ. FT.
		28 SQ. FT.
		14 SQ. FT.
		14 SQ. FT.
	− 76 SQ. FT.	
	244 SQ. FT.	

FIG. 13. Deducting Voids.

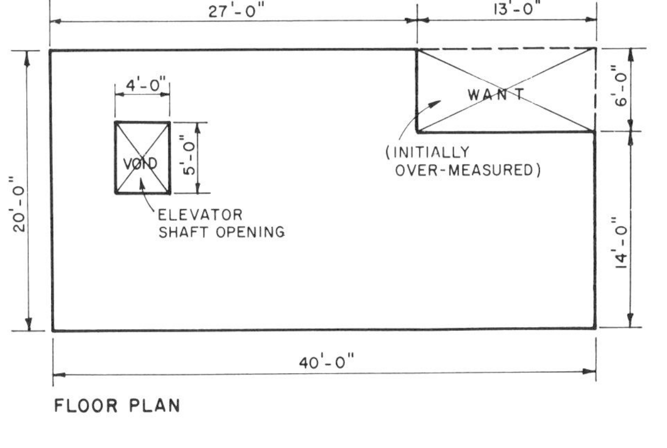

AREA = (40 x 20) - (13 x 6) - (5 x 4)
= 800 - 78 - 20
= 702 SQ. FT.

FIG. 14. Deducting Void and Want.

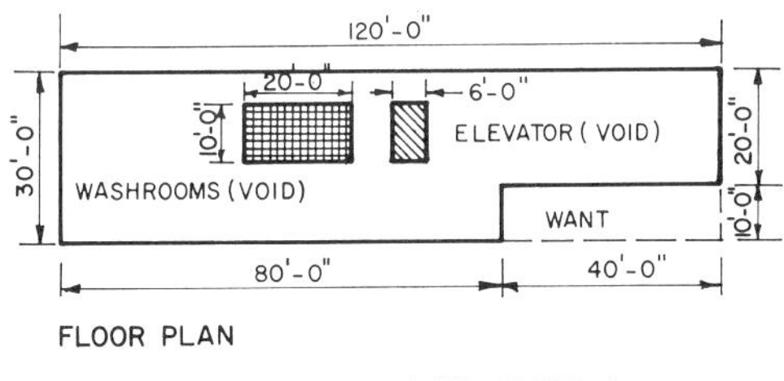

FLOOR FINISHES: VYNIL ASBESTOS TILING;
EXCEPT FOR CERAMIC TILING IN WASHROOMS

AREAS OF FLOOR FINISHES :

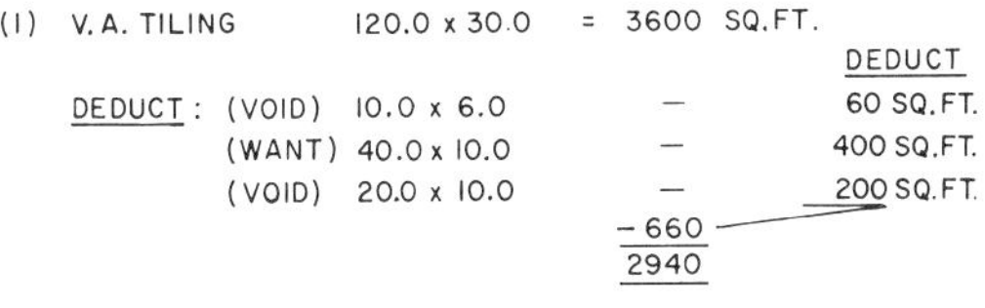

(1) V. A. TILING		120.0 x 30.0 =	3600 SQ. FT.	
				DEDUCT
DEDUCT:	(VOID)	10.0 x 6.0	—	60 SQ. FT.
	(WANT)	40.0 x 10.0	—	400 SQ. FT.
	(VOID)	20.0 x 10.0	—	200 SQ. FT.
			− 660	
			2940	

(2) CERAMIC TILING 20.0 x 10.0 = 200 SQ FT

FIG. 15. Deducting Voids and Want.

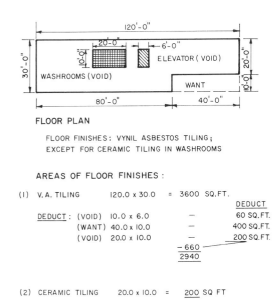

Other aspects of the description of items come from the Precepts of Measurement in Estimating, in Chapter 6. The application of these precepts requires that descriptions of the content and scope of each *item of work* be precise and clear. Loose and careless descriptions may result in bad measurement practices and inaccurate estimates.

Descriptions of items frequently contain dimensions, because not all dimensions are entered in the three dimension columns of an estimate sheet. Entering dimensions in an estimate is discussed below. For the moment it is sufficient to point out that descriptions usually do contain dimensions, and that dimensions in descriptions are written as they appear on the drawings, unlike those that are entered in the dimension columns (see Fig. 16).

In the figure, the example shows several items with none, or some, or all of their dimensions entered in the description instead of in the dimension columns adjoining. For example, no dimensions for the first item, 'Cont Ftgs' (Continuous Footings), are in the description; all three are in the dimension columns. For the next item 'Isolated Ftgs (less than One CY ea) size 4′ × 4′ × 1′3″D' all three dimensions are in the description and none are in the dimension columns, only the number of isolated footings. For the last item '12″Fnd Walls (below Main Floor)', one of its three dimensions is retained in the description until the total area of the 12-inch walls has been calculated. Then the third dimension (the thickness) is transferred from the description and applied to the total area to give the total cubic volume of the 12-inch walls.

Dimensions of Work

We live in a three-dimensional world, and the three dimensions of each *item of work* should be in an estimate, either explicitly or implicitly. Usually the dimensions are explicit in the item's description, or in one or more of the three dimension columns of the estimate sheet. Or one or more of the dimensions may be implicit in the item's description (see Fig. 17). In this example, for the first item (cont ftgs) all three dimensions are in the dimension columns, because the item measures volume, or is a *cube* item. For the next item, the 8-inch thick wall, there is one dimension in the description and two in the dimension columns. Although this item is subsequently measured as a *cube* item, it is expedient to measure it first as a superficial item, or as a *super,* to obtain the area of the

PRACTICAL "STANDARDIZED" FORMS FOR CONTRACTORS
Form 514 MFD. IN U.S.A.

GENERAL ESTIMATE

BUILDING: Sandown School
LOCATION: Sandown I.W.
ARCHITECTS: Charlesworth
SUBJECT: WORK BELOW MAIN FLOOR LEVEL –

ESTIMATE NO. 1
SHEET NO. 2 of 27 R
ESTIMATOR: KC
CHECKER: JB
DATE: June '72

FRANK R. WALKER CO., PUBLISHERS, CHICAGO

DESCRIPTION OF WORK	NO. PIECES	DIMENSIONS	EXTENSIONS	EXTENSIONS	TOTAL ESTIMATED QUANTITY	UNIT PRICE M'T'L	TOTAL ESTIMATED MATERIAL COST	UNIT PRICE LABOR	TOTAL ESTIMATE LABOR COST
REINF CONC (3500 psi/1½" agg)			Conc P.C. $19.60 cy delv'd						
Cont Ftgs	1/	600.00 x 2.00 x 1.00	1200						
8'-10' below grade	2/	20.50 x 2.00 x 1.00	82						
(allow 2% Waste)			1282	(÷27) =	47½ cy	@ 20.00	950	@ 3.00	143
Isolated Ftgs									
(< one cy ea) size									
4' x 4' x 1'3"D									
(allow 10% Waste)	5/4		20		20 No	@ 17.60	352	@ 13.00	260
Isolated Ftgs	4/	4.00 x 4.00	64						
(> one cy ea) in	6/	5.00 x 4.00	120						
S.Q. x 1'3"D	4/	5.00 x 5.00	100						
(allow 10% Waste)			284						
		(Cube)	x 1.25 D						
			355	(÷27) =	13 cy	@ 21.56	280	@ 11.00	143
12" Fnd Walls	1/	600.00 x 4.00	2400						
(below Main Floor)		20.50 x 4.00	164						
(allow 5% Waste)		(Wall Area)	2564						
		(Cube)	x 1.00 Th						
			2564	(÷27) =	95 cy	@ 20.58	1955	@ 4.00	380
				sub-totals (p. 2)			$ 3537		$ 926

f16

FIG. 16. An Example of Items of Work Entered on an Estimate Sheet.

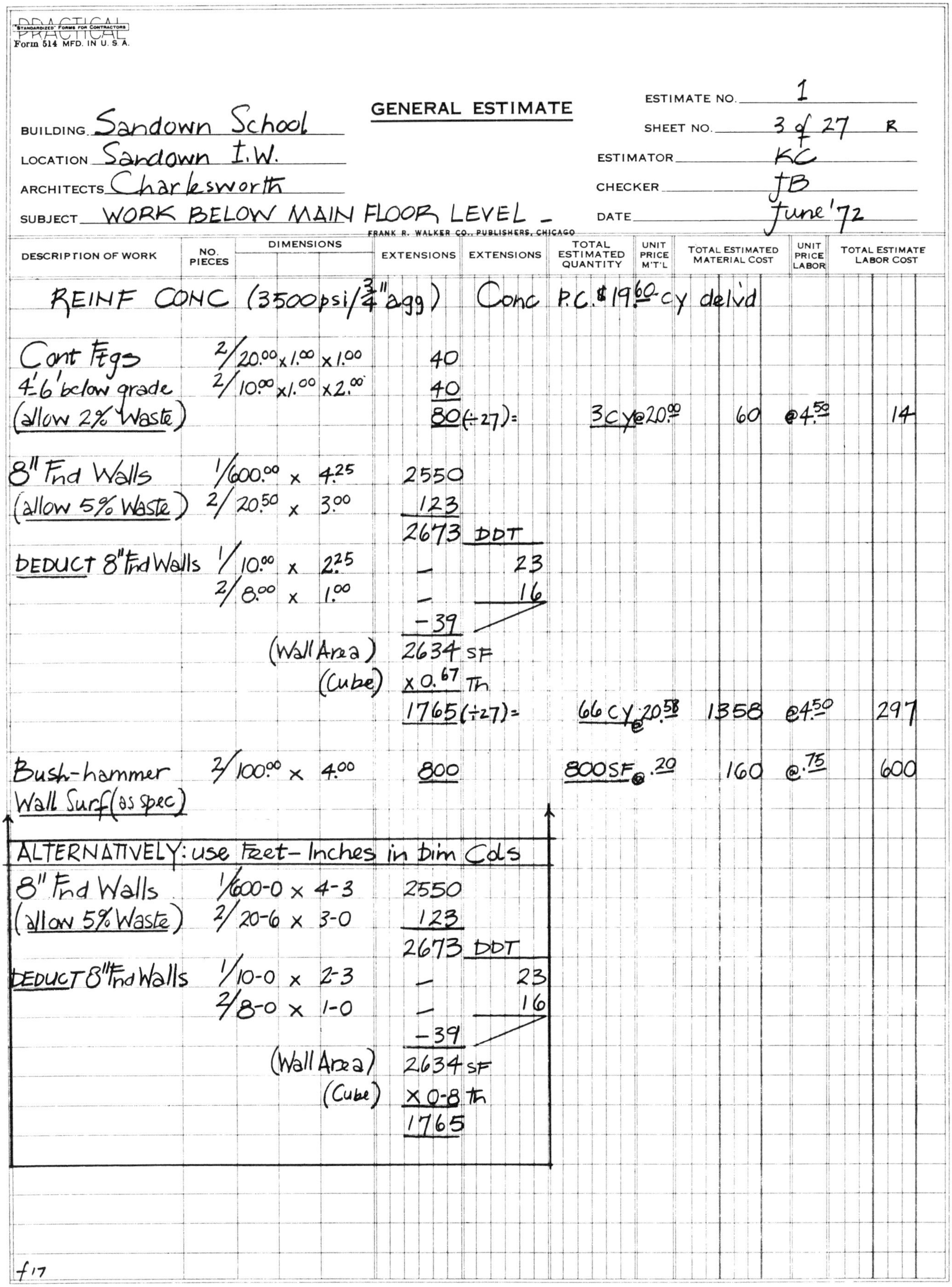

"STANDARDIZED" FORMS FOR CONTRACTORS
PRACTICAL
Form 514 MFD. IN U. S. A.

GENERAL ESTIMATE

BUILDING: Sandown School
LOCATION: Sandown I.W.
ARCHITECTS: Charlesworth
SUBJECT: WORK BELOW MAIN FLOOR LEVEL –

ESTIMATE NO.: 1
SHEET NO.: 3 of 27 R
ESTIMATOR: KC
CHECKER: JB
DATE: June '72

FRANK R. WALKER CO., PUBLISHERS, CHICAGO

DESCRIPTION OF WORK	NO. PIECES	DIMENSIONS	EXTENSIONS	EXTENSIONS	TOTAL ESTIMATED QUANTITY	UNIT PRICE M'T'L	TOTAL ESTIMATED MATERIAL COST	UNIT PRICE LABOR	TOTAL ESTIMATE LABOR COST
REINF CONC (3500psi/¾" agg) Conc P.C. $19.60 cy delvd									
Cont Ftgs	2/	20.00 x 1.00 x 1.00	40						
4'-6' below grade	2/	10.00 x 1.00 x 2.00	40						
(allow 2% Waste)			80 (÷27) =		3 cy	@ 20.00	60	@ 4.50	14
8" Fnd Walls	1/	600.00 x 4.25	2550						
(allow 5% Waste)	2/	20.50 x 3.00	123						
			2673	DDT					
DEDUCT 8" Fnd Walls	1/	10.00 x 2.25	–	23					
	2/	8.00 x 1.00	–	16					
			–39						
(Wall Area)			2634 SF						
(Cube)			x 0.67 Th						
			1765 (÷27) =		66 cy	@ 20.58	1358	@ 4.50	297
Bush-hammer Wall Surf (as spec)	2/	100.00 x 4.00	800		800 SF	@ .20	160	@ .75	600

ALTERNATIVELY: use Feet-Inches in Dim Cols

DESCRIPTION OF WORK	NO. PIECES	DIMENSIONS	EXTENSIONS	EXTENSIONS
8" Fnd Walls	1/	600-0 x 4-3	2550	
(allow 5% Waste)	2/	20-6 x 3-0	123	
			2673	DDT
DEDUCT 8" Fnd Walls	1/	10-0 x 2-3	–	23
	2/	8-0 x 1-0	–	16
			–39	
(Wall Area)			2634 SF	
(Cube)			x 0-8 Th	
			1765	

f17

FIG. 17. An Example of Dimensions Entered on an Estimate Sheet in Different Ways.

wall and thus to obtain the measurements of other related items, such as the wall forms. If the 8-inch thickness dimension were placed in the dimension column, the *super* area of the wall would not appear for future use. However, the three dimensions of the concrete wall are explicit as between the description and the dimension columns.

This is not so in the next item, 'bush-hammer wall surf.' Here, the two *super* dimensions of this tooled finish to concrete are explicit in the dimension columns, but the third dimension is not explicit. It is implicit in the description. The specifications may describe this item in detail, and it may be specified that the bush-hammering (in which a pneumatic tool with a special chisel-head is used to chip away the surface of the concrete to expose the aggregate) "shall remove from between 1/8 inches and 1/4 inches of the concrete wall surface." The third dimension is therefore implicit in the specified depth of the bush-hammering.

Similarly, for *painting work*, two dimensions are usually entered in the dimension columns, whereas the third dimension is implicit in the description. There, the third dimension would be more obvious, since the description should state the number of coats and the minimum thickness of a coat of paint.

Thus, the three dimensions of each item are required in an estimate, either explicitly or implicitly. If any are omitted, the estimate is incomplete and accurate pricing is not possible. In this way, both the "description" and the "dimensions" are one in describing each *item of work* in an estimate so as to enable it to be priced as accurately as possible.

The style of dimensions written in descriptions has already been mentioned. It should be as it is in the drawings, with the diacritical signs (above the figures) for feet and inches. But the style of writing dimensions in the dimension columns should be different from that in the descriptions. Dimensions in the dimension columns may be written as decimal fractions of a foot, or, more easily, simply as feet and inches (*duodecimals*) without the diacritical signs, as in the following table.

DIMENSIONS

In Descriptions (as on drawings)	*In Dimension Columns*	
	Decimal Fractions	feet-inches
2′ 0″	2.00	2–0
2′ 3″	2.25	2–3
3′ 10½″	3.875	3–10½
18′ 1″	18.083	18–1

Obviously, the easiest and simplest way is to enter dimensions in the columns as feet-inches. But, some students fear that such entries may be confused with decimal fractions of a foot. However, in practice there is generally no problem because decimal fractions of a foot are immediately recognizable because they all should be carried to at least two decimal places (except for 6 inches). If an electric calculator is to be used for calculating the quantities from the dimensions, the feet-inches can be converted to decimal fractions of a foot as they are entered into the calculator. In this way the estimator's task of entering dimensions is made easier, and the estimate sheets are simpler and easier to read.

Before calculators, quantities were often calculated manually by junior assistants using *duodecimals.* But the advent of calculators no longer makes their use necessary. However, as long as we continue to use feet and inches (and North America is the only place in the world where the changeover to the metric system is not yet complete) a knowledge of *duodecimals* may be useful to anyone who may find that he has to calculate quantities without the aid of a calculator.

In the examples in this book, the dimensions are entered in feet-inches in the dimension columns (with the exception of some in Fig. 17), and in the descriptions of items the dimensions are shown as they appear on the drawings.

Many dimensions required by an estimator are not shown on the drawings and must be calculated from those dimensions that are shown. These calculations should be shown in an estimate, either at the beginning as preliminary calculations or in the description column close to the description of the related item (see the examples in Chapter 9).

Nothing is more exasperating to a cost accountant, or to any other person who has to use an estimate compiled by another person, than finding dimensions in an estimate that are not on the drawings and for which there is no explanation in the estimate. Some estimators seem unable to grasp this fact, probably because of their limited view of an estimate—that it is simply a means to a *bid.* If an estimator knows and understands that his estimate will be used by others for other purposes, he will probably be more helpful in explaining how he has arrived at the assumptions, conclusions, and dimensions in the estimate that are not obvious from the *bidding documents.*

As estimator must make his estimate understandable to others by supplementing it with brief notes and explanatory calculations. Frequently, these calculations are simple enough to be done mentally by the estimator, and as a result they are too often omitted. The use of electric calculators also tends to cause estimators to neglect to enter calculations and subtotals in their estimates, much to the frustration of the readers of the estimates. An estimator will find that such notes and calculations often help him to avoid errors and to review and check an estimate, but primarily they are necessary to explain the estimate to others. All calculations that are obvious (even though they may be too long or complex for mental calculation) and that contribute nothing to explaining the reasoning in an estimate should not be shown. For example, in Fig. 18, the concrete footing length of 132 feet is not shown on the drawings, so it has to be calculated from the dimensions. This calculation should be shown in the estimate. No other calculations need to be shown, other than the extensions and the total quantity. Calculations that cannot be done mentally should be done by calculator or on a note pad, but no "rough work" should appear in an estimate.

By carefully examining the drawings and specifications before starting an estimate, an estimator is often able to reduce and simplify his work by grouping together similar items with common dimensions and calculating averages and totals and by tabulating information. This step is better done at the beginning of an estimate, and the estimator usually ignores the printed columns and column headings of the estimate sheet in doing preliminary calculations. Student estimators often have trouble with the printed columns on an estimate sheet. Some ignore them completely, whereas others use them scrupulously. The columns are intended to be an aid, not an obstacle. Without them, most estimates would be a puzzle, and student estimators should make an effort to use them properly. On the other hand, they should learn to ignore the printed columns when necessary.

The column between the "Description" and the "Dimensions" is headed "No. of Pieces" in the estimate sheets used in the examples. It is sometimes called the "timesing column" because the dimensions are "times"

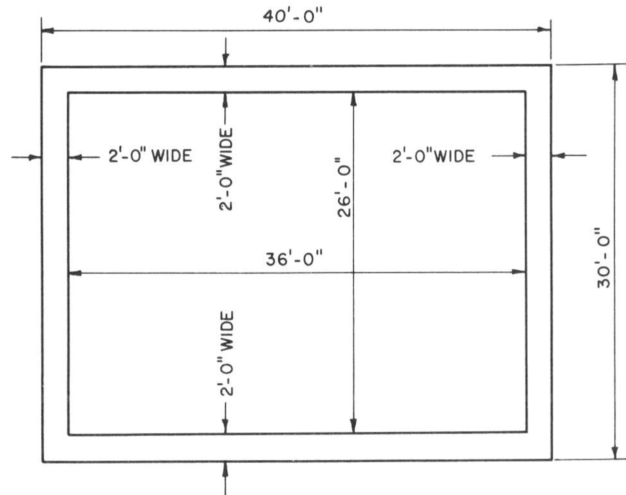

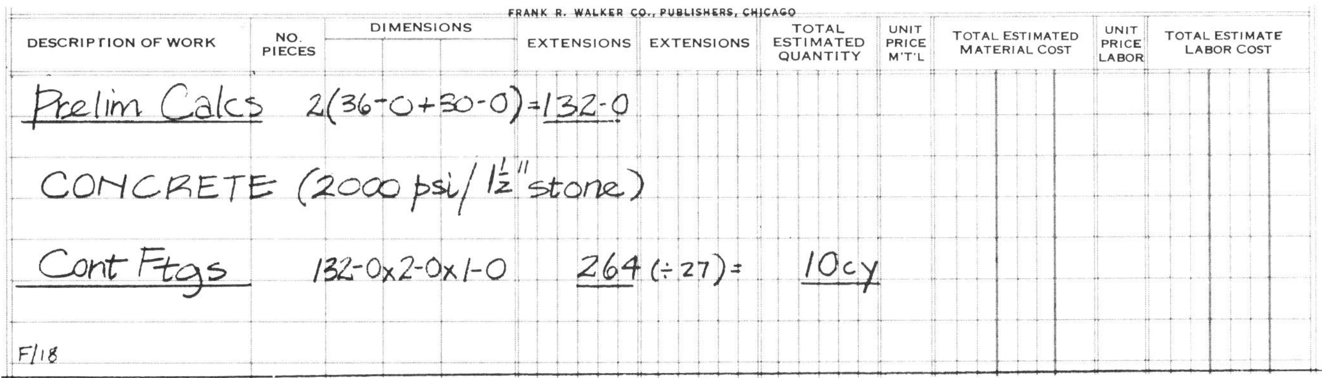

FRANK R. WALKER CO., PUBLISHERS, CHICAGO

DESCRIPTION OF WORK	NO. PIECES	DIMENSIONS	EXTENSIONS	EXTENSIONS	TOTAL ESTIMATED QUANTITY	UNIT PRICE M'T'L	TOTAL ESTIMATED MATERIAL COST	UNIT PRICE LABOR	TOTAL ESTIMATE LABOR COST
Prelim Calcs	2(36-0+30-0)=132-0								
CONCRETE (2000 psi/1½" stone)									
Cont Ftgs		132-0x2-0x1-0		264 (÷27)=	10cy				
F/18									

FIG. 18. Plan Showing Concrete Footings and Their Entry in the Estimate.

(multiplied by) the numbers entered there. Many estimators use a "timesing stroke" to separate a "timesing number" from the dimensions; thus: 2/ 2.0 × 4.0. Mathematically, this means 2 × 2 × 4; but in an estimate, 2/ 2.0 × 4.0 indicates something more. The entry 2/ 2.0 × 4.0 shows that there are two equal areas, both 2 feet × 4 feet. Otherwise, 2.0 × 2.0 × 4.0 indicates a volume 2 feet × 2 feet × 4 feet. Similarly, the "timesing stroke" is used in side calculations and in preliminary calculations to indicate multiplicity rather than multiplication. Timesing can be used most effectively to convey information. For example, in measuring the doors in an apartment block of forty suites, with an entrance door to each suite and ten suites on each of four floors, it is better to enter the dimension for entrance doors as 4/ 10/1 = 40, because this notation immediately conveys a picture to the reader of one entrance door × ten suites × four floors. Of course, the estimator could mentally calculate the total and enter it directly without showing the figures. But this defeats one of the purposes of the estimate: to convey information. Other examples of descriptive entries of dimensions are:

(1) 2/ ½/18.0 × 9.0 = 162 SF
(Two triangular areas, each with a base of 18 feet × a perpendicular height of 9 feet.)

(2) 1/ 3⅐/8.0 × 8.0 = 201 SF
(Area of a circle, 16 feet in diameter.)

(3) 3/ 3⅐/16.0 = 151 LF
(Three times the circumference of a circle 16 feet in diameter.

Simple fractions such as ½ must be written carefully so as to avoid confusion with 1/ 2/ (once times, twice times...).

You may ask, Why times a dimension by 1? Yet there are occasions when "once times" is significant if only to indicate that the dimension should not be timesed by any other figure, particularly if other items, before and after, are timesed by other figures. The "once times," properly used, can convey information that might not otherwise be known. This is the purpose of the conventions and techniques of entering dimensions: to convey information.

Sometimes it is necessary to go back to an item in an estimate to repeat an entry. This can be done by timesing the dimensions (if not already timesed) or by adding on the timesing figure—thus, 2.3/3⅐/16.0. (Originally, three times; now, 3 plus 2, or five times the perimeter of a circle 16 feet in diameter.) The 3/ times might have been erased and replaced by a 5/ times when the additional two items were found. But erasures are not always convenient or desirable, and sometimes the "3 plus two" tells more than "five." This is called "dotting on," but if there is any risk of confusion do not use it.

It is a useful practice to always enter dimensions in this order: horizontal dimensions first; vertical dimensions last. The third dimension in *cube* items should always be the vertical dimension of height/ depth/ thickness, as the case may be. Again, the use of a simple convention will convey more information.[2] Students of estimating who have some knowledge of computer languages and programs will probably not be conservative about using symbols in estimates; and when computers are used more in estimating, many more symbols will have to be created for handling and communicating the data.

Having entered the dimensions in the dimension column and in the description column when required, the next logical step is to enter the extension, or calculation, in the first extension column to the right of the three dimension columns. The extensions should be in whole numbers. As the dimensions are in feet and inches, so the extensions are in feet. It is not usually necessary to enter fractions of a foot in the extensions, and "rounding off" (up or down) to the nearest foot is usually satisfactory. With several extensions to an item, the extensions will usually more or less compensate. In some cases, there is another calculation required to arrive at the "total estimated quantity" of an item that also requires "rounding off" and that obviously affects the degree of accuracy required in the extensions preceding it.

In practice, the extensions are better left until all the *work*, or a major section of the *work*, has been "taken off" (measured) to avoid interrupting the estimator's train of thought. It is important for an estimator to have as few interruptions as possible while taking off quantities. This aspect of an estimator's work is the least understood. Estimating requires great concentration and a lively imagination, and measuring *work* should not be the detailed drudgery that some say it must be. The "taker off" constructs the *work* in his mind as he measures it, and at the same time he analyzes the *work* shown by the *designer* and visualizes the circumstances and conditions affecting it. From this mental process he describes the *items of work* and enters their dimensions. He has

[2] This is the common precept in measuring *work*. Some manufacturers give different orders of dimensions for their products to draw attention to the most critical (the variable) dimension of their product by stating it first. For example, in the case of concrete blocks with standard face dimensions of 16 inches long by 8 inches high, the variable third dimension (the thickness in the wall) is stated first.

translated the *bidding documents* into a concise, verbal, and numerical account. Some other person can do the extensions, for this imaginative process should not be interrupted by such routine tasks.

Quantities of Work

The unit of measurement in which an *item of work* is priced in an estimate should depend on several things, including:

1. The materials used in the *item of work*
2. the labor, plant, and equipment used in the *item of work*
3. the trade practices related to the *item of work*
4. the relative cost of a unit of the *item of work*
5. the appropriate "standard method of measurement"
6. the functions of computers in estimating

But, above all, the unit of measurement used for the total estimated quantity of an item should be that unit that most readily facilitates accurate pricing of the item in the estimate.

The units of measurement of many *items of work* are the same units in which the primary material of the item is bought and sold. For example, clay bricks are sold by the thousand, so clay brickwork is measured by some in square feet and then converted into units of a thousand bricks by a factor that varies according to the size of the brick, the width of the joint, and the number of bricks in a square foot area of wall of a stated thickness. Others prefer to simply measure brickwork in square feet (or yards), stating the thickness, the size of the brick, and the joint width in the description, leaving the further step of computing the total number of bricks until they are ordered for purchase, at which time an allowance for *waste* is added.

If one unit of measurement (U_2) is the product of another unit of measurement (U_1) and a factor (f), and $U_2 = U_1 \times f$, mathematically it does not matter which unit of measurement is used in the estimate (U_1 or U_2), if all the required information is given, including the factor (f). The various units of measurement for the most common *items of work* are given in Chapter 8.

Measuring *items of work* in the same units of measurement as are used for the materials may be satisfactory if different *unit prices* are used for similar material items of different dimensions. For example, the *unit price* "per thousand bricks" for 4-inch thick brick walls should be higher than the *unit price* for 8-inch thick brick walls, all else being equal. There does appear to be some risk of overlooking this matter when using the unit of measurement of the material rather than a superficial unit–say, square foot of wall area, with which the description of the wall's thickness becomes a much more obvious requirement. With an experienced estimator, either unit, per "thousand bricks" or per "square foot of wall area," will serve equally well; but since the *super* measurement has to be obtained first to calculate the number of bricks, and since the *super* measurement better describes the *item of work*, the original *super* unit of measurement is preferable.

The original *super* unit of measurement for masonry walls represents the *work* more clearly in the mind of an estimator. To speak of 100 SF of 8 inch thick concrete block wall conveys a clearer idea of the *quantity of work*, than to speak of 112½ concrete blocks (of 16-inches × 8-inches nominal face size with 3/8-inch wide joints). This is because the *work* consists of concrete blocks that, when built into the wall, lose their identity as separate units. The same point about using original units of measurement can be made for and illustrated with any other *item of work* for which the unit of measurement is sometimes that of its primary material, such as wood framed construction when measured in units of a thousand board feet; steel framed construction when measured in weight-units of tons, hundredweights, or pounds; and reinforced concrete when measured in units of a cubic yard. These *items of work* and their units of measurement are discussed further in Chapter 8.

It may be said in light of the preceding discussion that generally the most rational unit of measurement for any *item of work* is that unit in which the item is first measured from the drawings. That unit is the one that most aptly describes and represents the item in the minds of an estimator and others using the estimate.[3] The use of these units usually requires that one or more dimensions of the items be given in the item's description to provide the complete information about the item in the estimate.

As we shall see, trade practices and customs, and methods of measurement based on them, do not always call for the *work* to be measured in the most

[3] If we escape from the limited concept of an estimate as simply a means to a *bid* and see it as the building's design in another format—as *work* to be done for the *owner*—then we can see that the rational units of measurement of this *work* are not necessarily the same units used to measure and purchase the basic materials. Because in doing the *work* the materials are arranged and changed by skill, labor, and tools, and it is the *work* to be done that is to be measured and priced.

rational units, owing, in part, to compromises with the past, and the past emphasis on materials and their costs rather than labor. But now research is being conducted into the rationality of measurement and computers that require logical programming of construction management systems are being introduced, which means that there will have to be changes made in the methods of measurement. The coming of "industrialized construction" that began with the mass production of building components in the nineteenth century and the development of "systems building" that is the rationalization of construction processes require estimators to examine their methods and to replace some of them with new methods of measuring and estimating costs.

The costs of today's mass-produced building components do not originate from the same sources as the costs of similar components made by hand or by earlier production methods. For example, a flush door, size 1′6″ × 6′8″, may cost $7.90 each, and a similar but larger door, size 3′0″ × 6′6″, may cost $11.50 each. The smaller door therefore costs 79 cents per SF, and the larger door costs 59 cents per SF. Also, the installation costs of preparing and hanging a door will bear slight relationship to the door's size. This illustrates a fact about measuring doors and other such stock items; namely, that they should be and usually are enumerated and fully described. To measure them and price them in square feet would lead to inaccurate pricing. However, in the past, when most doors were specially made, published standard methods of measurement[4] called for doors to be measured in square feet. Now, all doors are enumerated and described.

But not all methods of measurement are so rational. Present standard methods of measurement require formwork for most cast-in-place concrete to be measured as *super* items, the measurements to be those of the actual surfaces of formwork in contact with concrete. Consider the case of formwork for the cast-in-place reinforced concrete stairs shown at (A) in Fig. 19. Suppose that the time required to build the forms for the stairs in (A) is 30 man-hours (say, 2 carpenters for 15 hours). What time will be required to build the forms for the stairs at (B), which are 50 percent wider than the stairs in (A)? They will not require 50 percent more time. The only extra labor required for (B) is in the erection of formwork. The *labor costs* of layout and cutting the material should be about the same in both cases.

This means that *unit prices* obtained through *cost accounting* for forming stairs 3′0″ wide cannot properly be used to price the formwork for stairs 4′6″ wide, if the unit of measurement is "square feet." But if the unit of measurement is, **"one staircase, 3′0″ wide by 12′6″ going × 10′0″ rise,"** the estimator can use the *unit price* for forming of, say, $210.00 each (30 hours @ $7.00 per hour) to estimate the cost of forming for a similar but wider staircase (which is 4′6″ wide) of, say, $230.00 each. An increase in cost of about 10 percent appears to be reasonable. An increase of 50 percent does not appear to be reasonable.

You may ask, Why then do the standard methods of measurement require such formwork to be measured as a *super* item instead of being described and enumerated? It seems that in the past it was necessary to measure all *work* in the standard units (e.g., feet and yards as *cubes, supers,* and *runs*) to combine and handle the many *items of work* in one job. By combining the quantities of similar *items of work,* the number of individual items was reduced, the reasoning probably being that the cost differences between similar items were too small to be significant.

For example, floor tiling installed in a number of rooms of various sizes and proportions of a project was and still is usually measured as one *super* item—the total of all the rooms' floor areas. This total quantity of floor tiling and the accompanying description of the item, say, "1/8 inch thick marbleized vinyl asbestos floor tiling," does not indicate either the room sizes and their proportions (length/breadth) or the straight cutting of floor tile usually required at the perimeters, all of which affect the costs of supplying and laying the floor tiling. The total amount of straight cutting would be a useful indication of certain costs; but straight cutting generally is not measured (its measurement is not generally required by the standard

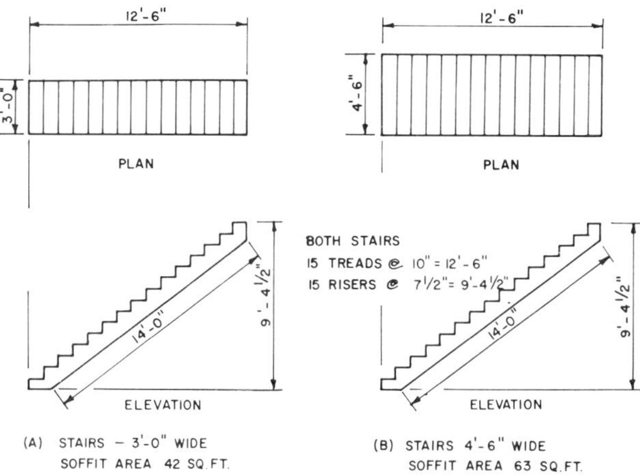

FIG. 19. Cast-in-Place Concrete Stairs to be Formed.

[4] Two standard methods of measurement currently published and in use are referred to in Chapter 8.

methods of measurement), and it is allowed for in the main *item of work*. Obviously, the way to price floor tiling accurately is to measure and price the *work* and the straight cutting in each room separately.[5] But this requires considerably more effort by the estimator, more paperwork and more calculating; and the extra accuracy has not been considered to be worth the extra effort when other more indeterminable parts of the estimate are of greater significance. Now, computers and electronic data processing (EDP) have potentially changed all this.

Computers and EDP can overcome some of the obstacles that in the past have kept us from estimating in greater detail and with greater accuracy, particularly in handling the large volume of data and calculations involved. If a computer can do the calculations and "print out" a copy of the estimate in a matter of minutes, why should we not estimate in greater detail? It is easier for an estimator if he does not have to write down the number of individual items and collect and compute and total them for pricing. Computer results can be more accurate, because more data can be utilized and stored for future use. The only remaining question without a clear and positive answer at present concerns economic feasibility. Is the use of computers in estimating economically feasible for the *contractor* and the *subcontractor*? This is a difficult question, and an answer cannot be given without research into the economics of bidding and computers. Probably at present, under the traditional system of competitive bidding for "lump sum" contracts (for a stipulated sum), computers would not be economical because the bidding system itself is at present fundamentally uneconomical in that most estimating work is non-productive.

Two things are happening to change this situation (1) the tendency to move away from the traditional system of competitive bidding for stipulated sum contracts, and (2) the continuing development and refinement of small, desktop computers that will eventually be within reach of most contracting firms. When building components were handmade by craftsmen, the methods of measurement reflected the materials and the labor used. When building components were mass produced in factories, the methods of measuring those items changed (or should have changed) to suit the new products. This change in methods of measurement is still going on, and soon the changes will be accelerated by something greater than mass production—mass cerebration through computers.

Quantities in an estimate are usually rounded-off, according to the nature of the item and the unit of measurement. There is no general rule other than commonsense that can be applied. Delusions of accuracy sometimes cause estimators to enter *total quantities of work* to two or three decimal places. There is a practical hazard in this unnecessary practice, because a decimal point can be easily overlooked, thereby possibly causing a major error. Generally, we are better off without fractions of any kind in an estimate. They are rarely necessary, because the costs of most *units of work* are not high and the tolerances of accuracy are usually sufficiently wide. The use of fractions depends on the costs of the item and the unit of measurement. If a fraction is necessary, use a simple fraction rather than a decimal fraction. The figures 19½ CY are unmistakable, but 19.5 CY can easily be mistaken for 195 CY. It seems that many are misled by a nebulous and mystical belief that, somehow, decimals must be more accurate, probably because they are often harder to use.

Mensuration of Work

Mensuration means measurement and the computation of measurements, and the student of estimating must have a knowledge of and a facility with mensuration at least as advanced as it is taught in high school. In addition, he should develop a practical comprehension of measurements and quantities. This ability comes with practice; but a conscious effort should be made to acquire it as early as possible.

Much of an estimator's work requires imagination, so that he can visualize *construction work* being done before it can be described and measured in the estimate. It is also necessary that he be able to visualize quantities of materials and *work*. For example, can you visualize a cubic yard; an excavated pit 3′0″ × 3′0″ × 3′0″; a cubic yard of gravel in a pile (about 6 feet across the base and about 3 feet high); a stack of twenty-seven bags of portland cement; or a stack of about 700 standard clay bricks? Can you visualize 1000 board feet of lumber; one-hundred 2″ × 6″ joists,

[5] A common trade practice is to measure areas of finishes "gross" by taking the actual dimensions of rooms up to the nearest multiple of the dimension of the product (tile, sheet) to be installed. This allows for waste material, but it does not necessarily allow for the *labor cost* of straight cutting at perimeters. An allowance for this *labor cost* may be made by pricing the gross quantity measured at a *unit price* for both the *material costs* and *labor costs*; but this is not necessarily accurate because *straight cutting and waste* depend on job and product dimensions, and the differences between them, which are always random and variable. Of course, the experienced estimator can usually make a reasonable judgement, but measuring net quantities and an item of straight cutting in each area is more consistent and rational.

each 10 feet long; 2000 linear feet of 1″ × 6″ boards; or twenty-five 2″ × 12″ boards, each 20 feet long? It is not too difficult to visualize quantities of materials in a heap or a stack. It is much more difficult to visualize them as *work installed* and incorporated in a building, such as the bricks in a wall and the joists in a framed floor or roof. The ability to visualize quantities of *work* comes in part from an ability to do mental computations, and the mind can learn to compute rapidly and without apparent effort. In fact, in mental calculations a strained and deliberate effort can be an obstacle, and it is often better to relax and rely on the brain and its reflexes after some training and practice in mental computation. Also, it seems that a complete reliance on mechanical means deprives the mind of a training that has a value beyond doing simple computations. However, we shall not make a general review of mensuration practices here, but rather shall look at some features that are of special interest and importance to estimators.

Integrating dimensions is one of these features. An estimator should bring together as many dimensions and parts of the same *items of work* as possible so that there is only one total dimension to be used. For example, in measuring partition walls, all the lengths of similar walls of the same height should be integrated into one total length, as in the following example: (note that all dimensions are entered here in feet-inches, or *duodecimals*).

Lengths	*Description*	*Dimensions*	*Quantity*
20.0	6″ Std conc		
60.0	blk ptns		
20.0	(two flrs)	2/126.0 × 8.0 =	2,016 SF
17.6			
8.6			
126.0			

It is not unusual to see experienced estimators enter dimensions as follows:

Conc Ftgs
200.0 × 2.0 × 1.6 = 600 CF
132.0 × 2.0 × 1.6 = 396
200.0 × 2.0 × 1.6 = 600
132.0 × 2.0 × 1.6 = 396
50.0 × 2.0 × 1.6 = 150
34.0 × 2.0 × 1.6 = 102
50.0 × 2.0 × 1.6 = 150
4.0 × 2.0 × 1.6 = 12
2.0 × 2.0 × 1.6 = 6
2412 CF

and so on, and on, and on, with each length of concrete footings separately measured and entered in a mechanical and thoughless way. No wonder some think estimating is tedious! By first collecting the lengths on the estimate sheet (not on a scratch pad) and by obtaining the total length of concrete footings, the dimensions may be entered thus:

Conc Ftgs 804.0 × 2.0 × 1.6 = 2412 CF

Or, if there is a reason for separate measurements, such as for two buildings or for two distinct parts of a building shown on the drawings, the estimator might enter two totals, thus:

200.0	50.0	Conc Ftgs 664.0 × 2.0 × 1.6 = 1992
132.0	34.0	140.0 × 2.0 × 1.6 = 420
332.0	50.0	2412 CF
× 2	4.0	
664.0	2.0	
	140.0	

The object is to simplify the estimate by brevity and to reduce the number of computations required to obtain the total quantity of the item without obscuring useful data. Only *items of work* with two of their three dimensions identical can be integrated by the addition of the various third dimensions. The previous 6-inch thick concrete block partitions are all 8 feet high, and their various lengths can be collected. The above footings are all 2 feet wide × 1½ feet deep, and their various lengths can be collected. Of course, all such collections must be carefully checked.

Averages are helpful, but they are often misunderstood and abused. For example, Fig. 20 shows the elevation of a concrete retaining wall.

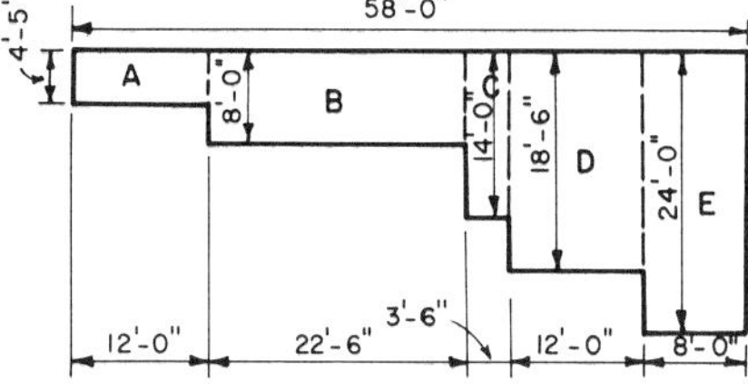

AREA OF WALL FACE = 696 SQ. FT.

FIG. 20. Elevation of a Retaining Wall.

The area of the wall is measured in feet-inches thus:

(A)	12.0 × 4.5 =	53 SF
(B)	22.6 × 8.0 =	180
(C)	3.6 × 14.0 =	49
(D)	12.0 × 18.6 =	222
(E)	8.0 × 24.0 =	192
	(58.0)	= 696 SF

Area = Length × Height ($A = L \times H$); then $H = A/L$
Average height = Area ÷ Length

$$= \frac{696.0}{58.0} = 12 \text{ feet (average height)}$$

This average height may be used for measuring other *items of work* in the reinforced concrete retaining wall. For example, the vertical steel rebars (reinforcing bars) may be measured as averaging 12 feet in length. The average number of rows of horizontal rebars can be calculated by dividing the average height of the wall by the spacing of the horizontal bars and adding an extra bar, because there is always one more bar than spaces. The formwork to the sides of the wall can be measured directly from the surface area of 696 SF already computed. The formwork quantity will be 2/696 SF = 1392 SF plus formwork to the ends, which average 14.2½ high (not 12.0).

Some may try to measure the wall area thus:

	4.5	12.0	58.0 (total length)
	8.0	22.6	× 13.9½ (average height)
	14.0	3.6	(799.11)
	18.6	12.0	800.0 SF
	24.0	8.0	
5)	68.11	58.0 (total length)	
	13.9½ (average height)		

This measurement is not correct, because the average height is not 13 feet 9½ inches, but 12 feet. The error is large (about 15 percent) and unacceptable.

If the wall's elevation were as in Fig. 21., the wall's area could be computed by first obtaining the average height, thus:

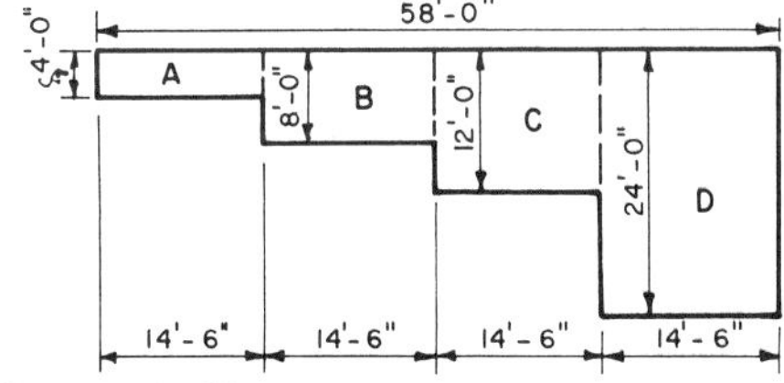

FIG. 21. Elevation of Another Retaining Wall.

	4.0	
	8.0	
	12.0	
	24.0	58.0 (total length)
4)	48.0	× 12.0 (average height)
	12.0 (average height)	696.0 SF

because in this case the four part lengths are all the same, and all the bays have two identical dimensions, the wall thickness and the bay length of 14 feet 6 inches.

Perimeters are often used to measure *work* in buildings, and the accurate computation of perimeters sometimes causes trouble. First, a simple example of a 2-feet-wide concrete footing. The total length of the 2-feet-wide concrete footing is 132 feet. This can be obtained in several ways, thus:

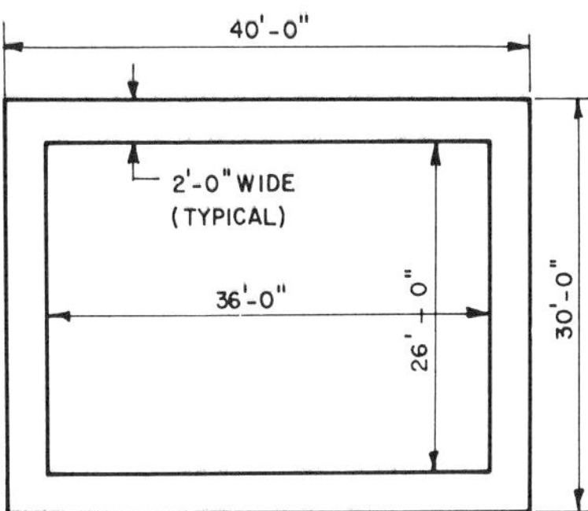

FIG. 22. Plan of Footings.

40.0	40.0	30.0	30.0	40.0
40.0 (*or*)	26.0 (*or*)	30.0 (*or*)	36.0 (*or*)	30.0
26.0	66.0	36.0	66.0	70.0
26.0	× 2	36.0	× 2	× 2
132.0	132.0	132.0	132.0	140.0
			(*less:* 4/2.0 =)	− 8.0
				132.0

These computations are all basically the same. In the last computation, the outside perimeter is computed first and 4/2.0 = 8.0 is deducted to obtain the true length of the footings. This is the best method.

The first four computations appear be more simple and logical, and you might ask, Why first calculate the outside perimeter (140.0) and then make a deduction? There are reasons that require explanation.

Many building plans are not simple rectangles, and computing their perimeters is not so obvious as it is in the example. Also, the outside perimeter is often a useful dimension to have, and it is convenient to compute it while computing the length of the footings. The length of the footings in Fig. 22 is 132 LF, and this is the length of the mean perimeter of the footings, the perimeter at the centerline of the footings as shown in Fig. 23. In Fig. 22, the outer perimeter length is 140 LF; the inner perimeter length is 124 LF; and the mean perimeter length is 132 LF, which

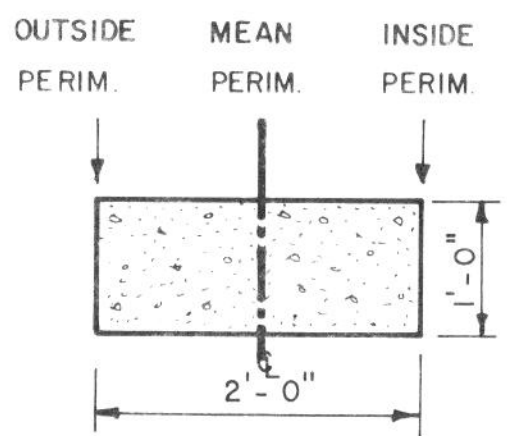

FIG. 23. Concrete Footing (In Section).

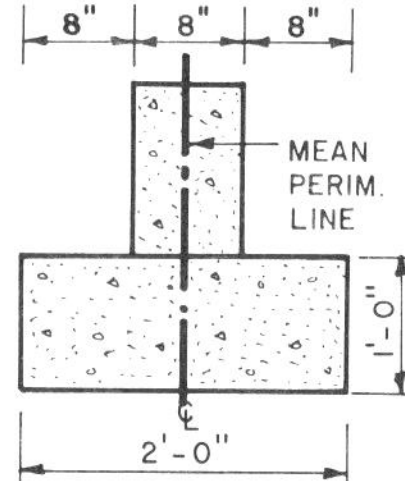

FIG. 24. Concrete Footing and Wall (In Section).

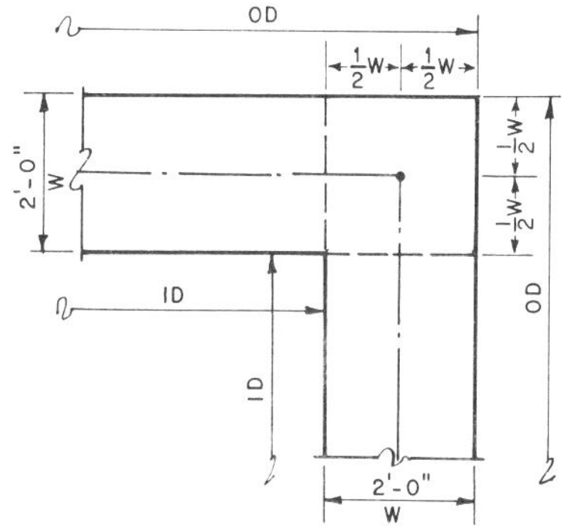

FIG. 25. Concrete Footing (In Plan) at Typical Corner.

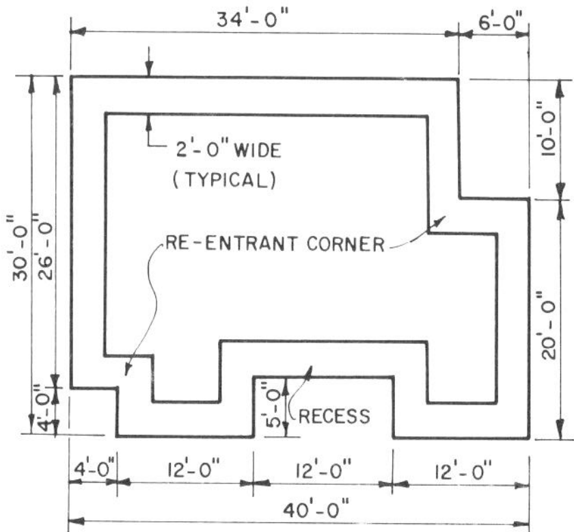

FIG. 26. Plan of Footings.

is the average (mean) of the other two perimeter lengths. The footings' mean perimeter (and true length) is also the true length of any other items symmetrical about the same mean perimeter line, such as the wall in Fig. 24. Since the 8″ foundation wall is centered on the footing, they both have the same mean perimeter and the same length.

Figure 25 shows a typical corner (in plan), and a 2-feet-wide concrete footing. It also shows that the inner dimensions (ID) are each less than the outer dimensions (OD) by the width of the footing, and that the difference between the outer and inner perimeters is twice times the footing width at each of four corners. From this it can be seen (and calculated) that the difference between the **mean perimeter** and either the inner or the outer perimeter is (plus or minus) twice times half the footing width at each of four corners. This is because there are two dimensions to be adjusted at every corner, each by half the width difference between the inside and the outside perimeters, to arrive at the **mean perimeter.**

Notice that some building plans do not show the dimensions of footings; only the plan and the dimensions of the exterior walls are shown. In those cases, the perimeters can be calculated from wall dimensions in exactly the same way; and the mean perimeter thus calculated can be used to measure the footings, as indicated in Fig. 24.

A less simple foundation plan, as in Fig. 26., can be approached in the same way, and the mean perimeter may be computed thus:

$$2 \times (40.0 + 30.0 + 5.0) - (4. \times 2.0) = 142 \text{ LF}$$

Notice that re-entrant corners, such as those indicated in Fig. 26, do not affect the length of a perimeter. However, the presence of a recess, such as is shown in Fig. 26 (size 12′0″ × 5′0″), does increase the length of a perimeter. Without the recess, the mean perimeter in Fig. 26 would be 132 LF, the same as in Fig. 22 despite the difference in shape. But the 5′0″-deep recess increases the mean perimeter by 2 × 5.0 = 10 feet. The rule for computing perimeters can be simply expressed:

> **Ignore all re-entrant corners, and add the overall length to the overall width of the building, add the depth of each recess, if any, and multiply the sum by 2. The result is the outside perimeter.**
> **To adjust an outside (inside) perimeter to the mean perimeter, deduct (add) four times the width of the footing, (or the thickness of the wall, as the case may be). The result is the mean perimeter.**

Notice that the adjustment is always four times the width or thickness—once for each 90-degree corner,

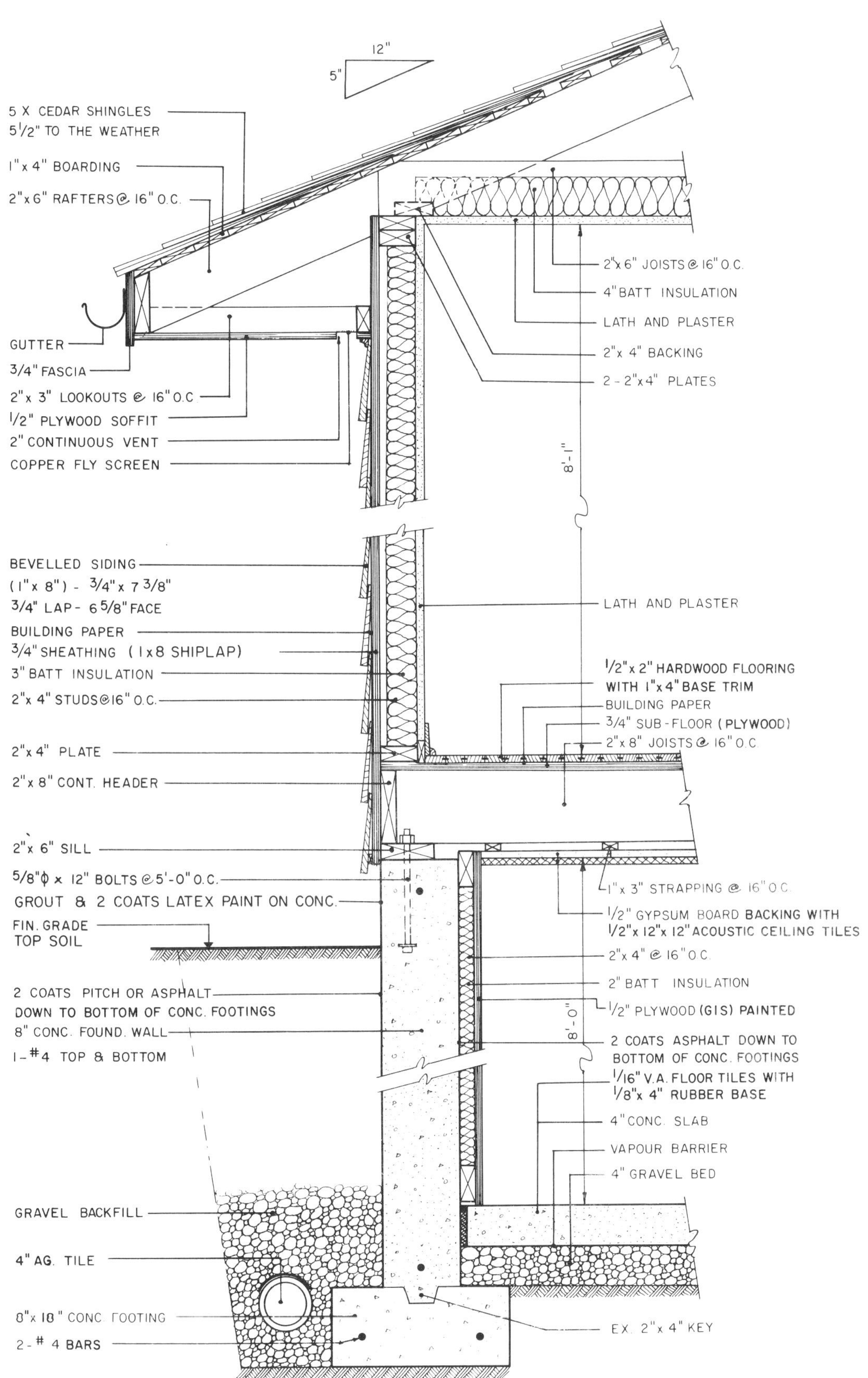
12"
5"
5 X CEDAR SHINGLES
5 1/2" TO THE WEATHER
1"x 4" BOARDING
2"x 6" RAFTERS @ 16" O.C.
GUTTER
3/4" FASCIA
2"x 3" LOOKOUTS @ 16" O.C.
1/2" PLYWOOD SOFFIT
2" CONTINUOUS VENT
COPPER FLY SCREEN
2"x 6" JOISTS @ 16" O.C.
4" BATT INSULATION
LATH AND PLASTER
2"x 4" BACKING
2 - 2"x 4" PLATES
8'-1"
BEVELLED SIDING
(1"x 8") - 3/4"x 7 3/8"
3/4" LAP - 6 5/8" FACE
BUILDING PAPER
3/4" SHEATHING (1x8 SHIPLAP)
3" BATT INSULATION
2"x 4" STUDS @ 16" O.C.
2"x 4" PLATE
2"x 8" CONT. HEADER
2"x 6" SILL
5/8"ϕ x 12" BOLTS @ 5'-0" O.C.
GROUT & 2 COATS LATEX PAINT ON CONC.
FIN. GRADE
TOP SOIL
2 COATS PITCH OR ASPHALT
DOWN TO BOTTOM OF CONC. FOOTINGS
8" CONC. FOUND. WALL
1-#4 TOP & BOTTOM
GRAVEL BACKFILL
4" AG. TILE
8"x 18" CONC. FOOTING
2-# 4 BARS
LATH AND PLASTER
1/2"x 2" HARDWOOD FLOORING
WITH 1"x 4" BASE TRIM
BUILDING PAPER
3/4" SUB-FLOOR (PLYWOOD)
2"x 8" JOISTS @ 16" O.C.
1"x 3" STRAPPING @ 16" O.C.
1/2" GYPSUM BOARD BACKING WITH
1/2"x 12"x 12" ACOUSTIC CEILING TILES
2"x 4" @ 16" O.C.
2" BATT INSULATION
1/2" PLYWOOD (G1S) PAINTED
8'-0"
2 COATS ASPHALT DOWN TO
BOTTOM OF CONC. FOOTINGS
1/16" V.A. FLOOR TILES WITH
1/8"x 4" RUBBER BASE
4" CONC. SLAB
VAPOUR BARRIER
4" GRAVEL BED
EX. 2"x 4" KEY

FIG. 27. A Typical Exterior Wall Section.

since the perimeter always passes through 360 degrees. This fact is not affected by re-entrant corners and recesses. Also, if a building's inside perimeter is calculated first, the adjustment of four times the width or thickness is **added** to give the mean perimeter.

It is often useful to be able to compute the length of other perimeters in addition to that of the mean perimeter. Looking at a section through a building's foundations and exterior walls indicates how the adjustment of perimeters can be simply used and expressed.

The difference in length between two perimeters is always four times, twice times, the horizontal distance between the perimeter lines.

In Fig. 22, the mean perimeter is 132 LF. The horizontal distance between the **mean perimeter line** and the line of the outer perimeter is 1 foot. Applying the above adjustment: 4 × 2 × horizontal distance of 1 foot = 8 LF; and (132 + 8) = 140 LF, the length of the **outer perimeter**. The length of the **inner perimeter** is, by the same adjustment (132 − 8) = 124 LF. The same adjustment is added or deducted, depending on whether the required perimeter is inside (shorter than), or outside (longer than), the original perimeter. This adjustment rule conforms with the rule for computing perimeters given above.

Consider the exterior wall section of a building and the number of *items of work* whose lengths are related to the building's perimeter. In Fig. 27, many such items have been indicated to show their number. The true length (perimeter) of each one of these items can be easily computed by making one adjustment (an addition or deduction) to a computed perimeter length.

Volumes of earthworks such as excavations and fillings are required for most estimates, and the accuracy of the measurements and the volumes depends on the quality and amount of available information, as well as the estimator's skill. It is easy to have delusions of accuracy in measuring earthworks; and an estimator should keep in mind the object of the measurement—to estimate the *costs of the work*—as well as the practical accuracy of the measurements and the relative importance of the quantities involved compared with other economic and physical conditions affecting the *costs of the work*.

Measured quantities probably are least significant in estimating the costs of earthworks, because there are so many other factors that are more important. These include:

1. the type and moisture content of the soil, which will determine the type and productivity of the excavating equipment and the degree of *swell and shrinkage* (increase and decrease in volume) that will occur when earth is excavated and later backfilled
2. the climate, location, and accessibility of the site, which will affect the type of equipment that can be used and the handling and disposal of excavated material on and off the site
3. the size and topography of the site, which will also affect the type and productivity of the excavating equipment as well as the amount of *work* to be done and material to be handled
4. the type and depth of excavation, which will also affect the type and productivity of equipment and the amount of *work* to be done and material to be handled.

These and other factors are often more important to the *costs of the work* of excavating than the measured quantities. Nevertheless, most estimators will not try to estimate the costs of earthworks without measured quantities, and many will make two estimates of costs —one by pricing quantities and another by estimating the equipment time and costs required to do the *work*.

Other considerations in measuring and pricing earthworks are the working space allowances to be made for other *work* in excavations, and the provisions to be made for *work* in stabilizing excavations.[6] The allowances for *swell and shrinkage* caused by excavating and consolidating backfilled materials should be included in the *unit prices* according to the Precept of Measurement to measure *work* net in place, as explained in Chapter 6. This leaves the actual measurement of the volumes of earthworks to simple solid geometry, in the first instance.

Excavations in trenches, basements, and the like, are measured simply by multiplying length times breadth times depth. Excavations in banks[7] are measured by multiplying the cross-section area by the length, if the cross-section is more or less constant throughout the length. When the volume to be excavated is enclosed by sides that are not parallel, the **Prismoidal Formula**[8] may be used, as illustrated in Fig. 28. In some cases the volume of a prismoid can be calculated to sufficient accuracy by $L \times C$, which in Fig. 28 (b) gives a volume of 90 cubic feet instead of the 93 1/3 cubic feet computed with the formula. For small works, the first would be sufficiently ac-

[6] These allowances and provisions for excavations are referred to in Chapter 8, Measuring Work: Particular.

[7] A bank refers to the face of an excavation at a change in level between two more or less horizontal planes, not to a financial institution.

[8] Appendix C contains explanations of this and other formulas for measuring areas and volumes.

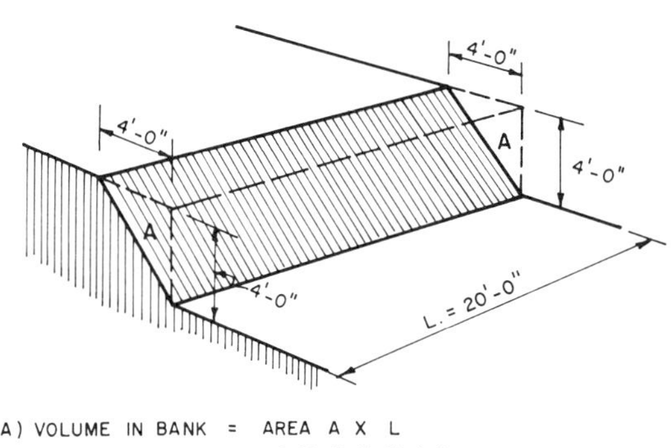

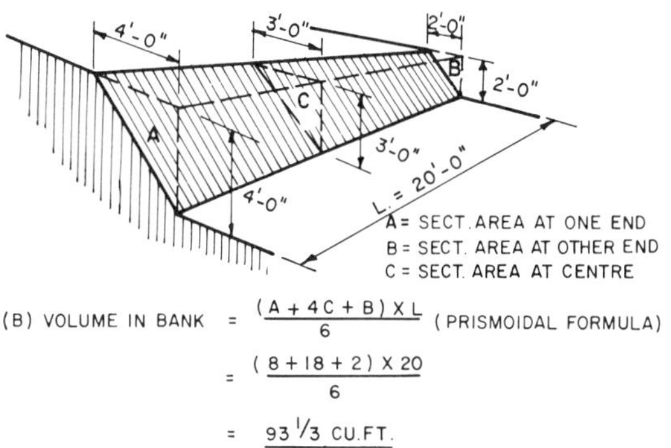

FIG. 28. Excavations in Banks.

curate. The Prismoidal Formula may be used to measure the volume of excavation over a sloping site or area, as in Fig. 29, where excavating is to extend down

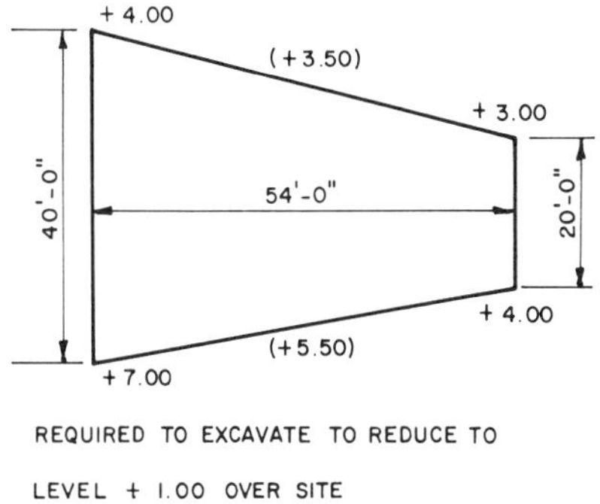

FIG. 29. Site Plan Showing Existing Levels.

to level + 1.00 over the site. Assuming even and regular slopes, we may interpolate the levels at the centers (in brackets). Using the Prismoidal Formula,

$$V = \frac{(A + 4C + B) \times L}{6}$$

$$A = 40 \times \frac{3 + 6}{2} = 180 \text{ SF}$$

$$B = 20 \times \frac{2 + 3}{2} = 50 \text{ SF}$$

$$C = 30 \times \frac{2.5 + 4.5}{2} = 105 \text{ SF}$$

Then,

$$V = \frac{(180 + 420 + 50) \times 54}{6} = 5850 \text{ CF}$$

If, as is commonly done, an average of the existing levels at the four corners is taken, the result would be slightly different:

4.0
7.0
3.0
4.0
4) 18.0 = 4.5 (Average existing level)
− 1.0 (Reduced level required)
3.5 (Average depth of excav)
V = 54.0 × 30.0 × 3.5 = 5670 CF

The difference is not large, about 3 percent, and this second method may be used for a small site or with each of the grid squares over a larger site, as explained below.

If a site has no parallel sides, it will be necessary to draw a "compensating line" parallel to another side in order to use the Prismoidal Formula, as in Fig. 30., Site "X."

The areas of irregularly-shaped sites with boundaries that are not straight lines can be calculated by first drawing compensating lines, as in Fig. 30, Site "Y" and then by dividing the equivalent area into triangles to calculate the area. The volume of earthworks can then be calculated by averaging the depths of excavation, as above, or, if two sides are more or less parallel, by using the Prismoidal Formula.

Some degree of approximation depending on the position of compensating lines is inevitable, but these methods can be sufficiently accurate for measuring earthworks and siteworks for which the *unit prices* are not high. The compensating lines are usually easily drawn in the optimum locations by using a transparent rule or square to keep the plan visible and to facilitate visual equalization of the compensating areas on either side of the line. A transparent plastic sheet

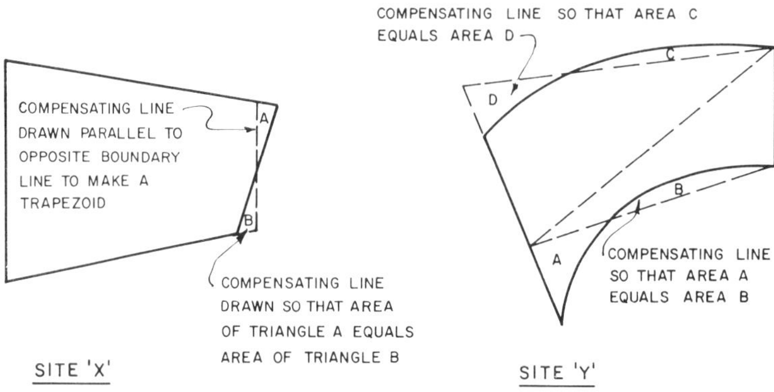

FIG. 30. Plans of Irregular-Shaped Sites with "Compensating Lines" Drawn.

with a printed grid of small squares makes equalization much easier. The squares and part-squares on either side of each compensating line are counted and balanced. Likewise, an irregular area can also be measured by counting all the squares and part-squares within its boundaries.

In Fig. 31., a more precise method of measuring irregular areas is shown. The site plan shows a parcel bounded on the west by a road, on the east by a river, on the south by a creek, and on the north by a boundary line that is not straight. Over the plan have been laid strips of a selected equal width. The area of the site can be found by measuring to scale, from the plan, the average length of each strip inside the boundaries and by multiplying the total length of the strips by the standard width. Irregular areas must be added or deducted as necessary, and as indicated. These irregular areas can sometimes be measured by using compensating lines, and by triangulation, or by dividing the area into smaller strips in the same way. Alternatively, and instead of using strips, grid squares can be drawn on the plan and used to measure areas, as indicated and described above. Measurements are made more accurate by using narrower strips or

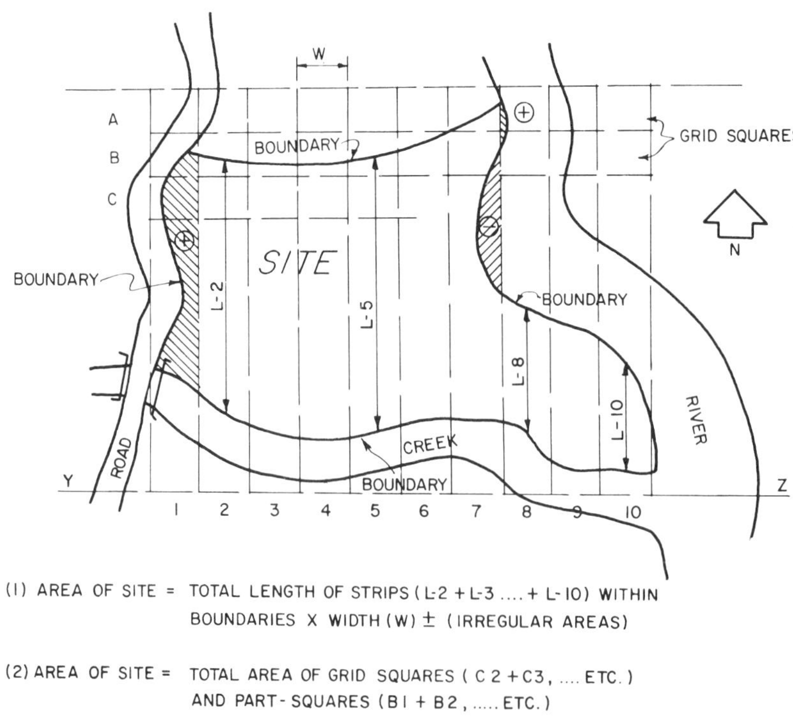

FIG. 31. Plan of Irregular-Shaped Site Divided into Strips (or Squares).

smaller grid-squares, the dimensions of which are selected for suitability and convenience.

Earthworks to finish the ground to specified levels and slopes by "cutting down the hills and filling in the valleys" requires other means of measurement that are based on those already described. The primary means is a grid laid over the site plan to artificially divide up the site into small and manageable areas. The grid may be on a transparent overlay sheet, or it may be drawn on the plan as previously explained. The size of the squares in the grid should be selected to ensure optimum accuracy, as explained below.

All measurements of earthworks by solid geometry are based on an essential assumption—that all the lines and planes are straight—because the solid geometry used is based on volumes bounded by straight lines and planes. The use of grids also involves the same application of solid geometry, and the same assumption is made. In Fig. 32 this assumption is illustrated

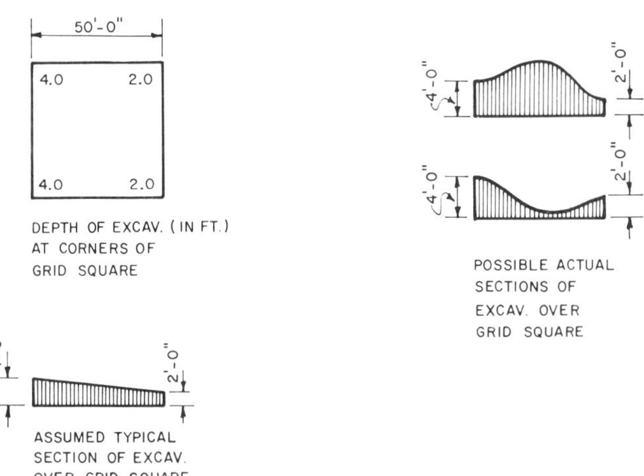

FIG. 32. Excavation Measurement Within a Typical Grid Square.

together with conditions that might actually exist. If the topography is gently rolling, a large grid square can be used; but if the topography is rough and broken, a smaller grid square should be used. If, in Fig. 32, a grid square of half the size were used, the actual profiles within the smaller grid squares would come much closer to the assumption that surface slopes with a grid square are straight slopes. The same applies to areas to be filled. The more variations and topographical features there are to affect the accuracy of excavation and filling measurements (cut and fill), the smaller must be the grid squares to enable more levels and dimensions to be used in computing the volumes of cut and fill.

Having selected the appropriate size of grid, and having overlaid the site plan with the grid, the estimator may apply one of several methods to calculate the volumes of cut and fill. If there are both cut and fill to be measured, it is first necessary to distinguish the areas of each on the site plan. This is done by plotting and drawing "cut and fill lines." A cut and fill line is a contour line, and a contour line is a line joining points on a plan that are at the same elevation. In this case, a cut and fill line is a contour line that joins those points on a plan at which elevation neither cut nor fill is required, because that particular elevation is the required "finished elevation."

The cut and fill lines are, therefore, the boundaries between those areas requiring excavation and those areas requiring fill, and in some grid squares (through which cut and fill lines pass) there will be some cut and some fill. Because of the shallow depths of cut and fill in the proximity of cut and fill lines, it is usually accurate enough to take part-squares (caused by cut and fill lines) to the nearest half of a square; more than three quarters of a square can be counted as a whole square in measuring areas of "cut" and "fill."

The following reproduction[9] compares three apparently different methods of measuring cut and fill over a grid square, or a site. The so-called "Four-point" uses a formula (the proof of which is given) that is useful when dealing with a large site and with several cut and fill lines, for it enables the estimator to simply collect the sum of the cuts and the sum of the fills and to apply the formula to these two totals. The formula is based on the necessary assumption stated earlier, that the slopes within grid squares are straight and even.

Earthwork Volumes, by

I. Prismoidal formula
II. Four-point method
III. Center-square method.

I. Prismoidal Formula:

$$V = \frac{L(A_1 + 4A_m + A_2)}{6}$$

V = Volume in Cubic Feet

L = Perpendicular distance between end planes (A_1 & A_2) in Feet

A_1 = Area one end plane in Square Feet

A_2 = Area other end plane in Square Feet

A_m = Area middle plane in Square Feet

[9] Reproduced by permission of Concosts Services, Ltd., *Construction costs reports,* Vancouver, B.C., Canada.

(**Note:** see application of formula below for an understanding of the above terms.)

II. Four-point Method:

Use formulas:—(all measurements in Feet)

$$V_C = \frac{L^2 \times (H_C^2)}{{}^*108 \times (H_C + H_f)} \;\&\; V_f = \frac{L^2 \times (H_f^2)}{{}^*108 \times (H_C + H_f)}$$

to calculate volumes of Cut and Fill in those grid squares divided by a cut and fill line.

V_C = Volume of Cut in Cubic Yards
H_C = Sum of Cuts on four corners of grid squares
V_f = Volume of Fill in Cubic Yards
H_f = Sum of Fills on four corners of grid squares
L = Length of side of grid squares (in ft.)
* 108 = 4 × 27: (4 × for four corners: and 27 × for cubic feet per cubic yard)

(**Note:** see application of formula and also the proof of the formula below.)

Applying:—**I. Prismoidal Formula** to a typical grid square — 100′ × 100′. with C = cuts F = fills as shown (in ft.) L = 100′: L^2 = 10,000 SF.

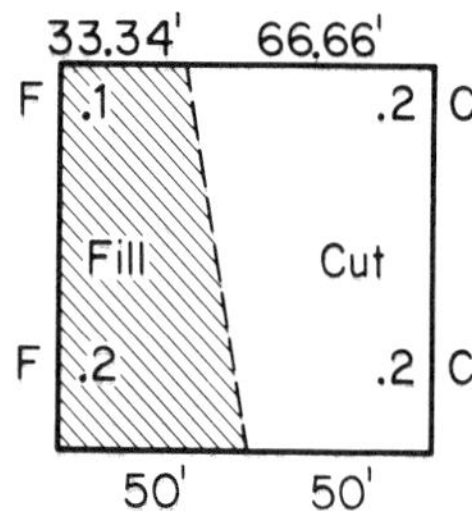

We must assume an even fall across grid square; then, proportion of C/F will be indicated by Cut/Fill line plotted as shown.

$$\text{Area of Fill} = \frac{50.00 + 33.34}{2} \times 100 = 4{,}167 \text{ SF}$$

$$\text{Area of Cut} = \frac{50.00 + 66.66}{2} \times 100 = 5{,}833 \text{ SF}$$

$$\text{Total Area} = 10{,}000 \text{ SF}$$

Volume of Fill:—considering end-planes A_1 and A_2 also, middle-plane A_m.

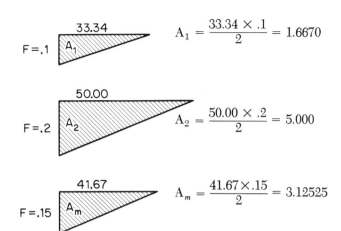

$$A_1 = \frac{33.34 \times .1}{2} = 1.6670$$

$$A_2 = \frac{50.00 \times .2}{2} = 5.000$$

$$A_m = \frac{41.67 \times .15}{2} = 3.12525$$

substituting in $V = \frac{L_1(A_1 + 4A_m + A_2)}{6}$

$$V = \frac{100(1.6670 + 4 \times 3.12525 + 5.00)}{6}$$

$$= 319.5 \text{ CF} = \underline{11.83 \text{ CY of Fill}}$$

Volume of Cut:—(as for Fill)

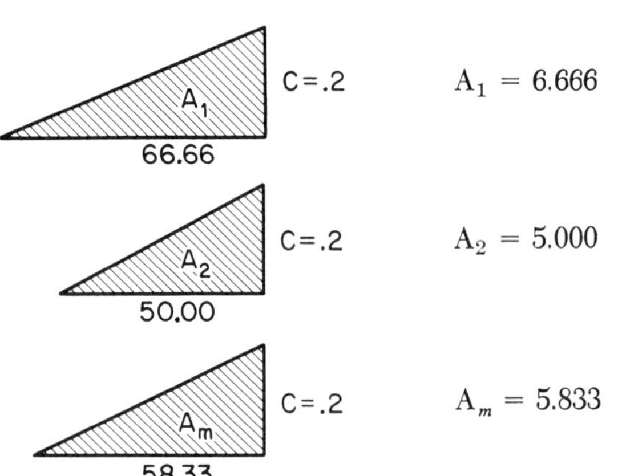

$A_1 = 6.666$

$A_2 = 5.000$

$A_m = 5.833$

substituting as above:

$$V = \frac{100(6.666 + 4 \times 5.833 + 5.000)}{6}$$

$$= \underline{21.60 \text{ CY of Cut}}$$

Applying: **II. Four-point Method:** substituting into formulas:—

$$V_f = \frac{10{,}000(.3^2)}{108(.4 + .3)} = \underline{11.90 \text{ CY of Fill}}$$

$$V_C = \frac{10{,}000(.4^2)}{108(.4 + .3)} = \underline{21.16 \text{ CY of Cut}}$$

Applying:—**III. Center-square Method** to the same example:

Area of Fill = 4,167 SF (as above)
Area of Cut = 5,833 SF (as above)
10,000 SF Total Area

Volume of Fill: calculate average depth of fill, thus:—

$$0.1 + 0.2 + 0.0 + 0.0 = 0.3 \text{ feet}$$

$$\frac{0.3}{4} = 0.075 \text{ feet average depth}$$

$$\text{Volume} = 4{,}167 \text{ SF} \times 0.075 = 312.5 \text{ CF} = 11.57 \text{ CY of Fill}$$

Volume of Cut: as above;

$$0.2 + 0.2 + 0.0 + 0.0 = 0.4 \text{ feet}$$

$$\frac{0.4}{4} = 0.10 \text{ feet average depth}$$

$$\text{Volume} = 5{,}833 \text{ SF} \times 0.10 = 583.3 \text{ CF} = 21.60 \text{ CY of Cut}$$

Note: Most grid squares would be either 'cut' or 'fill', and only one calculation per square would be necessary. The average depths of 'cut' and 'fill' are better separately tabulated and totaled; and the total depths multiplied by the total area of the squares, to give the total volumes in both cases.

Proof of Four-point Method Formulas

Taking $V_C = \frac{L^2}{108} \times \frac{(H_C^2)}{(H_C + H_f)}$; applied to the same grid square on page 84:— $L^2 = 10{,}000$ SF.

$$H_C + H_f = (.2 + .2) + (.1 + .2) = .7$$

Since location of Cut/Fill line in grid square (shown above) divides two sides of square proportionately, in each case, to the elevations at the corners (and again, this assumes a uniform slope at each side): a proportionate division of square's area (10,000 SF) may be made as follows:—

$$H_C + H_f = .7$$

$$\frac{10{,}000}{7} = 1{,}428.57 : \text{and by proportion,}$$

Area of Cut = 4 × 1,428.5 = 5,714.0 SF approx.
Area of Fill = 3 × 1,428.5 = 4,286.0 SF approx.
Total Area C/F = 10,000.0 SF

[NB: comparable Areas calculated for I. above.]

i.e. $\text{Area of Cut} = 10{,}000 \times \frac{H_C}{(H_C + H_f)}$

$\left[\text{Area of Fill} = 10{,}000 \times \frac{H_f}{(H_C + H_f)}\right]$

$$\therefore \quad V_C(\text{in CF}) = 10{,}000 \times \frac{H_C}{(H_C + H_f)} \times \frac{H_C}{4}$$

$$\therefore \quad V_C(\text{in CY}) = 10{,}000 \times \frac{H_C}{(H_C + H_f)} \times \frac{H_C}{4 \times 27}$$

$$V_C = \frac{10{,}000}{108} \times \frac{(H_C^2)}{(H_C + H_f)}: \text{ and substituting:—}$$

$$V_C = \frac{L^2}{108} \times \frac{(H_C^2)}{(H_C + H_f)}: \text{ and, similarly,}$$

$$V_f = \frac{L^2}{108} \times \frac{(H_f^2)}{(H_C + H_f)}$$

Questions and Topics for Discussion

1. What are the primary criteria for accuracy in measuring *construction work?*

2. What are the characteristics of *construction work* that must be included in the descriptions of *work* in an estimate?

3. Differentiate between *voids* and *wants* in measuring *work.*

4. Describe how an estimator can make his estimate more understandable to others, and why he should do so.

5. What are the primary factors that usually determine the units used in measuring *work,* and what criteria govern the selection of the most suitable units of measurement for all kinds of estimates?

6. Select and criticize the unit and method of measurement of one major class of *construction work*, according to the answer to the previous question.

7. Explain why the rule for adjusting one perimeter length to another perimeter length (for the same building) is always plus (or minus) four times, two times, the horizontal distance between the perimeter lines.

8. State and explain the primary reasons behind the statement: It is in estimating the costs of earthworks that measured quantities are of the least significance.

9. Explain why, in measuring cutting and filling over a site, the size of the overlaid grid squares should be selected according to the site's topography.

10. Define (a) a contour line; (b) a cut and fill line.

8

Measuring Work: Particular

Methods of measurement vary from place to place, according to local practices and customs. In this chapter, some of the more common *items of work*, the units and methods used to measure them, and two national standard methods of measurement will be examined:

1. **Method of Measurement of Construction Works,** published by the Canadian Institute of Quantity Surveyors, Toronto, Canada, 3rd ed. **(MM–CIQS).**
2. **Standard Method of Measurement of Building Works,** published by The Royal Institution of Chartered Surveyors and The National Federation of Building Trades Employers, London, England, 5th ed. **(SMM–RICS).** The "Fifth Edition–Metric" was published in 1968, and it is also referred to below **(SMM-RICS/M).**

The **SMM-RICS**, and its metric edition, are generally referred to only when the differences between **SMM-RICS** and **MM-CIQS** are significant, or when the **SMM-RICS** illustrates a point that is not in the other.

The **MM-CIQS** is published for use in North America. The **SMM-RICS** is published for use in Britain, where it is the basis of the "quantities method" in bidding, as described in Chapter 13. As such the **SMM-RICS** is of necessity, and by agreement between the joint publishing bodies, a comprehensive and detailed document, whereas the **MM-CIQS** is less detailed and still growing. Its primary, stated purpose is education.

References to both these methods of measurement are made by permission of the publishing bodies. Since both the publications contain much information and are both readily available, the selected references are brief, and refer only to certain major points. The publications themselves should be consulted for more complete information. Every student and practitioner of estimating should have copies of both these documents for information and guidance. The **MM-CIQS** is the first such publication in North America, and it is the beginning of what could become a North American method of measurement. The **SMM-RICS** is the first published method of measurement in the English language, is now in its fifth edition, and has almost fifty years of publishing and several centuries of measurement tradition and practice behind it, as is apparent from its scope and detail.

Alternative methods of measurement are indicated in this chapter for some items, and there are undoubtedly many other methods in use that may be known to readers. It is suggested that they compare them with those methods described below. If they remain convinced that their own methods are better, they are urged to tell us, so that we may have the opportunity to refer to the other methods in future editions of this book.

The order of the North American, "Uniform Construction Index," with its sixteen *standard divisions of construction work,* is followed here in reviewing the methods and units of measurement. This order is also used in the **MM-CIQS.** Units of measurement are generally given here in the "English System" (yards, feet, and inches). But in some cases the units are also indicated in the "metric system," as in **SMM-RICS/M,** so that the reader may become acquainted with the system of measurements we shall eventually be using in North America. If the unit of measurement in the English System is the yard or the foot, the corresponding metric unit would be the meter and its decimal fractions. Dimensions in inches are given in millimeters in the metric edition, **SMM-RICS/M.** (E.g., 6-inches is given as 150 millimeters, 12-inches as 300 millimeters, in the metric model.)

Before reviewing methods of measurement for the several *divisions of work,* we might take note of the general principles that are found in the **MM-CIQS**, and here is a synopsis.

1. Schedules of quantities shall briefly describe materials and workmanship but shall accurately represent the quantities of work to be executed.

2. The **MM-CIQS** is a definition of principle rather than an inflexible document.

3. Unless otherwise stated, all work shall be measured net as fixed in place.

4. In giving dimensions the order should be consistent and generally in the sequence of length, width, and height. Each item shall, unless otherwise stated, include conveyance and delivery, unloading, hoisting, all labor setting, fitting and fixing in position, lapping of materials, and straight cutting and waste. Circular work shall be given separately. All work in or under water shall be given separately. Any work required to be carried out in compressed air shall be given separately.

Similar general principles are to be found in the **SMM-RICS**, and both sets of general principles are worthy of attention in the original. They are generally adhered to in the later examples, except that *laps* in materials have been measured where they are required in some of the examples.

Methods of Measuring—General Requirements (Division 1)

Some of the general requirements for *construction work* are referred to under the heading of *Job Overhead Costs* in Chapter 5, and many of them do not require that *work* be measured in estimating their costs. Those general requirements that may require measurement, including such *work* as temporary fences, water lines, and drains, will be measured according to the methods described in the other fifteen *standard divisions of work.* Therefore, no special methods of measurement are stipulated for general requirements.

Methods of Measuring—Site Work (Division 2)

Demolitions are generally classified under two headings:

1. Mass demolitions of complete structures

2. Alterations, involving partial demolitions and *new work.*

In (1), there are usually few if any restrictions prescribed by the *owner*, whereas in (2), there may be many prescribed and inherent restrictions on the methods used. Both standard methods of measurement cover this *work* in detail.

Demolitions and alterations are among the most difficult tasks to estimate, because the conditions and methods are so variable. Mass demolitions are usually carried out by specialists who have their own methods and rely on their own experience. Alterations are a part of many projects for new construction, and they are, therefore, more likely to be a problem to any estimator.

One difficulty that alterations pose for an estimator is a frequent lack of information in the *bidding documents*. Few *designers* know how to write a good specification for alterations, perhaps because of the lack of opportunity. Ideally, alteration specifications are prepared from extensive field notes, and the specifications are virtually written on the site. Such specifications, if well written, enable the estimator to get into the mind of the writer and the result is *consensus ad idem*—a meeting of the minds—one of the essentials of a good contract. In alterations, the specifications are usually much more important than in most other kinds of *work*, because many aspects of *alteration work* can only be communicated through writing. Alteration specifications are essentially "performance specifications." The required result is specified; but the means of supporting, cutting, removing, and the like, are usually left to the *contractor* and are not prescribed by the *designer*.

New work in alterations is usually measured in the same way as other *new work*, but it should be specified as being "in alterations" and in "small quantities" if such is the case because these factors affect costs. The costs of matching *new work* with *existing work* must be allowed for, and attention should be drawn to these costs in the estimate so that they are not overlooked.

Check for the following things in estimating demolitions and alterations:

1. legal restrictions and special permits
2. location, time, and completion requirements, which restrict the use of certain methods of demolition and disposal
3. termination of existing services
4. the need to underpin other buildings on adjoining properties to avoid damage to them
5. requirements concerning removal (or otherwise) of existing substructures and services below ground
6. disposal of debris and parts, including feasibility of and restrictions on selling salvage
7. preservation of any specific parts that are to remain in place or are to be removed to storage for the *owner*
8. the effects of any regulations that will apply to the buildings once they are altered or improved.

Refer to **MM-CIQS** for the methods and units used in measuring buildings that are to be demolished.

Clearing the site of vegetation is generally measured in square yards, and includes trees of less than 12″ girth (at 5 feet above ground) in **MM-CIQS**. Trees of 12″ girth and more to be removed are enumerated in groups in **MM-CIQS** according to girth sizes; 12″ to 36″ girth; 36″ to 72″ girth; and so on, in 36″ multiples. (**SMM-RICS** refers to different sizes, but the method is the same.)[1]

Topsoil removal for later use is measured in cubic yards stating the depth removed in **MM-CIQS**. In **SMM-RICS**, it is measured in square yards stating the average depth. (**SMM-RICS/M** in square meters). Any subsequent treatment or handling of topsoil prior to reuse should be fully described.

Check for the following things in estimating site works:

1. disposal of organic and inorganic material
2. restrictions on burning
3. tree stumps to be removed and roots grubbed up
4. trees and other items to be preserved
5. dry stone walls to be built around trees to be preserved, and other similar protective and retaining *work*
6. ground water level and its effect on the *work*
7. unusual soil conditions, such as garbage or other imported fill, loose sand and peat, and other unusual ground conditions that would affect the *work* and the use of equipment.

Excavation is generally measured in cubic yards, net in place ("bank measure").[2] Allow for *swelling and shrinking* of excavated materials in *unit prices*.

All the different types of excavations should be measured separately. The differences arise out of such things as:

1. purpose of excavation; e.g., to reduce levels over site; for basements; trenches for continuous footings; isolated footings; trenches for pipes, curbs, and the like

[1] Large sites, highways, and rights of way to be cleared for transmission lines and pipelines are usually measured in acres. If all trees have to be removed, there is no need to measure and classify them; but if *unit prices* are to be used to price the removal of individual trees, they must be classified according to size. Removal of individual trees (at *unit prices*) is sometimes required outside the prescribed boundaries of area clearings to eliminate hazardous snags.

[2] "Net measurement" already has been explained. Here, the meaning is similar, but also different in that *excavation work* is different from other *work*. The net dimensions of an excavation are the minimum dimensions required by *other work* in the excavation, plus any "allowances" as described.

2. extent and depth of excavation
3. type and condition of materials to be excavated
4. ground water level and soil moisture conditions
5. machine or hand work
6. type of excavating equipment required and available, including need for drilling and blasting equipment
7. requirement and suitability of excavated materials for backfilling and for other purposes
8. locations of excavation on site, at general ground level, at basement level, at other levels
9. location of site; restrictions on work methods
10. location of possible borrow excavations, fill sources, and disposal sites.

If an *item of work* varies, or might vary, from another item for any cause or condition, it should be measured separately so that it can be priced at a different *unit price* if required. Allowances are usually made in measuring excavations for working spaces around footings and foundation walls, for erecting and removing forms, and for applying dampproofing, and the like. Usually, the allowance is 2 feet beyond basement walls, and 6-inches beyond footings. Allowances for sloping the sides of excavations for stability vary with the types of ground (see **MM-CIQS**). Otherwise, temporary shoring should be allowed for by measuring the area of the vertical faces of excavations that may require supporting.

Foundation trench excavations are generally measured in cubic yards. The purpose and starting level for each separate item of trenching should be stated; i.e., "surface trenches for foundations;" "basement trenches for foundations." Pipe trenches in **MM-CIQS** are also measured in cubic yards, with width allowances made according to the pipe size. The **SMM-RICS** method of measuring pipe trenches in linear yards, including all related *work* and stating the average depth to the nearest 6 inches, appears to be a better method in that volume of trench excavation is usually less significant than length.

Trimming and grading are usually measured as *super* items in square feet (**MM-CIQS**), or square yards (**SMM-RICS**). Machine excavations usually require some hand labor to finish the bottoms to required levels, since machines cannot always work to the required degree of accuracy. As concrete footings must be placed on undisturbed ground, it is frequently necessary for the last few inches of depth in trenches for footings to be removed by hand. All such *work* should be measured as separate items of hand trimming to bottoms of trench and other excavations. Other *super* items originating with machine excavations include trimming vertical and sloping surfaces, compacting filled and excavated surfaces, treating excavated surfaces for stability, and temporary shorings.

Temporary shoring costs include *labor costs* of erecting and removing the shores and the costs of renting and using the shores and shoring materials, such as timbers. Like formwork for cast-in-place concrete, temporary shoring is measured by the area of the excavation faces that are supported by the shoring. (Concrete formwork is measured by the area of the concrete faces in contact with formwork.)

Excavations are not shown on drawings, and *bidding documents* may specify the *work* only in a limited way, because the *designer* and the *owner* are not usually interested in how the excavations will be done, only that they be done satisfactorily, and that the *contractor* knows the extent of his responsibilities. A visit to the site by the estimator before he begins his work is essential for a realistic estimate. If he really wants to get the *work*, he may stay away from the site before bidding, but this practice is not recommended.

An estimator has to make many decisions about site works, and in this kind of *work* measurements and quantities are less significant than in most other work, because most of the costs of site works are *labor costs, equipment costs,* and *overhead costs,* and these are more difficult to estimate than *materials costs.* Added to this factor are all the possible problems and unexpected costs arising from ground and soil conditions, which vary according to location, climate, and season. For these reasons, *bids* for site works often vary widely, and it is not uncommon to see as much as a 100 percent difference between the high and low *bids* for a site works contract.

Backfilling of excavated materials is required in most projects. In some cases, however, because the excavated materials at the site are inadequate or unsuitable, fill material must be imported.[3] Like excavations, filling and backfilling are measured in cubic yards, net in place. In an estimate, the total net volumes of backfill are deducted from the total net volumes of excavations, and the surplus is the total net volume of material for disposal. If the result is a "minus quantity" (more backfill than excavation), the difference is the total net quantity of imported material required. The actual total quantity of fill required will be greater than the total net quantity calculated, depending on the type and moisture content of the fill material specified and used and the degree of compaction required. As already explained, it is better practice to always measure net quantities and to make

[3] Backfill means excavated native materials returned to the original excavation, as distinct from imported fill. But often the term "backfill" is loosely used to refer to both.

allowances for increases and decreases in volume due to *swelling and shrinking* in the *unit prices*. This is the only definite basis for estimating the quantities involved (any other basis is estimated and variable). The estimated allowances for *swelling and shrinking* can be checked through *cost accounting* by comparing actual gross quantities with measured net quantities.

Stone beds, such as gravel beds under concrete slabs and paving, are similar to other imported fill. If beds are not specified and priced for placing in layers of a stated maximum thickness, a separate *super* item for leveling and finishing the surface of such beds should be measured and priced. A 4-inch thick bed does not cost half as much (per unit area) as an 8-inch thick bed. It costs more than half as much, because both beds require the same *amount of work* at the surface. This is why some estimators prefer to measure such beds in square yards and to state the thickness (as in **SMM-RICS**). Alternatively, an estimator may use different *unit prices* per cubic yard for beds of different thicknesses, including the *work* on layers and surfaces.

In **Piling,** as in excavating, all the unique conditions affecting the *work* and its costs must be properly considered. The two published methods of measurement (**MM-CIQS** and **SMM-RICS**) both indicate in great detail the information required for proper cost estimates for *piling work*. The **MM-CIQS** contains comprehensive descriptions of most types of piling.

Site drainage includes piping and trenching. In the **MM-CIQS**, the piping is measured in linear feet and the trenching in cubic yards, as already described. The **SMM-RICS** combines both piping and trenching into one item, measured in linear yards (**SMM-RICS/M** in linear meters), including all related *items of work* such as grading bottoms, backfilling, and temporary shoring. In both methods, pipe fittings such as bends, tees, and junctions are enumerated as *extra over items* (extra costs of fittings over plain pipe, with no deductions made from the lengths of piping for the fittings). In both methods the piping, method of jointing and laying, and other related items must be completely described.

In trench excavations, the width of the trench may be determined by the piping, and by the depth of the trench. In some cases, the trench widths may be determined by the type of soil and other ground conditions, and whether or not the trench sides must be shored or cut back to a safe slope. In some cases, the trench width may be determined by the type of equipment used. The volume of trench excavations is, therefore, not always the most significant factor in trenching costs; and this may be one reason for measuring pipe trenches as linear items, combined with the piping and other related items.

The **SMM-RICS** requires pipe trenches to be described according to the depth range in successive steps of 5 feet (e.g., not exceeding 5 feet total depth; between 5 feet and 10 feet total depth, etc.) (**SMM-RICS/M** in 1.50 and 3 meters). These depths are also applied to other trench excavations in **SMM-RICS.** This practice appears to have come from the time when most trench excavations were done by hand, and 5 feet was a maximum throw for a laborer in a trench. Nevertheless, there appears to be some merit in describing trenching in this way, and there appears to be no reason for steps of other than 5 feet, unless laborers have other opinions.

Asphalt paving is measured in square yards, the thickness being stated in the description. Sub-grade materials, granular base courses, and the like, are measured in the same way. The **MM-CIQS** states that the required compaction shall be stated, and that the conversion to tonnage measurements shall be done in the pricing.

Local asphalt paving practices vary, and some *unit price contract documents* give only tonnage quantities, without stating the required thicknesses for base courses and pavings. At a recent roadbuilders' seminar, the superintendents present could not agree on the best method of measurement for *unit price contracts*; but most appeared to prefer tonnage measurement and payment, because it is easier to determine the quantities from the trucks' weigh-in slips. Accurate pricing and proper *cost accounting* would seem to require that paving areas should be measured and thicknesses should be stated, because thickness determines the area covered and the amount of *superficial work* required per ton.

Asphalt and other types of pavings should be classified according to type and intended use, and whether flat and level, or cambered, or laid to slopes. All these things affect the costs. Pavings to steep slopes should be identified and kept separate, as should pavings inside buildings and other pavings to be laid in restricted spaces or under unusual conditions. Anything that affects costs should be included and distinguished in the estimate.

Items in conjunction with asphalt and other types of paving, such as curbs, gutters, edge boards, and edge treatments, should be measured as *run* items in linear feet (or yards) and fully described according to dimensions and details and whether the item is for labor only, or for both labor and materials.

Natural conditions of soil, ground, water, location, and weather greatly affect the costs of all site works, and the estimator must make sure that as many of these factors as possible are known and considered and accounted for in the estimate, because any one of them could affect the costs much more than a discrepancy in measured quantities.

Methods of Measuring Concrete (Division 3)

Concrete is generally measured in cubic yards and no deductions are made for embedded reinforcing steel or for openings and voids of less than 1 cubic foot in volume (**MM-CIQS**). *Concrete work* is described according to mix specifications, location in the building (e.g., footings, walls, columns, beams, etc.), and according to any particular method of placing required, such as pumping. The **MM-CIQS** lists thirty-eight different classifications of concrete according to location in the building.

The **SMM-RICS** requires the sectional areas of such concrete members as beams and columns to be grouped and described as not exceeding 48 square inches (sectional area), over 48 but not exceeding 144 square inches, and over 144 square inches. (**SMM-RICS/M:** not exceeding 0.05 square meter; over 0.05 but not exceeding 0.10 square meter; and over 0.10 square meter). This size grouping is not required by **MM-CIQS.**

Plain, reinforced, cast-in-place, precast, and prestressed concrete should be so described and each measured separately, as should concrete in small quantities, continuously poured concrete, and concrete placed in special locations (such as underwater) and by special means. All these affect the *costs of the work* and must be identified and taken into account in an estimate.

Although *concrete work* is generally measured and priced in cubic yards, an estimator should not be too prone to directly measure all concrete as *cube* items. Rather, he should measure the concrete first in linear feet (with the concrete member's sectional area stated in the description), or in square feet (with the concrete's thickness stated in the description). The total quantities of each concrete item can be converted to cubic yards later in the estimate. By first measuring concrete footings in linear feet, and concrete walls and slabs in square feet, the estimator can obtain the reflected quantities of several other *items of work* related to the concrete. Such items as formwork, steel rebar, and cement finishing can often be measured directly from the concrete quantities. If all *concrete work* is immediately measured as a *cube* item, these other related *items of work* must then be measured separately and this requires extra effort.

Measure all concrete of the same type and mix together under the appropriate heading in the estimate. Such headings might be as follows, for example:

Plain Conc (3000 psi × 1 1/2″ stone)
R. Conc (3500 psi × 3/4″ stone)
R. Conc (3500 psi × 3/8″ stone)
Precast Conc (4000 psi × 3/4″ stone)

By collecting together like items under such headings an estimator facilitates subsequent pricing; and in an estimate for a large project, the estimator should start a new page for each major item with a different heading.

Both published methods of measurement state that voids of less than 1 cubic foot in volume shall not be deducted from concrete measured in cubic yards. (**SMM-RICS** states that concrete beds not more than 12 inches thick shall be measured separately in square yards (stating the thicknesses), and voids not exceeding 1 square foot shall not be deducted.)

Most estimators do not deduct small voids from beds and slabs; similarly, no deductions are usually made for other items passing through slabs, such as columns, pipes, and ducts. Any theoretical reduction in *materials costs* because of small voids is usually offset by increased *labor costs* incurred in working around the intruding item or the void.

In a reinforced concrete structure, the average amount of concrete displaced by reinforcing steel is usually from 1 to 2 percent of the total volume of reinforced concrete, depending on the type of reinforced concrete structure. In a high-rise apartment building, the average amount of steel per cubic yard of concrete is usually about 125 to 150 pounds. Taking the weight of steel to be about 480 pounds per cubic foot, this represents about 1/4 cubic foot of steel per cubic yard of concrete, or about 1 percent of the volume. In a reinforced concrete parking garage, the volume of steel rebar to concrete will usually be about 2 percent. Estimators should be conscious of the general magnitude of such relative quantities in construction projects so that they can make rational decisions about allowances for *waste.*

Although concrete structures are essentially monolithic, an estimator analyzes the structure and identifies and measures separate components such as the columns, beams, and slabs. Some junior estimators become confused about the sometimes artificial divisions among these monolithic components, and they may wonder where a column ends and a beam begins and where a beam ends and a slab begins. In fact, they are monolithic and, with the exception of columns, must be poured in place without joints, as in Fig. 33.

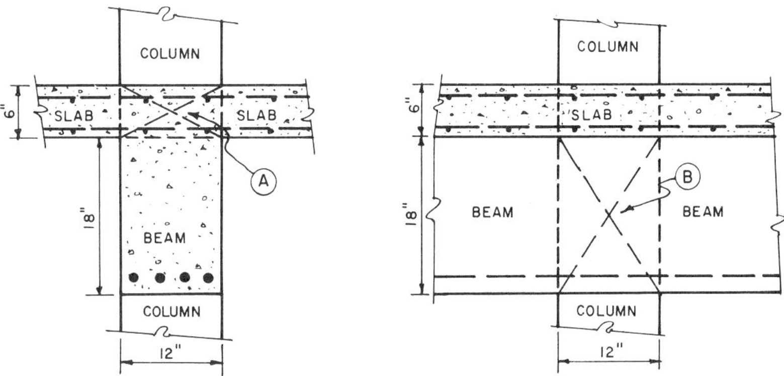

FIG. 33. Monolithic R. Concrete Columns, Beam, and Slab at Conjunction.

In Fig. 33, (A) might be measured as "slab" or as "beam." At (B), the concrete might be measured as "beam" or as "column." How is an estimator to decide what to do? The problem is more imagined than real, and an estimator may do whichever is expedient. **SMM-RICS** requires beams to be measured below the slab, and usually this will be the better method although it makes little difference in the estimate. The *concrete work* is to be measured so that it can be priced, to arrive at an estimate of costs. In an estimate, the estimator may have the following items with their *unit prices* for labor, materials, and equipment:

R. Conc (3500 psi × 3/4" stone)
Suspended slab (6" thick) .... $25.00 per cu. yd.
Beams .... $26.50 per cu. yd.
Columns .... $28.50 per cu. yd

The distribution of costs within these *unit prices* might be: *material costs* of ready-mixed concrete delivered to site, $20.00 per cu. yd.; *equipment costs,* $1.50 per cu. yd.; and *labor costs* placing concrete in slabs, $3.50 per cu. yd.; placing concrete in beams, $5.00 per cu. yd.; and placing concrete in columns, $7.00 per cu. yd. These comparative cost figures show that the total cost is not significantly affected, regardless of whether the concrete at (A) in Fig. 33 is called "slab" or "beam" or whether the concrete at (B) is called "beam" or "column," because the theoretical differences are only in the *labor costs* and these placing costs are a small part of the *total costs of concrete work.* For example, at (A) the difference between the *labor costs* for "slab" and "beam" is $1.50 per cubic yard, or less than $.03 per linear foot of beam.

Small items of *concrete work* should be enumerated and described rather than measured in volume, because the volume of concrete and the *labor costs* in placing the concrete are not always directly related. A small concrete footing of 1/2 cubic yard of concrete may cost almost as much to place as a concrete footing containing twice as much concrete. An estimator must be careful in pricing such items, because measurement by volume may obscure facts about costs, especially *labor costs.*

For example, Fig. 34 shows two concrete footings for steel columns. Footing (A) contains 1 cubic yard of concrete, and footing (B) 1½ cubic yards of concrete. If a project contains many such isolated footings, it may be better to measure them by enumeration and description and to arrive at a *unit price* for each type.

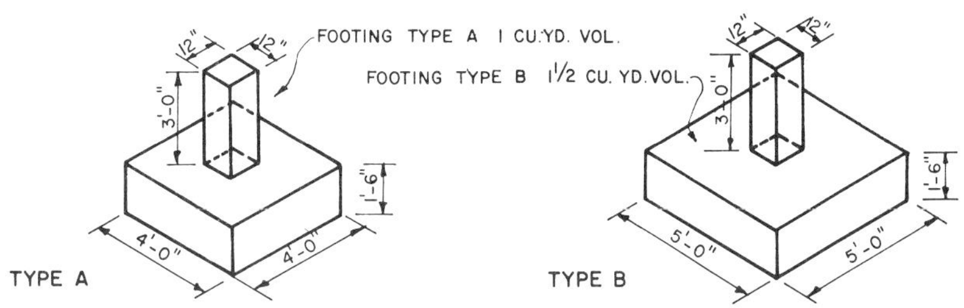

FIG. 34. Concrete Footings of Different Volumes.

To measure them all in cubic yards under one description, such as, "R. Conc (3500 psi × 3/4" stone); isolated ftgs," may result in inaccurate pricing, because the *labor costs* (per concrete footing) in placing both types may be about the same, despite the 50 percent difference in volume. Alternatively, the two types of footings (A) and (B) might be measured separately in cubic yards, with a different *unit price* for each type. But to estimate the *unit prices* (per cubic yard), it would be necessary to first calculate *unit prices* for a single concrete footing of each type. One may then question the sense and value of converting the *unit prices* per concrete footing to *unit prices* per cubic yard and run the risk of inaccurate pricing when using the price per cubic yard for different footings in a different job.

Concrete pan joist slabs and waffle slabs are best measured by the overall thickness of the slab, deducting the voids formed by the pans, as shown in Fig. 35. Long steel pans are used to form pan joist slabs, and square steel pans are used to form waffle slabs. Sometimes plastic pans are used. The deduction for the voids formed by steel pans of various standard sizes and shapes are published by the manufacturers and can be found in several handbooks and manuals.[4]

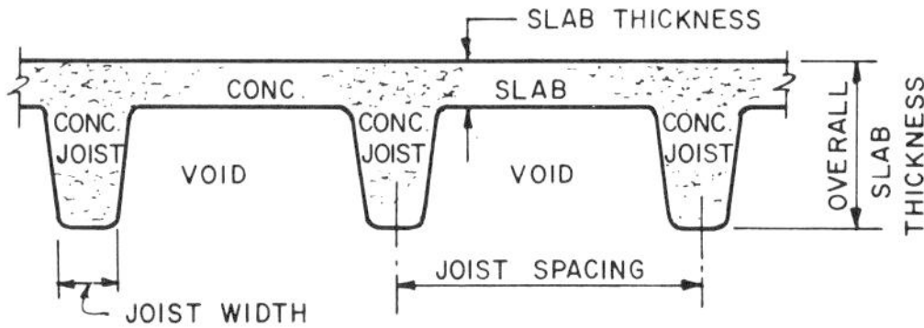

FIG. 35. Typical Section of a Reinforced Concrete Pan Joist or Waffle Slab.

In pan joint slabs, the joists (and the voids) go in one direction between slab beams. In waffle slabs, the joists go in two directions (at right angles) and the voids are square on plan.

Concrete curbs and similar items appear to be better measured as *runs* rather than *cubes*. Which is more descriptive?

(1) R. Conc (3500 psi × 3/4" stone) Curbs, size 6" × 12" 300LF

(2) R. Conc (3500 psi × 3/4" stone) Curbs, size 6" × 12" 5½CY

[4] Any estimate of concrete quantities based on such published data (for voids) should be carefully checked by *cost accounting*, since theory and practice are not always the same.

Can you visualize from (2) the amount of curb to be placed? Also, there is a risk of combining two or more items of concrete curbs of widely different sizes if they are measured as *cubes*, with resultant inaccuracies in pricing.

Formwork is a part of most *concrete work*, and formwork for cast-in-place concrete is a large part of its total costs. Formwork is unique in building construction because it is *temporary work*, and it is not incorporated into the building; therefore, it is rarely shown on the drawings. It is stripped from the hardened concrete and usually reused, and the number of uses largely determines the costs of form materials; but the *labor costs* are the biggest cost factor.

Formwork is generally measured in square feet (**MM-CIQ**) of the actual surfaces of the concrete structure in contact with and supported by formwork. Some students of estimating are troubled by this method and want to measure not only the superficial area of the formwork sheathing surfaces, but also all the formwork's supporting structure, including wales, struts, and all other material. They cannot at first understand how formwork can be estimated from the superficial "contact area."

Remembering that all measurement is done to enable *work* to be priced, let us consider the basic economics of formwork. The major part of formwork costs are *labor costs*. Usually, the design and intent is to reuse formwork as many times as possible. Even if formwork can only be used twice, the *labor costs* are still probably more than twice the *materials costs*. If the formwork can be used four times, the *labor costs* may be three to four times the *materials costs*. Some plywood forms may be used up to twenty or thirty times before they are worn out, and steel forms have a much longer life.

The estimator does not always know exactly which types of form sheathing and supporting systems may be used on a particular project. Therefore, the pricing formwork is primarily related to the "contact area" of the supported concrete surfaces, because it is this area that largely determines the *labor costs*, the major part of formwork costs. Whether the formwork requires 2.0 or 2.25 board feet of supporting lumber per square foot of "contact area" (per "contact foot") is a consideration, but it is not a major consideration if the supporting lumber can be used ten times over. The difference in costs of the two quantities of supporting lumber, above, would be about 1/4 of a cent per "contact foot," assuming ten uses of lumber that costs $.10 per board foot. More important factors in the economics of formwork are the number of uses possible and the efficiency in erecting and stripping formwork;

neither of which can be reflected in an estimate simply by measurement.

Deductions are not usually made in measuring formwork for:

1. intersections of concrete beams
2. intersections of beams with columns and walls
3. voids and openings in concrete of 100 square feet or less (**MM-CIQS**).

Allowances are not made in formwork measurements for overlaps and passings at angles and corners. All costs of supporting falsework and hardware, and the erection, oiling, transporting, and cleaning of formwork are made in the pricing, as are the allowances for reuse and waste of forming materials. Therefore, the measurement of formwork is relatively simple. More effort and skill is required in the pricing, as is often the case in estimating.

The easiest way to understand the basic economics of formwork is to reflect on the implications of renting prefabricated steel forms and erecting, stripping, and cleaning the steel forms before they are returned to the rental firm. In principle, the same types of costs are applicable to wood forms owned and used by a *contractor* as apply to rented steel forms, but the costs are more obvious in the second case. A *contractor* should allow for the costs of his own forms and their use in the same way as he has to allow for the costs of rented forms.

Formwork for concrete footings is sometimes left in place and not removed for reuse in the belief that it is sometimes more costly to remove forms for reuse than it is to abandon them. Lower-grade lumber can then be used for these forms. An estimate should indicate forms to be "left in." Formwork should be measured to the face of any "steps" (changes in level) in continuous footings and wherever else concrete is supported and retained by formwork. Formwork is not usually required over horizontal surfaces; but it is required over surfaces to a steep slope, depending on the specifications and thickness of the concrete.

Formwork for walls of different thicknesses should be kept separate, since the wall thickness affects the formwork costs. Generally, no distinction is made for walls of different heights; but an estimator should separate them if he decides that different conditions may create different costs, as he should in measuring *work* of any kind. If, subsequently, no difference in costs is estimated, the separate items can always be priced at the same *unit price,* nothing has been lost, and some useful information may still be shown.

Forms for piers, pedestals, and columns are generally measured in square feet; but formwork for circular columns is measured in linear feet stating the diameter, because this method better describes the tubular cardboard forms that are used. Small formwork items should be enumerated and described, as explained under Concrete, above.

Forms to beam soffits should be measured separately from forms to beams sides. The posts or shores supporting the undersides of beams should be replaced immediately after the forms have been removed. This allows the beam forms to be re-used as soon as possible, but primarily it provides the necessary support for the beams until the new concrete is sufficiently cured and strong enough to carry its own weight and the loads it bears. The same practice is followed with suspended concrete slabs. This reshoring of soffits is either allowed for in the *unit prices* for formwork to soffits of slabs and beams, or it is measured and priced as a separate *item of work.*

Formwork for the underside of suspended concrete slabs and for the sides and soffits of beams should be grouped and separated according to the heights of supports required. The **MM-CIQS** classifies them as not exceeding 10 feet high, 10 to 15 feet high, and so on, in 5-foot increments. Thus, different costs can be applied to formwork at different heights, because the height affects the type and size of supports required and the amount of labor required to erect and strip the forms. Formwork for slabs with minor differences in thickness is often grouped and priced together, but it may well be kept separate according to slab thicknesses. Formwork for unusually thick slabs should always be priced separately.

Formwork for pan joist slabs and waffle slabs may be measured in square feet, the same as formwork for other suspended concrete slabs. But it is not unusual for a *contractor* to erect the wood plank forms (at the undersides of the concrete joists) on posts or steel shores, and to sub-let the erection and removal of steel pan forms to a *subcontractor* who has the steel pan forms for hire. This method may require that the pan forms and the wood plank forms for the undersides of joists be measured separately. Both are better measured in linear feet, stating the other dimensions in the items' descriptions. Special end pans should be enumerated.

Forms for openings in walls and slabs are enumerated and described in terms of their sizes and thicknesses. The costs of forming one window opening, size 2′0″ × 3′0″, and another window opening, size 3′0″ × 4′0″, would not be very different; and to measure the perimeter (or area) of the formwork around these openings would not make for accurate pricing. It is the number of openings to be formed which is of primary significance.

Formwork for edges of slabs, for bulkheads at ends of walls (at vertical construction joints), and the like, are best measured in linear feet, stating the widths, although **MM-CIQS** requires these items to be in square feet. These items have to be measured first in linear feet before a *super* quantity can be calculated, and there appears to be no advantage in converting the *runs* to *supers*. In fact, by doing so, some information is lost. It is the total *run* of such items that primarily affects the costs. One hundred square feet of 6-inch wide forms will cost almost twice as much to erect as 100 square feet of 12-inch wide forms. The *labor costs* **per linear foot** would be almost the same in both cases, and only the *materials costs* would be more or less proportionate to the widths.

Screeds and edge forms at construction joints in concrete slabs should be measured as separate items, under formwork. Screeds may be measured in linear feet at indicated or assumed locations and at the slab perimeter, or the slab area can be used as basis for pricing screeds and similar items if actual costs per square foot of slabs have been obtained, based on more or less standard spacings for the items.

Formwork for grooves, sunk bands, sills, cornices, and other similar, continuous items, (either recessed or projecting), are generally measured in linear feet, stating the sizes. They are best described and priced as *extra over* the general formwork in which they occur. If they are over 24 inches in girth, **MM-CIQS** states that they shall be measured in square feet. Items of lesser girth are measured in linear feet because, again, the length is the dimension most significant to the costs, and the other dimensions make little difference.

Forms for sloping and battered faces should be so described and measured separately, indicating an average height for supports where appropriate. Forms for stringers and risers are generally measured in linear feet and also should be measured separately.

Reinforcing steel bars are generally measured by weight (in pounds, hundredweights, or tons) and should be measured separately and described according to:

1. **Type and grade of quality:** Bars may be plain or deformed; straight or spirally bent; steel may Grade 40, 50, 60 or 75; it may be billet, axle, or rail steel.

2. **Size (diameter):** The Concrete Reinforcing Steel Institute (CRSI) recommends in its manual of standard practice that each size should be priced separately (as does **SMM-RICS**); the **MM-CIQS** requires only that sizes #6 and under shall be separated according to each size, and that sizes above #6 be grouped together.[5]

3. **Length:** The **MM-CIQS** requires several groupings according to length because of different fabrication,[6] handling, and installing costs; the **SMM-RICS** calls for only bars over 30 feet long to be separated: (**SMM-RICS/M,** 10 meters long).

4. **Bending:** The CRSI recommends that estimates show the total quantity of bending separated into two classes (as *extra over* items):

a. **Heavy bending:** Bar sizes #4 through #18 that are bent at not more than six points; radius bent to one radius; and bending not otherwise defined.

b. **Light bending:** Includes all #3 bars that are bent and all stirrups and column ties and all bars #4 through #18 that are bent at more than six points, or bent in more than one plane, or radius bent with more than one radius in any one bar, or a combination of radius and other bending (radius bending being all bends having a radius of 12 inches or more to inside of bar). Notice that light bending costs more per ton than heavy bending. The terms may be misleading.

(The **MM-CIQS** requires several classifications of bending, according to the number of bends per bar. Often, the best and easiest practice is for an estimator to separate each size of bar and each type of bending.)

5. **Special fabrication:** The CRSI recommends that the following be measured separately.

a. bending to special tolerances
b. shearing to special tolerances
c. spirals
d. bending to less than recommended minimum radius
e. bars requiring hot bending
f. square (saw cut) ends
g. beveled ends
h. unusual bends or end preparations not otherwise defined.

Radial prefabrication of steel, for use around curved surfaces such as domes and tanks, is affected by the required radius and by the bar size; hence, some radial bending is done in the shop and other radial bending can be done at the site. Consult the CRSI manual for recommended estimating practices.

Hooks are not always required on deformed bars.

[5] Bars of #14 and #18 sizes cannot always be priced accurately if measured together with smaller bars down to #6 size, and the CRSI recommendation to separate all sizes is preferable. Each size has to be measured separately in the first place, anyway, and little is gained by combining sizes.

[6] "Fabrication" refers to fabricated bars, which includes bars cut to a specified length and bars cut and bent to a specified length and configuration.

Recommended hook sizes are given in the CRSI manual. Splices must be allowed for in measuring bars. The lengths of splices depend on location of splice, size of bar, concrete and steel strengths, and design method and stresses. Allowances must conform to the requirements of the *bidding documents.* Where not otherwise shown, lap splices in temperature steel in slabs and walls, and in bars in footings, can be taken as 24, 30, and 36 bar diameters for specified yield strengths of 40,000, 50,000, and 60,000 psi., respectively. But welded splices or splices made by mechanical means should be enumerated and described, as indicated in the drawings and specifications.

If lengths of bars are not specifically shown on the drawings, consult the CRSI manual for standard measuring practices. For example, the out-to-out dimensions of column ties are to be 3 inches less than the outside dimensions of the concrete column (to allow for coverage of steel).

Bar supports and spacers should be measured as separate items with reinforcing bars. If no details are given in the *bidding documents,* the CRSI manual lists recommendations concerning quantity and sizes of support bars, high chairs, spacer bars, and other bar supports. Individual chairs and spacers are enumerated. Continuous chairs and bolsters are given in linear feet.

Welded wire fabric is supplied in full rolls and sheets and is generally measured in square feet (**MMCIQS**, in squares) and described. Allowances must be made for *laps* as required.

Any estimator measuring steel reinforcing materials should have a copy of the CRSI manual and should memorize the more common and important standard estimating practices.

Precast concrete is easily measured. Panels, slabs, and the like, are fully described and measured in square feet. Beams, posts, curbs, and the like, are fully described and measured in linear feet. In all cases, the number of exposed faces and the number of units of each size should be stated. Caps, pads, bases, and similar items, are fully described and enumerated. Forms, molds, and steel rebars in precast concrete are not measured separately but are included in the description of the precast item. Adequate descriptions for pricing should also include:

1. dimensions and shapes of cross-sections, with sketches in the estimate if necessary; weights of units should be calculated and stated where significant
2. concrete specifications for both the exposed faces and the interiors of units if different specifications are indicated
3. nature and extent of all surface finishes and treatments
4. reinforcing steel specifications, sizes, and amounts
5. bedding and fixing requirements
6. descriptions of any inserts.

Sizes and weights of units are important in handling and transportation, and this information should not be lost in the measurement process. In many projects it is the handling, transporting, and erecting of precast concrete that causes the most trouble and poses the greatest risk of financial loss.

Most *precast concrete work* is supplied and sometimes supplied and installed by specialist precast manufacturers, and a *general contractor's* estimator would usually measure the *work* as described above. Only a precast manufacturer's estimator would need to measure the *work* in greater detail, separating forms, reinforcing steel, and the other component items as in measuring *cast-in-place concrete work.*

Hollow block and concrete suspended slab construction may consist of many different arrangements of clay or concrete hollow blocks, with precast or cast-in-place concrete joists and beams and with cast-in-place concrete fill and topping. Generally, this *work* is measured in square feet with a detailed description, including:

1. overall thickness of slab
2. sizes and types of blocks, tiles, joists, and other components
3. spacing of joists, etc.
4. specifications for cast-in-place concrete beams, joists, and toppings
5. size and type of any tiles to soffits
6. finishes to soffits and toppings
7. particulars of any concrete end-filling of hollow blocks and any other similar *work* at edges.

Such items as concrete filling to ends of blocks should be measured separately in linear feet, and described. Forms and reinforcing steel should also be measured separately.

Special concrete work should be adequately described for accurate pricing, including the heights to which pneumatically placed concrete must be pumped, the requirements for cold-weather placing, use of admixtures, special cements and aggregates; and any other requirements that may affect the *costs of the work.* Methods of measurement state the **minimum requirements** for accurate pricing, and the estimator should always include more information if he thinks that it is necessary.

Cement finishes include many different *items of work* performed on portland cement concrete; all are generally measured in square feet, using either the dimensions of the concrete or, in many cases, the formwork as the basis. These items include:

1. rough screeded finishes
2. wood floated finishes
3. steel troweled finishes
4. applications of various hardeners
5. applications of coloring agents
6. non-slip and textured finishes
7. grouted and rubbed surfaces
8. ground or polished surfaces
9. cement concrete toppings
10. applications for curing, hardening, sealing and wearing.

The estimator should be careful to properly interpret the *designer's* requirements in the *bidding documents*, because costs for such items can vary greatly.

Many finish items are not easily specified, and samples are not always available during bidding. For example, grouting and sacking (rubbing) concrete surfaces straight from the forms can mean different things to different people. To the *contractor* it may mean: fill up with cement grout and smooth off by rubbing with a sack or a sponge all visible cavities in the concrete surfaces. To the *designer* it may mean the production of a smooth, dense concrete surface not unlike terrazzo. Despite the advances made in writing specifications, it is still sometimes necessary to estimate according to a *designer's* reputed requirements, or expectations, rather than what is in the specifications, particularly in the matter of workmanship. Yet this should not be; and to help remove this source of misunderstanding and dispute, *designers* should have models and representative samples of unusual items and finishes made available for bidders to examine before submitting their *bids*. Specifications should clearly identify that which is required. In return, bidders should examine any models and samples and obtain through addenda any necessary clarification of the *bidding documents* before submitting their *bids*.

Methods of Measuring Masonry (Division 4)

Unit Masonry is measured by two basic methods, and one is an extension of the other. The first is as a *super* item, as measured from the drawings in square feet, square yards, or square meters. The second is as a *number* item, obtained from the *super* quantity multiplied by a factor that depends on the size of the masonry unit and the joint. The **SMM-RICS** calls for the former, and the **MM-CIQS** calls for the latter method.

The advantage of measuring masonry as a *number* item is that it facilitates ordering the units, although an allowance for *waste* (from cutting and breakage) must first be made to the estimated quantities, which should be measured net, in the first instance, as previously explained. The **MM-CIQS** states that the mortar shall be measured separately in cubic yards, describing its composition. This method essentially involves measuring materials, and it is not the preferred method.

The advantages of measuring masonry as a *super* item are that it requires no further computation of the *super* quantities measured from the drawings until the materials are to be ordered; and, more important, it retains the *quantities of work* in a form that is more easily comprehended. One-hundred square feet of 4 inch thick wall is much more comprehensible than 650 masonry units of a specified size and joint width. With the first measurement, one can visualize a wall, say, 12½ feet long × 8 feet high, or 10 feet × 10 feet. Either way, a more or less accurate image can be formed. With only the number of units given, the most one can visualize is, perhaps, a stack of masonry units.

Masonry work usually includes scaffolding. The **MM-CIQS** describes *masonry work* as also generally including cutting for pipes, ducts, and conduits, and cutting around, against, and to the underside of concrete and steel members. The **SMM-RICS** generally requires all cutting to be measured separately, and it differentiates between "rough cutting" (where concealed) and "fair cutting" (at exposed faces). Fair cutting per linear foot may cost as much as a square foot of *masonry work*, and it should always be estimated.

Unit masonry work should be measured separately and described according to:

1. **type and size of masonry unit,** usually the nominal dimensions are stated
2. **type and width of mortar joints,** including composition of mortar and any special method of jointing (such as "face shell bedding")
3. **type of bond,** in which units are laid
4. **thickness of wall or veneer,** the **MM-CIQS** is not explicit about wall thickness, but it is better to measure all masonry separately according to thickness
5. **type of construction,** such as veneer, facework, backing to other masonry, partitions, walls, cavity walls, fire protection for steelwork, piers, columns, supports, chimney stacks, and linings

6. **location of masonry,** such as in exterior and interior walls, filling to existing openings, and masonry starting at elevated levels (such as on beams) and requiring additional scaffolding below

7. **surface treatment,** including the pointing of joints, any requirements about selecting units and the workmanship at exposed faces, and any requirements about cleaning, treating, or coating surfaces.

Cut Stone is generally measured in square feet and fully described according to:

1. **kind of stone,** including any requirements concerning source, quarry, texture, bed, and finish to exposed faces (eg. sawn, tooled, etc.)
2. **thickness of stone,** stating an average or minimum thickness (bed width) where necessary
3. **arrangement of stones,** such as coursed or uncoursed, and any requirements concerning bedding, bonding, or pattern
4. **type and width of mortar joints,** including composition of mortar
5. **type of construction,** such as veneer, walls, piers, columns, supports, and chimney stacks
6. **location of masonry,** as described for unit masonry above
7. **surface treatment,** as described for unit masonry above.

Coping, sills, and the like, are generally measured in linear feet and the item's section, lengths of pieces, method of bedding, and the like, are fully described.

Field stone is generally measured in the same way as cut stone. This kind of *work* is not easily described and specified, and the estimator should take care to ensure that he is measuring and pricing exactly what is required by the *bidding documents.* Any samples or examples of the required stonework referred to in the specifications should be examined before estimating, otherwise it may be very difficult to properly estimate the *labor costs.*

Centering for masonry is *temporary work,* usually done by carpenters to support masonry during its construction and until it is completed, cured, and self-supporting. The most common example is centering for an arch, from which the name of this *work* is derived. Curved supports are laid out from the center of the arch, as in Fig. 36.

Like formwork, centering is generally measured by the actual masonry surface to be supported. It is described according to:

1. the nature of the *work* to be supported (e.g., brickwork, blockwork)
2. the shape of the *work* (e.g., flat, semicircular).

In most cases, centering is best measured by enumeration and description. The omission of centering from an estimate may be quite serious, for it can be a significant part of masonry costs.

Arches should be measured the mean length on the vertical face, as a *run* item; and the width of all faces, the number of arches of each type and size, and the *arch work* should be fully described, including any cutting or other labor on masonry units in the arch; whether or not joints are tapered and their width; any cutting of masonry along the length and at the ends of the arch; and any pointing, finishing, and other treatment to arch surfaces. Or arches may be measured by enumerating and fully describing them.

The **MM-CIQS** calls for "decorative patterns and features" in unit masonry to be measured in square feet and classified as "extra labor and material," and described. However, in some instances, some features such as decorative bands and courses may be better measured in linear feet, and with the other two dimensions stated in the items' descriptions. Many masonry items, including arches, are conveniently measured and priced as *extra over* items (extra costs of labor and materials), which means that in measuring these items no deductions have to be made from the measurements of the general *masonry work* in which the items occur.

Masonry is of many types, and it contains many *items of work* that create mainly *labor costs,* which must be allowed for in an estimate. The **SMM-RICS** deals with the measurement of masonry in great detail, and most "labors" on masonry are measured separately from and additional to the main masonry items. Some estimators question such detail, and simpler methods of measurement may be adequate in many cases. Nevertheless, the **SMM-RICS** provides an excellent check list for the estimator of *masonry work,* and the estimator should always measure in as much

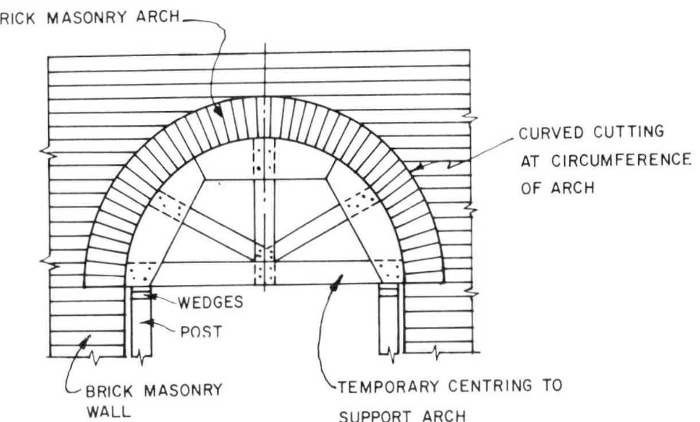

FIG. 36. Centering for a Masonry Arch.

detail as the nature of the *work* and its costs require. It is by ignoring the various "labor" items, such as cutting, and plumbing at corners, that estimates are often made too low.

Methods of Measuring Metals (Division 5)

Structural metals are generally measured by weight in pounds, hundredweights, or tons. The **MM-CIQS** calls for structural steel in pounds, or short tons of 2,000 pounds. (The **SMM-RICS** uses hundredweights of 112 pounds, and the **SMM-RICS/M**, kilograms.)

Similar steel sections are usually divided into weight groups. **MM-CIQS** requires column members to be separated according to the section weights per linear foot; over 60 pounds per LF; from 17 to 60 pounds; and less than 17 pounds per LF. (The **SMM-RICS** goes into greater detail for weight groups of several types of steel sections.)

The national steel institutes in the U.S.A. and Canada[7] refer to computing weights for *unit-price bids* and present a method to be used unless the *bidding documents* require actual scale weights or some other method of calculation. This standard method assumes the weight of steel to be 0.2833 of a pound per cubic inch. Weights of shapes, bars, and hollow structural sections shall be computed from detailed shop drawings and from manufacturer's published weights. No deductions shall be made for holes or for material removed in fabrication, and no allowances shall be added to the finished dimensions for cutting, milling, or planing.

Weights of plates and slabs are computed by the theoretical rectangular dimensions of plates and slabs from which the finished pieces shown on the shop drawings can be cut. No allowances are added for burning, cutting, and trimming. Weight allowances for rivets and welds and for painting and galvanizing should be made according to the tables given in the steel institutes' publications, to which the estimator will need to refer for this and other steelwork data.

[7] American Institute of Steel Construction, Inc., 101 Park Avenue, New York, N.Y. 10017, U.S.A.; Canadian Institute of Steel Construction, 1815 Yonge Street, Toronto, Ontario, Canada.

The method of computation described does not result in actual weights of fabricated steel. It is based on quantities that can be readily computed, and any greater degree of precision is seldom warranted if the parties involved are each cognizant of the unit weights to which the bid unit price applies. The institutes' publications should be read for a full and detailed explanation.

The steel institutes' handbooks are essential to an estimator of structural steelwork.

The **SMM-RICS** follows a useful practice of describing steelwork as "framed steelwork" when: "The majority of the steelwork is extensively and accurately jointed together to act as a frame. A small amount of unframed steelwork in a steel framed building would be treated as framed. A small amount of framed steelwork (e.g., two stanchions and a beam) in an unframed building would be unframed." The reason for this practice is the usual differences in cost (per unit weight) of these two classes of steelwork because of the quantities of material used and the *overhead costs* of the steel fabricator and erector, which are higher (per weight unit of steel) on the smaller quantities of unframed steelwork. The classifications "framed" and "unframed" should be defined if used.

Steelwork should be grouped in an estimate according to:

1. **Grade of steel required,** according to the specifications

2. **Type of steel section or member,** such as;

a. beams and columns	f. pipe columns
b. purlins	g. baseplates
c. trusses	h. bolts
d. open-web joists	i. rivets
e. built-up members	j. welds

and so on, according to the steel member's section, components, and function

3. **Method of site connection,** such as, riveted, bolted, turned-bolted, welded

4. **Location of steelwork,** such as connected to *existing work* or steelwork in locations involving extra or special handling, where appropriate (but not usually necessary for describing *work* in a new steel-framed building).

In addition, the "fabrication classification" of each item must be identified in the estimate, such as:

1. **Beams:** plain or punched, at web or flange, or both; framed horizontal, with connections; cut or coped; with cover plates, welded, or riveted; framed sloping and punched, or with connections; curved or bent with connections

2. **Columns:** plain, WF, or H sections; round sections; with cover plates, welded or riveted; built-up sections, welded or riveted

3. **Girders:** light or heavy sections, welded or riveted

4. **Trusses:** light or heavy sections, welded or riveted.

Shop cleaning and painting must be allowed for as required. This is commonly priced at a cost per ton, and sometimes as a percentage of the total weight/cost.

To accurately estimate erection costs of steelwork, it is necessary to know the numbers, sizes, and weights of pieces requiring erection by crane and by other means, in addition to the total weight of steel. Also, it is necessary to know the numbers of rods, stringers, bolts, rivets, and the like. Any steel member requiring further assembly or fabrication in the field prior to erection should be enumerated and described. Any field painting should be indicated and described.

As an alternative method of measurement, **MM-CIQS** says that: "each piece of steelwork may be enumerated and described in detail as to size, length, and weight." This method provides more detailed information in the estimate, and it is a better method if the estimator's pricing information is sufficiently detailed and accurate to use the information. As explained before, grouping items in an estimate is a convenience based on the assumption that greater detail is not necessary for pricing, although it may be needed later for other purposes such as ordering, fabricating, and erecting. As estimating and *cost accounting* techniques improve, and as data processing systems are utilized to handle large amounts of data, this grouping of items (and the obscuring of data) will not be necessary, and estimating will be more accurate.

Structural steel can be measured only by a systematic take-off of all members. A good knowledge of steel construction and terminology is essential, because steelwork drawings indicate steel members by single lines and symbols. Special estimating forms for steelwork are used to tabulate the various types of steel members with the many labors required, such as punching, drilling, and cutting. These forms vary from place to place, and some typical formats are shown in Fig. 37.

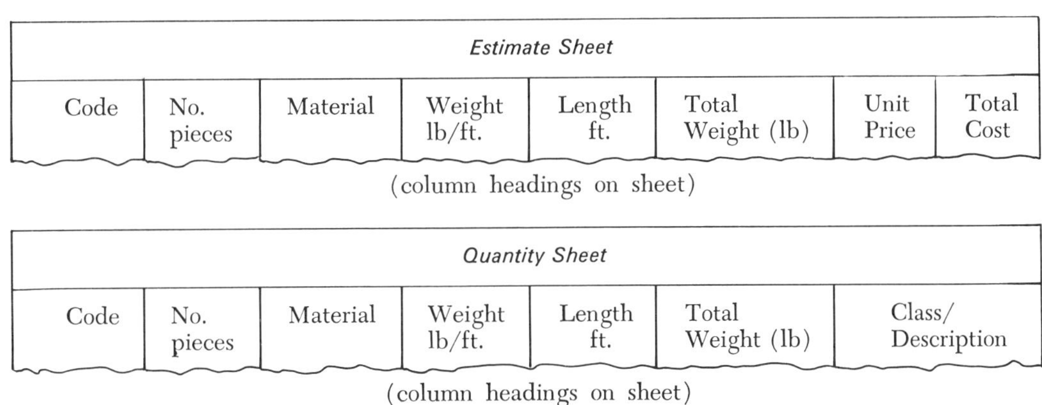

Estimate Sheet

Code	No. pieces	Material	Weight lb/ft.	Length ft.	Total Weight (lb)	Unit Price	Total Cost

(column headings on sheet)

Quantity Sheet

Code	No. pieces	Material	Weight lb/ft.	Length ft.	Total Weight (lb)	Class/ Description

(column headings on sheet)

FIG. 37. Samples of Forms Used in Estimating Structural Steelwork.

A common and logical sequence for measurement of steelwork is:

1. grillages, and other work at footings
2. columns, including bases and caps
3. beams, of all kinds
4. plate girders
5. trusses.

Many start the "take-off" at the top left-hand corner of the steelwork plan and work across and down the plan. All beams in one direction (say, west to east) are taken-off first before turning the plan sideways to take-off the beams in the other direction (say, north to south).

Beam connections are best enumerated and described (as seated or framed), with the weights of the connections' materials entered in the estimate. Not all connections are shown in detail on the drawings, and the estimator may have to decide on the types of connections to be allowed for in the estimate by selecting them from the standard beam connections tabulated and described in the steel institutes' manuals. Some steel fabricators have their own tables of connections for estimating. Short spans and heavy sections (such as 12 and 14 WF sections) require heavier connections than usual, and the maximum reaction values (in kips) and minimum spans for standard connections are tabulated in the steel manuals to assist the estimator in selecting the proper connections for beams and columns. A less accurate method is to add a percentage for connections to the total weight of main members.

If the steelwork is properly measured and recorded in detail in the estimate, it seems that the only reason for making several weight groups, or one total weight group, of the entire steelwork, is to simplify pricing, because there presumably is insufficient detailed cost information on steel fabrication and erection to price the steelwork in greater detail. A detailed take-off ensures more accurate pricing if the cost data is available. But, the necessary cost data will not be available unless *cost accounting* is done, and this first requires a detailed take-off. So, in order to achieve more accurate pricing through *cost accounting*, it is necessary to start with more detailed estimates. For this reason, the **MM-CIQS** alternative method of measuring steelwork by enumerating and describing each piece in detail in terms of size, length, and weight is preferable.

Miscellaneous metals should be fully described according to:

1. **kind,** such as steel, aluminum, copper, bronze
2. **grade and quality,** such as alloy composition; carbon content (of steel); strength; manufacturing method (e.g., hot rolled and cold rolled steel); tempering and artificial aging treatments; mill and other finishes; metals for special purposes, such as alloys with the necessary characteristics for working, welding, and resisting corrosion
3. **gauge, thickness, or substance,** such as sheet, plate, bars, rolled and extruded sections; drawn wire; drop forgings; casting; machine products
4. **type of metalwork,** including such items as structural connections for timbers; matwell frames; straps, hangers, brackets, and the like; floor plates and duct covers; gratings and grills; access doors and similar items; gates, handrails, railings, and balustrades; ladders and staircases; sheet metalwork in such items as coverings to doors and counter-tops, linings, and panels; *work* in wire mesh or expanded metals; anchors, bolts, and the like
5. **method of fabrication and installation,** such as bolted, welded, or brazed; hot or cold bending
6. **type of finish,** such as mill finish; priming and painting; galvanizing; electro-plating; enameling; anodizing; etching; polishing; and the like.

Generally, miscellaneous metals should be kept separate according to all such classifications and should be measured in square feet, linear feet, or enumerated, as appropriate, and fully described. This *work* is generally "custom made" (as opposed to standard products), and adequate descriptions are essential since simple references to catalogues of standard products are not usually possible. Descriptions should include the weights of items, where appropriate; also, the number of items (if grouped and measured together), and the methods of fixing and installation.

The practice of measuring miscellaneous metals by weight is often not conducive to accurate pricing, because the *labor costs* of fabrication and installation are often not directly related to the weight, although the *material costs* usually are. It is therefore possible to price this *work* accurately by weight only if accurate and valid cost data (related to the weights) are available from other *work* that is almost identical and has similar proportions of *material costs* and *labor costs*.

Items of work such as labors on metal items, which create additional *labor costs* or *material costs,* should always be measured and described. Such items may often be included with the main item, but in some cases it is more convenient and more accurate to measure them separately. Such items might include raking and curved cutting on mesh, sheets, and plates; rounded and scrolled ends, angles, bends, ramps, wreaths, and the like, of handrails; hemmed and beaded edges of sheet metal; machined and polished edges; fillet welds and spot welds; holes (drilled, cut, countersunk, tapped); and special screwing, bolting, and other fixing methods.

Miscellaneous metals, custom made for a job, are often fabricated and supplied by a *subcontractor* (not a *supplier,* because the *work* is specially designed and made for the job, and is not a standard and stock item), and it is often installed by the *contractor,* depending on the type of *work.* However, no matter who does what, somebody has to estimate the costs, as with precast concrete and other building components.

Methods of Measuring Carpentry (Division 6)

Rough carpentry is permanent woodwork generally of a structural nature and usually concealed in the finished building. It does not include temporary woodwork, such as formwork and centering. There are two basic methods of measuring this *work,* and one is an extension of the other. One, by which all framing and other lumber is measured in linear feet. The other method uses the board measure, and units of 1000 board feet,[8] in which case it is first measured from the drawings in linear feet and then converted to board feet. Each of these two methods is widely used in estimating according to local practices, and both are given in the **MM-CIQS**.

It is not easy to find a good argument for the board measure, other than custom. The same can be said for the practice of measuring some framing lum-

[8] A "board foot" is one-twelfth of a cubic foot, calculated from the nominal (not actual) dimensions of the lumber; i.e., a 12-inch length of 2 × 6 lumber. It is the unit of "board measure".

ber in cubic feet (**SMM-RICS**). The board foot and the cubic foot are both units of volume by which lumber is bought and sold. Measurements in linear feet are simpler for estimating, and the quantities are more comprehensible.

Board foot quantities can be particularly misleading in estimating *framing labor costs.* There is less labor required per thousand board feet for framing 2 × 12 joists than for 2 × 10 joists, and 1 × 4 strapping requires almost twice as much labor per thousand board feet as 2 × 4 strapping at the same spacing. Such facts are overlooked by some estimators, in part because of the use of the board foot measure in estimating, which can be misleading.

Framing and other lumber should be separated according to:

1. **species,** according to the appropriate rules of lumber manufacturers
2. **grade,** according to the appropriate rules of lumber manufacturers
3. **surfaces,** dressed or undressed, and whether or not specially treated
4. **dimensions,** using nominal dimensions for dressed (surfaced) lumber and actual lengths in even 2-foot increments.

Species and grades should be identified according to recognized local grading rules, which also govern the definitions of nominal and actual dimensions according to whether or not the lumber's surfaces are dressed, and the dressing allowances.

Lumber specified to have a special treatment such as air or kiln drying, or a preservative treatment of any kind, should be measured separately. Also, lengths over 14 feet should be separated in even 2-foot increments, because of the increased costs for these longer lengths, according to **MM-CIQS.**[9]

[9] It is a common practice for estimators to measure construction lumber up to the nearest 2-foot increment; i.e., lengths greater than 10 feet and up to and including 12 feet are measured as 12 feet, because lumber is usually sold in even lengths. But it is important that an estimator should record and consider all the facts that affect costs, and therefore it would seem that the better way to measure lumber is net, in linear feet, stating the estimated percentage of *waste,* if any.

Minor items that can utilize available offcuts of lumber, such as bridging, girts, and firestopping, can be described as "labor only" items if the estimator knows that offcuts will be available from major items. And often it is cheaper to overlap joists that are longer than necessary than it is to cut them. However, if the estimator can foresee utilization of the additional material unavoidably purchased, he can measure it and indicate less or no *waste* in his description of the *work.* Job prices and conditions vary, and an estimator must measure so as to price the *work* as accurately as possible, using standard methods and estimating precepts as a guide.

In addition, framing and other lumber should be separated according to its use and location in the building. Plates, studs, joists, lintels, beams (solid or built-up), rafters (common, hip, or valley), ridges, collars, ties, purlins, and the like, should all be estimated separately. Also, short studs (less than 8 feet) and long studs (over 8 feet) should be separated in even 2-foot increments. All other items of framing also should be separated and described. Plates and studs in prefabricated (i.e., tilt-up) framed walls may be grouped together; but if they are, any additional labors to plates and sills, such as bedding, leveling, drilling, and bolting, should be measured separately and described as extra labor items.

Lumber costs increase according to section and length, and larger sections and longer lengths are generally more expensive. *Labor costs* vary according to type of *framing work* and location. Roof framing is more expensive than framing in floors and walls. Supporting partitions with short studs, in attics and crawl-spaces, are more expensive than ordinary partition walls. Hence, these and other such items must be measured separately and priced differently. Remember, it is not only the lumber that is measured, but also the *work.* Studs include nailing the ends to plates. Hip and valley rafters include cutting, fitting, and nailing the ends of common rafters against them, on both sides. Each piece of lumber is installed, and both the lumber and the installation must be described, and eventually priced.

Items such as bridging, girts, fire-stopping, and the like, should be measured in linear feet over the joists (or studs, etc.) and so described. Such items consist mostly of *labor costs* and the material content is minor, if not insignificant. Mismeasurement and poor descriptions of *work* lead to inaccurate pricing; and board foot measurements of this type of *work* are usually misleading and often incomprehensible, because there is no relationship between the board foot quantities and the *labor costs* of items such as cross-bridging.

Wood decking and shiplap sheathing are best measured in square feet, with walls, roof, and floors measured separately and fully described. Board foot measurements require allowances to be made in the measurements for the difference between actual sizes of surfaced (dressed) lumber and the nominal sizes, as well as allowances for laps, or tongues and grooves. With *super* quantities measured net, in square feet, such allowances are made in the *unit prices.*

Sheathing paper is measured in square feet, with an allowance included for *laps,* preferably in the *unit price.* Plywood sheathing and underlay are also measured in square feet, and the description should refer to specie, grade, surface, and thickness. The methods of jointing, fixing, and supporting edges of plywood sheathing should also be stated. Raking and circular cutting should be measured in linear feet. Other common rough carpentry items such as blockings, furrings, grounds, bucks, nailers, bearers, and the like, are all measured separately in linear feet and described according to purpose and location, as well as specie, grade, surface, and dimensions. The methods of fixing and the surfaces to which items are fixed should also be described unless they are usual and commonplace.

Methods of framing and fixing, other than with spikes or nails, should always be fully described and the *work* kept separate. Spikes and nails are measured separately by weight or allowed for in the pricing. Other rough hardware items such as anchor bolts, joist hangers, connectors, and the like, are enumerated and described.

Prefabricated work should be enumerated by components, each fully described. An estimate should contain headings for any *prefabricated work*, such as:

> "The following in one prefabricated roof truss ("type A") 24 feet span × 8 feet rise"

followed by a detailed measurement and pricing of the *work* in one such truss. This procedure can be used to isolate and estimate in detail the costs of any building components and any other parts of a job requiring special consideration by the estimator, such as unusual framing and other *work*.

Finish carpentry is woodwork that is exposed in and around the finished building and is, therefore, selected and installed for its appearance and finish. It may or may not be structural in nature. It includes all *carpentry work* that is not classed as rough carpentry, and also the installation of millwork (custom woodwork) that is made in a factory and delivered to the site. It also includes many items that are not of wood, such as certain finish hardware, plastic laminates, and trim of various kinds. It does not include special products such as glued-laminated timbers and wood flooring: wood doors and windows are included in Division 8 of the Uniform Construction Index. The distinctions among *sections of work* in Division 6—Carpentry are not always clear, but this is a problem to be solved first by the specification writer and then by the estimator.

Trim is measured in linear feet, and each type, specie, and pattern should be separate, although **MM-CIQS** allows some grouping of wood trim according to widths. Soffits should be measured in square feet (**MM-CIQS**), but a linear quantity with the width stated is often more descriptive. For a similar situation see the next item—Shelving.

Shelving is generally measured in linear feet. **MM-CIQS** states that shelving over 12 inches wide should be measured in square feet. It is convenient to group them together if there are many widths to be measured, but the estimator should always use his discretion. It would be fallacious to measure 200 linear feet of 13-inch wide shelving and 200 linear feet of 24-inch wide shelving in square feet together and to price them at the same *unit price*. Of course, both could be measured separately in square feet and priced at different *unit prices*; but since the way to ensure different *unit prices* for different items is to measure them separately, the easier way is to measure them as *run* items, stating the widths in the descriptions.

Paneling and siding are both generally measured in square feet and fully described, including full details of joints, tongues and grooves, laps, and any requirements relating to joints and their layout. Raking and circular cutting should be measured separately in linear feet. Straight cutting is generally deemed to be included with the main items in most kinds of *work* in all trades. But there is no reason why straight cutting should not be measured and priced separately if required; and this should be done if there is much straight cutting, and if accurate pricing is not otherwise possible.

Cutting of all kinds includes not only the *labor costs* of cutting, but also the additional *material costs* of *waste* caused by the cutting. This *waste* should be allowed for in the prices, and it varies in amount according to job dimensions and the size of the piece or sheet of material used. With plywood paneling and large sheet goods, it may be a considerable amount, particularly if a specified joint layout requires a minimum number of joints. Mitred corners, narrow widths (less than 12 inches wide), trim, and similar items should each be measured separately in linear feet and described.

Partitions that are custom made should be measured in linear feet and fully described, with doorways and borrowed lights enumerated. Partition costs do not vary in direct proportion to height and area; and, therefore, partitions cannot be accurately estimated from *super* quantities of partitions of different heights grouped together. Each height should be separated, thus rendering linear measurement logical.

Staircases are a typical example of the change from traditional woodwork to factory production. Methods of construction are changing radically, and methods of measurement must follow. The **SMM-RICS** describes the method for measuring wood staircases in some detail. Each part is to be measured separately and described under an appropriate heading, such as

"The following *work* in one staircase, 3 feet 3 inches wide, 10 feet going (length), and 8 feet 1½ inches total rise." The **SMM-RICS** also states elsewhere that, "standard units" (of woodwork) shall be enumerated and described. Cannot a prefabricated staircase be called a "standard unit"? Perhaps not always; but the same reasons for measuring by enumeration would apply to many a prefabricated staircase. If a project requires several staircases with each one identical, then their similarity to a standard unit increases. Therefore, factory-made wood staircases should be enumerated and described. The **MM-CIQS** approaches this method by stating that "treads and risers shall be combined, fully described and enumerated." (But it also requires strings and carriages to be measured in linear feet.) An estimator should decide on the method of measurement according to the method of manufacture so as to facilitate accurate pricing.

Custom woodwork, sometimes known as "millwork," is similar to finish carpentry except that it is custom made in a millwork shop or factory and delivered to the site partially or fully assembled for installation. The installation is usually included in Finish Carpentry.

Cabinetwork is generally enumerated and described by the *general contractor's* estimator, who will probably obtain *bids* from millwork firms on supplying the cabinetwork. He will estimate only the costs of installation. The millwork manufacturer's estimator, however, will have to estimate custom cabinetwork in detail. This requires an analysis and listing in his estimate of all the materials to be individually priced to estimate the *material costs.* The *labor costs* may then be estimated as a percentage of the *material costs* (about 50 percent), or the *labor costs* may be more accurately estimated by a detailed analysis of the *work* and by the *cost accounting techniques* used in factory production.[10] To these costs must be added factory and other *overhead costs,* sales taxes, shipping costs, and *profit.* Other custom woodwork items, such as custom paneling and trim and certain items described under Finishing Carpentry, which in some cases may be part of the custom woodwork supplied to a job, should be measured and priced as described above.

[10] There is a fundamental difference between *cost accounting* as applied by other industries to mass production and *cost accounting* as applied to construction, and the difference stems from the fact that the products are different. Buildings are generally unique and complex and their construction takes months or years, whereas the products of most other industries are mass produced in a relatively short time. A week's production on a construction site may be hardly perceptible at times, whereas a week's production of units in a factory may be counted in tens or thousands. But as construction moves from the site into the factory, so the methods of management, including *cost accounting,* will need to change.

Glued-laminated timbers are usually a separate *section* of the *Carpentry division.* There are several methods of measurement used for these timbers. Rectangular sections can be measured and priced on a linear foot basis, but generally all estimates of glued-laminated timbers are basically made on volume or volume converted to weight by a pounds per cubic foot factor. Volume measurements are either in cubic feet or board feet. On the Pacific Coast, where "glulam" timbers are manufactured and widely used, the board foot measure was common in the past; but it has now been replaced by the weight measure by some manufacturers.

To measure the number of board feet in a rectangular glulam beam, first ascertain the number of 1 5/8-inch or 1 1/2-inch laminations in the beam's section. (The depth of the beam will be a multiple of either 1 5/8 inches or 1 1/2 inches.) Convert the actual size of the lamination to its nominal size. The nominal width is the nearest multiple of 2 inches above the actual width (e.g., 7 inches actual width, 8 inches nominal width); the laminations' nominal thickness is 2 inches (for both 1 5/8 and 1 1/2 inch actual thicknesses). For example, a rectangular beam of 9 inches × 39 inches actual section is made up of twenty-six laminations, each of 2 inches × 10 inches nominal size. This equals 43 1/3 board feet per linear foot of beam. To the net quantity of board feet in the beam add 20 percent for trimming and jointing: e.g., a 9-inch × 39-inch beam, 24 feet long, would require 1,250 board feet of 2 × 10 lumber: (43 1/3 × 24) + 20% = 1250 board feet (to the nearest 10 BF).

Curved glulam timbers are made from laminations of down to 1/4 inch thickness; depending upon the radius of curvature required. The smaller the radius the thinner the laminations required.

Glulam timbers should be described according to:

1. **form:** straight; tapered; curved; or other
2. **function:** simple beams or purlins; cantilevered beams; arches (including two hinged arches, such as the foundation arch and the tied arch; and three hinged arches, such as the parabolic arch, the A-frame arch, the radial arch, etc.); trusses; rigid frames; and posts
3. **section size:** may be variable, depending on function and form, such as in tapered members
4. **strength grade:** according to function, such as 24f (bending); 18c (compression); 26t (tension);
5. **service condition:** "dry" (at or below 16 percent moisture content in timber when installed and in service); or "wet" (above 16 percent moisture content when installed and in service), which determines the type of ad-

hesive required in manufacturing, according to appropriate published standards

6. **appearance quality:** the lowest quality, for industrial use; the intermediate quality, for "paint" finish; and the top "architectural" appearance quality, according to appropriate published standards

7. **protection requirements:** including the application of penetrating sealer, or sealer coating; bundle wrapping, or individual wrapping; and any preservative treatment.

For transportation and erection, the weight and size of individual timbers may be critical, and erection should be measured separately with all pieces enumerated and described, including the dimensions and the weight of each piece. Which leads one to question why glulam timbers should not be always measured by enumeration and description, in the same way as described in the **MM-CIQS** as an alternative method of measuring structural steel members, in Division 5.

Steel hardware for connections is best enumerated and described, stating the weight. Sometimes it is all grouped together and estimated by weight, but the precepts given above for estimating miscellaneous metals also apply here.

Methods of Measuring Moisture Protection (Division 7)

Dampproofing and **waterproofing** are often misused terms. "Waterproofing" should properly be used in reference to *work below ground* that is intended to seal a building against water under static pressure, whereas "dampproofing" should be used in reference to other moisture protection where ground water pressure is not expected. With bituminous materials, waterproofing usually includes membranes, whereas dampproofing usually does not.

The **MM-CIQS** requires this *work* to be measured in square feet, with no deductions made for openings of less than 40 square feet. The *work* should be separated according to the type of surface to be treated, such as masonry, or concrete, vertical or horizontal. *Special work,* such as extra plies at corners, curbs, and the like, should be measured in linear feet and described. *Special work* to seal around pipes, and the like, should be enumerated and described. Some bituminous waterproofing is similar to bituminous built-up roofing (see below) and should be measured in like manner. In any event, the larger unit of measurement used for roofing would be more appropriate.

Roofing comes in many types and varieties, but most of them have certain basic characteristics in common. Since it is usually not possible to roof a building with a single continuous piece of material, most manufactured roof coverings are lapped in such ways as to make one, two, three (or more) plies of roof covering, as required for protection. This is true of built-up roofing, wood shingles, tiles, and slates, all of which are laid to various *laps* according to the number of plies, or thicknesses, of roof covering required.

All roof coverings are measured as *super* items, usually in "Squares." A Square contains 100 square feet. This is a traditional unit of measurement for roofing in many parts of North America, and the Square is the unit used by the **MM-CIQS**. (The **SMM-RICS** uses the square yard; the **SMM-RICS/M** uses the square meter.)

Roofing is measured net over the area covered by the roofing. No allowances are added to the measurement for *laps,* which are part of the roof covering's specifications and description and are allowed for in the pricing. (**SMM-RICS** calls for seams, drips, and the like, in sheet metal roofing such as copper, to be measured and added to the roofing measurements.)

Roofing descriptions should describe the surfaces to receive the roofing. *Work* to flat, sloping, and circular (on plan or elevation) surfaces should be separated. The degree of slope, or pitch, should be indicated. The **MM-CIQS** calls for the height above grade at which the *work* will be executed to be stated. It also states that no deductions should be made for openings in roofing of less than 10 square feet in size.

Such items as underlayment or double courses at eaves, and cutting and extra items of labor and materials at verges, ridges, hips, and valleys, should be each measured separately in linear feet and fully described. Similarly, in built-up roofing, extra felts at cants, curbs, eaves, turning into reglets, and the like, should each be measured separately from the roofing in linear feet. Isolated *items of work* such as dressing roofing felts into scuppers and hoppers, making gum pockets, and *work* around hatches, roof lights, and the like, should be enumerated and fully described.

Sheet metal flashings are best measured in linear feet and fully described, stating the width, or girth, of the metal. This *work* should be separated according to:

1. type of sheet metal
2. weight, gauge, or thickness of sheet metal
3. type of *work* (e.g., gravel stop, eaves flashing, valley flashing, hip flashing, gutter, etc.), and the type of *roofing work* with which it is associated.

Such items as joints, fixings, underlayment, and calking should be included in the items' descriptions.

Flashings to such items as hoppers, vents, and pipes should be enumerated and described separately.

If inspection of moisture protection and a warranty of the *work* is required under the contract, the requirements should be indicated and priced in the estimate. Inspection is generally quoted at a *unit price* per Square, depending on quantity and type of *work*, type of inspection required, and the location. Cut tests are usually quoted at a *unit price* per test, or on the basis of one test to so many Squares of roofing. All such requirements as these should be entered and priced as separate items in an estimate.

Methods of Measuring Doors, Windows, and Glass (Division 8)

Doors and windows are generally enumerated and fully described. The **MM-CIQS** requires that glass and glazing done by window manufacturers be separately measured the same way as other glass and glazing, but identified with the windows. This would also presumably apply to glazed doors.

Glass and glazing is generally measured in square feet stating the sizes of the lights (panes). The **MM-CIQS** says "light sizes shall be given in separate groupings in accordance with the manufacturer's break sizes," and that "each light [shall be] dimensioned in inches (to the next larger even figure)." The **SMM-RICS** follows a similar method, but says that sheet glass and rolled and cast glass should be classified as follows:

1. panes not exceeding 1 square foot stating the number of panes
2. panes over 1 but not exceeding 4 square feet
3. panes over 4 but not exceeding 8 square feet
4. panes over 8 square feet.

These size-groups are a convenience to reduce the number of separately measured items. Panes (lights) having one or both dimensions "over the manufacturer's normal maximum" should be given separately in square feet stating the size of the pane (**SMM-RICS**).

Glass and glazing work should be classified according to:

1. kind, type, and thickness (or weight) of glass
2. size and shape of light (pane)
3. method of glazing and glazing materials used
4. kind and pattern of bead or other means to secure glass and its method of installation (e.g., snap-in, screwed)
5. kind and type of window or other opening to receive glass and whether glass to be installed at site or in factory
6. location of glazing (e.g., inside or outside of frame) and position from which site glazing is to be done (e.g., from inside or outside).

There are many kinds and types of glass and several qualities of each. There are also several thicknesses (nominal) and weights for most types. Glass prices vary considerably according to size and total quantity. Consequently, careful measurement and description are important.

Light sizes are measured up to the nearest multiple of 2 inches in the **MM-CIQS** (and to the nearest inch in the **SMM-RICS**). Shapes other than rectangular should be measured separately to the nearest rectangular area necessary, and fully described.

The **MM-CIQS** requires the method of glazing to be fully described and measured separately in linear feet. Glazing materials and methods vary greatly in quality and cost, and the only way to price this *work* accurately is to measure the glazing and the glass separately. Interior and exterior glass and glazing should be separated. If scaffolding or other equipment is necessary for site glazing from the outside, it should be allowed for by describing and separating the *work*, and pricing it accordingly.

Labor on glass, such as raking (angular) and circular cutting, and beveled or polished edges, should be measured separately in linear inches (or feet), and fully described. Drilled holes are to be enumerated. Any *work* on glass surfaces, such as cleaning, acid etching, sand-blasting, and the like, should be measured in square feet and each portion enumerated, Narrow strips of such *work* should be measured as *runs*, stating the widths of the strips. Such *work* done in special or isolated shapes and areas, or at the site, should be enumerated and fully described.

Mirrors should be enumerated and fully described, stating the quality, size, and method of installation. Labor on edges, drilled holes, special backings, and all other items should be included in the description.

Other work similar to glass and glazing should be measured in the same way. This includes vitrified products used as wall linings and claddings, glass doors, and patented glazing systems.

Finish hardware is an important part of the *work* in this *division*. It can amount to a significant part of the total costs of a building, and it often consists of a great variety of items, each of which may have numerous alternatives in style, quality, and finish. Because finish hardware is often so complex, its supply may be covered in a contract by a "cash allowance" prescribed by the *designer*. This practice is further considered in Chapter 11, under Division 8.

Finish hardware is measured simply by enumeration and description. Descriptions must be complete in all respects, and a good knowledge of finish hardware is essential. Hardware for major projects is usually measured by a hardware specialist. However, although the hardware may be supplied by a hardware firm, it is usually installed by the *prime contractor* whose estimator must measure and allow for the installation in his estimate. If the supply is covered by a cash allowance, the estimator may have little to aid him in estimating installation costs other than the amount of the cash allowance and any experience he may have obtained from similar jobs. Usually the specifications are not much help, because the *designer* himself does not know at that time exactly what hardware will be purchased. Nevertheless, the *designer* should indicate in the *bidding documents* all that he can about the finish hardware: its types and its scope; even if only in general terms.

The estimator will measure by enumeration and estimate the costs of hanging doors of the various types; and this estimate will embrace much of the finish hardware. Hardware on metal windows is usually supplied and installed by the window specialist. Finish hardware on cabinets and other custom woodwork may be installed by the millwork firm. Some *designers* include miscellaneous fixtures and accessories with finish hardware under one cash allowance, and this makes measurement for estimating installation costs rather difficult. (Such items not installed on doors and windows should be specified in Division 10.)

Methods of Measuring Finishes (Division 9)

Generally, finishes are not difficult to measure. Most *finish work* is measured as *super* items in square feet or square yards. (The **SMM-RICS/M** uses square meters.) In several trades, *work* in narrow widths (of less than 12 inches) is measured as *runs,* in linear feet, stating the widths. Items such as beads and trim are also measured in linear feet.

The **SMM-RICS** (and the **SMM-RICS/M**) contains a section entitled. "Plasterwork and Other Floor, Wall and Ceiling Finishings," which recognizes the similarities between them. The general requirements for this section of **SMM-RICS** are that:

(a) Work shall be grouped according to the kind of material and each group with its associated labors shall be given under an appropriate heading.

(b) Work and labors on external surfaces shall be so described.

(c) Work and labors executed overhand[11] shall be so described.

(d) Work and labors in repairs shall be so described and the preparation of old surfaces to receive such work shall be given in the description.

(e) Work on old surfaces shall be so described stating the thickness of any dubbing.[12]

(f) Work in isolated areas not exceeding one square yard [one square metre] each shall be so described.

(g) Work to a pattern or in more than one color shall be so described stating the nature thereof.

(h) Curved work, conical work and spherical work shall be each so described stating the radius. Elliptical work and other work curved to more than one radius shall be so described stating the radii. Curved labors shall be so described irrespective of the radius.

(i) Temporary rules, temporary supports to risers and the like, temporary screeds and templets shall be deemed to be included with the items.

Generally, all *finish work* to horizontal and vertical surfaces is measured separately. Measurements are made of the area of the *work* in contact with the base on which it is done. Rough cutting to straight lines is generally not measured separately and is deemed to be included in the general item. Fair cutting to straight lines, (e.g., where a finish abuts another finish surface and the cutting at the joint is exposed) and all raking cutting are generally measured as *run* items.

Lath and plaster (and stucco) is generally measured in square yards. The **MM-CIQS** makes no distinction between lath and plaster to plain areas and that to narrow widths, although it does require that lath and plaster to beams, pilasters, and columns each be classified séparately. In some other finish trades **MM-CIQS** does separate all *work* in widths of less than 12 inches. Generally, this distinction is made by **MM-CIQS** in trades in which the basic material is a dry board or sheet, and is not plastic or wet like plaster. (The **SMM-RICS** requires all *finish work* in widths of less than 12 inches to be measured separately as *run* items.)

Lath and plaster to wall, ceilings, solid partitions, beams, pilasters, and columns are each measured separately and described. *Work* above a certain height (**MM-CIQS**, 14 feet; **SMM-RICS**, 11 feet;) is to be

[11] "Overhand" refers to *work* done on a surface that the workman cannot directly face while working; e.g., *work* to the outer face of a parapet wall performed by workmen standing on the roof behind the parapet.

[12] "Dubbing" refers to *preparatory work* to an old surface to receive *new work*; e.g., patching and leveling an old floor with a filler material to receive new resilient flooring.

kept separate so that the additional scaffolding costs can be allowed for in the pricing.

In the **MM-CIQS,** no deductions are to be made from the measurements of lath and plaster for openings under 40 square feet. (The **SMM-RICS** says openings under 4 square feet; the **SMM-RICS/M** says openings under 0.50 square meter.) This means that under the **MM-CIQS**, most doors and window openings should not be deducted, and this is customary practice with many estimators. But not all estimators agree with this practice, and some deduct all door and window openings from their measurements and measure the *work* net.

In some buildings, openings are a considerable proportion of the total wall areas to be plastered. If they represent, say, 20 percent of the gross area, how realistic and how useful is the measured quantity of *work* if the openings have not been deducted? An argument against deducting the openings is that the *additional work* measured (by not deducting the openings) allows for the *additional labor costs* possibly created by the openings; that is, the additional labor caused by the interruption of the plain areas by openings and in working up to and around the openings. Logically, this argument should also apply to large openings (of over 40 square feet) for which deduction are made, and the estimator should then measure a labor item around the perimeters of such openings because no other allowance for any additional labor is made. But this is not usually done.

Generally, it is not good practice to substitute one item for another in estimating, as stated in the Precept of Measurement, Chapter 6. Therefore, it is probably better practice to deduct all openings and to measure any *additional work* created by the openings. Very small openings may be ignored (as in **SMM-RICS**) because the costs involved are insignificant. Also, it is better practice to measure separately all *work* in narrow widths (less than 12 inches), because *work* containing a high proportion of narrow widths will normally cost more than *work* that contains a lower proportion of narrow widths or none at all.

A strong and continuing argument is made by some against the separate measurement of narrow widths and of *finish work* to surfaces such as beams and pilasters. The argument states that since the costs of such *items of work* cannot be segregated by *cost accounting,* there is no point in measuring them separately, because there is no factual basis on which to price them differently. A wall with pilasters, or a ceiling with beams, usually costs more to plaster than a plain wall or ceiling of equivalent area without pilasters or beams. It seems reasonable that the more *work* there is to pilasters or beams in a given area, the higher the *unit price* will be. But because of the nature of the *work,* it is true that the actual *costs of work* to the pilasters or beams practically cannot be isolated from the total costs of plastering the wall or ceiling. What then should an estimator do?

In Chapter 6, the statement was made that generally speaking there is no reason for an estimator to measure *work* in greater detail than *cost accounting* procedures require, because the details of the estimate can never be verified by *cost accounting*. However, it was pointed out at the same time that there are exceptions to this general rule, and it is in such an exception that we can find the answer to the above question.

In an effort to move towards more accurate estimates through a greater knowledge of costs, an estimator must make allowances for all the costs that he can foresee. This means, for example, that in estimating *plaster work* to walls or ceilings containing pilasters or beams, if the estimator believes that 1 square yard of plaster applied in narrow widths (such as to faces of pilasters or to beam soffits) is more expensive than 1 square yard of plaster applied to the plain area of a wall or a ceiling, he must somehow make allowances for the extra cost.

Some estimators try to take into account the *extra costs of work* arising from corners, changes in plane and surface, edges, and the like, by separately measuring and pricing items such as arrises, angles, edges, and similar features. The **SMM-RICS** calls for the separate measurement of such labors (as they are called) on finishes, whereas, following common trade estimating practices, the **MM-CIQS** does not. However, the general principles of **MM-CIQS** state that, "if it is in the interest of accurate and practical estimating, he [the estimator] may give more detailed information than is demanded by strict adherence to the document." Thus, an estimator should try to make his estimate as accurate as possible by measuring whatever he believes is necessary or desirable to represent the *work* to be done in order to create the best possible basis for pricing the estimate and accounting for the costs when the *work* is done. If features such as narrow widths and arrises appear to make *work* more expensive, then they are better measured as *separate items of work* and priced accordingly.

There are two ways in which an estimator can estimate and allow for the extra costs of plastering a ceiling with beams. Let us assume that from recent experience and through *cost accounting* an estimator knows that *lath and plaster work* will probably cost $10.00 per square yard in plain, flat areas of ceilings without beams. He can make an intuitive judgment and say that, since this *work* contains some plaster to beams, it will cost, say, an extra 5 percent, or $10.50 per square yard, without knowing exactly what proportion of the *work* is in plain areas and what proportion is in sides and soffits of beams. Alternatively, he can measure separately the sides and soffits of the beams, and the arrises. He can price the plain areas at $10.00 per square yard, and price the *work* to beams at, say $12.00 per square yard, allowing an estimated 20 percent extra cost. He can also price the arrises at so much per linear foot.

Either way, the estimate should be subsequently checked after the *plaster work* is finished and cost accounted. If the estimator chooses the first way of pricing the *work* he will gain little from the *cost accounting*. He may know generally whether or not his judgment to price the plaster at $10.50 per square yard was good or bad; but because he does not have a quantitative basis for his judgment, he cannot carry his cost experience forward to use it in making future judgments about similar cases involving different quantities. If he chooses the second and more analytical way to price the *plaster work*, he is better able to accumulate cost data and experience, and each estimate and cost account will bring him closer to a better knowledge of costs.

This is the argument for measuring in greater detail rather than in less in any trade—that it often makes it possible to obtain more knowledge about the *costs of work*. One argument against it, however, is that it may tend to make measurement an end in itself, and this must be guarded against, as with all good things.

Gypsum drywall is measured in square feet under **MM-CIQS,** with *work* less than 12 inches wide measured separately (in square feet), and with no deductions made for openings of less than 20 square feet. This is presumably to conform to certain trade practices. Square cutting is not measured under **MM-CIQS**. (**SMM-RICS** requires the measurement of square cutting around openings and it requires openings of more than 4 square feet to be deducted. Cutting around smaller openings is measured by enumeration.)

It should be pointed out that some estimators measure drywall by enumerating the required boards. For example, a room 12′6″ × 10′0″, with an 8′0″ high ceiling, might require:

Walls:	4 boards, size 4′ × 13′ =	208 SF
	4 boards, size 4′ × 10′ =	160
(boards installed horizontally)		368
Ceiling:	3 boards, size 4′ × 13′ =	156
		524 SF

The net areas are:

Walls:	45.0 × 8.0	= 360 SF
Ceiling:	12.6 × 10.0	= 125
		485 SF

The difference between 524 and 485 is about 8 percent *waste,* and additional *waste* might occur at door and window openings, depending on their locations and on different layouts and boards.

If a drywall estimator is not concerned with the amount of *waste,* he might always measure by enumerating the different size boards in this way rather than by the net *super* areas; but there are reasons why an estimator should want to know how much *waste* occurs. Drywallers are sometimes employed on a piecework basis, and this may encourage the workmen to follow more wasteful but faster methods of cutting and applying the boards. Similarly, excessive *waste* may occur in contracts in which the wallboard is supplied by the *owner* and applied under a labor contract.

The amounts of straight cutting and *waste* are related, and if a drywall *contractor* is concerned with productivity and the costs of materials and labor he will want to know the amounts of cutting and *waste* in each job. It is always necessary to measure the net area to find the amount of *waste,* and unless this is done no accurate knowledge of *waste* can be obtained. This is why it is a Precept of Measurement to measure *work* net in place, as explained in Chapter 6.

Ceramic tiling is measured in square feet under **MM-CIQS,** with no deductions made for voids of less than 4 square feet. Coved internal angles, rounded external angles, and tile bases are measured in linear feet. The **MM-CIQS** does not require that straight cutting be measured, but again it should be measured if the accuracy of the estimate requires it.

The primary variations among the **MM-CIQS** methods of measuring different finishes are in the minimum deductions for openings and in the measurement of narrow widths, and these can be found in the **MM-CIQS** publication. Hopefully, as the use of the **MM-CIQS** becomes more widespread, there will develop a greater uniformity and rationality in all methods of measurement; and an active interest in the development and practical use of this publication is recommended as a means to this end.

Other finishes, such as terrazzo, stone veneer, acoustical treatments, wood flooring, resilient flooring, painting, wall coverings, and special floorings and coatings are all measured in ways similar to those described above. However, different classes of finishes require different considerations by estimators; and since all measurement leads to pricing, we must look to the pricing of *work* for further aspects of measurement. Chapter 11—Pricing Finishes (Division 9) is, therefore, a further source of information on the measurement of finishes of all kinds.

Methods of Measuring Specialties and Similar Work (Divisions 10, 11, 12, 13 and 14)

These *divisions of work* include:

Division 10—Specialties
Division 11—Equipment
Division 12—Furnishings
Division 13—Special construction
Division 14—Conveying systems.

Most of the *items of work* in these *divisions* are typically manufactured away from the site and, in many cases, they are installed by a specialist firm. In some instances they will be dealt with in the contract and be estimated like finish hardware (*Division* 8). Generally, their measurement will simply consist of enumeration together with a detailed description or a reference to an item's specifications in the *bidding documents.* However, such description may be inadequate for a proper estimate of installation costs, and the estimator should obtain other information, such as the item's overall size and gross weight. Installation may include unloading, hoisting, and moving several times, as well as temporary storage, protection, and security. All of these factors give rise to costs that must be recognized and estimated; and the installation of an unusual piece of equipment, supplied by others, may involve only a few anchor bolts. But the total costs of installation may amount to hundreds of dollars because of difficulties in access and handling.

Any *work* in these *divisions* requiring more analytical methods of measurement, including construction and finishes for special rooms and structures, such as studios, radiation centers, and conservatories (which are all usually specified in *Division 13*—Special Construction), can usually be measured by utilizing appropriate methods for *similar work* in other *divisions.*

Certain *sections of work* are included in the **MM-CIQS**:

(*Division 10*) Chalkboards and tackboards
Toilet and shower compartments
Demountable partitions
Storage shelving (metal)

(*Division 11*) Food service equipment
Laboratory equipment

(*Division 12*) Blinds and shades
Carpets and mats
Drapery and curtains

(*Division 13*) Radiation protection

(*Division 14*) Elevators.

Other *sections of work* will be published in the future and sent to purchasers of the CIQS document.

Methods of Measuring Mechanical Items (Division 15)

It is usual for special "General Requirements for Mechanical Trades" to be stated in *bidding documents,* and an estimator should note in his estimate any of these requirements that may give rise to costs. Most of these will be in the nature of *job overhead costs,* including such items as:

Record drawings (as built)
Operating instructions
Temporary and trial usage
Detail drawings
Wall plates and access doors
Valve tags and charts
Electric wiring and starters
Guarantees and warranties.

These are some of the requirements indicated in Division 15 of **MM-CIQS**.

Plumbing and drainage should be measured with each system separated in the estimate under headings such as are listed in **MM-CIQS**; e.g., Outside Services, Inside Buried Piping, Water Piping, Gas System, and so on.

Outside services are usually measured from property lines up to a perimeter line from 3 to 5 feet outside the structure to connect to inside buried piping measured up to the same perimeter line outside the building. The **MM-CIQS** designates the limits of the various other systems in this way, using a standard distance of 5 feet outside the structure.

Generally, all piping is measured in linear feet, and all fitting, valves, and the like are enumerated. No deductions are made from pipe lengths for fittings, etc., except in the case of pipes 10 inches in diameter and over (**MM-CIQS**). The measurement itself is simple, if the estimator has the knowledge and ability to read and interpret the drawings and specifications. All *items of work* must be fully described, including

Ductwork

Ductwork generally	**S26**	For rules relating to Section S generally see Clause S1 hereof.
Ducting	**S27**	**(a)** Ducting (measured over all ducting-fittings, short running lengths and branches) shall be given in linear yards stating the size. Stiffeners shall be given in the description. **(b)** Flexible ducting and extensible ducting (measured as fully extended) shall each be so described. **(c)** Curved ducting shall be so described stating the mean radius. **(d)** Lining ducting internally with acoustic or protective material shall be given in the description of the ducting.
Ducting-joints	**S28**	**(a)** Providing materials, heat, bolts, nuts, washers and everything else necessary for making joints in ducting shall be deemed to be included with the items. **(b)** Joints in the running length shall be given in the description of the ducting stating the method of jointing. **(c)** Special connections and special joints in ducting shall each be enumerated separately (except where given in the description of another enumerated item) stating the size and kind of ducting concerned and the method of jointing. Classification shall be as follows:— (i) Connections between ducting of differing materials. (ii) Connections between ducting and equipment stating the nature of the equipment. (iii) Isolated joints which differ from those given in the description of the ducting (eg flanged joints in ducts having spigot and socket joints generally). (iv) Flexible connections between ducting and plant.
Ducting-fittings	**S29**	Ducting-fittings (eg stop-ends; bends; offsets; diminishing pieces; change-of-section pieces; junction pieces; nozzle outlets) shall each be enumerated separately as extra over the ducting in which they occur. Where there is a preponderance of fittings (eg in plant rooms), they shall each be enumerated separately as individual items. Cutting and jointing ducting to the fittings and forming openings in ducting for branches shall be deemed to be included with the ducting.
Ducting-supports	**S30**	**(a)** Particulars of the method of fixing shall be given in the description of ducting supports as Clause S1(g) hereof. Cutting and pinning ends of ducting supports shall be given as Clause S117 hereof. **(b)** Components for supporting ducting (eg brackets; hangers) shall be enumerated stating the size of the component and the size of the ducting. Spring-compensated components shall be so described stating the loading and the movement to be accommodated.
Dampers, ducting-turns, access-doors and openings	**S31**	Manually-operated regulating dampers and louvres in ducts, ducting-turns and mid-feathers, test-holes and covers, access-doors and openings in ducting (other than for branches) shall each be enumerated separately as extra over the ducting in which they occur. Forming and stiffening openings in ducting shall be given in the description.
Shutters, grilles, diffusers and equalisers	**S32**	Louvre-and-butterfly back-draught shutters, grilles, diffusers, diffusers with special dampers, deflectors, equalisers and anti-smudge rings shall each be enumerated separately stating the method of jointing. Forming and stiffening openings in ducting shall be given in the description.
Cowls and terminals	**S33**	**(a)** Cowls, terminals and the like shall each be enumerated separately stating the size and type thereof, the size of ducting to which it is attached and the method of jointing thereto. **(b)** Flashing-plates, weathering-aprons, cravats and the like shall each be enumerated separately stating the size thereof and the size of the ducting passing through.
Roof-ventilators	**S34**	**(a)** Roof-ventilators shall be enumerated stating the type, the rated output of air in cubic feet per minute, the temperature, the minimum static resistance, the maximum outlet velocity, the minimum efficiency and the number, type and size of the connections for ducts. Prime movers, drives and guards shall be given in the description stating the number, the type and the size. **(b)** Weathering-aprons, flashing-plates and the like where not forming an integral part of the unit shall each be enumerated separately stating the type and the size.
Fire dampers	**S35**	Fire dampers shall be enumerated stating the type and the size. Fusible links and integral operating-gear shall be given in the description.
Special dampers	**S36**	Damper units other than fire dampers shall be enumerated stating the type and the size. Operating-motors with linkage shall be given in the description.
Extract hoods	**S37**	Extract hoods shall be enumerated stating the type and the size.

FIG. 38. Ductwork
from pages 74 & 75 of the "Standard Method of Measurement of Building Works"; Fifth Edition (reproduced by permission of the Royal Institution of Chartered Surveyors).

pipes, fittings, etc., and the methods of installing and jointing.

Heating and air conditioning is similarly measured by separating systems, and with all *work* in boiler and equipment rooms measured separately. *Ventilation work* is measured by systems such as Supply Air, Return Air, General Exhaust, etc., as listed in **MM-CIQS**. Ducts are measured in linear feet, stating dimensions of cross-sections of ducts, types and gauges of materials, methods of jointing, and types of hangers to be used. The **MM-CIQS** also explains how elbows and branches in ducts shall be measured, in linear feet. The **SMM-RICS** has similar requirements; and it sets down in detail the features of equipment and other items that should be listed for accurate pricing. It is, therefore, a useful check list for the estimator who is pricing his own measured quantities, and for mechanical specification writers. The reproduction of the section on ductwork from the fifth edition of **SMM-RICS** in Fig. 38 indicates the kind of detailed information required for the accurate pricing of *mechanical work.*

Ductwork is still estimated by an average *unit price* per pound by some estimators. Yet it is practically impossible to properly take into account the variations from job to job in duct sizes and in the numbers of duct fittings when estimating *labor costs* from the total weight of ductwork. The best way to estimate ductwork is to measure it like piping, stating the weight per linear foot for each size and type of duct and the size and weight of each enumerated duct fitting. In this way, all the data is retained in the estimate to facilitate both accurate pricing and *cost accounting*, both of which go hand in hand.

Mechanical thermal insulation for piping is measured the same way as piping, in linear feet with fittings enumerated. Thermal insulation for ductwork should be measured as a *super* item, in square feet, according to **MM-CIQS**, stating type, thickness, and finish. But, again, the insulation could be measured more simply, and the *work* priced more accurately, if insulation to ductwork were to be measured the same as the ductwork—in linear feet (stating the dimensions), with fittings enumerated. *Material costs* for linear items can readily be estimated because they are usually directly related to the cross-section's dimensions. *Labor costs* often do not reflect this direct relationship to a cross-section's dimensions, and different *unit prices* must be established for different sizes. This requires separation according to size in the estimate and in the subsequent *cost accounting.*

To facilitate both the measurement and *cost accounting* of mechanical systems, most *mechanical work* (including insulation) can be measured along the following lines:

1. *The Linear Systems*
   a. vertical stacks
   b. horizontal mains
   c. branches to terminal units (from stacks and from mains)
2. *The Terminal Units*
   a. plumbing and other fixtures
   b. equipment (plumbing, heating, cooling, etc.)
   c. other terminal units connected to a linear system (sprinkler heads, registers, grilles, etc.).

Because these parts elements are basic to most mechanical systems, most systems can be broken down into these two major divisions: the terminal units and the linear systems to and from them.

The linear systems (of piping and ductwork) include the fittings, such as elbows and tees, and hangers, brackets, and other related items. By subdividing the linear systems into stacks, mains, and branches, measurement often can be shortened and simplified, and pricing accuracy can be increased because cost data can be obtained and allocated to these subdivisions of *work.* It is not usually feasible to obtain *labor costs* for individual fittings and lengths of pipe by normal *cost accounting* methods. However, the linear systems should be measured according to the standard methods described above, with all fittings, hangers, brackets, and the like, described and enumerated so that they can all be individually and separately priced according to the system described in Chapter 11.

The terminal units (connected to the linear systems) include all those items that are not part of the linear systems that serve them, and thus they usually identify the linear systems connected to them.

Both linear systems and terminal units in such locations as mechanical rooms and boiler rooms, and in any other locations such as would affect the *costs of the work,* should be so identified and measured separately so that appropriate allowances can be made in the pricing for any additional *labor costs* required owing to space restrictions and complexities of the *work.*

Methods of Measuring Electrical Items (Division 16)

Generally, that which has been said about mechanical items also applies to electrical items, particularly with regard to general requirements and to the measurement of linear systems and terminal units. The

MM-CIQS lists some forty sections of *electrical work* as a guide for the estimator, and the **SMM-RICS** presents a very detailed set of measurement rules.

Electrical work, as with *mechanical work,* is not difficult to measure for the estimator who has a good knowledge of these installations. In both instances, measurement includes making a complete bill of materials, preferably under appropriate headings, as indicated, so that the various parts of the *work* can be identified and priced according to the circumstances and the data available. But unlike some other kinds of *construction work, mechanical* and *electrical work* are not always illustrated and described in the *bidding documents* in sufficient detail for an estimator to be able to simply measure what is shown. In the first place, it is not practically possible for the *designer* to indicate every junction box and pipe fitting on the drawings, or every conduit and pipe line in its precise location in the building; and linear systems often are shown diagramatically and only the terminal units are located. In the second place, mechanical and electrical installations are subject to government regulations and official inspection, and sometimes the *bidding documents* depend heavily on codes and regulations to state the requirements for the *work.* As a result, many details of *mechanical* and *electrical work* are determined by the estimator, who measures what he believes the installations will require rather than what is explicit in the *bidding documents.*

Measurement for Pricing

In measuring *construction work* the purpose of measurement must be remembered: to obtain information and quantities so that the *costs of the work* can be estimated. So much measurement (even that done according to published methods) is not always entirely appropriate and rational, often because it is done by traditional methods that are no longer relevant, because methods of construction have changed and because construction costs are no longer constituted as they were in the past.

Measurement should be simpler, but much more detailed. And this becomes feasible with the use of EDP and computers. It is becoming more usual to measure "pieces" by enumeration, such as measuring structural steelwork with each piece fully described and identified instead of measuring the total gross weight of steel members of all kinds. But this simpler measurement requires that each piece shall be fully described, with all its dimensions, its weight, and its relationships to the other pieces, or parts, included in its description. And more complex and detailed descriptions and the use of computers in turn require estimating languages consisting of new symbols for brevity and precision to be devised and used by estimators.

Questions and Topics for Discussion

1. Present an argument for measuring trench excavations in linear yards rather than in cubic yards.

2. Explain in detail why it is usually necessary to measure certain *super items of work* in conjunction with *cube* items of excavating, giving reasons based on both mensuration principles and construction practices.

3. Explain in detail why concrete formwork is measured by its "contact area," with particular reference to pricing.

4. Describe, with the aid of a sketch, how you would proceed in measuring concrete in a panjoist slab.

5. Explain the two usual methods of measuring unit masonry, and give the advantages and disadvantages of both.

6. Recognizing the fact that unit masonry will cost more if it has more corners, how might an estimator allow for the additional *labor costs* arising from corners in a building?

7. Explain the two methods of measuring structural steelwork given in **MM-CIQS,** and present an argument for one method in preference to the other.

8. State four primary characteristics that should determine the classification and grouping of structural steelwork.

9. Explain why it is better to measure wood strapping as a *run* item, instead of in board feet; and explain the use of another unit of measurement for this item that would be preferable to board feet.

10. What are "narrow widths" of *finish work,* and explain how and why they should be taken into account in an estimate.

9

Measurement Examples and Exercises

The examples given in this chapter are based on three sets of drawings contained in this text; namely, the: 1. Preliminary drawings, 2. Residence drawings, and 3. Warehouse drawings. There is also a fourth set, the Apartment Block drawings, from which no measurement examples have been taken, but which is included for measurement exercises, as explained below.

These sets of drawings were selected and arranged so that the examples taken from them are progressively more complicated, and so that there is a wide variety of *construction work* included. In addition, measurement exercises may be taken from the second, third, and fourth sets of drawings; and if the examples from earlier drawings have been studied and understood, exercises from the later drawings should not be too difficult.

Each one of the Preliminary Examples is taken from a Preliminary

Drawing, and each of these examples is complete in itself in that it includes all the *work* illustrated. However, the Residence Examples embrace only part of the total *work* shown in the Residence Drawings; and it is the same with the Warehouse Examples. This means that the remaining *work* in the Residence and Warehouse not already measured in examples can be measured as measurement exercises, and examples of most of the types of *work* in these exercises can be found in earlier measurement examples. No examples have been taken from the Apartment Block Drawings, and all the *work* shown in those drawings may be measured in exercises. All the various types of *work* in the Apartment Block have been measured in previous examples taken from the other three sets of drawings to prepare for these exercises.

For example, the measurement of cast-in-place concrete and formwork is shown in all the Preliminary Examples, and steel rebar is included in some of them. The *concrete work* in each of the Residence, the Warehouse, and the Apartment Block can be measured in exercises; and each one is increasingly complex. Similarly, some wood framing is measured in a Preliminary Example followed by a larger framing example taken from the Residence; and the framing in the Apartment Block can be measured as a major exercise after studying the examples of framing measurement.

Teaching experience shows that one of the best ways to learn and to practice the measurement of *work* is to first study measurement examples, then to measure the *work* without reference to the example, and finally, to compare the measurement exercise done with the measurement example. Studying the examples without actually measuring the *work* is not very fruitful.

Most *work* can be measured in more than one way, and the methods used in the examples should be compared with the other methods available. All the measurement examples should be reworked by the student, and all examples and exercises should be priced using local and current *unit prices*. In this way, two things should be accomplished. The significance of the methods of measurement and the degrees of accuracy used should become clearer, and a familiarity with representative *costs of work* will develop.

With a growing knowledge and understanding of estimating and *cost accounting* precepts and practices, there will come also an understanding of the limitations of learning practical estimating from a book. The hypothetical construction job illustrated and used in an example or an exercise is always limited by an absence of some of the facts. The Preliminary Examples have been contrived to show certain estimating methods and techniques, some of which cannot be properly demonstrated to show their full value to the estimator simply because they are short and limited examples such as are necessary in a textbook. For example, the technique of measuring the *work* in cast-in-place concrete walls (whereby the wall's thickness is indicated in the item's description and is finally applied to the total wall area to obtain the volume of concrete, whereas the total wall area is also used to measure the forms) is particularly useful with large projects for which there will probably be many sets of concrete wall dimensions to be entered in the estimate. In estimates for large projects this minor technique can save a considerable amount of time and effort; but this fact may not be immediately apparent from a brief example in which the technique has been used and shown.

However, examples and exercises are essential to learning practical *work* measurement, and the examples given can provide an almost unlimited supply of practice exercises if the dimensions are changed. At the end of the commentary on some of the examples given below, alternative dimensions are provided so that the examples can be reworked with different dimensions to measure different quantities of *work*. Since it is impossible to present complete information and data for hypothetical examples of *work* and its measurement, a few general observations and assumptions are made here.

Excavation and fill (and other *site work*), more than most other types of *work*, require specific information from the site for estimating the *costs of work*. The type of ground, the soil conditions, the types of local fill material available, the locations for the disposal of spoil—these are a few of the critical facts that must be known, and that cannot be provided for each example and exercise. Even the method of measuring excavation is affected by such things; and in those examples that include such *work*, more or less normal conditions have been assumed—if such exist. And unless otherwise indicated, the site may be assumed to be flat, and the soil and ground conditions ordinary.

Concrete is simply specified for strength and maximum aggregate size, and these details are given in the examples.

Formwork presents no measurement problems, because only the "contact area" is measured and most of the problems occur in the pricing. Formwork is required at "construction joints" (joints required by stopping and starting concrete pouring) where continuous concrete placing is not possible, as is usual on large jobs. Because small jobs are shown in the examples, generally no such forms have been measured; but they should be measured in the exercises.

Steel rebar has many possible variations in design and quality, and consequently there can be as many variations in its measurement. In all the examples, the following is assumed to apply, unless otherwise indicated:

Grade of steel . . . . #60
Maximum length of bars . . . . 30 ft
Length of splices (laps) . . . . 36 × bar dia
Concrete cover to steel, generally . . . . 1½ in.
Concrete cover in concrete footings . . . . . . 3 in.

These allowances are sometimes increased or decreased by small amounts to "round-off" a dimension in some measurement examples.

For large projects it is necessary to measure all the carrying bars, chairs, and other rebar accessories required. But in smaller and simpler projects, few special accessories may be required, and the minor costs of any such items may be covered by a small allowance in the estimate.

Bending of rebar is described as "light" (L. Bending) or "heavy" (H. Bending), as explained in Chapter 8. Bending should be measured as an *extra over item;* i.e., the extra cost of bending over and above the costs of supplying and installing the rebar.

Finally, experience shows that everyone will estimate a project differently, because estimating deals in probabilities not only of costs and productivity, but also of work methods and procedures. There is really only one ultimate test: How close is the estimate to the actual costs? All the facts are never available to an estimator, for if they were it would not be a question of probabilities but of certainties; and therefore estimating is to some extent always a matter of opinions and differences. But experience also shows that if a project (such as those in the following examples) is measured and priced according to the precepts and practices of good estimating, and with the same basic cost data, the results of any number of estimates will usually be sufficiently close to assure those who made them that their results are generally satisfactory, with possibly one or two exceptions. And in most cases the exceptions will result either from an obvious mistake or from a very different point of view concerning how the *work* might be done, which might be quite valid.

Preliminary Examples Generally

The Preliminary Drawings, unlike the others, are not taken from actual construction jobs. Instead, they were specially contrived for instruction and to illustrate certain precepts and practices of estimating. They do not necessarily represent good design and construction practice; and, in a few instances, unusual designs and drawing practices have been used to better illustrate a point or to facilitate an estimating practice. These are indicated and explained in the commentaries.

For brevity's sake, the commentaries do not usually repeat explanations, and consequently, the earlier commentaries contain more explanations and are longer than the later ones. These first estimating examples are simple and brief, but they do illustrate measurement techniques that are more effective when used for larger and more complex projects. It will help to understand why certain things are done if this is remembered. It is necessary to use simple examples to illustrate estimating techniques in a text, even though the techniques are most useful when applied to more complex cases. Therefore, the first impression on looking at some of these examples may be that they contain too much information in the form of explanatory calculations and notes. However, we are first concerned with good estimating practices and their general application. Besides, rules cannot be broken until they have been learned.

Preliminary Example

Foundations

Estimating Forms. There are many types of forms available. The "General Estimate" forms used here are ideal for small jobs with only a few sets of dims to be entered against each item in the estimate. For larger jobs, in which items may have many sets of dims, the "Quantity Sheets" used in later examples are preferable; because with only one or two items per sheet, "pricing columns" cannot be fully utilized and it is better to use "Recapitulation Sheets" to which the "total quantities" of the *items of work* are transferred from the "Quantity Sheets."[1] All of these estimating forms, and many others, may be obtained from the Frank R. Walker Company, 5030 N. Harlem Avenue, Chicago, Illinois, 60656., whose estimating publications are listed in the Bibliography. Similar forms are also published by the Robert Snow Means Company, Inc., whose annual publication, *Building Construction Cost Data,* is also listed in the Bibliography.

Each page of an estimate should carry the job title

[1] Many of the examples in this chapter have been done on Quantity Sheets in order to make the slightly reduced reproductions easier to read. However, in practice, General Estimate Sheets would be more suitable because the quantities can be measured and priced on the same sheets.

PRACTICAL
Form 514 MFD. IN U. S. A.

GENERAL ESTIMATE

BUILDING PRELIMINARY
LOCATION EXAMPLE
ARCHITECTS
SUBJECT FOUNDATIONS

ESTIMATE NO. P-1
SHEET NO. 1 of 2
ESTIMATOR KC
CHECKER AB
DATE 1975

FRANK R. WALKER CO., PUBLISHERS, CHICAGO

DESCRIPTION OF WORK	NO. PIECES	DIMENSIONS	EXTENSIONS	EXTENSIONS	TOTAL ESTIMATED QUANTITY	UNIT PRICE M'T'L	TOTAL ESTIMATED MATERIAL COST	UNIT PRICE LABOR	TOTAL ESTIMATE LABOR COST
prelim. calcs.		55-0	slab dims -			recess width - 10-0			
3/15-0 = 45-0		30-0		55-0 x 30-0				add 2/2 = 2-8	
10-0		4-0	less 2/3/8" =	-2-8 -2-8				12-8	
55-0	2/89-0 =	178-0		52-4 x 27-4					
	less 4/2/1-0 =	-8-0							
	MP =	170-0							
CONCRETE (2500 psi - 3/4" stone)									
R. mix Conc @ $21.00 cy delv'd.									
Ftgs, cont		170-0 x 2-0 x 1-0	340		13 cy	22.00	286	7.00	91
Fnd Walls x 8"t		170-0 x 4-0	680 (wall area)						
		cube	x 0-8 =	453	17 cy	22.00	374	9.00	153
Slab o/grd x 4"t		52-4 x 27-4	1430	DDT					
DDT (wants)		15-0 x 10-0	-	150					
		12-8 x 4-0	-	51					
			-201						
			1229 (slab area)						
		cube	x 0-4 =	407	15 cy	21.00	315	8.00	120
FORMS (ERECT & STRIP)									
Form Ftgs, cont	2/170-0 x	1-0	340		340 SFCA	25¢	85	75¢	255
Form Fnd Walls	2/680 SF (wall area)				1360 SFCA	25¢	340	70¢	952
Slab Screeds x 4"	3/27-4		82						
30-0 20-0 / 2/2 = -2-8 -2-8	2/23-4		47						
27-4 17-4	2/17-4		35		164 LF	7¢	12	15¢	25
27-4 - 4-0 = 23-4									
Form Key, 2x4	1/170-0		170		170 LF	10¢	17	20¢	34
					TOTALS (1)	$	1,429	$	1,630

and other information as indicated by the printed headings, because later confusion can be disastrous (when, perhaps, the estimate contains fifty or a hundred pages and some parts have been estimated twice to compare costs).

Entries on estimating forms should be neat and clear and free from superfluity. If a pen is used, mistakes must be crossed out and a new entry made. Generally, it is better to use a pencil, preferably a propelling pencil and an HB lead to ensure a reasonably consistent dark line.

Dims in descriptions are written as they appear on the dwgs,[2] but dims in the dim cols are written as decimals or feet-inches but without the diacritical signs for feet and inches. All entries are in feet (and fractions of a foot) and need no further identification except the total estimated quantities that are identified (for pricing) as *cube, super, run,* or *number* items. Otherwise, use letter abbreviations for units of measurement elsewhere in the estimate (i.e., SF, CY) only if there is a risk of confusion.

Prelim Calcs. Check the dims on each side of the plans to eliminate any errors made by the draftsman. Calculate MP (mean perim) of conc ftgs and fnd walls by first finding the "outside perim" from dims shown. (Wall dims are usually given, but not dims for conc ftgs. Here the conc ftgs are dimensioned to simplify and clarify the use of the MP.) The MP is calculated by the rule given in Chapter 7; that is, by deducting 4 times the conc ftg width from the outside perim, which is the same as applying the rule for the adjustment of perimeters; i.e., deduct (or add) 4 × 2 × the horizontal distance between the perimeters, which in this case is 4 × 2 × (½ × 2 ft), which equals 4 times the conc ftg's width of 2 ft.

Since the conc slab dims are not shown in the dwg, they are calculated together with the length of the *want* from the slab (its width is 4 ft, as shown).

Any calcs can be performed at the outset in this way if the estimator can anticipate needing them; and, in particular, dims that have a general application and dims that require several calculations are better obtained by prelim calcs. The cols of the estimating form can be ignored when it is convenient to do so, such as for prelim calcs.

Concrete. This heading is used so that conc items can be grouped beneath it, and their descriptions can be shortened. Grouping similar *items of work* also facilitates pricing.

2 The abbreviations used in this chapter are in the List of Abbreviations preceding Chapter 1. Their use in these commentaries reflects their use in the estimating examples.

Ftgs, cont. The volume of the continuous conc ftgs is found from the length (MP) × width × thickness, as shown. The measured quantity is the net actual quantity required and shown on the dwgs. Any allowance required for *waste* (from spillage, enlarged formwork, etc.) will be made in the *unit price* for materials. The measured quantity of 340 CF is divided by 27, and the result is taken up to the nearest cubic yard.

Fnd Walls. Such walls are described as 8 in. thick so that the total wall area can first be calculated for future use. This practice also reduces the amount of calculation required if there are many sets of dims in an estimate.

Slab on Grd. As for walls, the *super* quantity is found first, and the *cube* quantity, later. The *wants* are deducted (DDT) by totaling all deductions in the second extension column to the right of the dims under a heading "DDT." Some estimators use "Omit."

Forms. Again, a heading is used over a group of like items. Only the "contact area" (area of conc surfaces in contact with form faces) is measured, and allowances for the use and *waste* of materials are made in the *unit prices. Labor costs* are the major factor in pricing formwork, and it is not necessary to measure in detail such items as sheeting, support lumber, ties, and form hardware, as explained in Chapter 8.

Most formwork dims can be taken directly from the dims for conc, and for this reason many estimators measure each item of formwork immediately following the appropriate item of *conc work.* This practice may save a little time, but this minor advantage may be lost during the pricing. It is easier to price like items together and not have to alternate from *conc work* to *formwork.*

Slab Screeds. Slab screeds are not really formwork, but they are similar in use and costs because they are temporary and, like formwork, are placed before conc is placed. Conc slabs are "struck off" level with the tops of the screeds installed for this purpose by a straight-edged board, or by a length of steel pipe. Screeds for conc slabs on grd may be lengths of dimension lumber supported on wooden pegs driven into the fill under the conc slab, and spaced about 10 ft apart. Alternatively, the screeds may be lengths of steel pipe temporarily set down on pegs or chairs that have been fixed and leveled. The screeds are moved as required, and the pegs or chairs are pulled out or left embedded in the slab. Their locations must be assumed by the estimator. The *material costs* are minor, and most of the costs involve labor for installing and leveling the screeds. Some estimators prefer to relate the cost of

PRACTICAL
"STANDARDIZED" FORMS FOR CONTRACTORS
Form 514 MFD. IN U.S.A.

GENERAL ESTIMATE

BUILDING PRELIMINARY
LOCATION EXAMPLE
ARCHITECTS
SUBJECT FOUNDATIONS

ESTIMATE NO. P-1
SHEET NO. 2 of 2
ESTIMATOR KC
CHECKER AB
DATE 1975

FRANK R. WALKER CO., PUBLISHERS, CHICAGO

DESCRIPTION OF WORK	NO. PIECES	DIMENSIONS	EXTENSIONS	EXTENSIONS	TOTAL ESTIMATED QUANTITY	UNIT PRICE M'T'L	TOTAL ESTIMATED MATERIAL COST	UNIT PRICE LABOR	TOTAL ESTIMATE LABOR COST
REBAR (GRADE #60)									
#4 in Ftgs, cont	4/	178-0 (str)	712						
(36/½" = 1-6) (laps)	4/	1-6 (")	6						
			718	x 0.668 lb/ft =	480 LB	20¢	96	7¢	34
CONCRETE SUNDRIES									
Steel Trowel Slab		(slab area =)	1229		1230 SF	1¢	12	11¢	135
Cure Slab		(slab area =)	1229		1230 SF	1¢	12	2¢	25
Wpf Membrane u/slab (incl. laps)									
		(slab area =)	1229						
ADD (for laps & edges)		15% of 1229 SF	185						
			1414		1420 SF	7½¢	107	1½¢	21
Gravel Bed u/slab x 5"t (allow 20% shrinkage)									
Clear Gravel $4.00 cy delv'd		(slab area =)	1229						
DDT (wante Ftgs)	1/	164-8 x 0-8	−110						
			1119						
		cube	x 0-5	466	17 CY	5.00	85	5.00	85
				sub-totals (2)			312		300
				(1)			1,429		1,630
				TOTALS			$ 1,741		$ 1,930

4" CONC SLAB
5" GRAVEL BED
DEDUCTION OF GRAVEL BED x 8" WIDE DISPLACED BY FTGS
℄ ℄
8" (from ℄ - ℄)

this *item of work* to the slab area and do not measure the screed lengths.

Edge forms are required for some conc slabs, both at the slab's perim and at intermediate construction joints in the case of large conc slabs that are poured and placed in alternate panels. These edge forms (which are sometimes of steel shaped to form a tongued and grooved joint) are different from slab screeds, which are solely for leveling.

Control joints in which vertical strips of joint filling material are installed at a slab's perim, and sometimes at intermediate lines in the slab to enable the slab to move and to control the cracks that usually occur in a slab soon after placing, are also different from both screeds and edge forms. Each should be measured separately. Sometimes, the control joint strips at a slab's perim can be installed and leveled and used as a screed. This technique is shown in a later example.

Form Key. This item does not require a deduction for conc ftgs, as it is filled with conc when the fnd walls above are placed. The *labor costs* (as with all formwork) are the largest element in pricing keys; the size of the key does not affect its cost very much. It is often a 2 × 4, beveled to facilitate removal.

Rebar. The bars are measured at the outside perim of the conc ftgs to allow for crossing at corners. (The outside perim has not been reduced here by the few inches of conc cover required over the ends of the bars.) *Laps* have been measured for the one ftg length in excess of 30 ft, according to the assumptions about rebar stated above. The total length is multiplied by the weight per foot, and the total quantity is taken up to the nearest 10 lb. No bending is required or measured.

Conc Sundries. Conc sundries includes all other *items of work* related to cast-in-place conc, including curing and finishing. Some estimators also include here such items as gravel beds, whereas others include them with excav and fill.

Curing conc slabs is an *item of work*, and as such it is often measured (and cost accounted) separately.

Wpf Membrane under Slab. This item includes about 10 percent for laps at edges (4 in. for every 36 in. width), and about 5 percent for turned-up edges at the conc slab's perim. Alternatively, the simpler method (recommended by **MM-CIQS**) would be to measure it net, using the slab area, with an allowance for the *laps in the unit price*. However, the necessary allowances will vary slightly from job to job. In a larger job this item would be measured elsewhere, such as under Moisture Protection.

Gravel Bed. The deduction here is for that part displaced by the inner projection of the conc ftgs, the length of which is calculated from the MP (170.0) less 4 × 2 × the horizontal distance from the perim (center) line of the projection to the MP line (according to the rule given). The estimator makes an allowance in the *unit price* (and in ordering the gravel) for *shrinkage* owing to consolidation; and the quantities are measured net, as explained before.

Some estimators measure a separate *super* item for the compacting of gravel beds and measure the gravel in CY because the cost of compaction relates more directly to the area of a bed rather than to its volume (i.e., a 6-in. thick bed would not require 20 percent more compaction than a 5-in. thick gravel bed). Similarly, finishing a conc slab is measured separately from the conc placing.

This first example has been priced to show the

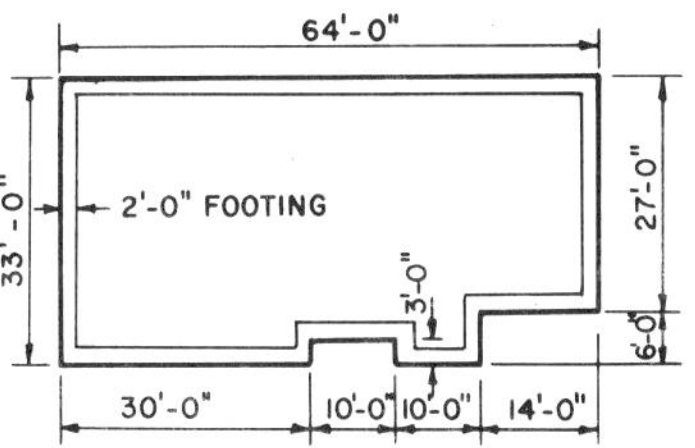

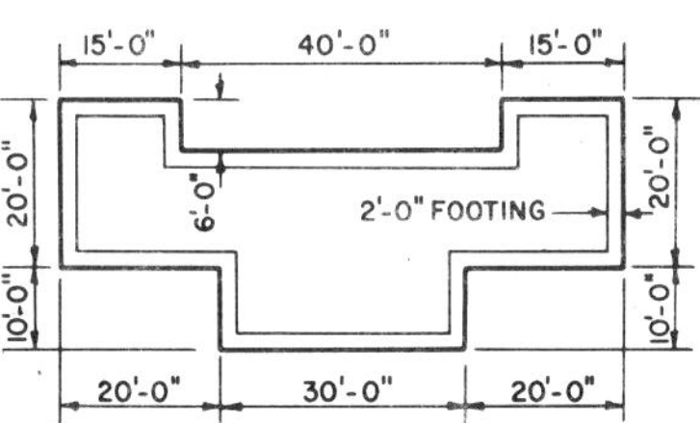

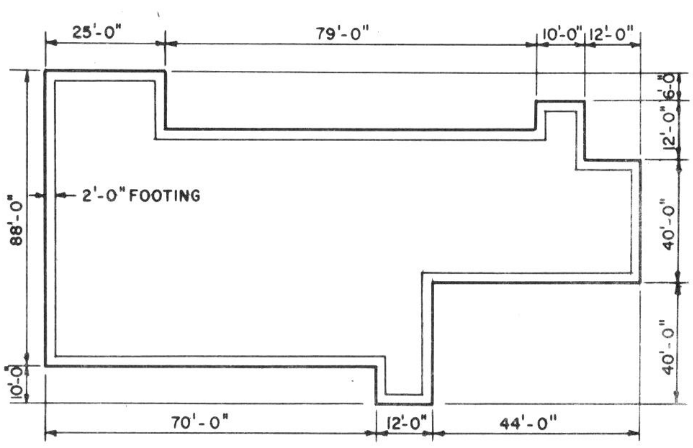

complete method and use of the estimating form. No *equipment costs* and no *overhead costs and profit* have been added. Each page is totaled and a summary of page totals is made so that any errors are not carried from page to page.

Suggested outline plans with dims for measurement exercises based on this example are shown on page 121. As in the example, dims are to outside faces of footings.

Preliminary Example

Foundations and Framing

Slab on Grd. The dims on the dwg are used, and the over-measurement is deducted as a *want*; i.e., a strip 16 in. wide, the length of which is calculated from the MP. Alternatively, the slab can be measured as in the last example by first calculating its actual dims.

Slab Screeds. Slab screeds are measured here by the slab area, unlike in the previous example. The better method is the one that yields the most valid *unit price*. A *run* item is better for pricing the *material costs*. But the *labor costs* are greater, and only cost data will show the better method of measuring for labor. If screeds are usually spaced about the same distance apart, there will be a relationship between *run* and *super* quantities; and the *super* item is certainly easier to measure. Thus, the estimator must decide how to measure items to ensure accurate pricing and *cost accounting*.

Exp Joint and Seal. This joint is really a "separation joint" and is commonly misnamed because the filler material used is commonly referred to as "expansion joint" material. This type of joint is not usually found in residential construction, but any such *work* at a slab perim would be measured in the same way. The *unit price* should include both the joint filler strip and the sealing compound poured into place at the top edge. The space for the compound may be formed by a ½ in. × ½ in. removable wood strip fixed over the top edge of the filler strip, and this strip may then be used as a slab screed at the perim.

Dpf Fnds. Dampproofing the conc fnd walls and ftgs is described as requiring 1 gallon of asphalt per coat per Square (Sq) (of 100 sq ft) to enable the item to be priced.

Wpf Membrane. This time the turned-up edges are measured, since the perim dim is readily available; and only the *laps* are added as a percentage. By slightly varying the methods of measurement, the validity of allowances for *laps* and the like can be periodically checked and confirmed, and revised if necessary. Too often, estimators look for, find, and employ forever and without question allowances and formulas and "rules of thumb." Alternatively, the membrane could be measured net, as explained before.

Framing. This heading is used, as before, to group like items and to shorten their descriptions.

Stud Walls. The 2 × 4[3] lumber is measured in LF and converted to BF. Some estimators leave the quantity in LF, which is simpler and adequate. It depends on local custom. The single 2 × 4 sill at the bottom and the double 2 × 4 plates at the top are often measured together with the studs (measured as 4 ft. because probably two will be cut from one 8-ft stud). The plates and studs combined are measured here as one *item of work*, but there is additional labor required between the conc fnd walls and the bottom 2 × 4 sill. This is measured here as an *extra over item* to allow for the *labor costs* of squaring and leveling. Alternatively, the sill (wall plate) and the studs can be measured separately (as required by **MM-CIQS**), with short studs separated from longer studs; this procedure should be followed for a larger project.

The number of studs (at 16 in. spacing) is calculated by dividing the total length of the walls by the spacing, to which are added three additional studs for each of the fourteen corners. If the length of a wall is not a multiple of the spacing, an extra stud is required; so, on this basis, all three studs at each corner are added, assuming that none has already been allowed for at the corners by dividing the wall length by the stud spacing. This procedure may cause a slight over-measurement if some wall lengths are in fact multiples of the spacing; but usually in framing some extra studs are required. It should be remembered that specified spacings are always the maximum spacings permitted, and that closer spacing is permitted and, indeed, is frequently required by the *work*. Additional studs (and lintels) would also be required for any openings.

A common "rule of thumb" in measuring residential –type framing is to "allow one stud per foot of wall–

[3] This is the nominal size of the lumber—2 in. by 4 in. The actual size after planing used to be 1 5/8 in. by 3 5/8 in. Now the standard actual size is 1 1/2 in. by 3 1/2 in. when surfaced (by planer) and dried. The standard sizes were revised in 1971.

PRACTICAL "STANDARDIZED" FORMS FOR CONTRACTORS
Form 514 MFD. IN U. S. A.

GENERAL ESTIMATE

BUILDING PRELIMINARY
LOCATION EXAMPLE
ARCHITECTS
SUBJECT FOUNDATIONS & FRAMING

ESTIMATE NO. P-2
SHEET NO. 1 of 4
ESTIMATOR KC
CHECKER AB
DATE 1972

FRANK R. WALKER CO., PUBLISHERS, CHICAGO

DESCRIPTION OF WORK	NO. PIECES	DIMENSIONS	EXTENSIONS	EXTENSIONS	TOTAL ESTIMATED QUANTITY	UNIT PRICE M'T'L	TOTAL ESTIMATED MATERIAL COST	UNIT PRICE LABOR	TOTAL ESTIMATE LABOR COST
prelim calcs	28-0 5-0 4-0 37-0	3/ 33-0 4-0 37-0	24-0 2-0 26-0	7-0 6-0 13-0 26-0	37-0 26-0 2-0 2/65-0 = 130-0 less: 4/2-0 = -8-0 MP = 122-0		Fnd Wall Ht — 3-4 + 0-8 4-0		
CONCRETE (2500 psi - 3/4" stone unless o/wise indicated)									
Ftgs cont (1½" stone)		122-0 x 2-0 x 0-9	183		7 CY				
Fnd Walls x 8"		122-0 x 4-0	488 (wall area) x 0-8 = 325		12 CY				
Slab o/grd x 4"		37-0 x 26-0	962	DDT					
DDT (wants)		5-0 x 7-0	—	35					
		4-0 x 13-0	—	52					
		11-0 x 2-0	—	22					
2/8" = 1-4		4-0 x 2-0	—	8					
(122-0 + 4/2/4") = 124-8		124-8 x 1-4	—	166					
			-283						
		(slab area) =	679						
			x 0-4 = 226		9 CY				
FORMS (erect and Strip)									
Form Ftgs		2/122-0 x 0-9			183 SFCA				
Form Fnd Walls		2/488 sf (wall area)			976 SFCA				
Form Small Rebate		1/124-8			125 LF				
Slab Screeds x 4"		(slab area) =			680 SF				

length, for studs at 16 inches over centers." This rule is fairly accurate for ordinary house design and construction, in which the houses do not have more than six corners. The studs measured in this way provide for additional studs at corners and at openings, but the rule cannot be applied to other frame construction. Such rules are useful if properly applied. The trouble is that novices think that these rules are among the infallible secrets of the estimating trade, and they abuse them. The only valuable rules are those rules found and proven by the estimator who uses them. Estimating has many gimmicks and short cuts to success, and none of them is worth much. Only those rules and precepts that have proven to be consistently valid within a certain set of conditions should be used. It is a matter of testing and proving by *cost accounting.*

Shiplap Shthg. This sheathing is measured net over the area to be covered, with an allowance made for *laps* and for the loss in width in dressing the board when measuring in BF. The nominal width of the boards is 8 in.; but the actual overall width is 7½ in., which includes a ⅜-in. rebate to make the *lap.* This leaves an effective width of 7⅛ in. out of the original 8 in. on which the board measure is based. The difference between 7⅛ in. and 8 in. is about 12½ percent, which is the amount allowed in the measurement for *laps* in converting to a board measure. For a shiplap board 6 in. wide, the allowance should be about 16 percent. An allowance for *waste* of about 5 to 10 percent should be made for end-cutting in the *unit price.* It is simpler, however, to measure sheathing net in SF (according to **MM-CIQS**) and to add for both *laps* and *waste* in the *unit price.*

Excav and Drains. The tendency is to measure this *work* first, because it is performed first at the site. But there are several advantages in measuring it last. The amount of excav for a building is determined in part by the sizes and depths of the conc fnds and ftgs, and these quantities are known better after the *conc work* has been measured. Sometimes an estimator cannot inspect a site before starting an estimate, and he has to wait for information from test holes and the like before he knows how the *excav work* should be measured. So he has to begin by measuring other *work.*

Remove Topsoil. Topsoil (also called loam and vegetable soil) is usually handled and measured separately from other soil because of its value in landscaping. The item in the example assumes that the topsoil is removed from over the building area (without the *wants* considered) and piled on the site for future use, part of which will be backfilled around the building later.

Machine Bulk Excav. Machine Bulk Excav is measured here down to the underside of the gravel bed, which is 3 in. above the bottom of the conc ftgs. The calculation of the depth (2.10) is shown. The area of the excav allows a working space 2 ft wide outside the conc fnd walls (actually 2 ft 2 in., in going to the nearest foot). The dwg shows the limit of the excav in the typical section, but this limit is not normally shown on dwgs. Only the major *wants* are deducted from the area of the excav. The area and method of excav would in fact depend on the soil and ground conditions. In hard ground a minimum amount would be taken out; in soft ground the soil at the re-entrant corners might collapse and have to be removed.

Hand Excav Trench Bottom. The additional depth of 3 in. under the conc ftgs and the drains is assumed to require removal by hand, to finish to an accurate level under the conc ftgs and to accurate grades under the drains. The width of 4 ft includes an allowance of 6 in. for the installation and removal of the inner forms to conc ftgs. The *hand work* under the conc ftgs and the drains would be done at different times, and might be measured as separate items. The same applies to the next item, which is directly related; but as the quantity here is small, combining the similar *work* under the conc ftgs and the drains in this case will not reduce the accuracy of the estimate.

Trim Trench Bottom Level. Most *excav work* requires additional labor at the newly exposed surfaces, and the method of measurement (**MM-CIQS**) calls for:

> An item of trimming and grading [to be] measured in square feet under the following headings:
> (a) Trimming to bottoms of excavation.
> (b) Trimming to vertical face of excavation. (If required when excavated face is vertical.)
> (c) Trimming and grading to form sloping banks.
> (d) Trimming to all faces of rock shall be kept separate.

Trim under Gravel Bed. This item is similar to the preceding one, but is performed for a different reason and probably at a different cost. This item involves finishing off after the machine excavates to a rough grade so that the gravel bed will be of the proper thickness. Since the trimming will probably be done

PRACTICAL "STANDARDIZED" FORMS FOR CONTRACTORS
Form 514 MFD. IN U. S. A.

GENERAL ESTIMATE

BUILDING PRELIMINARY — ESTIMATE NO. P-2
LOCATION EXAMPLE — SHEET NO. 2 of 4
ARCHITECTS — ESTIMATOR KC
CHECKER AB
SUBJECT FOUNDATIONS & FRAMING — DATE 1972

FRANK R. WALKER CO., PUBLISHERS, CHICAGO

DESCRIPTION OF WORK	NO. PIECES	DIMENSIONS	EXTENSIONS	EXTENSIONS	TOTAL ESTIMATED QUANTITY	UNIT PRICE M'T'L	TOTAL ESTIMATED MATERIAL COST	UNIT PRICE LABOR	TOTAL ESTIMATE LABOR COST
CONCRETE SUNDRIES		4/2/4" =	122-0 −2-8 = 119-4						
Exp Joint & Seal (as detail/specs page –)					120 LF				
Steel Trowel Slab (slab area) =					680 SF				
Cure Slab (slab area) =					680 SF				
DpF Fnds wi 2 cts Asphalt (1 gall per ct/per Sq)									
124-8	1/	124-8 × 3-4	416						
130-0	1/	130-0 × 0-9	98						
2) 254-8 = 127-4	1/	127-4 × 0-8	85 =	599	600 SF				
Wpf Membrane u/Slab (slab area) =			679						
ADD (edges)	1/	120-0 × 0-4	40						
ADD (laps)		10% × 720 SF	72 =	791	800 SF				
FRAMING (D fir "Construction" grade)									
2x4 Stud Walls (4'0" high approx)									
124-0 / 1-4 = 93 (plates)	3/	124-0	372						
+14/3 = 135	135/	4-0	540						
			912 × 2/3 BF		600 BF				
Extra Lab to Wall Plates					124 LF				
1/2" φ Anchor Bolts set in Conc (124-0/5-0 + 7)					28 No				
1x8 Shiplap Shthg diagonal on Walls (allow 5% waste)									
4-4 +0-1	1/	124-8 × 4-5	551						
ADD (laps)		12½% × 551 SF	+ 69 × 1 BF		620 BF				

PRACTICAL "STANDARDIZED" FORMS FOR CONTRACTORS
Form 514 MFD. IN U. S. A.

GENERAL ESTIMATE

BUILDING: PRELIMINARY
LOCATION: EXAMPLE
ARCHITECTS:
SUBJECT: FOUNDATIONS & FRAMING

ESTIMATE NO.: P-2
SHEET NO.: 3 of 4
ESTIMATOR: KC
CHECKER: AB
DATE: 1972

FRANK R. WALKER CO., PUBLISHERS, CHICAGO

DESCRIPTION OF WORK	NO. PIECES	DIMENSIONS	EXTENSIONS	EXTENSIONS	TOTAL ESTIMATED QUANTITY	UNIT PRICE M'T'L	TOTAL ESTIMATED MATERIAL COST	UNIT PRICE LABOR	TOTAL ESTIMATED LABOR COST
EXCAV & DRAINS									
Remove Topsoil avg 12" dp & stockpile on site									
37-0 x 26-0									
2/1-6 = +3-0 +3-0									
40-0 x 29-0		40-0 x 29-0 x 1-0	1160		43 CY				
Machine Bulk Excav Bmt x 2'10" dp									
		40-0 x 29-0	1160	DDT					
DDT (wants)		5-0 x 7-0	–	35					
+3-4		4-0 x 13-0	–	52					
-1-0									
+0-9			−87						
-0-3		(excav area) =	1073						
2-10 dp			x 2-10 =	3040	113 CY				
Hand Excav Trench Bottom									
(1-6 + 2-0 + 0-6) = 4-0		122-0 x 4-0 x 0-3	122		5 CY				
Trim Trench Bottom Level									
		122-0 x 4-0	488		490 SF				
Trim under Gravel Bed									
(122-0 − 4/2/8") = 116-8		(slab area) =	679						
DDT (want @ perim)		116-8 x 0-8	−78 =	601	600 SF				
Gravel Bed w/slab x 6" (allow 20% shrinkage)									
122-0									
4/2/1-3 = −10-0		(trim area) =	601						
112-0			x 0-6						
			300						
ADD (extra @ inside ftgs)		112-0 x 0-6 x 0-3	+14 =	314	12 CY				

by hand, it might not be practical to always treat it as an item separate from hand excav, and the two might be combined in one item. Some estimators usually measure the bottom few inches of depth as "hand excav, including trimming." But in some ground the trimming may, in fact, involve compaction using a pneumatic tool. Once again, measurement of *excav work* depends on the ground and the soil conditions and the way in which the *work* is done; and again it is clear that measured quantities of *excav work* are of limited value to an estimator when it comes to pricing.

Gravel Bed. Additional gravel is measured against the inner face of the conc ftgs where the excav allowance for formwork was made. If *work* on the surface of the gravel bed is required to be measured separately, an item such as that shown might be measured.

Drain Tile. The length is found by using the rule for adjusting perimeters.

Extra (for) Bends. Drain tile is not deducted where bends are measured because the bends are measured as an *extra over item.* This is the usual way of measuring (and pricing) fittings in small-diameter pipes of all kinds.

Drainage Rock. This item (over drains) is usually washed and screened material containing no fine material that would block the drain and is different from and much more expensive than ordinary gravel fill, which is often used as dug straight from the pit or river bed.

Remove Surplus Excav Native Mat. The **MM-CIQS** says:

> All excavation items are deemed to include for dumping on site in spoil heaps. When all excavation and backfill quantities are known and equated the excess excavation shall constitute an item of disposal of spoil. If the backfill and grading exceed the excavation quantities, then an item of borrow fill shall be taken, the type of which shall be specified.

This procedure has not been followed here with the topsoil on the assumption that the balance will be used later for landscaping on the site.

Suggested outline plans with dims for measurement exercises based on this example are given in the next column above. As in the example, dims are to outside faces of footings in the first plan; but in the others, dims are to outside faces of conc fnd walls, which is the more common practice.

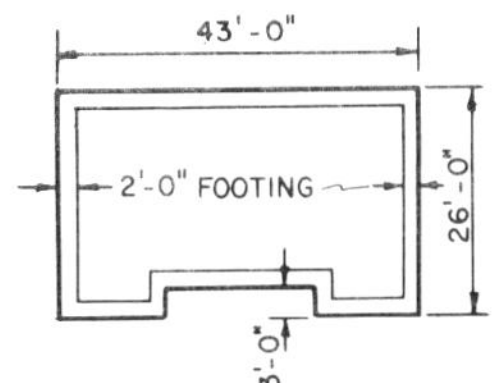

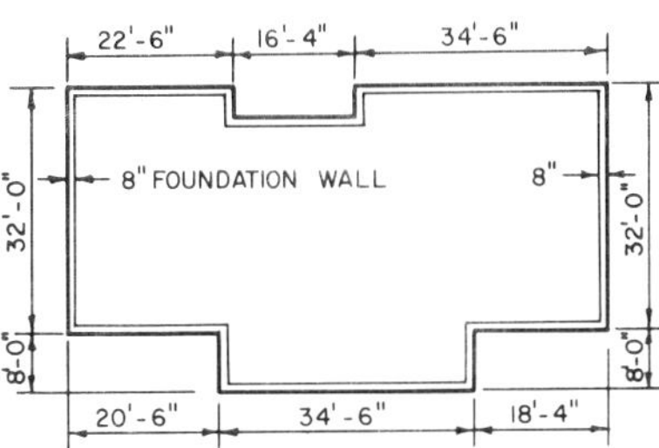

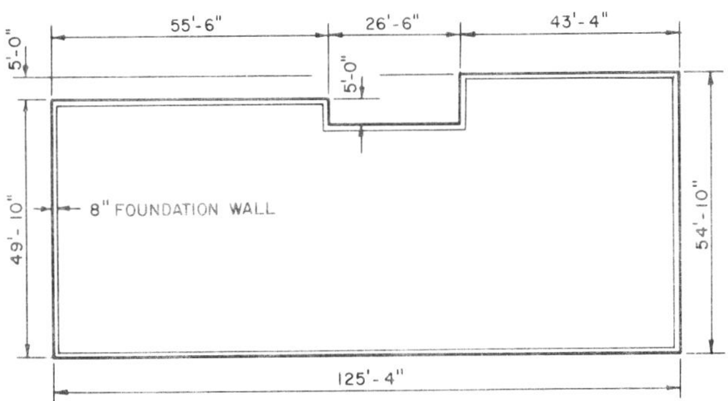

Preliminary Example

Swimming Pool

Prelim Calcs. These include calculating the MP of the pool walls, the avg wall ht's at the sides and at the two ends, and the lengths of the three sections of the bottom slab. These are each represented by the hypotenuse of a right-angle triangle, each with a 20-ft base and with varying perpendicular heights (H). The length (L) of the hypotenuse is calculated by $L^2 = 20^2 + H^2$ (to the nearest inch), and the avg length (L) of the three sections is found so that one set of slab dims can be "timesed" (multiplied) by three.

PRACTICAL "STANDARDIZED" FORMS FOR CONTRACTORS
Form 514 MFD. IN U. S. A.

GENERAL ESTIMATE

BUILDING PRELIMINARY
LOCATION EXAMPLE
ARCHITECTS
SUBJECT FOUNDATIONS & FRAMING

ESTIMATE NO. P-2
SHEET NO. 4 of 4
ESTIMATOR KC
CHECKER AB
DATE 1972

FRANK R. WALKER CO., PUBLISHERS, CHICAGO

DESCRIPTION OF WORK	NO. PIECES	DIMENSIONS	EXTENSIONS	EXTENSIONS	TOTAL ESTIMATED QUANTITY	UNIT PRICE M'T'L	TOTAL ESTIMATED MATERIAL COST	UNIT PRICE LABOR	TOTAL ESTIMATED LABOR COST
EXCAV & DRAINS (cont)									
4"ϕ Agric Drain Tile @ Ftgs									
(130-0 + 4/2/6")		1/134-0			134 LF				
Extra 4" Bends (90°)	14/1				14 No				
Drainage Rock over D. Tile (allow 15% shrinkage)									
		134-0 x 1-6 x 0-9	151						
		134-0 x 2-2 x 0-9	218						
			369		14 CY				
Backfill @ Ftgs (allow 20% shrinkage)									
(4-1 − 1-6 + 1-0) = 1-7		134-0 x 2-2 x 1-7	460		17 CY				
(1-6 + 0-8) = 2-2									
Backfill topsoil avg 12" dp (allow 25% shrinkage)									
		134-0 x 2-2	290						
		5-0 x 7-0	35						
		4-0 x 13-0	52						
		4-0 x 2-0	8						
		8-0 x 2-0	16						
			401						
			x 1-0 =	401	15 CY				
Remove Surplus Excav Native Mat (allow 30% Swell)									
		Amount Mach Excav =	113 CY						
		Amount Backfill =	−17		96 CY				

[NB: assumes that balance of topsoil remains on site for landscaping.]

PRACTICAL "STANDARDIZED" FORMS FOR CONTRACTORS
Form 514 MFD. IN U. S. A.

GENERAL ESTIMATE

BUILDING PRELIMINARY
LOCATION EXAMPLE
ARCHITECTS
SUBJECT SWIM POOL

ESTIMATE NO. P-3
SHEET NO. 1 of 3
ESTIMATOR KC
CHECKER AB
DATE 1972

FRANK R. WALKER CO., PUBLISHERS, CHICAGO

DESCRIPTION OF WORK	NO. PIECES	DIMENSIONS	EXTENSIONS	EXTENSIONS	TOTAL ESTIMATED QUANTITY	UNIT PRICE M'T'L	TOTAL ESTIMATED MATERIAL COST	UNIT PRICE LABOR	TOTAL ESTIMATE LABOR COST

prelim calcs

61-4
31-4
2/ 92-8 = 185-4
less 4/8" = -2-8
MP = 182-8

Pool side walls
11-6 11-6 4-6
8-6 4-6 3-6
2) 20-0 16-0 8-0
10-0 8-0 4-0

10-0
8-0
4-0
3) 22-0
7-4

Pool end walls
8-6
3-6
2) 12-0
6-0 (AVG)

Sloping Slab Lengths

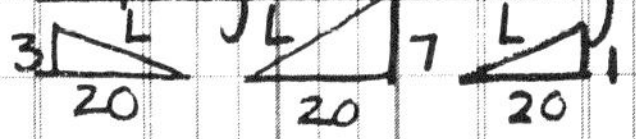

L = 20-3 = 21-2 = 20-0

20-3
21-2
20-0
3) 61-5
L = 20-6 (AVG)

CONCRETE (3500 psi - 3/4" stone)

Pool Slab o/Grd (ZERO SLUMP) x 6"
3/20-6 x 31-4 1927
(@ under end walls) 2/0-8 x 31-4 42
(slab area) = 1969
x 0-6 = 985 37 CY

Pool Walls x 8"
2/3/20-0 x 7-4 880
2/31-4 x 6-0 376
(wall area) = 1256
x 0-8 = 837 31 CY

FORMS (erect and strip)

Form Slab Edges x 6" 2/3/20-6 123
2/31-4 63
2/2/0-8 3 = 189 190 LF

Slab Screeds x 6" 5/31-4 157
2/3/20-6 123
2/2/0-8 3 = 283 280 LF

Form Pool Walls 2/1256 sf (wall area) 2510 SFCA

RC & W on last item 2/2/3/20-6 246 250 LF

Similarly, the end walls have an avg ht, and the dims are twice "timesed" (multiplied by 2).[4] If a drawing is accurate (and many prints are not), sometimes dims may be scaled. However, calculation is generally preferable, with scaling used as an approximate means of checking the calculation.

Forms. It is assumed that the forms of the slab perim will also be used as slab screeds, and only intermediate slab screeds have been measured. Forms for conc walls are measured as before, and item of *"raking cutting and waste"* (R C & W) is also measured to allow for the additional costs of labor and *waste* in cutting the bottom edges of wall forms to slopes on both sides of both the side walls.

Many estimators do not know how to deal with R C & W, even though they recognize that it must involve extra costs for labor and materials. Some ignore it, whereas others add an allowance; but it should be measured and a *unit price* should be analyzed for each item.

The forming of the gutter recess, and the small recess (for perim paving), and the projection, are measured as *extra over items,* primarily as a basis for the additional *labor costs* and also for the additional use and *waste* of lumber inserted to form the recesses and used as supports for the projection. They are measured as *run* items because their lengths are the most significant dims for the *labor costs,* as explained before. No adjustment to the volume of conc walls has been made for these features, because the projection almost balances with the recess and a precise deduction would be only about 3 CF which can be ignored.

Rebar. First, some prelim calcs have to be made to determine the avg conc wall ht for all four sides. This average can be found by dividing the conc wall area by the length (the MP), and the result can be used to calculate the avg length of the vertical rebars and the avg number of horizontal rebars in the four sides.

The avg length of vertical rebars is found by deducting the top conc cover of 1½ in., adding 3 in. into the slab at the bottom, and adding 6 in. and 24 in. for the top and bottom hooks, respectively. The number of vertical bars is calculated by dividing the MP by the spacing (6 in.) and adding additional bars for the corners (one each) and for the wall lengths (one each), similar to the way wall studs were measured.

[4] The term "to times" (to multiply), as in "timesed by three," is used by many estimators to express a useful convention: i.e., "3/20.6 × 31.4," in which a distinction is made between multiplying dimensions by other dimensions and multiplying dimensions by a number (in the "Number of Pieces" column to the left of the dimension column) in the estimate sheet. The use and value of this convention in estimating was explained in Chapter 7.

Remember that indicated spacings are always maximum spacings, and that extra members (bars) are required for less than maximum spaces (caused by job dims that are not multiples of the spacing) and at corners and ends.

The length of the horizontal rebars in the four sides is calculated by adjusting the MP to obtain a perim length which is 2 in. inside the outer faces of the walls, for conc cover. This gives the maximum length, as obviously some horizontal bars are shorter than others because of the sloping bottom. But, this perim length multiplied by the average number of horizontal bars in the four sides will yield a sufficiently accurate result, and the avg number of horizontal bars is found by dividing the avg wall ht (6.10½) by the spacing (12 in.) plus one bar extra (over the number of spaces). *Laps* have been measured; two per bar in the long sides, and one per bar in the ends, assuming a maximum bar length of 30 ft, as previously stated.

The lengths of rebar in the slab are measured by allowing one *lap* in the shorter bars less 3 in. cover at each end, and two *laps* in the longer bars less 3 in. cover at each end. The numbers of bars in both directions are found by dividing the slab dims (less 2 × 3 in. for conc cover at edges) by the rebar spacing of 9 in. in both cases. Because the bars are equally spaced in both directions, the total length of the longer bars (2719 LF) is almost the same as the total length of the shorter bars (2688 LF) in the other direction. If bar spacings are identical both ways, the total lengths of bars each way should be the same, except for minor differences resulting from *laps* and from dims that are not exact multiples of the spacing.

If the slab rebar is measured by "bar length per unit area," the result is similar; thus:

With bars at 9 in. spacing both ways there are 2⅔ *LF* of bar per *SF* of slab area.

$$\text{Slab Area} \times 2\tfrac{2}{3}\,LF = 1969 \times 2\tfrac{2}{3}\,LF = 5251\,LF$$

To which must be added an extra bar at two adjacent edges:

$$(32.0 + 64.8 = 96.8) = 5251 + 97 = 5348\,LF$$

To which must be added an allowance for *laps* (at every 30 ft):

$$(5348 \div 30) \times 14\text{ in.} = 5348 + 208 = 5556\,LF$$

"STANDARDIZED" FORMS FOR CONTRACTORS PRACTICAL
Form 514 MFD. IN U. S. A.

GENERAL ESTIMATE

BUILDING PRELIMINARY
LOCATION EXAMPLE
ARCHITECTS
SUBJECT SWIM POOL

ESTIMATE NO. P-3
SHEET NO. 2 of 3
ESTIMATOR KC
CHECKER AB
DATE 1972

FRANK R. WALKER CO., PUBLISHERS, CHICAGO

DESCRIPTION OF WORK	NO. PIECES	DIMENSIONS	EXTENSIONS	EXTENSIONS	TOTAL ESTIMATED QUANTITY	UNIT PRICE M'T'L	TOTAL ESTIMATED MATERIAL COST	UNIT PRICE LABOR	TOTAL ESTIMATE LABOR COST
FORMS (cont)				4/2/6½" 185-4 -4-4					
Extra Lab & Mat to Form in Walls:–				181-0					
Recess 1½"x6"		1/181-0		181 LF					
Recess 1½"x1½"		1/186-0		186 LF					
Projection 1½"x4½"		1/186-0		186 LF					
REBAR (grade #60)									
Avg Wall Ht = Area / MP = 1256 SF / 182-8 = 6-10½ (AVG)									
#3 Bars in Slab									
(31-4 - 2/3") + 1.2 = 32-0; 62-4 / 0-9 = 83 +1 = 84		84/32-0 (str)	2688						
(62-4 + 3/1-2) = 64-8; 30-10 / 0-9 = 41 +1 = 42		42/64-8 (str)	2719	= 5407					
#3 Bars in Walls									
(6-10½ - 0-1½ + 0-3 + 0-6 + 2-0) = 9-6; 182-8 / 0-6 = 365 (+4+4) = 373		2/373/9-6 (Vert Bent)	7087						
(182-8 + 4/2/2") = 184-0; 6-10½ / 1-0 = 7 +1 = 8		2/8/184-0 (Horiz Str)	2944						
(36 x 3/8") (laps)		2/8/6/1-2 (Horiz str)	112	= 10,143					
				15,550					
TOTAL #3 Bars in Pool				x 0.376 lb =	5850 lb				
Extra L. Bending		2/373/9-6	7087 x 0.376 lb =		2670 lb				

[NB: No "CONCRETE SUNDRIES" included here.]

PRACTICAL "STANDARDIZED" FORMS FOR CONTRACTORS
Form 514 MFD. IN U. S. A.

GENERAL ESTIMATE

BUILDING: PRELIMINARY
LOCATION: EXAMPLE
ARCHITECTS:
SUBJECT: SWIM POOL

ESTIMATE NO.: P-3
SHEET NO.: 3 of 3
ESTIMATOR: KC
CHECKER: AB
DATE: 1972

FRANK R. WALKER CO., PUBLISHERS, CHICAGO

DESCRIPTION OF WORK	NO. PIECES	DIMENSIONS	EXTENSIONS	EXTENSIONS	TOTAL ESTIMATED QUANTITY	UNIT PRICE M'T'L	TOTAL ESTIMATED MATERIAL COST	UNIT PRICE LABOR	TOTAL ESTIMATE LABOR COST
EXCAV & FILL					2/2-0 =	61-4 x +4-0 65-4 x	31-4 +4-0 35-4	AVG HT LONG SIDES = 7-4 SLAB = +0-6 7-10	
Remove Topsoil (NIL)					—				
Machine Excav Pool		65-4 x 35-4 x 7-10	18,083		670 CY				
Remove Excav Mat off Site (allow 25% swell) (none backfilled)					670 CY				
Hand Trim & Ram Bottom to Slopes (slab area) =					1970 SF				
Temp Shoring to Sides Excav avg 7'10" dp 2x(65-4 + 35-4)	1/201-4 x 7-10				1580 SF				
Ditto, 12" Extra Ht at Top (above ground)	1/201-4				200 LF				
Fill Imported Pit Run Gravel around Pool Walls, & Ram in 12" layers, as spec (allow 20% shrinkage) 4/2/1-0 = 201-4 -8-0 193-4		193-4 x 2-0 x 7-10	3029		112 CY				

[NB: No gravel bed under conc slab required.]

This quantity is about 150 LF (= 50 lb) greater than that in the estimate, primarily because no adjustment has been made for conc cover at slab edges. This method of measuring rebar by "bar length per unit area" is useful and accurate if the extra bars at two adjacent edges are added (because there is always one more bar than the number of spaces); and the method can just as easily be used when the spacings are not the same both ways. (In some instances it is convenient to use a unit area of more than 1 SF, which is also a multiple of the two spacing dims.) Measurement by this method and measurement by the other (calculating bar lengths and numbers of bars) generally produce results within 2 or 3 percent of each other; differences arise from differences between job dims and multiples of the spacing dims and from the absence of adjustments to areas for the conc cover to rebar, as indicated above.

Extra L. Bending. This type of bending is measured for the bent vertical bars in the conc walls, the assumption being that the conc slab is placed first and that the vertical rebar in the walls is bent and placed before the slab is poured. Alternatively, cranked bars would have to be placed in the slab with about 15 in. upstanding, and the vertical rebars would have to be spliced to the cranked bars. This procedure would probably prove to be more expensive because of the *laps,* although there would then be less steel to be handled in the light bending. Often, one finds that the costs of two reasonable but different ways of doing the same thing are about the same, which means that if an estimator selects one good method for his estimate, the costs will usually be valid even if the method is changed.

Excav and Fill. The prelim calcs add a working space 2 ft wide around the outside of the pool to obtain the dims of the excav. It is assumed that all excav mat is removed and that imported gravel fill is to be used. The absence of topsoil is noted to indicate that it has not been overlooked.

Temp Shoring. This *work* is measured when extra excav to form sloping banks is not measured. The measured quantity is a basis for pricing the costs of supporting the sides of the excav, or it provides a contingency for any *extra excav work* required if shoring is not done and a collapse occurs. The **MM-CIQS** requires an additional foot of depth to be added to allow for the shoring to project at the top. By measuring this additional 12 in. of height as a separate *run* item, more information about the *work* is available from the estimate; i.e., the actual face area of excav to be shored and the length and avg depth of the excav. Information that would be otherwise obscured by adding the extra ht to the general area of shoring and timbering is thereby retained.

The **MM-CIQS** requires that shoring and timbering be measured only to depths of more than 4 ft, and then only if extra excav to form sloping banks is not measured. As with conc formwork, only the contact areas (supported faces) are measured for shoring, because the costs are mostly *labor costs* and (like formwork) the use and *waste* of lumber are calculated and allowed for in the *unit price.*

An exercise may be done from the dwg of this example, but with inside dims as shown below.

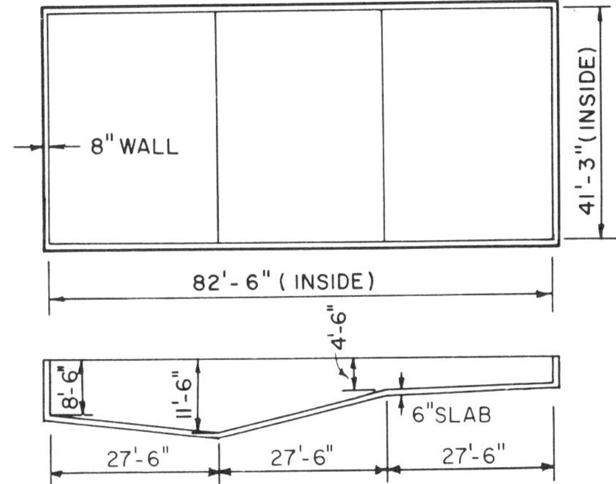

In addition, a dwg is included with the dwg for the example for an entirely different pool for a measurement exercise.

Preliminary Example

Concrete Pan Joist Slab

Concrete. Although the perim beams and the slab will be placed at the same time so that they are monolithic, the beams and the slab are measured separately, as required by the methods of measurement (**MM-CIQS** and **SMM-RICS**). This provides an opportunity to obtain cost data and to price the two items—beams and slab—at different *unit prices* for labor, if such differences can and have been established. Some estimators claim that such distinctions are impractical and

PRACTICAL "STANDARDIZED" FORMS FOR CONTRACTORS
Form 514 MFD. IN U. S. A.

GENERAL ESTIMATE

BUILDING PRELIMINARY EXAMPLE
LOCATION EXAMPLE
ARCHITECTS
SUBJECT PAN JOIST SLAB

ESTIMATE NO. P-4
SHEET NO. 1 of 2
ESTIMATOR KC
CHECKER AB
DATE 1972

FRANK R. WALKER CO., PUBLISHERS, CHICAGO

DESCRIPTION OF WORK	NO. PIECES	DIMENSIONS	EXTENSIONS	EXTENSIONS	TOTAL ESTIMATED QUANTITY	UNIT PRICE M'T'L	TOTAL ESTIMATED MATERIAL COST	UNIT PRICE LABOR	TOTAL ESTIMATE LABOR COST
Prelim Calcs MP(Beams) = 2(45-6 + 21-9) + 4/1-0 = 138-6						Avg Depth Beams	2.9 + 2.7 / 2 = 2.8 (avg)		
CONCRETE (3500 psi – 3/4" stone)									
Perim Beams		138-6 x 1-0 x 2-8	369		14 CY				
Pan Joist Slab x 15"		45-6 x 21-9	989½ x 1-3 = 1237						
DOT (@ pans)		15/21-6 (x 2.45 CF/LF)	= –790						
			447		17 CY				
FORMS (erect and strip) btwn 10 & 15 ft above deck									
Form Beam Soffits		1/138-6 x 1-0			139 SFCA				
Form Beam Sides (out) 138-6 + 4/2/0-6 = 142-6		1/142-6 x 2-9			392 SFCA				
Form Beam Sides x 7" (In)		1/134-6			135 LF				
Ditto x 9" (In)		1/134-6			135 LF				
Form Joist Soffits x 6"		16/21-6			344 LF				
Form Soffits x 1½"		2/45-6 (next perim beams)			91 LF				
Form Pan Joists wi 30" x 12" Pans		15/21-6			323 LF				
Extra for End Pans		2/15			30 NO				
Slab Screeds x 3"		(slab area) =			990 SF				

unnecessary, because separate and different *unit prices* for two such closely related *items of work* cannot be substantiated. But if the distinction between such items is not made in estimates, more accurate costs never will be obtained. Measuring such items separately ensures better *cost accounting* results and cost analyses.

Pan Joist Slab. Many students see this item as a thin conc slab supported by conc joists and think that it should be measured that way. But structurally, and for a better aspect of the *work* to be measured, it should be seen as a 15-in. thick conc slab with the superfluous concrete displaced by pans.[5] The volume displaced by the pans is given, and similar figures for standard steel pans of different sizes can be obtained from manufacturers' tables and some estimating data handbooks. However, as pans are used and dented their displacement may vary and *cost accounting* will help to show how much.

Form Beam Soffits. Although the dwg shows masonry below the beam, it is unlikely that the conc would be placed on the masonry and therefore soffit forms would be required. The standard methods of measurement call for formwork for sides and soffits of beams to be measured together as a *super* item. But many estimators prefer to measure soffits separately, because they have to be reshored after the forms are stripped, or else the forms for soffits are left in place for several days after the forms for the sides have been stripped. Either way, there is an additional cost for forming soffits. However, it is a general principle of standard methods of measurement that they are not inflexible, and that providing the method of measurement is made clear in the estimate, estimators and quantity surveyors may measure in greater detail and adopt special methods of measurement which are not standard to better enable the *costs of work* to be estimated. With this in mind, the soffits are better measured separately from the sides.

Form Beam Sides. There are some complications involved in measuring and estimating this item, because of the junction between beams and slab. The first question is, Should there be a deduction of formwork at the inner side of the beam, at the 15-in.-deep slab? According to usual practice (and **SMM-RICS**), formwork is not deducted at intersections of beams with walls or columns or other beams. But in this case a deduction is justified. It can also be argued that a *super* "contact area" does not directly reflect the costs of formwork for the perim beams, and that it would be more realistic to price all the *work* as a *run* item; i.e., "form perim beam sides (two) average 32 in. (with no deduction for adjoining 15 in. deep slab; with soffit measured separately)." *Cost accounting* would probably show that formwork costs (per linear foot) would not vary significantly with minor variations in beam depths. It can be argued that the presence of the adjoining slab would not decrease the costs of forming the beam sides (by displacing a portion of the beam forms) and that, therefore, no deduction of side forms is necessary. This illustrates a fundamental fact of estimating: that *work* should be measured so that the quantities of *work* can be priced, and that methods of measurement should be determined primarily by the measurements that can best be used to account for the *costs of work.* In the example, the formwork to the beam sides has been separated between outer and inner faces, and inner faces have been measured as *runs.* But such *work* cannot be accurately priced without considering the locations of the *work.* For example, the 7-in.-high face should cost less than the 9-in.-high face below the slab, not because of the minor difference in height but because of the location.

Form Joist Soffits. Similarly, this item is better measured as a *run* item. Joist soffits 8 in. wide would not cost one-third more (per linear foot) than joist soffits 6 in. wide. These forms and their supports are often erected and stripped by the *contractor,* whereas the next *item of work* is often done by a *subcontractor.*

Form Joists wi Pans. Some estimators measure the gross slab area overall (excluding any slab beams), whereas others measure only the actual plan area of the pans (15/21.6 × 2.6), and still others measure the pans as a *run* item, stating the pan width. Although the first method is common, the other two are preferable, particularly the last method, for the reasons already given for measuring such *work* as *run* items. In no case is the two-dimensional contact area of the pans measured (i.e., the actual pan surface in contact with concrete), which in fact is contrary to the previously stated general principle for measuring formwork. The reason for this is that measurement of the pans' contact area would be too complicated and the result would not increase the accuracy of estimating. This is significant, and it raises other questions about the

[5] From the measured quantities, it can be seen that the 15-in. pan joist slab shown contains less concrete than a 6-in.-thick flat slab of the same area. But by increasing the effective depth of the slab, by using a thinner floor slab and joists, the concrete and the steel rebar are used more efficiently in the structure.

PRACTICAL
Form 514 MFD. IN U.S.A.

GENERAL ESTIMATE

BUILDING: PRELIMINARY
LOCATION: EXAMPLE
ARCHITECTS:
SUBJECT: PAN JOIST SLAB

ESTIMATE NO. P-4
SHEET NO. 2 of 2
ESTIMATOR: KC
CHECKER: AB
DATE: 1972

FRANK R. WALKER CO., PUBLISHERS, CHICAGO

DESCRIPTION OF WORK	NO. PIECES	DIMENSIONS		EXTENSIONS	EXTENSIONS	TOTAL ESTIMATED QUANTITY	UNIT PRICE M'T'L	TOTAL ESTIMATED MATERIAL COST	UNIT PRICE LABOR	TOTAL ESTIMATE LABOR COST
REBAR (grade #60) up to 30 ft lengths										
#7 Bars (in J1)	14/22-9	(str)		319	x 2·044 lb	650 lb				
#8 Bars (in J1)	14/25-6	(bnt)		357						
(in B1)	4/138-6	(str)		554						
36 x 1" = 3-0 (laps)	4/2/3-0	(")		24						
3L3 = 6-0 (corner splices)	4/4/6-0	(bnt)		96						
				1031	x 2·670 lb	2755 lb				
#5 Bars (in B1)	2/138-6	(str)		277						
36 x 5/8" = 2.0 (laps)	2/2/2-0	(")		8						
2L2 = 4-0 (corner splices)	4/2/4-0	(bnt)		32						
(138-6)/(1-6) = 92 (+4+4) = 100	2(1·9+0·4½)+0-9 = 5·0			317	x 1·043 lb	330 lb				
#4 Bars (in B1)	100/5-0	(stirrups)		500						
9⟌24	100/2-9	(bnt)		275						
((46-6)/(1-6) = 31 +1 = 32)	32/22-9	(str)		728						
((22-9)/(1-6) = 16 +1 = 17)	17/46-6	(")		791						
36 x ½" = 1-6 (laps)	17/1-6	(")		26						
				2320	x 0·668 lb	1550 lb				
Extra for Bending:-										
H. Bending	14/25-6 x 2·670 lb			953						
	4/4/6-0 x 2·670			256						
	4/2/4-0 x 1·043			33						
	100/2-9 x 0·668			184		1430 lb				
L. Bending	100/5-0 x 0·668					340 lb				
Allow for Conc Blk Bar Supports						Item				

[NB: No "CONCRETE SUNDRIES" taken here.]

validity of other methods of measurement and units of *work* that are commonly used.

Rebar. Rebar is often more easily measured for a large project than for a small project, because for large projects the structural engineer often annotates or schedules every structural rebar by size, length, and location on the dwgs, whereas for smaller jobs the information is often somewhat sparse. Problems incurred in measuring rebar often arise from requirements not shown on the dwgs; although they are often implicit in the contract through a reference to a standard or a code of practice. Therefore, the estimator must know of all the usual requirements for rebars, including such items as non-structural carrying bars that are not usually shown on the dwgs but are required for installing repetitive structural bars that are wired at the required spacing to the carrying bars for secured placing and support.

The conc joists against the perim beams (at the two shorter sides) are not required to be reinforced, according to the CRSI Code of Practice. There are, therefore, fourteen #7 bars in conc joists with 6 in. at each end into the perim beam. There are also fourteen #8 bars (bent) in joists. In the perim beams the #8 bars (straight) are the MP length (average), and one *lap* is measured for each of the two longer sides (× 4 bars). At corners, #8 bars for corner splices (bent to 90 degrees) are measured at twice the *lap* length. Alternatively, and according to the Code, the continuous bars may themselves be bent to 90 degrees to splice at corners, but the separate splice bars as measured here would probably be easier and cheaper. The #5 bars in the perim beams are similarly measured.

The #4 stirrups are measured as 1.9 deep × 0.9 wide and with two 4½-in. hooks at the top.[6] The 2.9 long cranked #4 bars are in conjunction with the stirrups. The #4 bars in the slab are measured 6 in. into the beams at each end. The total length of #4 bars 46.6 long is higher than the total length of #4 bars 22.9 long, because whereas 46.6 is a multiple of the 18-in. spacing, 22.9 is not; and an additional bar 46.6 long must be allowed for the part space of 3 in. If the slab steel is calculated on the basis of "bar length per unit area," as follows, the result is almost the same.

Taking a square yard (36 in. × 36 in.) as the unit area, it would contain four bars each 36 in. long; a total of 12 *LF*, which equals 1⅓ *LF* per *SF*.

Reinf Area × 1⅓ *LF* = (46.6 × 22.9) × 1⅓ *LF*
= 1058 *SF* × 1⅓ *LF* = 1411 *LF*

add: extra bars at two edges = +69
add: extra bar one way (for one dim not multiple of spacing) . . . . = +47

Total: 1527 *LF*

The total measured by the other method in the estimate is 1519 LF. The longer bars are assumed to require one *lap* per bar. If in fact they are in one length (of 46½ ft) the extra length measured will help to offset the higher handling costs.

Extra for Bending. The weights of bent bars are measured so that the additional cost may be estimated. "Heavy" bending (the cheaper of the two) includes all the bent bars in this example except the stirrups, which are always classed as "light" bending. It is helpful to describe all bars according to whether they are bent or straight when they are measured, and in larger projects they are better separated as different items to avoid confusion and to simplify the measurement of rebar and bending.

The **MM-CIQS** requires bars to be grouped according to length and the number of bends per bar. It also says that bars larger than #6 shall not be separated according to bar size; but the increasing use of large bars (#14 and #18) makes it desirable to do so, as recommended by the CRSI Code. The differences between the two methods of measurement are not great. The important thing is that the method of measurement used is clear and appropriate to the *work* and the locality; and the clearest method is to measure separately every bar that is different in bar size and bending and to group them according to lengths.

Bar supports, such as steel bolsters and chairs, are usually measured as *number* items, except continuous chairs, and the like, which are measured as *run* items in 5 ft and 10 ft lengths. If such items are required by the *work*, they should be specified and indicated by standard symbols (see CRSI Code). Alternatively, plastic or precast conc bar supports are used, either plain, or with wires, or with dowels of rebar, particularly in smaller projects; and it is assumed that these are to be used in these Preliminary Examples. Costs for plastic or conc supports are not high, and they can be established by CA according to the type of job and the rebar quantities. Therefore, these small accessories need not be measured in detail and can be priced by allowing a suitable sum.

[6] Stirrups are usually measured in width and height as 3 in. less than the outside width and height of the conc beam. Column ties are similarly measured. For other standard practices, see Chapter 8 and the CRSI Code of Practice.

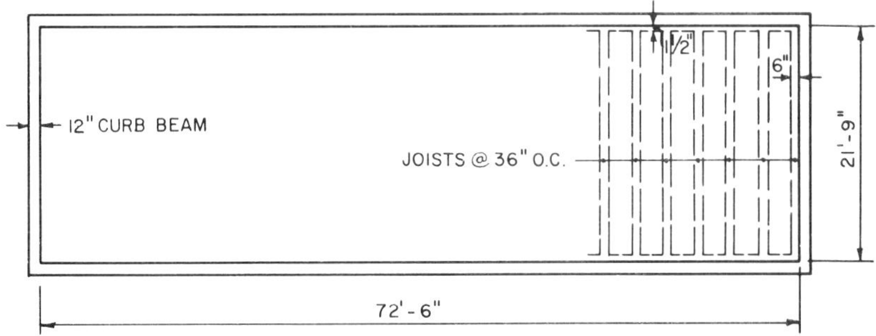

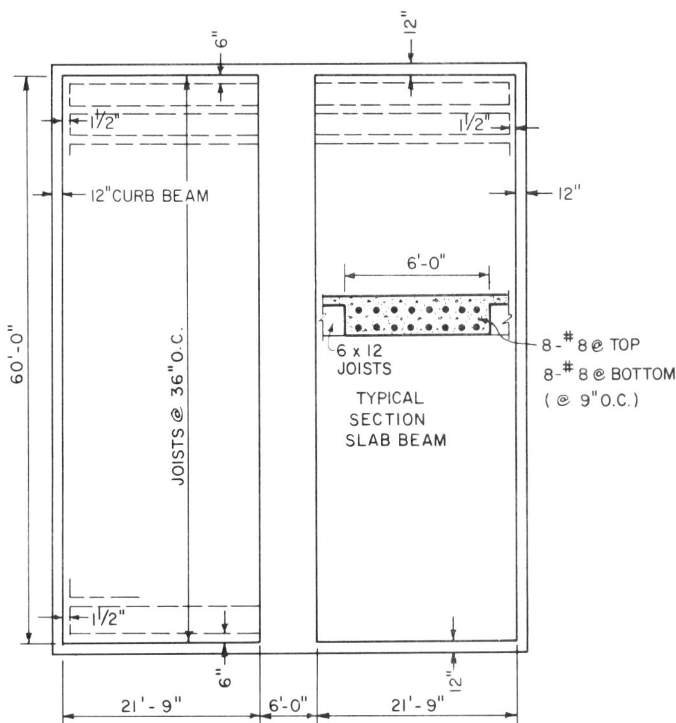

Exercises may be done from the dwg of the example but with the different dims shown above.

Preliminary Example

Concrete Building

Prelim Calcs. Preliminary calculations include the MP and the wall ht, which is taken up to the underside of the roof slab because the wall would probably be placed that way. The avg ht of the plinth, sill, and fascia projs is calculated so that they can be grouped and measured together. The slab dims are also calculated here, which is usually necessary even when the dims are shown for walls rather than for ftgs, as in the example.

Extra Lab and Mat in Projs to Walls. These items are measured separately so that they can be priced at a higher *unit price* than the general conc wall item. There is undoubtedly additional time and effort required in ensuring that the cast-in-place conc does fill the forms of the projs. If it does not, there will be some patching to do when the proj forms are stripped, and this risk should be allowed for. In addition, the precept is to measure and price *basic items of work* whenever possible because cost data for *basic items* is more likely to be available. By measuring these projs separately, the 8-in. conc walls are retained as a *basic item of work,* whereas projs are not *basic items.*

Susp Roof Slab. Susp Roof Slab is measured as extending over the conc walls as it would probably be placed —monolithic with the fascia proj so that there would be no construction joint in the fascia. With all struc-

PRACTICAL "STANDARDIZED" FORMS FOR CONTRACTORS
Form 514 MFD. IN U.S.A.

GENERAL ESTIMATE

BUILDING: PRELIMINARY
LOCATION: EXAMPLE
ARCHITECTS:
SUBJECT: CONCRETE BUILDING

ESTIMATE NO.: P-5
SHEET NO.: 1 of 7
ESTIMATOR: KC
CHECKER: AB
DATE: 1972

FRANK R. WALKER CO., PUBLISHERS, CHICAGO

DESCRIPTION OF WORK	NO. PIECES	DIMENSIONS	EXTENSIONS	EXTENSIONS	TOTAL ESTIMATED QUANTITY	UNIT PRICE M'T'L	TOTAL ESTIMATED MATERIAL COST	UNIT PRICE LABOR	TOTAL ESTIMATE LABOR COST

prelim calcs

6-0; 20-0 + 14-0 = 40-0; 20-0 + 10-0 = 30-0

40-0; 30-0; 4-0; 2/74-0 = 148-0

less 4/2/1-0 = -8-0

MP = 140-0

Wall Ht: 8-0; 0-6; 0-6; 3-6; 12-6; -1-0; 11-6 High

Wall Projs: plinth 15"; sill 3; fascia 12; 3)30 = 10" AVG; + 1/2"(slope); 10 1/2" (AVG Ht.)

slab dims: 40-0 x 30-0; less 2/1-4 = -2-8 -2-8; 37-4 x 27-4

slab perim: 37-4; 27-4; 4-0; 68-8; x2; 137-4

CONCRETE (3000psi – 3/4" stone – except as noted)

Ftgs cont (1 1/2" stone) 140-0 x 2-0 x 1-0 — 280 — 10 1/2 CY

Walls x 8": 140-0 x 11-6 — 1610; DDT

DDT (openings): 5/5-0 x 4-0 — 100; 1/3-0 x 7-0 — 21; −121

4/2/6" = +4-0; 140-0 + 4-0 = 144-0

(net wall area): 1489 x 0-8 = 993 — 37 CY

Extra Lab & Mat in Projs to Walls: 3/144-0 x 0-4 x 0-10 1/2 — 126 — 5 CY

Slab on Grd x 6": 37-4 x 27-4 — 1021; DDT

DDT (wants): 14-0 x 10-0 — 140; 6-0 x 10-0 — 60; (10-0 + 2/1-4) = 12-8; 12-8 x 4-0 — 51; −251

(slab area) = 770 x 0-6 = 385 — 14 CY

Susp Roof Slab x 6": (slab area) = 770

ADD (@ over walls): 140-0 x 0-8 — 93

(roof slab area) = 863 x 0-6 = 432 — 16 CY

Roof Beams 12" x 12": 1/37-4 — 37; (20-0 − 2/1-4) = 17-4; 1/17-4 — 18; 55 x 1-0 x 1-0 = 55 — 2 CY

tural conc, the separation of different parts of monolithic walls, slabs, and beams for purposes of measurement is somewhat arbitrary, but the effect is not significant.

Roof Beams. These are better measured as a *run* item at first, with the volume calculated from the total length. (The value of this procedure is less obvious when the beam is 12 in. × 12 in.) The total length of beams (and columns) of the same size can be used for measuring forms and (when adjusted) for rebar.

Form Walls. The additional quantity (71 SF) at slab edges is a quantity required to theoretically extend the wall forms up to the top of the slab and behind the fascia proj so that the forming of the fascia proj can be measured and priced as an *extra over item,* as is the forming of the other projs. No deduction is made from the forms for the openings. (The **MM-CIQS** calls for no deductions for openings of 100 SF or less.)

Extra Lab and Mat to Form Wall Projections. These are measured as *run* items and described, because to measure them as *super* items (length × girth), and to combine them into one *super* item would make accurate pricing impossible. (**MM-CIQS** calls for such items to be measured as *super* items if over 24 in. in girth.) The *labor costs* of these items will not vary in proportion to their girth, but rather in proportion to their length. The *material costs* will be affected by the girth, but not much, because most of the *material costs* are in the top and bottom (and in the external supports), and it is only the form sheathing between that will vary with the vertical dim of the projection.

Form Openings. These openings are measured as *number* items to more accurately estimate the *labor costs.* Variations in opening sizes affect the costs slightly; more important are the number and the types of openings.

Slab Screeds. Slab screeds are required on grade, but suspended slabs (placed on leveled forms) require only movable screeds, such as a length of steel pipe laid on chairs, which do not require laborious installation with pegs, and leveling. The costs of these is therefore almost negligible and can be included with the tools and equipment used for placing the concrete.

Concrete Sundries. These *items of work* on conc surfaces are inadequately described for accurate pricing, and the job specs would have to be referred to again by the estimator. Spec page references might be included in such descriptions. No deductions have been made for openings, but if they were, the *work* on reveals (jambs) should be measured; and if the openings were numerous, it would be better to measure them. Corners and arrises often require special care and attention, particularly with bush-hammering, which may be another good reason for measuring reveals (and deducting openings) if they are numerous.

Rebar. The #4 bars in the conc ftgs are measured at the outside perim of the ftgs, assuming they cross at the corners. The conc wall ht is divided by the spacing to give thirteen horizontal rows of bars. These are assumed to be spliced by bent #4 bars at corners, as in the last example. The length of the vertical bars in walls is equal to the wall ht (11.6), and their number is calculated by dividing the MP by the bar spacing, to which are added twelve extra bars for the number of wall lengths, plus an addition of twelve extra bars at corners. There is an equal number of dowels in conc ftgs, joined to the vertical bars in the walls above. It is not possible to measure exactly how many bars are eliminated by the openings, so the minimum number is deducted and the usual #5 bars around openings are measured.

The #4 bars in the roof are measured "per unit area," as explained previously, and *laps* are measured for the few bars that are over 30 ft long. (This is to be consistent with the assumption made regarding the maximum bar length; in fact, bars 40 ft long would probably be used.)[7]

The rebar in beams is taken 6 in. into the supporting walls. The two beam lengths are added for convenience. The cranked #7 bars are measured by adding an additional 5 in. to each of the bar lengths (for the portions at 45 degrees), plus 10 in. at each end for a hook; two such bars are in each beam. No *laps* are allowed for, as noted. Stirrups are measured as 3 in. less than the beam dims, plus 4 in. hooks at the ends. Heavy and light bending are identified and measured as before.

The welded wire fabric (WWF) in the conc slab on grd requires a side-lap of at least 2 in. and an end-lap of at least 8 in. (one mesh + 2 in.). Since, however, it may be cheaper to make bigger *laps* than to cut the WWF, the amount for *laps* varies slightly. The

[7] The **MM-CIQS** has not always been strictly followed here. For example, the **MM-CIQS** calls for rebar to be classified by lengths, as stated therein; and for brevity and simplicity this has not been done here. The larger and the more complex the *work* being measured, the more useful further classifications become. However, these examples of quantity surveys are intended to reflect some of the methods in general use as well as published standard methods; and, in some cases, alternatives to both of these.

PRACTICAL "STANDARDIZED" FORMS FOR CONTRACTORS
Form 514 MFD. IN U.S.A.

GENERAL ESTIMATE

BUILDING: PRELIMINARY
LOCATION: EXAMPLE
ARCHITECTS:
SUBJECT: CONCRETE BUILDING

ESTIMATE NO.: P-5
SHEET NO.: 2 of 7
ESTIMATOR: KC
CHECKER: AB
DATE: 1972

FRANK R. WALKER CO., PUBLISHERS, CHICAGO

DESCRIPTION OF WORK	NO. PIECES	DIMENSIONS	EXTENSIONS	EXTENSIONS	TOTAL ESTIMATED QUANTITY	UNIT PRICE M'T'L	TOTAL ESTIMATED MATERIAL COST	UNIT PRICE LABOR	TOTAL ESTIMATED LABOR COST
FORMS (erect and strip)									
Form Ftgs	2/	140-0 x 1-0	280		280 SFCA				
Form Key 2x4	1/	140-0			140 LF				
Form Walls	2/	1610 SF (gross wall area)	3220						
(140-0 + 4/2/4") (@slab edges)	1/	142-8 x 0-6	71						
			3291		3290 SFCA				
(140-0 + 4/2/6") = 144-0									
Extra Lab & Mat to Form Wall Projections (4")									
Plinth 4" x 15"	1/	144-0			144 LF				
Sill 4" x 3"	1/	144-0			144 LF				
Fascia 4" x 12"	1/	144-0			144 LF				
Form Wdw Openings (5'x4')	5/1				5 NO				
Form Door Openig (3'x7')	1/1				1 NO				
Form Beam Sides	2/	55-0 x 1-0			110 SFCA				
Form Beam Soffits	1/	55-0 x 1-0			55 SFCA				
Form Susp Slab (<10' h.) (slab 0/grd area) =			770						
DDT (@ Beam Soffits)			-55		715 SFCA				
Slab Screeds x 6"	1/	137-4	138						
(20-0 − 2/1-4)	2·1/	17-4	52						
(16-0 − 2/1-4)	2/	13-4	27						
			217		220 LF				
[NB: Slab screeds for Roof not incl.]									

PRACTICAL "STANDARDIZED" FORMS FOR CONTRACTORS
Form 514 MFD. IN U. S. A.

GENERAL ESTIMATE

BUILDING PRELIMINARY
LOCATION EXAMPLE
ARCHITECTS
SUBJECT CONCRETE BUILDING

ESTIMATE NO. P-5
SHEET NO. 3 of 7
ESTIMATOR KC
CHECKER AB
DATE 1972

FRANK R. WALKER CO., PUBLISHERS, CHICAGO

DESCRIPTION OF WORK	NO. PIECES	DIMENSIONS	EXTENSIONS	EXTENSIONS	TOTAL ESTIMATED QUANTITY	UNIT PRICE M'T'L	TOTAL ESTIMATED MATERIAL COST	UNIT PRICE LABOR	TOTAL ESTIMATE LABOR COST
CONCRETE SUNDRIES									
Cut Back Ties & Grout Holes Flush									
(as specs page –) (form walls area)					3290 SF				
(140-0 − 4/2/4") = 137-4									
Sack-rub Finish Walls (above grd) (as specs p–)									
(inside)	1	137-4 × 8-0	1099						
(outside on projs)									
(1-0 + 0-3 + 0-5) = 1-8	2	144-0 × 0-4	96						
(144-0 + 4/2/2") = 145-4	1	145-4 × 1-8	242						
			1437		1440 SF				
Bush-hammer Finish Walls (above grd) (as specs p–)									
(8-6 − (1-1 + 0-4)) = 7-1	1	142-8 × 7-1	1011		1010 SF				
Wood Float Finish Slab (roof slab area) =					860 SF				
Steel Trowel Finish Slab (flr slab area) =					770 SF				
Ditto, Sloping tops Projs × 4" wide									
(not sack-rubbed)	3	144-0	432		430 LF				
Cure Slabs		(roof slab area) =	860						
		(flr slab area) =	770		1630 SF				
Control Joints & Seal (@ slab perim) (as specs p–)									
	1	137-4			140 LF				
Gravel Bed u/Slab × 6" (allow for 20% shrinkage)									
		(flr slab area) =	770						
			× 0-6 =	385	14 CY				

PRACTICAL
Form 514 MFD. IN U.S.A.

GENERAL ESTIMATE

BUILDING PRELIMINARY
LOCATION EXAMPLE
ARCHITECTS
SUBJECT CONCRETE BUILDING

ESTIMATE NO. P-5
SHEET NO. 4 of 7
ESTIMATOR KC
CHECKER AB
DATE 1972

FRANK R. WALKER CO., PUBLISHERS, CHICAGO

DESCRIPTION OF WORK	NO. PIECES	DIMENSIONS			EXTENSIONS	EXTENSIONS	TOTAL ESTIMATED QUANTITY	UNIT PRICE M'T'L	TOTAL ESTIMATED MATERIAL COST	UNIT PRICE LABOR	TOTAL ESTIMATED LABOR COST
REBAR (grade #60) up to 30' lengths											
#4 Bars (in Ftgs)	5	148-0		(str)	740 × 0.668 lb		500 lb				
#4 Bars (in Walls)	2/13/1	142-0		(str)	3692						
	2/1	18/11-6		"	2714						
					6406						
9/24 dowels in ftgs/walls	2/118	2-9		(bnt)	649						
18/18 corner splices (outside only)	12/1/13	3-0		"	468						
					7523	DDT					
DDT (@ openings)	5/2/4	5-0			–	200					
	5/2/3	4-0			–	120					
	1/2/7	3-0			–	42					
	1/2/2	7-0			–	28					
					−390						
					7133 × 0.668 lb		4770 lb				
#4 Bars (in RF Slab) (slab area 863 SF × $3\frac{1}{3}$ LF/SF)					= 2877						
18-8 (extra @ edges) (slab perim) −4-0 / 148; 14-8 / 1-6 = 10, +1 = 11 (laps)	11/1	-6		(str) "	= 145; 17						
					3039 × 0.668 lb		2030 lb				
#5 Bars (in Walls)											
2(5-6+4-6) = 20-0 (arnd openings)	5/2	20-0		(str)	200						
2(7-6)+3-6 = 18-6	1/2	18-6		"	37						
					237 × 1.043 lb		250 lb				
#5 Bars (in Beams)											
2/= 37-4 + 1-0 = 38-4; 38-4 + 17-0 + 1-0 = 56-4	2	57-0		(str)	114 × 1.043 lb		120 lb				

PRACTICAL "STANDARDIZED" FORMS FOR CONTRACTORS
Form 514 MFD. IN U. S. A.

GENERAL ESTIMATE

BUILDING PRELIMINARY
LOCATION EXAMPLE
ARCHITECTS
SUBJECT CONCRETE BUILDING

ESTIMATE NO. P-5
SHEET NO. 5 of 7
ESTIMATOR KC
CHECKER AB
DATE 1972

FRANK R. WALKER CO., PUBLISHERS, CHICAGO

DESCRIPTION OF WORK	NO. PIECES	DIMENSIONS		EXTENSIONS	EXTENSIONS	TOTAL ESTIMATED QUANTITY	UNIT PRICE M'T'L	TOTAL ESTIMATED MATERIAL COST	UNIT PRICE LABOR	TOTAL ESTIMATE LABOR COST
REBAR (cont) up to 30' lengths										
#7 Bars (in Beams)		3/57-0 (str)		171						
37-4 +2(0-5+0-10 hook)		2/40-0 (bnt) ✓		80						
17-4 +2(0-5+0-10 hook)		2/20-0 (") ✓		40						
				291 x 2.044 lb		600 lb				
#3 Bars (stirrups)										
37-4/1-6 = 25 +1 =		1/26 (bnt)		26						
17-4/1-6 = 12 +1 =		1/13 "		13						
(extra c ends)	2/2/2	"		8						
2(15+4)+9=47" (say) 4-0 long		(No stirrups) =		47						
				x 4-0 long						
				188 x 0.376 lb		70 lb				
Extra for Bending:-	LF	lb/FT								
H Bending	649	x 0.668	=	746						
	468	x "								
	80	x 2.044	=	245						
	40	x "		991		1000 lb				
L Bending	188	x 0.376 =				70 lb				
6"x6" x 10/10 Gauge WWF in Slab o/grd										
		(slab area) =		770						
ADD (laps)		5% x 770 sf =		+38 =		810 SF				
Allow for Bar Supports						Item				

PRACTICAL "STANDARDIZED" FORMS FOR CONTRACTORS
Form 514 MFD. IN U.S.A.

GENERAL ESTIMATE

BUILDING PRELIMINARY
LOCATION EXAMPLE
ARCHITECTS
SUBJECT CONCRETE BUILDING

ESTIMATE NO. P-5
SHEET NO. 6 of 7
ESTIMATOR KC
CHECKER AB
DATE 1972

FRANK R. WALKER CO., PUBLISHERS, CHICAGO

DESCRIPTION OF WORK	NO. PIECES	DIMENSIONS	EXTENSIONS	EXTENSIONS	TOTAL ESTIMATED QUANTITY	UNIT PRICE M'T'L	TOTAL ESTIMATED MATERIAL COST	UNIT PRICE LABOR	TOTAL ESTIMATE LABOR COST
MOISTURE PROTECTION									
Dpf Conc Below Grd wi 2cts Asphalt (1gal per ct/per Sq.)									
(140-0+4/2/4") =	1/	142-8 x 2-2	309						
(140-0+4/2/8") =	1/	145-4 x 0-8	97						
(140-0+4/2/12") =	1/	148-0 x 1-0	148		550 SF				
Wpf Membrane u/Slab (slab area) =			770						
(laps)		10% x 770SF =	77						
(edges)	1/	137-4 x 0-6 =	68		920 SF				
SITE WORK		(2/1-6 = 40-0 x 30-0 +3-0 +3-0 = 43-0 x 33-0)							
Excav to Remove Topsoil avg 6" dp & stockpile on site									
		43-0 x 33-0	1419 x 0-6 = 710		26 CY				
Excav Trenches at Surface									
(140-0+4/2/5") = (2-0+0-6+1-4) = 3-10		143-4 x 3-10 x 3-6	1925		71 CY				
Temp Shoring to Trench Sides									
	2/	143-4 x 3-6	1003						
(@6" topsoil Remvd)	1/	143-4 x 0-6	72		1075 SF				
Ditto, 12" Extra H't at Top									
(above ground)	2/	143-4	287		290 LF				
Hand Trim Trench Bottom under Ftgs									
	1/	140-0 x 2-0			280 SF				
[NB: Trimming under Drain taken with that item.]									
Hand trim under Gravel Bed (G. Bed area) =					770 SF				

PRACTICAL "STANDARDIZED" FORMS FOR CONTRACTORS
Form 514 MFD IN U.S.A.

GENERAL ESTIMATE

BUILDING PRELIMINARY
LOCATION EXAMPLE
ARCHITECTS
SUBJECT CONCRETE BUILDING

ESTIMATE NO. P-5
SHEET NO. 7 of 7
ESTIMATOR KC
CHECKER AB
DATE 1972

FRANK R. WALKER CO., PUBLISHERS, CHICAGO

DESCRIPTION OF WORK	NO. PIECES	DIMENSIONS	EXTENSIONS	EXTENSIONS	TOTAL ESTIMATED QUANTITY	UNIT PRICE M'T'L	TOTAL ESTIMATED MATERIAL COST	UNIT PRICE LABOR	TOTAL ESTIMATE LABOR COST

SITE WORK (cont)

4"ϕ Agric Drain Tile @ Ftgs incl trimming under to slopes
(148-0 + 4/2/9") = 1/154-0 — 154 LF

Extra 4" Bends (90°) 12/1 — 12 No

1" Drain Gravel over D Tile (allow 15% shrinkage)
1/154-0 x 1-4 x 1-0 = 205 — 8 CY

Backfill Trenches @ Ftgs (allow 20% shrinkage)
(trench excav) = 1925 DDT
DDT (wants) (ftgs) (conc ftgs) — 280
(fnd walls) 140-0 x 0-8 x 2-6 — 233
(plinth) 144-0 x 0-3 x 0-3 — 12
(D.Tile gravel) (D. gravel) = 205
−730
1195 — 44 CY

Remove Spoil from Site (allow 30% swell)
44 + 27 = 71 cy (Excav) (amount not Backfill) = 730 CF — 27 CY

Backfill Topsoil avg 6" dp (allow 25% shrinkage)
(area removed) = 1419 SF DDT
DDT (wants) (roof slab area) = — 863
(@ plinth) 1/144-0 x 0-4 — 48
−911
508
x 0-6 = 254 — 10 CY

unit price for this item is quite low, and this influences the degree of accuracy required.

Excav Trenches. This item is described as "at surface" to indicate the level at which excav starts. Some use the description "surface trenches" (or "basement trenches," as the case may be.) The trenches are excav to 24 in. outside the conc fnd walls for the working space, and 6 in. inside the conc ftgs for ftg forms.

Temp Shoring to Trench Sides. This item is similar to the item measured in the example of the swimming pool, and the need for it here depends on ground conditions. The extra 12-in. height is measured separately as a *run* item, as before.

Hand Trim Trench Bottom under Ftgs. This item is measured as the width of the ftgs, for the *work* can be done between the erected ftg forms. The trimming (to falls) under the tile drains is usually done when the drains are laid and may therefore be included with the drains in one *item of work.* This is another way of measuring these items, which were measured differently in the second example.

Backfill Trenches. Backfill trenches is measured by applying an important precept of measurement; i.e., to measure *work* overall and to correct the over-measurement by deducting the *wants.* The quantities of backfill and removal of spoil (surplus excav material) should equal the quantity of trench excav because they are all net quantities by "bank measure;" the *shrinkage* and *swell* allowances are to be made in the *unit prices.*

Exercises may be done from the dwg of this example but with different dims, as shown below.

Residence Example

Generally

The Residence Drawings are of a house designed by its owner, an architect, Mr. E. Kuckein of Vancouver, B.C. This house was chosen because of the excellence and simplicity of the design and the dwgs, and because the house contains the kinds of *work* required for these examples and for the estimating exercises to be done. The dwgs are reproduced here essentially as they were when the construction contract was made and when the *work* was done.

The house is shown in the two photographs on page 148, and the designer described it as follows:

> The lot size of 50 feet by 120 feet, the setback requirements, and the orientation dictated the general form of the house. No attempt was made to adhere to accepted standards of design or of construction. The problem was to design a one family house of high aesthetic value at minimum cost, in an area where none of the existing houses had to be considered. The most economical finish materials were chosen: gypsum wall board for the interior, and stucco for the exterior. For this reason, balloon framing with wood joists and studs was selected and the house was designed to eliminate the effects of wood shrinkage. As a result, three years after completion no cracking in the finish materials is apparent.

The measurement examples of the Residence generally involve different kinds of *work* from those mea-

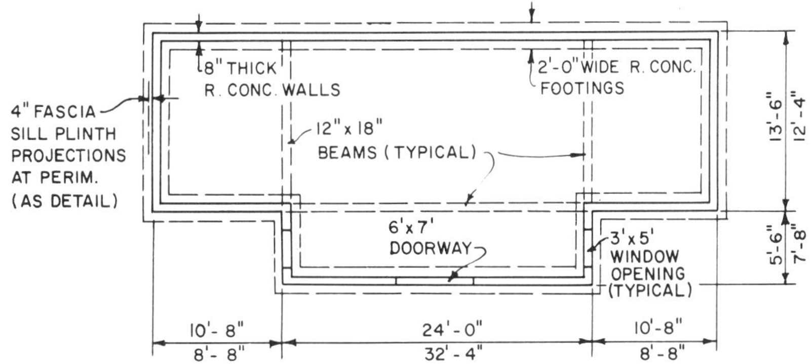

NOTE: DETAIL ALL AS ON LARGE DRAWING OF CONCRETE BUILDING

EXERCISES (1) USE DIMENSIONS ABOVE LINES
(2) USE DIMENSIONS BELOW LINES

sured in previous examples, which means that the *site work, excavation,* and *concrete work* of the Residence will be measured as exercises.

Although the following examples are set down in standard specification order, which is approximately the same as the order in which the *work* would be executed, estimators often measure in a different order. Many find it is better to first measure the major finishes such as stucco, drywall, floor coverings, and painting. In this way, the estimator soon gets a knowledge of the building's layout and main features, which then helps him to measure the structure of the building beneath the finishes. Therefore, the later examples of the Residence might be studied first, leaving the measurement examples involving framing and metals until later.

Residence Example

Masonry

The amount of masonry in the Residence is small, but masonry chimney stacks often present a measurement problem because of the voids in the stack and because of the usually different exposed and concealed *masonry work.* Often, too, dwgs of stacks are vague and incomplete, showing the exposed masonry, the fireplace, and the stack above the roof, but nothing in between to explain how the stack size is reduced, how the flue location is changed as the stack goes up, and how the throat is to be formed over the fireplace.

The measurement of masonry in chimneys was the

PRACTICAL
Standardized Forms for Contractors
Form 516 MFD IN U.S.A.

QUANTITY SHEET

PROJECT RESIDENCE — ESTIMATOR KC — ESTIMATE NO. R-1
LOCATION — EXTENSIONS KC — SHEET NO. 1 of 2
ARCHITECT ENGINEER — CHECKED — DATE 1972
CLASSIFICATION MASONRY (CHIMNEY STACK)

DESCRIPTION | NO. | DIMENSIONS | ESTIMATED QUANTITY | UNIT

THE FOLLOWING IN ONE BRICK MASONRY CHIMNEY STACK & FIREPLACE:-

4" Bk Perim at Floor – 2/6-8 (4-0 + 2-8) = 13-4; less 4/4" = 1-4; 12-0; less opening −2-8; 9-4

4" Bk Perim above Opening: 4-0 (or) 9-4; 2/= +3-4; 2/= −2-0; 7-4; 7-4

4" Bk Perim above Roof: 4/1-9 = 7-0; less 4/4" = −1-4; 5-8

Stack Ht: bk paving = 0-1; 8-7½; 7-6; roof = 1-2; above roof = 3-0; 20-4½

20-4½; −2-6 open'g; 17-10½; −4-0 above clg; 13-10½

Selected Facing Bricks (P.C. \$130.00 per M delvd) nominal standard size 8 x 2¼ x 3¾ in. (6⅓ units per SF) wi ½ in. concave tooled joints in Cmt mtr & running bond (as specs) (allow 3% Waste for Facings)

4" Bkwk pointed one side

1/	9-4	x	2-6	23	
1/	7-4	x	13-10½	102	
1/	5-8	x	3-8	21	
				146 x 6⅓ bks	925 bks

4-0 (above clg)
−0-4 (pcc. cap)
3-8

8" x 8" Bkwk Piers (attached) at stack corners pointed 2/sides & wi 2/angles

2/13-10½ — 28 x 9 bks — 250 bks

Common Bricks (P.C. \$90.00 per M) as above (but not pointed) (allow 5% Waste)

4" Bkwk wi raked jts (as key for stucco, by others)

1/2-8 x 13-10½ — 37 x 6⅓ bks — 235 bks

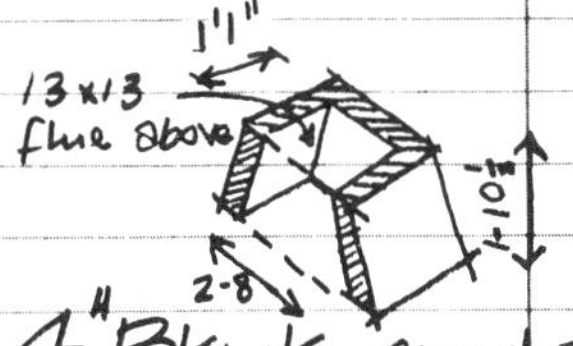

(sketch: 13 x 13 flue above; 1'1"; 2-8; 1-10½)

4" Bkwk arnd smoke chamber & flue (3 sides)

1/	4-9	x	1-10½	9	
1/	3-11	x	12-0	47	
				56 x 6⅓ bks	355 bks

2-8; 1-1; 2)3-9; 1-10½ (avg); 2/= 2-2; 2/= 0-8; 4-8½ (avg)

3/1-1 = 3-3; 2/ = +0-8; 3-11

13-10½; −1-10½; 12-0

FRANK R. WALKER CO., PUBLISHERS, CHICAGO

PRACTICAL "STANDARDIZED" FORMS FOR CONTRACTORS
Form 516 MFD IN U.S.A.

QUANTITY SHEET

PROJECT: RESIDENCE — ESTIMATOR: KC — ESTIMATE NO. R-1

LOCATION — EXTENSIONS: KC — SHEET NO. 2 of 2

ARCHITECT ENGINEER — CHECKED — DATE: 1972

CLASSIFICATION: MASONRY (CHIMNEY STACK)

DESCRIPTION	NO.	DIMENSIONS				ESTIMATED QUANTITY	UNIT
CHIMNEY STACK (cont)							
4" Standard Firebricks in fireplace, wi ½ in. flush joints in fireclay mtr (as specs) (allow 3% Waste)							
(hearth)	1/	3-4	x	2-0	7		
(2/2-0 + 2-8) (sides)	1/	6-8	x	2-6	17		
					24 x 6⅓ bks =	155	bks
Steel Throat & Damper Unit size 38½" x 13" x 4½" h - o/all (as specs) built-in & incl rough bk fill to form smoke shelf & incl Parging smoke chamber above wi cmt-lime mtr about ½ in. thick	1/					1	No.
2½" x 2½" x ¼" M. Steel Lintel x 3'4" long built in over f'place opening	1/					1	No.
13" x 13" (o/all size) terracotta Flue Liners							
(20-4½ – (2-6 + 1-10½))	1/	16-0				16	LF
4" thick Precast Conc Chimney Cap, size 1'9" x 1'9", wi weathered top face & 13" x 13" flue hole, set on bk stack	1/					1	No.
Mortar (1:3) } Scaffolding }		include in unit prices for items above —				—	

(End of Stack & Fireplace)

FRANK R. WALKER CO., PUBLISHERS, CHICAGO

subject of some early construction disputes, and formalized methods of measurement subsequently appeared as the disputes were settled. One measurer wrote in the seventeenth century that "the truest way was to measure [chimneys] as a Solid, and deduct the Vacancies"—meaning the voids,[8] which appears to be perfectly simple and logical. But, it seems, there are good arguments for other methods, and other methods are used.

The **MM-CIQS** requires masonry in chimney stacks to be measured separately, the masonry units being enumerated (bricks, per thousand). The **SMM-RICS** requires much the same, except that brickwork is required to be measured in square yards, stating the thickness; but "brickwork of two-brick thickness and over. . .shall be reduced to one-brick and given separately in square yards," which would generally be the case with chimney stacks. The **SMM-RICS** also says that no deductions shall be made for flues wherein the voids and the *work displaced* (by flues) do not together exceed 3 square feet in sectional area; the **MM-CIQS** wisely avoids any mention of deductions for flues. Obviously, anyone measuring chimney stacks must tread carefully, and it seems that the seventeenth-century measurer had the simplest and clearest idea: measure that which is solid, and deduct the voids. It could be argued, however, that an even simpler method would be to measure the stack as a *run* item (in LF), describing the flues and stating their sizes and the overall size of the stack. There does not appear to be any advantage in following the published standard methods, for chimney stacks are not usually numerous in any project and it would be much easier to *cost account* for a *run* item.

Special work of this nature should be placed under a descriptive heading so that it may be priced as such; i.e., one brick masonry chimney stack and fireplace. A sub-heading describes the basic material to be used: the facing bricks, which have had a prime cost (per thousand) applied to them in the contract by the *designer*, so as to leave his choice within that price range open. The mortar, the joints, and the bond must also be fully described.

Prelim calcs are for the o/a ht, the ht above fireplace, and the ht above fireplace and below ceiling, since there is a different plan section at each level. The perim is calculated at each of the three levels.

The bkwk of facings is measured in SF, indicating thickness and finish. The attached piers are more easily priced as a *run* item. Both quantities can easily be converted to the number of bricks, as shown. The factor varies with the size of the bricks.

The bkwk of firebricks in the fireplace is measured likewise, followed by the supply and installation of the prefabricated steel throat and damper unit described in the specifications. This installation requires some additional *masonry work* around and above the throat, the costs of which are mostly *labor costs* and the *material costs* of which are not large and are fairly constant according to the size of the unit. Consequently, this *work* is best included and priced with the damper unit. Attempting to measure this part of the *work* more accurately in order to price the masonry materials more precisely would simply give the appearance of accuracy.

The flue liners start above the smoke chamber (above the damper unit) and extend up through the precast conc chimney cap, measured later. Common bkwk (not of facing bricks) is measured to three sides, the fourth side of facings having already been measured.

If the bkwk is measured as a *super* (or *run*) item, the cost of mortar is usually allowed for in the *unit price.* If the bkwk is measured by the number of masonry units, the mortar can still be allowed for in the same way. But **MM-CIQS** calls for masonry mortar to be measured separately, with the net quantity required (no allowance for *waste* to be made in the measured quantities) in cubic yards. The quantity of mortar can be calculated according to the type and quantity of masonry units from the same data that would be used to calculate the mortar costs for the *unit price.* Considering the relatively minor costs of mortar, and the very large and variable amount of *waste* usually incurred, it would seem simpler and adequate not to measure masonry mortar separately. Scaffolding for *masonry work* is not always measured, and the costs are often allowed for in the pricing. Yet in some kinds of masonry, the costs of scaffolding are as much as the costs of mortar.

Residence Example

Metals: Miscellaneous and Structural

In such a small job all metals can be measured together. The measurement is simple, but full descriptions of the *work* are essential for proper pricing. Each item is enumerated and described; but with

[8] F.M.L. Thompson, *Chartered Surveyors: the Growth of a Profession* (London: Routledge and Kegan Paul Limited, 1968). Chapter Four, The Origins of Quantity Surveying, refers to Venterus Mandey's *Marrow of Measuring,* of 1682, in which the author, a professional measurer, is "much exercised by chimneys."

PRACTICAL
STANDARDIZED FORMS FOR CONTRACTORS
Form 516 MFD IN U.S.A.

QUANTITY SHEET

PROJECT RESIDENCE	ESTIMATOR KC	ESTIMATE NO. R-1
LOCATION	EXTENSIONS KC	SHEET NO. 1 of 2
ARCHITECT ENGINEER	CHECKED	DATE 1972

CLASSIFICATION METALS: MISCELLANEOUS & STRUCTURAL

DESCRIPTION	NO.	DIMENSIONS	ESTIMATED QUANTITY	UNIT
THE FOLLOWING IN STEEL BEAM WITH CONNECTIONS:—				
10 WF 45 x 17'6" long (=787½ lb) Beam installed in wood framed upper floor —	1/	17-6 x 45 lb = 787½ lb	(lb 787½)	1 No.
¾"φ x 10" long m.s Drift Pins welded to beam flanges at ends	2/1	(x 1¼ lb)	(2½)	2 No.
¼"x4"x6" x 4" long m.s Angle Brackets 3/drilled (for ⅜"φ bolts m/s)	4/2	(x 3 lb)	(24)	8 No.
⅜"φ m.s Bolts x 6" long with head, nut, & washers (for brackets m/s)	4/4	(x ¼ lb)	(4) (818)	16 No.
Drill Web of 10 WF 45 Beam (for ⅜"φ bolts m/s)	4/2			8 No.
Drill Flanges of 10 WF 45 Beam (for r/carp lag-bolts to plates m/s)	4/2			8 No.
(End of Steel Beam)		approx total weight	(820 lb)	
¾"φ x 10" long m.s Drift Pin a.b (installed in conc slab for 6x8 wood post over)	1/	(x 1¼ lb)	(1 lb)	1 No.

FRANK R. WALKER CO., PUBLISHERS, CHICAGO

PRACTICAL "STANDARDIZED" FORMS FOR CONTRACTORS
Form 516 MFD IN U.S.A

QUANTITY SHEET

PROJECT RESIDENCE | ESTIMATOR KC | ESTIMATE NO. R-1

LOCATION | EXTENSIONS KC | SHEET NO. 2 of 2

ARCHITECT ENGINEER | CHECKED | DATE 1972

CLASSIFICATION METALS (cont)

DESCRIPTION	NO.	DIMENSIONS		ESTIMATED QUANTITY	UNIT
			b-fwd	(lb 1)	
3/8"x 3½" m.s Flat Bar Anchors x 3'6" long hooked (180°) one end & 4/drilled for & incl 3/8" ϕ x 6" long m.s bolts wi head, nut & washers; set in conc fnd & bolted to wood framing (as Specs and Detail "F.")	5/2	(x 17 lb)		(lb 170)	10 No.
2"x3" (x 6 lb) Steel Channel Stiffeners x 7'0" long 3/drilled for & incl 3/8"ϕ x 8" long m.s bolts wi head, nut & washers; set in conc flr slab & bolted btwn partn studs (at ends of free-standing partn)	2/1	(x 43 lb)		(85)	2 No.
¼"x 1"x 1" m.s Angle Supports x 10" long for wood Stair Treads 4/ drilled for wood screws	2/12	(x 1½ lb)		(40)	24 No.
				(296)	
		(End of Misc. Metals)	approx total weight	(295 lb)	

[NB: Steel Chimney Throat & Damper Unit & lintel taken with Masonry.]

FRANK R. WALKER CO., PUBLISHERS, CHICAGO

many more items, some similar items might be grouped together and measured by weight, stating the number of items in each group.

Each item is described as installed; but in some estimates, such as that of a *miscellaneous metals subcontractor*, the items would be "supply only," and installation would have to be estimated separately by the *contractor* or by other *subcontractors*.

The steel beam is measured first, followed by the ancillary items—the drift pins, brackets, and bolts to connect the beam and the wood frame. The dwg detail of the steel beam to column connection indicates a pair of brackets to "every 4th floor joist," which does not exactly indicate how many are required. Since the beam is pinned to the wood columns at one end, three pair would appear to be adequate; but the *designer* might argue for four. Each pair requires four bolts, which in this case are more easily measured separately instead of with the brackets.

Flat bar anchors are shown in Detail "F", and here the bolts can be easily measured with the item.

The 2″ × 3″ steel channels bolted between two studs at the ends of the free-standing partition (on the GF) are to be set in the conc slab (suitably thickened) to give rigidity to the partition.

Weights are needed to price the *material costs* of these items, and it is useful to remember that steel plate ¼ in. thick weighs 10.2 lb per SF, from which weight the weights of most steel flats, angles, and plates can be calculated with sufficient accuracy for most items.

Other metal items required for this job should be included in other *sections of work*. Rough and finish hardware for carpentry are measured with the carpentry items. Reinforcement is measured with concrete or masonry, wherever it occurs. Misc metals generally include only custom made metal items, and not standard products. Structural metals are generally of standard sections cut to the specific lengths as required by the job, with the necessary connections.

Residence Example

Rough Carpentry Framing

Rough carpentry frmg usually includes sheathing to walls and decks, and it is usually better to begin by measuring the sheathing and then to measure the wood frmg beneath. In this way, the estimator becomes familiar with the layout and the dims he will use in measuring the more complicated frmg later.

Because of this building's design, there is little or no advantage in calculating and using a wall perimeter to measure the frmg, and so the total length of extr walls is collected from the dims. Sheathing to walls, floor, and roof deck are each measured separately. The number of exposed ends of walls makes measuring the sheathing to these ends (in n.w.'s, as a *run* item) worthwhile. Likewise, the n.w.'s of sheathing behind parapets is measured as a *run* item. Usually such items can be described as less than 3 in. wide, 3 to 6 in. wide, 6 to 9 in. wide, and less than 12 in. wide because exact width is not significant to cost and in this way similar widths can be conveniently grouped together. Cants and sheathing paper, both attached to the plywood sheathing, are measured next.

The distinction between "rough" and "finish" carpentry is not always absolutely clear; thus, items that are left out of one *section* to be measured in the other are sometimes overlooked. Therefore, it is wise to make notes of such items as the exposed ½ in. ply cladding. Rough carpentry is taken here to be *rough, concealed work*; and all woodwork that is exposed and finished is taken as finish carpentry. Nevertheless, exterior siding is often classified as rough carpentry.

The lengths of extr and intr walls are collected and the numbers of studs calculated by dividing by the stud-spacing (16 in.) and by adding extra studs for each wall length (one); for short returned lengths of extr wall (four, assuming that a maximum of seven are needed and that three have already been measured); for corners (three); and for junctions (two extra studs in extr walls, and one extra stud in intr walls). Framing methods and the numbers of studs used at such locations vary; for example, four studs are sometimes used at corners in extr walls.

Extr wall frmg is measured separately because the extr walls are often constructed differently from intr walls. One bottom plate and a double top plate require three times the extr wall length. Intr walls require two or three times the wall length for plates, depending on a single or double top plate. Here, double top plates have been measured. Extra intr wall studs are measured at doorways, and extra material is also measured at the bathtub. Intr finishes and fixtures always require some extra frmg material, and this must be allowed for.

Fixing plates to conc is measured as a separate item. Girths in stud walls are also measured separately because of their relatively high *labor cost*.

The **MM-CIQS** calls for all frmg lumber to be measured separately if over 14 ft long, and to be grouped in 2-ft stages. Studs are required to be grouped separately in lengths less than 8 ft and in lengths 8 to 12 ft. These categories relate to lumber costs, but their

PRACTICAL "Standardized" Forms for Contractors
Form 516 MFD IN U.S.A.

QUANTITY SHEET

PROJECT	RESIDENCE	ESTIMATOR	KC
		ESTIMATE NO.	R-1
LOCATION		EXTENSIONS	KC
		SHEET NO.	1 of 6
ARCHITECT ENGINEER		CHECKED	
		DATE	1972
CLASSIFICATION	ROUGH CARPENTRY FRAMING (SHEATHING)		

Prelim Calcs

Dims – U. Floor		Roof				Ext Walls (collection)				Height	
(W) 13-6 (N)	5-8	15-2 (E)	22-0 (S)	15-2 (N)	5-8	a'rnd Pool	30-0	√133-0	13-6	8-7½	L.Flr
8-4	8-4	½/= 0-2	8-0	10-10	8-4	13-4	14-0	4/= 8-0	8-4	7-6	U.Flr
21-6	2/= 4-8	15-4	30-0	5-4	14-0	1-8	2-0	2/2/= 12-0	1-0 21-10	1-2	Roof
43-4	18-8 ✓	2-0	-5-0	31-4		8-0	23-0	2/2/= 10-0	0-2 = 1-2	0-8½	Parapt
2/= -0-8	2/= -0-8	17-4	25-0	3-0	2-6	23-0	21-8	163-0	20-8	18-0	
42-8 ✓	18-0 ✓		½/= -0-4	0-4	0-4	21-6	21-8				
			24-8	3-4	2-10	½/= 0-2	20-8				
						21-8	133-0				

DOUGLAS FIR PLYWOOD SHEATHING (EXTERIOR GRADE)

DESCRIPTION	NO.	DIMENSIONS			ESTIMATED QUANTITY	UNIT
½" Select Sheathing (T&G) to Floor (allow 10% Waste)						
	1/	42-8 ×	18-0	768		
	1/	17-4 ×	8-4	144		
				912		
DDT (@ S/well)	1/	8-9 ×	3-0	-26 DDT		
				886	890	SF
½" Select Sheathing (T&G) to Roof (allow 10% Waste)						
	1/	42-8 ×	18-0	768		
	1/	24-8 ×	17-4	428		
	1/	14-0 ×	2-10	40		
	1/	31-4 ×	3-4	104		
				1340	1340	SF
5/16" Standard Sheathing to Extr Walls (allow 7½% Waste)						
	1/	163-0 ×	18-0	2934		
(@ Garage door jambs)	2/	0-4 ×	6-6	4		
				2938	2940	SF
5/16" Ditto in N.W <6" wide (wall ends)						
	1·4/	18-0		90	90	LF
5/16" Ditto in N.W <9" wide (parapets)						
(42-8 + 2-10 + 3-4)	2/	48-10		98		
(18-0 + 17-4)	2/	35-4		70		
	2/	3-0		6		
				174		
DDT (want) (31-4 + 17-0 + 14-0)	1/	62-4		-62 DDT.	112	LF

FRANK R. WALKER CO., PUBLISHERS, CHICAGO

PRACTICAL "STANDARDIZED" FORMS FOR CONTRACTORS
Form 516 MFD IN U.S.A.

QUANTITY SHEET

PROJECT RESIDENCE — ESTIMATOR KC — ESTIMATE NO. R-1
LOCATION — EXTENSIONS KC — SHEET NO. 2 of 6
ARCHITECT ENGINEER — CHECKED — DATE 1972
CLASSIFICATION ROUGH CARPENTRY FRAMING (SHEATHING)

DESCRIPTION	NO.	DIMENSIONS			ESTIMATED QUANTITY	UNIT
4x4 D.fir 'Constr' grade Roof Cants (on sheathing) (allow 5% Waste)						
		(Length of parapets) =		112		
ADD (@ E. overhangs)		1/62-4		62		
(@ arnd c. stack)		1/9-4		10	184	LF
Standard (black) Building Paper to Extr Wall Sheathing (Laps measd)						
(allow 10% Waste)		(area ply wall shthg) =		2938		
ADD (@ N.W's)		1/ 90-0 x	0-6	45		
		1/ 112-0 x	0-9	84		
2-3 4-10 7-1 ✓ (@ ply siding-N)		1/ 14-0 x	7-1	99		
		1/ 17-0 x	3-4	57		
2-3 1-1 3-4 ✓ (@ ditto -S)		1/ 31-4 x	3-4	104		
				3327		
ADD (laps)	approx	20% x 3327 SF =		673	4000	SF
1"x2" Thickener planted on u/s proj ply shthg at top stair tread approx 4' long					1	No
[NB: ½" Ply Siding (exposed) measd as 'Finish Carpentry.']						
W. RED CEDAR 'CONSTRUCTION' GRADE						
2x4 Framing for planting pots supports (N. Elev) in 16' lengths (Waste measd)						
		8/14-0		112		
		5/2-0		10	128	LF
1x6 Bevel Siding as parapet capping (allow 5% Waste)						
		(Length of parapets)			112	LF

PRACTICAL Form 516 MFD IN U.S.A.

QUANTITY SHEET

PROJECT: RESIDENCE — ESTIMATOR: KC — ESTIMATE NO. R-1

LOCATION — EXTENSIONS: KC — SHEET NO. 3 of 6

ARCHITECT ENGINEER — CHECKED — DATE: 1972

CLASSIFICATION: ROUGH CARPENTRY FRAMING (WALLS)

DESCRIPTION	NO.	DIMENSIONS	ESTIMATED QUANTITY	UNIT

Prelim Calcs

Extr Dims
- (E) 22-0+8-0 = 30-0
- (N) 14-0 − 3/4" = 13-4
- (E) 21-6 + 2 1/4" = 21-8
- (W) 13-8+3-10 +5-4+20-6 = 43-4

Extr Wall Perim
- (E) 30-0
- (N) 13-4
- (E) 21-8
- (W) 43-4
- 2/2-0 = 8-0
- 2/3-0 = 6-0
- 2/2-6 = 5-0
- 1/2-0 = 2-0
- (129-4)
- Total (say) 130-0

No Studs
- 129-4 / 1-4 = 97
- W. lengths = 13
- S. lengths 5/4 = 20
- corners 6/3 = 18
- junctns 5/2 = 10
- window 1/2 = 2
- Total 160 No

Int Walls (collect)
- (GF) 2/ = 36-0 (e-w)
- 3/ = 15-0 (n-s)
- 1/ = 3-0 (@ furnace)
- 2/ = 3-0 (@ HW Tank)
- 57-0
- (GF-6½' Wall) 28-6
- (stack) −4-0
- (24-6)
- 25-0

No Studs
- 57-0 / 1-4 = 43
- w. lengths = 8
- junc 7/1 = 7
- (58)
- Total 60 No
- 24-6 / 1-4 = 17
- 4/2 = 8
- 25 No

(UF)
- 2/ = 36-0 (e-w)
- 2/ = 3-0 "
- 1/ = 7-0 "
- 2/ = 26-0 (n-s)
- 1/ = 8-0 "
- 1/ = 21-0 "
- 2/ = 5-0 (closets)
- Total 106-0

No Studs
- 106-0 / 1-4 = 80
- wall lengths = 11
- junctions 9/1 = 9
- corners 4/3 = 12
- (112)
- Total 110 No

D. FIR "CONSTRUCTION" GRADE

2×4 Extr Wall Frmg in 18' & 20' lengths (Waste incl)

Description	No.	Dimensions	Extension	Estimated Quantity
(plates)	3/	130-0	390	
(studs)	160/	18-0	2880	
			3270 (× 2/3 BF)	3300 LF (2200 BF)

2×4 Girths btwn extr wall studs (Waste incl)

Description	No.	Dimensions	Extension	Estimated Quantity
(2/3-0 + 2/2-6 + 1/2-0)	1/	13-0		13 LF

2×4 Intr Wall Frmg in 18' lengths (Waste incl)

Description	No.	Dimensions	Extension	Estimated Quantity
(GF plates)	3/	57-0	171	
"	3/	25-0	75	
(UF plates)	3/	106-0	318	
(@ Gallery Balustr)	3/	12-0	36	
			600 (× 2/3 BF)	600 LF (400 BF)

2×4 Intr Wall Frmg in 8' lengths (Waste incl)

Description	No.	Dimensions	Extension	Estimated Quantity
(① 25 for 6½' wall) (GF studs)	25 ① 60/	8-0	680	
(extra for GF col @ Beam)	6/	8-0	48	
(UF studs)	110/	8-0	880	
(extra @ B. Tub)	4/	8-0	32	
(extra @ Doorways)	15/2/	8-0	240	
(extra @ top plates for drywall ceilings)	2/2/	36-0	144	
	1-1/2/	8-0	32	
			2056 (× 2/3 BF)	2050 LF (1370 BF)

① EXTRA (PARALLEL TO JOISTS)

FRANK R. WALKER CO., PUBLISHERS, CHICAGO

PRACTICAL "Standardized" Forms for Contractors
Form 516 MFD IN U.S.A.

QUANTITY SHEET

PROJECT RESIDENCE — ESTIMATOR KC — ESTIMATE NO. R-1

LOCATION — EXTENSIONS KC — SHEET NO. 4 of 6

ARCHITECT ENGINEER — CHECKED — DATE 1972

CLASSIFICATION ROUGH CARPENTRY FRAMING (WALLS)(FIXTURES)

DESCRIPTION	NO.	DIMENSIONS		ESTIMATED QUANTITY	UNIT
D.FIR "CONSTRUCTION" GRADE					
Extra Lab & Mat Fastening 2x4 Wall Plates to conc wi & incl ½"φ A. Bolts @ 6' o/c	1	130-0		130	LF
Ditto 2x4 Plates to conc slab wi & incl Power Fasteners @ 4' o/c					
(57-0 + 25-0)	1	82-0		80	LF
2x4 Extr Wall Frmg in Panels (for ½" Ply Siding m/s) in 14' lengths (Waste incl)					
	2/4	14-0	112		
	14	5-0	70		
	14	2-0	28		
			210 (x ⅔ BF)	210	LF
				(140	BF)
2x3 Intr Wall Frmg at Stairs in 10' & 12' lengths (Waste incl)					
	2/3	10-0	60		
	2/8	12-0	192		
	2/2	12-0	48		
			300 (x ½ BF)	300	LF
				(150	BF)
2x3 Ditto at Kitchen Cabinets & Fixtures in 12' lengths (Waste incl)					
(W. Elev plates)	6	12-0	72		
(" studs)	22	6-0	132		
(" hatch)	3/2	4-0	24		
(@ Range & Frig)	2/6	2-0	24		
(" " studs)	2/4	6-0	48		
			300 (x ½ BF)	300	LF
				(150	BF)
2x8 (full ht) Frmg in Bases to K. Cabinets (Waste incl)					
(8-0 + 12-0)	1	20-0	20 (x 1⅓ BF)	20	LF
				(30	BF)
2x12 (full ht) Frmg in Tops to K. Cabinets (Waste incl)					
	2	12-0	24 (x 2 BF)	25	LF
				(50	BF)
3x12 Frmd Stairs Stringers in 12' lengths (Waste incl) incl ends to slope					
	2	12-0	24 (x 3 BF)	25	LF
				(75	BF)
2x12 "C & Better" V.G. Stepping in Stairs treads in 6' lengths (on Brkts m/s)					
	12	3-0	36 (x 2 BF)	36	LF
				(72	BF)
Extra Lab Selecting & Installing Clear Exposed 2x4 Studs at Doorways (as detail)					
(both jambs meas'd.)	15	2		30	No
[NB: Exposed Wood Posts & Mullions measd as "Finish Carpentry".]					

FRANK R. WALKER CO., PUBLISHERS, CHICAGO

PRACTICAL "STANDARDIZED" FORMS FOR CONTRACTORS
Form 516 MFD IN U.S.A.

QUANTITY SHEET

PROJECT RESIDENCE | ESTIMATOR KC | ESTIMATE NO. R-1
LOCATION | EXTENSIONS KC | SHEET NO. 5 of 6
ARCHITECT ENGINEER | CHECKED | DATE 1972
CLASSIFICATION ROUGH CARPENTRY FRAMING (DECKS)

DESCRIPTION	NO.	DIMENSIONS			ESTIMATED QUANTITY	UNIT
U.Floor No Joists						
13-6 (43-4)/(1-4) = 33		(8-8)/(1-4) = 7		Roof (25-0)/(1-4) = 19	Joist Lngth 2-4	
8-4 +1		2/1 8-4 +1 38		30-0 +1 38	5-8	
21-6 (extra) +4		2/=0-4 (extra) +4 12		−5-0 (extra) +3 23	8-4	
43-4 38		8-8 12 50 No Joists		25-0 23 61 No Joists	2-4	
					(18-8)	
					20-0 long	
D.FIR "CONSTRUCTION" GRADE						
2x12 Floor Joists in 20' lengths (Waste incl)						
	50	20-0	1000 (x2 BF)		1000	LF
					(2000	BF)
2x12 Roof Joists in 20' lengths (Waste incl)						
	61	20-0	1220			
(@ S. Overhang)	3/2	20-0	120			
(@ N. Overhang)	3/1	20-0	60			
(Lam. beams-roof)	3	20-0	60			
(extra for blockings)	2	20-0	40			
			1500 (x2 BF)		1500	LF
					(3000	BF)
Extra Lab (only) in Solid 2x12 Blockings to btwn ends of joists (using offcuts)						
(flr and roof)	2/2	43-0	172			
(roof overhangs)	2/2	3-0	12			
(ditto)	1/2	2-6	5			
(@ Steel Beam - N. end)	2	4-0	8			
			197		200	LF
Extra Lab Cutting & Fitting 2x12 Joists' Ends to Steel Beam						
(17-6)/(1-4) = 13 +1 = 14	3 @ 14	1	17		17	No
Extra Lab & Joist Hangers Trimming 2x12 Joists at Lam. Beams (O/s meas'd)						
(10-0 + 6-0)	2	16-0	32		32	LF
Ditto Trimming at Stairwell, 9' x 3'	1				1	No
Ditto Trimming at Stack, 4' x 2'8"	1				1	No
X-Bridging to 2x12 Joists at 16" o/c (meas'd over joists)						
(43-4 + 8-0) (Floor)	2	51-4	103			
(43-4 + 3-0 + 2-6) (Roof)	2	48-10	98			
(30-0 − 5-0) + 3-0 (")	2	28-0	56			
			257		260	LF

FRANK R. WALKER CO., PUBLISHERS, CHICAGO

PRACTICAL STANDARDIZED FORMS FOR CONTRACTORS
Form 516 MFD IN U.S.A.

QUANTITY SHEET

PROJECT RESIDENCE | ESTIMATOR KC | ESTIMATE NO. R-1

LOCATION | EXTENSIONS KC | SHEET NO. 6 of 6

ARCHITECT ENGINEER | CHECKED | DATE 1972

CLASSIFICATION ROUGH CARPENTRY FRAMING (SUNDRIES)

DESCRIPTION	NO.	DIMENSIONS		ESTIMATED QUANTITY	UNIT
D. FIR "CONSTRUCTION" GRADE					
1x6 Ledger Boards set flush into studs @ 16" o/c for joists (Waste 5%)					
(flr and roof)	2/2/	43-0	172		
(roof overhangs)	2/2/	3-0	12		
(ditto)	1/2/	2-6	5		
			189 (x ½ BF)	190	LF
				(95	BF)
Ex-2x12 Headers behind fascias in 16' & 18' lengths (Waste incl)					
(16-0+18-0) (N&S)	2/	34-0	68 (x 2 BF)	70	LF
				(140	BF)
2x12 Blockings btwn mullions in 16' & 18' lengths (Waste incl)					
(S)	2/	32-0	64		
(N)	2/	18-0	36		
			100 (x 2 BF)	100	LF
				(200	BF)
2x10 Headers behind valances in 16' & 18' lengths (Waste incl)					
(N&S)	2/2/	34-0	136 (x 1⅔ BF)	140	LF
				(230	BF)
2x4 Blockings on 4/sides {blockings / headers} in 16' & 18' lengths (Waste incl)					
(32-0 + 18-0) (on 2x12)	1/	50-0	50		
(on 2x10)	2/2/	34-0	136		
			186 (x ⅔ BF)	190	LF
				(125	BF)
2x4 Frmd Furrings to eaves soffites in short lengths (Waste incl)					
2x (31-0 / 1-4) = 48; 1·3/2 = +8 (say) 60	60/	2-6	150 (x ⅔ BF)	150	LF
				(100	BF)
2x4 Blockings lag-bolted to drilled steel beam in 14' lengths (Waste incl)					
	2/	14-0	28 (x ⅔ BF)	30	LF
				(20	BF)
1x3 Blocking/Furring on lower 2x4 (Waste incl)					
1x3	1/	14-0	14 (x ¼ BF)	15	LF
				(4	BF)
2x6 Blockings in Valences in 18' lengths (Waste incl)					
(S)	2/	36-0	72		
(N)	2/	18-0	36		
			108 (x 1 BF)	110	LF
				(110	BF)
Rough Hardware (include with items of work)				—	

FRANK R. WALKER CO., PUBLISHERS, CHICAGO

rigid application is not always practical or necessary, and in small jobs separation as shown in the example is adequate.

Estimated total quantities of frmg are shown in LF, and (in parentheses) in BF. Usually, one or the other unit of measurement is used, and the linear unit is preferable for reasons already given.

The 2 × 3 frmg material at the stairs and kitchen cabinets is similarly measured, as is the isolated 2 × 4 frmg for the ext ply panels on the north elevation. This kind of frmg requires more material per unit area than ordinary walls, and adequate quantities must be measured. Also, both these items should be priced higher than ordinary wall frmg because of the arrangements and locations of this *work*. No frmg details are indicated here, and it has been assumed that the 2 × 4 wall frmg for ply siding panels would be prefabricated frames. An estimator often has to make such assumptions before he can measure *work* that is not fully detailed.

The numbers of floor and roof joists required are calculated from the dims divided by the joist spacing, with extra joists added for double joists and for trimming around stairwell and stack. The joist lengths are taken up to the nearest even dim (20 ft), with the knowledge that offcuts can be used for solid blockings between joists. Two additional 20-ft lengths are included for solid blocking that cannot be obtained from offcuts, although they may not be required; but it is easy to under-measure these requirements. Additional labors on ends of joists and trimming are described and enumerated.

Cross-bridging can be measured as a *run* item, or enumerated in sets, one set per space but never in BF. Ledgers are described as "set into studs" to remind the estimator of the labor involved when pricing. Likewise, locations of blockings are indicated to facilitate pricing later. Fixing blockings to a steel beam is much different from fixing them to a wood frmg member.

Rough hardware for carpentry includes nails, screws, spikes, joists hangers, and the like. Usually, it is not necessary to measure quantities of rough hardware, because the common items are better allowed for in the *unit prices*, as shown in Chapter 11.

Residence Example

Finish Carpentry

The exposed plywood cladding in panels to extr walls would be regarded by some estimators as Rough Carpentry, but the fact that it is exposed and that the joints have to be properly laid out and carefully made means that it requires finish and that it should be priced accordingly and differently from plywood sheathing. The *intr work* is similarly measured, but kept separate. Drilling vent holes is described on the dwgs "as directed," which means that the estimator cannot measure the *work*. In the case of a major item so described, the *owner/designer* should be asked for more information. If it is not available, the estimator can make an allowance to cover the probable costs or he can refuse to submit a *bid*. If, as in this case, it is a minor item, the estimator can make an allowance without too much risk.

The hardwood ply paneling is shown with no horizontal joints, which means that about two thirds of the paneling will require sheets 12 ft long, more expensive than the standard 8-ft-long sheets.

Raking cutting and *waste* is measured as a basis for pricing the additional costs. Another way would be to measure the material gross (instead of net) and to measure and price the extra labor cutting to the required angle. Cutting slots in the hardwood ply for the concealed tread angles is an expensive labor item that must be included.

The hardwood ply paneling in the kitchen cabinets is similar to the paneling by the stairs, and both items could be grouped together because the extra labor items in each case have been identified and measured separately. Thus, the paneling in the cabinets requires mitred corners (not required at the stairs) which is measured separately; the extra labor involved in making the serving hatch door is also measured separately.

The Douglas fir finish material is described as "C" grade, vertical grain, which means it is kiln dried and of high quality (next to "B" grade, which is the highest). Some estimators would measure an extra labor item for the shaped ends to the 2 × 8 handrails. But in a small job, with a small number of different items, the estimator can sometimes allow for such things in his pricing without detailed measurement.

Measurement of base trim (of any kind) can be tedious and confusing. The simplest way is to use the total lengths of walls—once times extr walls, and twice times intr walls—and deduct the major *wants*. Whenever possible other trim should also be measured by utilizing dims of other *work* to which the trim is attached.

Finish carpentry usually includes the installation (by carpenters) of millwork and other fixtures supplied to the job by a *subcontractor* or a *supplier*. In this example it is assumed that the wood sash, wood

PRACTICAL "STANDARDIZED" FORMS FOR CONTRACTORS Form 516 MFD IN U.S.A.

QUANTITY SHEET

PROJECT RESIDENCE — ESTIMATOR KC — ESTIMATE NO. R-1

LOCATION — EXTENSIONS KC — SHEET NO. 1 of 7

ARCHITECT ENGINEER — CHECKED — DATE 1972

CLASSIFICATION FINISH CARPENTRY (PLYWOOD)

DESCRIPTION	NO.	DIMENSIONS					ESTIMATED QUANTITY	UNIT
D.FIR PLYWOOD (EXTERIOR GRADE) GOOD ONE SIDE (G1S) unless otherwise shown								
½" Ply Panels to Extr Walls (Fascias & Siding) wi exposed vertical 'veed' joints as indicated in Elevations (& as in Details #1/3 & #2/3) (allow Waste 25% on net quantity)						SHEETS (4'x8')		
(15-2+10-10+5-4)=31-4	(S) 1	31-4	x	2-3	71	4		
(4+11½+9½+2")=2-3	1	31-4	x	1-1	34			
(5-8+8-4)=14-0	(N) 1	14-0	x	2-3	32	4		
(*dim scaled)(14-0+3-0)=17-0	1	14-0	x	4-10*	68			
(ASSUME CONCEALED	1	17-0	x	2-3	38	2		
JOINTS IN PANELS OVER	1	17-0	x	1-1	19			
8FT LONG)					262	(10)	262	SF
½" Ply Panel (as above) 14" x 27" over Window (west) ✓							1	NO
Extra Lab making concealed x-tongued vert joints in ½" Ply Panels								
	(N) 1	2-3			2¼			
(S) ⊙	(N) 1⊙1	1-1			2¼ LF		(3 No) 4½	LF
½" Ply Intr Wall Panels wi concealed x-tongued vert joints (Waste 12½%)								
(as Detail #3/3)	(S) 1	31-4	x	1-2	37			
	(N) 1	17-0	x	1-2	20			
					57	(2)	57	SF
½" Ply (Good/Solid grade) in Valences x 11" wide wi x-tngd jts (Waste 12½%)								
(as Detail #1/3)	(S) 1	35-4			36			
	(N) 1	18-0			18			
					54		54	SF
½" Ply Eaves Soffites (on 2x4 furr'g m/s) (Waste 15%)								
(as Detail #1/3)	(S) 1	31-4	x	2-9	86			
	(N) 1	17-0	x	2-9	47			
	(N) 1	14-0	x	2-3	32			
					165		165	SF
Extra Lab Drilling Vent Holes in ½" Ply Eaves "as directed"								Item

[NB: ½" Ply Panel in Window (west) supplied with Window (in Div. 8).]

FRANK R. WALKER CO., PUBLISHERS, CHICAGO

PRACTICAL "STANDARDIZED" FORMS FOR CONTRACTORS
Form 516 MFD IN U.S.A.

QUANTITY SHEET

PROJECT RESIDENCE — ESTIMATOR KC — ESTIMATE NO. R-1

LOCATION — EXTENSIONS KC — SHEET NO. 2 of 7

ARCHITECT ENGINEER — CHECKED — DATE 1972

CLASSIFICATION FINISH CARPENTRY (PLYWOOD)

DESCRIPTION	NO.	DIMENSIONS				ESTIMATED QUANTITY	UNIT
AMERICAN BLACK WALNUT PLYWOOD ("CUSTOM" GRADE) as Specs							
1/2" Walnut Ply Panelling (at Stairs) (ex- 4x12 sheets) (67% of net area) wi exposed vertical "Veed" joints as indicated (Waste 20% on net)							
(7-6+1-1+3-0)=11-7	2/2/	8-8	x	11-7	402		
					SHEETS (4'x8') = 4; DDT 113 (4'x12') 4; 17		
DDT (wants)	3/3/1/2/	7-6	x	7-6	–		
(@ flr joists)	2/	8-8	x	1-0	–		
					-130		
					272	272	SF
Extra Lab in X-tongued end-matched joints in 1/2" Ply x 7 1/2" long	2/1					2	No
Extra Lab in cutting Slots in 1/2" Ply for 2x12 Treads & Sinkings for 1/4" thick Steel Angle brackets behind	2/12					24	No
RC & W on 1/2" Walnut Ply (@ btm edges to slope)	2/2/	10-9			43	43	LF
1/2" Walnut Ply Panelling at Kitchen Cabinets & Fixtures (as Details) (Waste 50%)							
(2-1+1-5+2-4)=(5-10) 6-0	1/	8-0	x	6-0	48		
	2/2/	1-3	x	6-0	30		
	2/2/	0-4	x	6-0	8		
	2/2/	0-4	x	2-0	3		
	2/2/	0-4	x	1-3	2		
					91	91	SF
Extra Lab to Mitered Corners in 1/2" Ply Panelling	4 06/	6-0			60	60	LF
Extra Lab Cutting Finishing & Hanging Hatch Door Panel 4'x1'6" incl Brass Piano Hinge & Ball Catch & Finger Pull	1/					1	No
1x2 Walnut Edge Trim (to match) Hatch Door	2/	4-0			8		
	2/	1-6			3	11	LF
1/2" D.FIR PLY (G1S) at Bases & Tops of K. Cabinets wi mitered corners x 8" & 12" wide	2/2/	10-3			41	41	LF

FRANK R. WALKER CO., PUBLISHERS, CHICAGO

PRACTICAL "STANDARDIZED" FORMS FOR CONTRACTORS
Form 516 MFD IN U.S.A.

QUANTITY SHEET

PROJECT RESIDENCE — ESTIMATOR KC — ESTIMATE NO. R-1

LOCATION — EXTENSIONS KC — SHEET NO. 3 of 7

ARCHITECT ENGINEER — CHECKED — DATE 1972

CLASSIFICATION FINISH CARPENTRY (TRIM)

DESCRIPTION	NO.	DIMENSIONS			ESTIMATED QUANTITY	UNIT
D. FIR FINISH ("C" V.G. GRADE)						
2x8 Handrails (on brackets m/s) incl shaped ends (Waste nil)						
	2/	12-0	24		24	LF
3x8 Rail (at Stairwell) framed in at ends (Waste incl)						
	1/	4-0	4		4	LF
3/4"x5 1/4" Double Dressed Cap (at Stairwell) incl miters & ends (Waste 4%)						
(8-9+10-9+11-7 1/2+4-1 1/2)	2/	35-3	71		71	LF
1/2"x1 1/2" Base Trim (at floors/walls) incl miters & ends (Waste incl)						
(length extr walls)	2/	130-0	260			
(" intr ")	2/	57-0	114			
(" " ")	2/	106-0	212			
(31-0+14-0+14-0)(@wdws)	2/	59-0	118			
			704	DDT		
DDT (Wants)(@ pool)	2/	24-0	–	48		
(@ Garage)	2/	38-0	–	76		
(@ S. Lengths)	2/	3-0	–	6		
(")	2/	2-6	–	5		
(@ L Room)	2/	13-0	–	26		
(@ Doorways)		(Nil)	-161			
			543		550	LF

FRANK R. WALKER CO., PUBLISHERS, CHICAGO

"STANDARDIZED" FORMS FOR CONTRACTORS
PRACTICAL
Form 516 MFD IN U.S.A.

QUANTITY SHEET

PROJECT RESIDENCE — ESTIMATOR KC — ESTIMATE NO. R-1
LOCATION — EXTENSIONS KC — SHEET NO. 4 of 7
ARCHITECT ENGINEER — CHECKED — DATE 1972
CLASSIFICATION FINISH CARPENTRY (WINDOW WALLS)

DESCRIPTION	NO.	DIMENSIONS		ESTIMATED QUANTITY	UNIT
DOUGLAS FIR (GRADES AS INDICATED)					
6x8 Structural grade Posts in 8' lengths (Waste incl)					
(8-7½+7-6)-(0-10+0-7½)	(S)	2/8-0	16	16	LF
2x6 Structural grade Mullions in 16' lengths (Waste incl)					
(8-7½+7-6)-0-7½	(S)	4/16-0	64		
	(N)	2/16-0	32		
			96	96	LF
2x4 Structural grade Frames in 16' lengths at walls (Waste incl)					
	(N&S)	2/2/16-0	64	64	LF
2x2 Structural grade Trim in 16' lengths planted on posts & mullions (Waste incl)					
	2·4·	1/16-0	112	112	LF
½"x1¾" "C" V.G. Stops in 6' & 8' lengths planted on (Waste incl)					
(@ Doors) (N)⊙(S)	1·4/	2/8-0	80		
	1·4/	1/3-0	15	95	LF
¾"x1¾" Ditto in 8' lengths planted on (Waste incl)					
(@Glass) (jambs only)	(N&S) 10/	2/2/8-0	320	320	LF
¾"x1¾" Ditto in 16' & 18' lengths planted on (Waste incl)					
(@Glass) (for sills & heads)	(N&S) 3/	2/32-0	192		
	3/	2/18-0	108		
			300	300	LF
½"x2½" Trim in 18' lengths planted on inside (Waste incl)					
(@ wdw heads & Valences)	(S)	1/36-0	36		
	(N)	2/18-0	36	72	LF
½"x2¼" Door Stops in 8' lengths planted on at intr doorways (Waste incl)					
(jambs only)	2·15/	2/8-0	272	272	LF

FRANK R. WALKER CO., PUBLISHERS, CHICAGO

PRACTICAL "STANDARDIZED" FORMS FOR CONTRACTORS
Form 516 MFD IN U.S.A.

QUANTITY SHEET

PROJECT RESIDENCE — ESTIMATOR KC — ESTIMATE NO. R-1
LOCATION — EXTENSIONS KC — SHEET NO. 5 of 7
ARCHITECT ENGINEER — CHECKED — DATE 1972
CLASSIFICATION FINISH CARPENTRY (WDW WALLS) (INSTALLATION)

DESCRIPTION	NO.	DIMENSIONS		ESTIMATED QUANTITY	UNIT
D. FIR FINISH ("C" V.G GRADE)					
1 5/8" x 5 1/4" Sill (as Detail #3/3) in 16' & 18' lengths framed to mullions (Waste incl)					
	(S)	1/32-0	32		
	(N)	1/18-0	18	50	LF
1 5/8" x 7 1/4" Sill (as Detail # 3/3) ditto (ditto)					
	(S)	1/32-0	32		
	(N)	1/18-0	18	50	LF
1 5/8" x 7 1/4" x 14" long ditto (ditto) (@ W. Wdw)		1/1		1	NO
3/4" x 3 1/2" Wdw Stool in 14' length (Waste incl)					
	(N)	1/14-0	14	14	LF
INSTALLATION OF WOOD WINDOWS & DOORS (SUPPLY IN DIV. 8)					
Install Wood Windows in 5 lights 14'0" x 3'10" o/a (N)				1	NO
Ditto 1'1" x 15'6" o/a (W)				1	NO
Hang Wood Doors with hardware to wood frames as follows:— (NB: Intr Doors have matching transom panels over)					
Door (A) 3'0" x 7'4" x 1 3/4" Solid		1/1 (entrance)	1	1	NO
" & Panel (B) 2'8" x 6'8" x 1 3/8" Hollow		5/1 (intr)	5	5	NO
Door (C) 3'0" x 7'4" x 1 3/4" Glazed		4/1 (patio)	4	4	NO
" & Panel (D) 2'6" x 6'8" x 1 3/8" Hollow		4/1 (intr)	4	4	NO
" " (E) 2'0" x 6'8" x 1 3/8" "		3/1 (")	3	3	NO
" " (F) 3'0" x 6'8" x 1 3/8" "		2/1 (")	2	2	NO
" " (G) 2'10" x 6'8" x 1 3/4" Solid		2/1 (")	2	2	NO
" " (HWT) 1'6" x 6'8" x 1 3/4" "		1/1 (")	1	1	NO
O/Head Garage Door 14'0" x 6'6" x 1 3/4" Solid			1	1	NO
Install Glass Louvres in Alum Frames (Supply in Div. 8) to wood frames single lights each 3'0" x 6'8" o/a				3	NO

FRANK R. WALKER CO., PUBLISHERS, CHICAGO

PRACTICAL "STANDARDIZED" FORMS FOR CONTRACTORS Form 516 MFD IN U.S.A.

QUANTITY SHEET

PROJECT RESIDENCE — ESTIMATOR KC — ESTIMATE NO. R-1

LOCATION — EXTENSIONS KC — SHEET NO. 6 of 7

ARCHITECT ENGINEER — CHECKED — DATE 1972

CLASSIFICATION FINISH CARPENTRY (INSTALLATION)

DESCRIPTION	NO.	DIMENSIONS	ESTIMATED QUANTITY	UNIT
Install Millwork Cabinets (supplied by Millwork Subcontractor)				
Sink Counter 8'0"x2'1"x 2'4" o/a inc 17"h Splashback wi returned ends (to frmd walls & panels m/s)			1	No
Upper Wall Cpb'ds (over sink) 8'0"x1'0"x2'1" o/a (ditto)			1	No
Ditto (over frig & stove m/s) 8'0"x1'0"x2'1" o/a (ditto)			1	No
Counter Unit (btwn frig & stove) 2'9"x2'1"x2'4" o/a inc 17"h Splashback (wi returned end) extending 5'2" long (behind stove)			1	No
Vanity Counter 4'0"x 1'8" x 2'4" o/a (in Bathroom)			1	No
Ditto 4'8"x1'8"x 2'4" o/a (in Bathroom)			1	No
Medicine Cabinets approx 2'x2' o/a (in Bathroom)			2	No
Install Bathroom Accessories (supplied by others) incl rough bearers				
Install Toilet Paper Holders (flush type)			3	No
Install Towel Rails			3	No
Install Soap Dish (flush type)			1	No

FRANK R. WALKER CO., PUBLISHERS, CHICAGO

PRACTICAL "STANDARDIZED" FORMS FOR CONTRACTORS
Form 516 MFD IN U.S.A.

QUANTITY SHEET

PROJECT RESIDENCE — ESTIMATOR KC — ESTIMATE NO. R-1
LOCATION — EXTENSIONS KC — SHEET NO. 7 of 7
ARCHITECT ENGINEER — CHECKED — DATE 1972
CLASSIFICATION FINISH CARPENTRY (SHELVING)

DESCRIPTION	NO.	DIMENSIONS	ESTIMATED QUANTITY	UNIT
3/4" D.Fir Ply (Good/Solid) Closet Shelves wi solid clear D. fir Front Edging and 1x2 Wall Bearers (to three edges) and wi Chrome-finish Closet Track wi small Eye Hangers (2 per foot of track-length), as follows:—				
Shelf 1'6" w x 4'6" long	1	(BR # 2)	1	No
Shelf 1'6" w x 6'6" long	1	(BR # 3)	1	No
Shelf 1'6" w x 7'6" long	1	(BR # 1)	1	No
Shelf 1'6" w x 3'0" long	1	(Coats)	1	No
Linen Closet Shelves 1'0" × 2'6" long of 1/2" × 1" W. Red Cedar Frame and Battens spaced 1/2" apart on & incl 1" × 2" D. fir Wall Bearers (to three edges) 4/1			4	No
Allowance for Lab & Mat to supply & install minor items of Trim required (and not indicated)				Item

FRANK R. WALKER CO., PUBLISHERS, CHICAGO

doors, millwork fixtures (cabinet work), and small bathroom fixtures (mirrors, medicine cabinet, and the like) are supplied to the job in this way, and only the installation is measured in this estimate.

Hanging doors includes installing hardware (assumed to be supplied through a cash allowance in the contract), and the cabinetwork is assumed to be supplied with all its finish hardware.

Shelving in this job is simple and probably will be made up on the job. Measurement is equally simple. In larger jobs requiring quantities of shelving, there would be some advantage in measuring it as *run* or a *super* item, depending on the widths. If it is not more than 12 in. wide, it is measured as a *run* item in the **MM-CIQS**.

Residence Example

Moisture Protection

The built-up roofing and the roof insulation is measured from the upper floor plan, from which the o/a dims inside parapets are computed, and from the roof plan shown on the site plan. Measurements are taken of the areas covered by *roofing work*.

The roofing felts are taken up over the cant strips, up the inside face of parapets, and nailed at the upper edges under the metal flshgs. Such *work* is better measured as a *run* item, as in the example. The metal flshgs are fully described, and their girths are stated. The number of breaks (bends) in the girth of each kind of metal flshg is sometimes stated, particularly when the flshg is to be formed to an unfamiliar profile to make more accurate pricing possible. Isolated metal flshgs to vent pipes and the like are described and enumerated.

If the project spec requires moisture protection to be guaranteed, following inspection of the *work* in progress, or subsequent testing, or both, an item should be noted for pricing the costs.

Residence Example

Finishes

Stucco

The dims are collected and added together in the preliminary calculations and, for clarity, they are identified N, S, E, or W, according to the elevation. The o/a vertical dim is also computed from the dims shown on the dwgs.

The extr stucco is described "as spec" (as specified). The full description of all *items of work* cannot always be entered on estimate sheets because of their length, and in such cases the description should contain the words "as spec" so that the estimator is reminded to look at the specified requirements when pricing the item. The narrow widths of stucco are described according to width, as before, and measured as *run* items.

Metal beads and expansion joint trim are measured; and although such items are not always indicated or fully described on dwgs, they may be required by the specs, or by the class of *work*.

Drywall

Intr dims for the finishes are computed first, and the perims of the rooms and intr areas are calculated below the items' descriptions, which contain the appropriate wall hts, later multiplied by the aggregated perims to give the areas of drywall finish to walls. In this way the dim entries and calculations are kept to a minimum, and the estimate sheet is uncluttered and easy to read. Sub-totals are entered in an "extension column" to enable the items to be grand totaled. Deductions are dealt with in the same way.

Narrow widths and metal beads are described and measured as *run* items, as before. Dims of clgs are calculated so as to measure the finish to clgs over the ptn walls and opngs, and the *wants* are deducted. To simplify measuring, the clg area over the GF—storage—WC—furnace—coats area is totally deducted (ddt—18.4 × 5.4), and the actual clg area (17.4 × 5.0) is added to the overall area measured above. The high LR clg (above 14 ft) is measured separately and so described because of additional scaffolding costs. Double wallboard in the garage clg (for fire isolation) is so described and measured separately.

Although *designer's* dwgs often describe *work* by using trade names such as "Gyproc," the specs usually make it clear that other equivalent products are acceptable. For this reason, and unless only one particular product is to be used in a job, it is better to avoid the use of trade names.

Measuring the total gross quantity of drywall required, in 4-ft-wide boards of the required lengths, will enable the amount of *waste* to be estimated. Assume that boards are installed vertically to walls, with a minimum length of 8-ft.

PRACTICAL "STANDARDIZED" FORMS FOR CONTRACTORS
Form 516 MFD IN U.S.A.

QUANTITY SHEET

PROJECT: RESIDENCE — ESTIMATOR: KC — ESTIMATE NO. R-1
LOCATION — EXTENSIONS: KC — SHEET NO. 1 of 2
ARCHITECT ENGINEER: MOISTURE PROTECTION — CHECKED — DATE: 1972
CLASSIFICATION: (ROOFING & SHEET METAL)

DESCRIPTION	NO.	DIMENSIONS				ESTIMATED QUANTITY	UNIT
Prelim Calcs - Dims		(Overhangs)		(Want)			
(E-W) 15-2 (N-S)	13-6	35-4	5-8	(E-W) 14-0	21-6		
10-10	8-4	2/=-4-0	8-4	3-4	-0-6 = 4" 2"		
5-4	21-6	31-4	14-0	17-4	21-0		
2/2-0 = 4-0	43-4	3-0	2-6		-3-0		
35-4	2/0-4 = 0-8	+0-4	+0-4		18-0		
	42-8	3-4	2-10				

4 Ply Tar & Gravel B/up Roofing on & incl 1" thick Rigid Fiberboard Insulation on Vapor Barrier (all as Specs pp —)

DESCRIPTION	NO.	DIMENSIONS				ESTIMATED QUANTITY	UNIT
	1/	35-4	×	42-8	1508		
(@ S. O/hang)	1/	31-4	×	3-4	104		
(@ N. ")	1/	14-0	×	2-10	40		
					1652		
DDT (Want)	1/	17-4	×	18-0	- 312 DDT		
					1340	13½	SQ

4 Ply Ditto on Vertical Faces of Parapet etc. × 8" high above Roof incl nailing Top Edges (as Specs p —)

DESCRIPTION	NO.	DIMENSIONS				ESTIMATED QUANTITY	UNIT
(E & W)	2/	42-8			85		
(S. O/hang)	2/	3-4			7		
@ above Entr ⊙ (N. O/hangs)	1·2/	3-4			10		
(")	2/	2-10			6		
(@ returned ends para)	2/2/	2-0			8		
(arnd chmny)	4/	1-9			7		
					123	123	LF

(NB: 4 Ply Ditto over Cants measd in general Roof Area)

Cut & Fit B/up Roofing to Roof Drains (m/s)

DESCRIPTION	NO.	DIMENSIONS				ESTIMATED QUANTITY	UNIT
	4/	1				4	No

FRANK R. WALKER CO., PUBLISHERS, CHICAGO

PRACTICAL "STANDARDIZED" FORMS FOR CONTRACTORS
Form 516 MFD IN U.S.A.

QUANTITY SHEET

PROJECT	RESIDENCE	ESTIMATOR	KC	ESTIMATE NO.	R-1
LOCATION		EXTENSIONS	KC	SHEET NO.	2 of 2
ARCHITECT ENGINEER	MOISTURE PROTECTION	CHECKED		DATE	1972
CLASSIFICATION	(ROOFING & SHEET METAL)				

DESCRIPTION	NO.	DIMENSIONS		ESTIMATED QUANTITY	UNIT
26 GA GALV STEEL SHEET METAL FLASHINGS (all as Details & Specs)					
Cap Flashings to parapets x 12" girth wi 2-Hemmed Edges & installed wi Clips max 8'0" apart (as Detail #4/3)					
(123-0 – 7-0) (As Rfg)	1/	116-0	116		
(corners)	4·4/	0-6	4	120	LF
Counter Flashings to parapets x 9" girth wi Hemmed Btm Edge & Top Edge Nailed & wi Clips a.b (as Ditto)					
	(As Cap Flshgs)			120	LF
Ditto to chmny stack x 9" girth wi Ditto & wi top Edge turned into Bkwk & wi Clips a.b					
	4/	1-9		7	LF
Cant Flashings x 9" girth wi 2-Hemmed Edges & Clips a.b					
(17-4 + 14-0)	2/	31-4		63	LF
Window Head Flashings x 6" girth wi Hemmed Btm Edge & hand to carpenters for installation under Ply (as Detail #2/3)					
(S)	1/	31-4	31		
(N)	1/	17-0	17		
(W)	1/	1-6	2	50	LF
3 LB LEAD SHEET FLASHINGS (as Specs)					
3"ϕ Vent Pipe Flashing wi 12"x12" Base & 8" high upstand				1	No
4"ϕ Ditto				1	No
Allowance for Roofing Inspection, Testing & Guarantee (as Specs)				Item	
[NB: Roof Drains & Pipes in Div. 15.]					

FRANK R. WALKER CO., PUBLISHERS, CHICAGO

PRACTICAL "Standardized" Forms for Contractors
Form 516 MFD IN U.S.A.

QUANTITY SHEET

PROJECT RESIDENCE — ESTIMATOR KC — ESTIMATE NO. R-1
LOCATION — EXTENSIONS KC — SHEET NO. 1 of 1
ARCHITECT ENGINEER FINISHES — CHECKED — DATE 1972
CLASSIFICATION STUCCO (EXTR & INTR)

DESCRIPTION	NO.	DIMENSIONS			ESTIMATED QUANTITY	UNIT
Prelim Calcs - Dims			Collect Dims	Height		
(N)(8-0 − 0-4) = 7-8	(E) 22-0	(W) 13-6 (E) 30-0	↓132-4	8-7½ (G. Floor)		
(14-0 − 2/0-4) = 13-4	+8-0	+8-4 (N) 14-0	4/2-0 = 8-0	7-6 (U. Floor)		
(inside @ Screen Wl) 1-8 / 22-8	30-0	21-10 " 2-0	2/2/3-0 = 12-0	1-1½ (Roof)		
	(E) 21-6	1-0 / 0-2 − 1-2 " 22-8	2/2/2-6 = 10-0	0-8 (Parapet)		
	½/ −0-2	20-8 (E) 21-4	TOTAL 162-4	17-11		
	21-4	21-6 (W) 20-8		+ 0-1 (@ Btm)		
		½/ = +0-2 21-8		TOTAL 18-0		
		21-8 132-4				

Extr Stucco Plaster wi 1" Galv Mesh Reinf & self-furring nails to walls (as Spec.)

	1/162-4	×	18-0	2922		
(Reveals @ Gar. Door)	2/0-4	×	6-6	4		
				2926 (÷9)	325	SY

Ditto < 3" wide to Reveals

(@ W. Wdw)	2/18-0	36 (÷3)	12 LY

Ditto 3" to 6" wide to N.W's

(@ Wall ends)	5/18-0	90 (÷3)	30 LY

Intr Stucco Plaster on Brick Chmny Face Panel in S.Q (as Spec)

1/2-8	×	13-4	36 (÷9)	4 SY

Zinc Alloy Stucco Trim (as Spec)

Expansion Joints	4/18-0	72	72 LF
Corner Beads	5/2/18-0	180	
(⊙ @ W. Wdw)(Walls)	2·6/18-0	144	324 LF
Edge Beads (@ Btm)	1/163-0	163	
(2-8 + 13-4 + 6-6) =	2/22-6	45	208 LF

FRANK R. WALKER CO., PUBLISHERS, CHICAGO

PRACTICAL "STANDARDIZED" FORMS FOR CONTRACTORS
Form 516 MFD IN U.S.A.

QUANTITY SHEET

PROJECT: RESIDENCE — ESTIMATOR: KC — ESTIMATE NO. R-1
LOCATION — EXTENSIONS: KC — SHEET NO. 1 of 4
ARCHITECT ENGINEER: FINISHES — CHECKED — DATE: 1972
CLASSIFICATION: DRYWALL (WALLS)

DESCRIPTION	NO.	DIMENSIONS							ESTIMATED QUANTITY	UNIT

Prelim calcs Dims

(GF) 15-2, 2/= 16-2, 2/2-4= 4-8, 36-0, 2/= −0-8, 35-4 ✓ | 22-0, 2/= −0-8, 21-4, 3-11, 1/2 = −0-2, 3-9 | 5-4, 2/= −0-4, 5-0, 11-0, 2/= −0-4, 10-8 | 3-1, 1/2 = −0-2, 2-11, 20-6, −0-6 4/2, 20-0 | 14-0, 2/= 4-8, 18-8, 2/= −0-8, 18-0 | 6½' Partn: 35-4, −3-0, −3-8, =28-8, −4-0, 24-8

(UF) 10-10, 1-3, 12-1, 2/= 0-4, 11-9 | 13-6, −0-6 4/2, 13-0, 5-4, 2-4, 7-8, −0-6 4/2, 7-2 | 13-6, −0-6, 13-0, −7-10, 5-2, 21-6, −0-6 4/2, 21-0 | 8-4, 2-4, 10-8, −0-6 4/3, 10-2 | 5-8, 2-4, 8-0, −0-6 4/3, 7-6 | Closets: 2/= 6-0, 2/= 0-2, 6-2; 35-4, −7-6, 27-10 | = 4-0, = 0-2, 4-2; 13-6, 8-4, 21-10, −0-6 = 2/4, 21-4

½" Gypsum Drywall boards wi taped & filled joints (as spec) to Walls (x 7'6" high) (allow 25% Waste)

Description	No.	Dimensions		LF	(4'x8' shts)	
(GF)						
LR/K. 2(35-4 + 21-4)	1/	113-4	(x 7-6)	113	2(9+6) =	30 No
WC. 2(3-9 + 5-0)	1/	17-6		18	2(1+1) =	4
Stor. 2(10-8 + 5-0)	1/	31-4		31	2(3+1) =	8
Furn. 2(2-11 + 2-6)	1/	10-10		11	2(1+1) =	4
Clos. 2(2-11 + 2-2)	1/	10-2		10	2(1+1) =	4
HWT. 2(1-6 + 1-6)	1/	6-0		6	offcuts =	0
Gar. 2(18-0 + 20-0)	1/	76-0		76	2(5+5) =	20
				265		70 No
(UF)						
BR1. 2(11-9 + 13-0)	1/	49-6		50	2(3+4) =	14
DR. 2(7-2 + 7-6)	1/	29-4		29	2(2+2) =	8
Bath. 2(7-2 + 5-2)	1/	24-8		25	2(2+1) =	6
Bath. 2(7-2 + 8-0)	1/	30-4		30	2(2+2) =	8
Lin. 2(2-6 + 1-3)	1/	7-6		8	2(1+0) =	2
BR2. 2(21-0 + 10-2)	1/	62-4		62	2(6+2) =	16
BR3. 2(21-0 + 7-6)	1/	57-0		57	2(5+2) =	14
Clos. 2(6-2 + 2-0)	1/	16-4		16	2(2+½) =	5
Clos. 2(4-2 + 2-0)	1/	12-4		13	2(1+½) =	3
UG. LR. 2(27-10 + 21-4)	1/	98-4		98	2(7+6) =	26
				653 LF		172 No
				X 7-6 high		
				4898 SF		
ADD (6½' Partn)	2/	24-8	x 6-6	321 SF	2(7)	14
(@ stairwell).	2/	3-8	x 1-0	8	offcuts	0
2/=26-0, 2-2 (extra ht in LR)	1/	28-2	x 1-0	28	(use 10' sheets)	0 (adjusted later)
(to balustrade)	2/	10-6	x 3-0	63	2(1½)	3
	1/	17-0	x 1-0	17		
			(TOTALS)	5335 SF	TOTAL	189 No

(carry forward to next sheet)

FRANK R. WALKER CO., PUBLISHERS, CHICAGO

PRACTICAL
Form 516 MFD IN U.S.A.

QUANTITY SHEET

PROJECT	RESIDENCE	ESTIMATOR KC	ESTIMATE NO. R-1
LOCATION		EXTENSIONS KC	SHEET NO. 2 of 4
ARCHITECT ENGINEER	FINISHES	CHECKED	DATE 1972
CLASSIFICATION	DRYWALL (WALLS)		

DESCRIPTION	NO.	DIMENSIONS				ESTIMATED QUANTITY	UNIT
Prelim Calcs - Dims							
15-2 14-0		5-8					
16-2 3-4		8-4					
31-4 17-4		14-0					
2/2 = 0-4 -0-6 = 4/2	2/2/2" =	-0-8					
31-0 16-10		13-4					
DDT 1/2" Gypsum Drywall a.b (x 7'6" high)					(4'x8' shts)		
(GF) (@ S. Wdws)	1/	31-0	(x 7-6)	31 LF	7 No		
(@ N. Wdws)	1/	16-10		17	4		
(@ Int. Doors)	2.3/2/	2-8		27 (2.3/2/1/2) =	5		
(@ W. Wdw)	1/	1-0		1	0		
(UF)				76	16 No		
(@ S. Wdws)	1/	31-0		31	7		
(@ W. Wdw)	1/	1-0		1	0		
(@ N. Wdw BRs)	1/	13-4		13	3		
(@ N. Wdw UG)	1/	16-10		17	4		
(@ Int. Doors)	11/2/	2-8		59 (11/2/1/2)	11		
				197 LF	41 No		
				x 7-6 high			
				1478 SF			
(Garage Door)	1/	14-0	x 6-6	91	3		
				1569 SF DDT	44 No DDT		
		(brought forward) TOTAL		5335	TOTAL 189		
				-1569 DDT	-44 DDT		
				3766	145 (incl 8 No 10' sheets)		
1/2" Gypsum Drywall boards wi taped & filled joints (as spec) to Walls (allow 20% Waste)						3770	SF
					(4'x8' sheets	137	No)
					(4'x10' sheets	8	No)

Waste Calcs

sheets 137 (4'x8') =	4384 SF
8 (4'x10') =	320 SF
Gross Quantity =	4704
Net " =	-3766
Waste =	938 SF

$\frac{938}{3766\text{SF}} \times \frac{100}{1} = 25\%$

FRANK R. WALKER CO., PUBLISHERS, CHICAGO

PRACTICAL "STANDARDIZED" FORMS FOR CONTRACTORS
Form 516 MFD IN U.S.A.

QUANTITY SHEET

PROJECT	RESIDENCE	ESTIMATOR	KC	ESTIMATE NO.	R-1
LOCATION		EXTENSIONS	KC	SHEET NO.	3 of 4
ARCHITECT ENGINEER	FINISHES	CHECKED		DATE	1972
CLASSIFICATION	DRYWALL (WALLS)				

DESCRIPTION	NO.	DIMENSIONS		ESTIMATED QUANTITY	UNIT
½" Gypsum Drywall to Walls a.b 3" to 6" wide (Lab only)					
(GF) (top 6½' partn)	1/	24-8	25		
(ends ")	2/	6-6	13		
(reveals W. Wdw)	2/	7-6	15		
			53		
(UF)					
(reveals W. Wdw)	2/	7-6	15		
(top balustrade)	1/	10-6	11		
	2/2/	3-0	12		
			91	90	LF
Galv Steel Corner Beads to ½" Drywall (as spec)					
(GF) (@ 3"-6" widths)	2/	24-8	49		
"	2/2/	6-6	26		
"	1/2/	7-6	15		
(@ wall corners)	2/	7-6	15		
(@ stairwell)	2/	3-8	7		
			112		
(UF)					
(@ 3"-6" widths)	1/2/	7-6	15		
"	2/1/	10-6	21		
"	2/2/2/	3-0	24		
(@ wall corners)	2·4·2·4/	7-6	90		
	1/	1-0	1		
(17+13) (over LR)	1/	30-0	30		
			293	300	LF
Galv Steel Edge Beads to ½" Drywall, (as spec)					
(GF) (exposed edges)	6/	7-6	45		
(next chmny stack)	2/2/	6-6	26		
			71		
(UF)					
(exposed edges)	1·5/	7-6	45		
		1-0	1		
			117	120	LF
Ditto, (ditto) at Door Jambs (as detail Dwg sht. 1)					
(c Int. Doors)	17/2/2/	7-6	510		
Types: B, 5 E, 3 / D, 4 F, 2 / HWT, 1 G, 2	2/	6-6	13	520	LF

[NB: Extra scaffolding for high walls in LR taken later.]

FRANK R. WALKER CO., PUBLISHERS, CHICAGO

PRACTICAL
STANDARDIZED FORMS FOR CONTRACTORS
Form 516 MFD IN U.S.A.

QUANTITY SHEET

PROJECT RESIDENCE
ESTIMATOR KC
ESTIMATE NO. R-1
LOCATION
EXTENSIONS KC
SHEET NO. 4 of 4
ARCHITECT ENGINEER FINISHES
CHECKED
DATE 1972
CLASSIFICATION DRYWALL (CEILINGS)

DESCRIPTION	NO.	DIMENSIONS								ESTIMATED QUANTITY	UNIT

Prelim Calcs-Dims

(GF) 13-8 WC 3-9 14-0 22-0 14-0 (UF) 13-6 10-10 15-2 13-6 | UF Partitions 42-8
15-2 3-10 Storage 10-8 3-4 2/= -0-8 2/= 4-0 8-4 5-4 2-4 -0-6 4" = 3" | 2/= 36-0
16-2 5-4 Furn 2-11 17-4 21-4 0-4 21-6 2-4 17-6 13-0 | 2/= 4-8 closets
2/= 4-0 22-10 17-4 -0-4 22-4 18-4 43-4 18-6 2" 4 = -0-6 | 7-2 shower
35-4 -0-6 5-4 17-0 -21-4 2/= -0-8 -0-6 17-0 | 1-2
22-4 2/= -0-4 1-0 42-8 → 18-0 → | 91-8
5-0

½" Gypsum Drywall boards wi taped & filled joints (as spec) to Ceilings (allow 12% Waste)

					4' wide sheets	
(GF) (LR, K, etc)	1/35-4	× 22-4	789		(5/14' + 4/10' + 5/8') =	150
(UF) (BR's, etc)	1/42-8	× 18-0	768		(5/14' + 5/12' + 10/8') =	210
(U. Gallery)	1/17-4	× 8-4	144		(partially 4/8') =	32
(over Doors)	3 @ 6/3-0	× 0-4	9		(no extra)	–
ADD (Net Area @ Storage, WC, Furn, etc)	1/17-4	× 5-0	87 *		(")	–
			1797			392'
DDT ½" Drywall (Wants)				DDT		×4'w
(GF) (@ N. Wdw & Entr)	1/17-0	× 1-0	–	17		1568 SF (GROSS)
(@ Storage, WC, etc)	1/18-4	× 5-4	–	98 *		
(over LR)	1/17-0	× 13-0	–	221		(14'×4' sheets - 10 No)
(@ Stairwell)	1/8-9	× 3-8	–	32		(12'×4' " 5 No)
(UF) (@ Partn Walls)	1/91-8	× 0-4	–	31		(10'×4' " 4 No)
			-399			(8'×4' " 19 No)
			1398			1400 SF

½" Ditto 16' 1½" high above G. Floor incl extra scaffolding (allow 27% Waste)

					(14'×4' sheets - 5 No)
1/17-0	× 13-0	221	(5/14')		220 SF

Double ⅜" Gypsum Drywall boards wi exposed joints taped & filled (as spec) to Ceiling of Garage (allow 11% Waste)

				(10'×4' sheets - 10 No)
1/20-0	× 18-0	360	(5/2/10')	(10'×4' backing sheets - 10 No)
				360 SF

* to deduct Partitions around Storage, WC, Furnace, & Coats.

FRANK R. WALKER CO., PUBLISHERS, CHICAGO

PRACTICAL
"STANDARDIZED" FORMS FOR CONTRACTORS
Form 516 MFD IN U.S.A.

QUANTITY SHEET

PROJECT	RESIDENCE	ESTIMATOR	KC	ESTIMATE NO.	R-1
LOCATION		EXTENSIONS	KC	SHEET NO.	1 of 2
ARCHITECT ENGINEER	FINISHES	CHECKED		DATE	1972
CLASSIFICATION	FLOOR COVERINGS				

DESCRIPTION | NO. | DIMENSIONS | ESTIMATED QUANTITY | UNIT

Prelim calcs - Dims

(GF) 16-2 22-0 | 13-8 22-4 | 15-2 | collect Walls
15-2 2/-0-8 | 3-10 -21-4 | 2-0 | 18-0
2/2-0= 4-0 21-4 | 5-4 1-0 | 17-2 | 2/= 10-0
35-4 | 22-10 | -0-2 | 3-0
4" 6"= -0-6 | 17-0 | 31-0
22-4 | 3/= 8-0 (doors)
23-0

1" thick x 9" x 4½" Brick Paviors bedded & flush jointed in P cmt mtr on conc floor slab (as spec) (SC & Waste m/s)

(GF-except Garage)	1/35-4	x	22-4	789		
					DDT	
DDT (Wants)(@ Entr)	1/17-0	x	1-0	—	17	
(@ LR. conc paving)	1/9-0	x	7-0	—	63	
(@ chmny stack)	1/4-0	x	2-8	—	11	
(@ Storage)	1/10-8	x	5-0	—	53	
(@ Furnace)	1/2-11	x	2-6	—	7	
(@ Walls)	1/23-0	x	0-4	—	8	
				-159		
				630		630 SF

Straight Cutting & Waste to 1" Bk Paviors (against walls)

	2/35-4	71
	2/22-4	45
(9-0+7-0)	2/16-0	32
(4-0+2-8)	2/6-8	13
(24-6+3-6+1-0)	2/30-0	60
	2/5-4	11
		232	230 LF

1" P Cement Fine Topping on conc floor slab steel trwld smooth

1/ 9-0 x 7-0 63 — 63 SF

FRANK R. WALKER CO., PUBLISHERS, CHICAGO

Floor Coverings

Floor coverings are easily measured after the clgs because their dims and quantities are similar, and the quantities used for one serve as a check on the quantities required for the other:

> Total clgs area (excluding garage) is 1398 + 221 = 1619 SF; the total area of flr coverings is 630 + 63 + 855 = 1548 SF, which is a difference of 71 SF. This difference is made up of the ddts from the brick flr area of 71 SF (11 + 53 + 7).

Descriptions of the floor items are abbreviated, and reference is made to the specs, as before. These items may have specific requirements in respect to adhesives, cleaning, waxing, and maintenance, which must be accounted for in the pricing. If it is necessary, the estimator should write fuller descriptions in the estimate to ensure that all the costs are included.

Painting (Exterior)

Painting and other decoration and coating *work* is usually measured last so that the dims and quantities of the measured *work* to be painted can be used. The measurements should be as accurate as the requirements of the items and their *unit prices* dictate. Most painting (two coats) will cost less than $.25 per SF. Besides, many other factors affect the costs much more than minor variations in quantity.

Where possible, the measured areas of other *work* such as cladding, paneling, stucco, drywall, and the like, are used to measure the quantities of painting required. Windows and doors are measured overall, with no deductions for glass. Painting on different kinds of surfaces is measured separately, and generally all surfaces are measured as though flat.

Allowances for surfaces that are not flat because of natural variations, applied moldings, and other features, are better calculated in the *unit prices*, and any allowances established from experience should be indicated in the items' descriptions for this purpose. The **MM-CIQS**, however, calls for such allowances to be made in the measurements.

Paint to be applied to surfaces of less than 12 in. girth is better measured and priced in LF. Isolated items to be painted are described and enumerated.

Having reviewed the examples for measuring finishes, it might be well to mention again that many estimators and quantity surveyors measure the *superficial work* such as roof coverings and finishes first, and then measure the structure and frame. This is done to first acquire a knowledge of the building and its layout before measuring the more complex *items of work*, such as the wood frame.

Residence Exercises

All the *work* not measured in the examples, including site work, excavation, concrete (including formwork), and interior painting can be measured as separate exercises. Finally, the entire project might be remeasured and priced (at current and local *unit prices*) as a complete estimating exercise; and at the same time, variations in cladding and finishes might be made and estimated by first making section-drawings showing the variations. For example, 4-in.-thick brick veneer might be substituted for the exterior stucco finish, or masonry exterior walls might be substituted for framing, each illustrated by a simple section-drawing, to provide more exercises.

Warehouse Example

Generally

The Warehouse Drawings are of a building that was designed and built a few years ago for a plywood distributor; and although it has a few features that are peculiar to the site, it is generally typical of numerous warehouses with attached offices designed and built for distributors.

The measurement examples taken from the Warehouse Drawings were selected to demonstrate the measurement of some types of *work* not previously shown. The *work* not measured in the examples, such as the site works, excavation and fill, foundations, miscellaneous metals, framing, finish carpentry, moisture protection, and finishes, can be measured as exercises. Similar *work* has been measured in previous examples.

Warehouse Example

Masonry (Concrete Blockwork)

As explained before, the **MM-CIQS** follows a common trade practice by stating that unit masonry shall

PRACTICAL "STANDARDIZED" FORMS FOR CONTRACTORS
Form 516 MFD IN U.S.A.

QUANTITY SHEET

PROJECT RESIDENCE — ESTIMATOR KC — ESTIMATE NO. R-1
LOCATION — EXTENSIONS KC — SHEET NO. 2 of 2
ARCHITECT ENGINEER FINISHES — CHECKED — DATE 1972
CLASSIFICATION FLOOR COVERINGS

DESCRIPTION	NO.	DIMENSIONS					ESTIMATED QUANTITY	UNIT
Prelim Calcs-Dims								
(UF) 10-10 13-6	15-2	35-4						
5-4 8-4	10-10	-18-0						
2-4 21-6	5-4	17-4						
4" 18-6 43-4 2/=	4-8	15-2						
2"=-0-6 2/=-0-8	36-0	2-0						
18-0 42-8 2/=-	0-8	0-2						
	35-4	17-4						
0.080" thick Vinyl Asbestos Floor Tiling (one color) installed on ply subfloor m/s. (allow 10% Waste)								
(UF) (B'rms, Baths, etc)	1/	42-8	x	18-0	768			
(u. Gallery)	1/	17-4	x	8-4	144			
(doorways)	6/	3-0	x	0-4	6			
					918			
DDT (Wants)						DDT		
(@ partn. walls)	1/	91-8	x	0-4	–	31		
(@ stairwell)	1/	8-9	x	3-8	–	32		
[NB: no deducts made for plumbing fixtures.]					-63			
					855		860	SF
Extra over last item for Waterproof Adhesive								
(Bathrooms)	1/	7-2	x	5-2	37			
	1/	7-2	x	8-0	57			
					94		100	SF

FRANK R. WALKER CO., PUBLISHERS, CHICAGO

PRACTICAL "STANDARDIZED" FORMS FOR CONTRACTORS
Form 516 MFD IN U.S.A.

QUANTITY SHEET

PROJECT RESIDENCE — ESTIMATOR KC — ESTIMATE NO. R-1
LOCATION — EXTENSIONS KC — SHEET NO. 1 of 2
ARCHITECT ENGINEER FINISHES — CHECKED — DATE 1972
CLASSIFICATION PAINTING (EXTERIOR)

DESCRIPTION	NO.	DIMENSIONS			ESTIMATED QUANTITY	UNIT
PRIMER & 2 COATS EXTR OIL PAINT ON PLYWOOD (as spec)						
Isolated Panels / Fascias up to 18' above ground						
	(as in FIN CARP p.1)		262 (÷9)		30	SY
Small Ditto (approx 2 sf each) up to ditto						
	(as in ditto)		2		2	No
Eaves Soffites (x 2'3" & 2'9" wide) about 15'6" above ground						
	(as in ditto)		165 (÷9)		20	SY
PRIMER & 2 COATS EXTR OIL PAINT ON WDW & DR FRMS IN WDW WALLS						
Wdw Frms (one edge next glass) 3"–6" wide						
(S)	2/2/	31-4	125			
(N)	2/2/	17-4	68			
(Avg jamb (S)	2/6/2/	7-0	168			
7' high) (N)	2/3/2/	7-0	84			
			445	DDT		
DDT (wants)	(Wdw Frms <3" wide)		–	130		
	(Dr Frms 3"–6" wide)		–	50		
	(Dr Frms <3" wide)		–	20		
	(Dr Frms 6"–9" wide)		–	15		
			– 215		230	LF
Wdw Frms (one edge next glass) less than 3" wide						
(Heads Frms) (S)	2/	31-4	63			
(N)	2/	17-4	35			
(Jambs @ Walls) (S & UF N)	2/2/	7-0	28			
" (GF N)	1/1/	7-0	7			
2/ = 14-0 +3-0			133		130	LF
Dr Frms 3"–6" wide	1·4/	17-0	85	DDT		
DDT (wants)	(Dr Frms <3" wide)		–	21		
	(Dr Frms 6"–9" wide)		–	14		
			– 35		50	LF
Dr Frms less than 3" wide (Jambs @ Walls)	1·2/	7-0	21		20	LF
Dr Frms 6"–9" wide (B/s mullion post)	2/	7-0	14		15	LF

FRANK R. WALKER CO., PUBLISHERS, CHICAGO

PRACTICAL "STANDARDIZED" FORMS FOR CONTRACTORS
Form 516 MFD IN U.S.A.

QUANTITY SHEET

PROJECT RESIDENCE — ESTIMATOR KC — ESTIMATE NO. R-1
LOCATION — EXTENSIONS KC — SHEET NO. 2 of 2
ARCHITECT ENGINEER FINISHES — CHECKED — DATE 1972
CLASSIFICATION PAINTING (EXTERIOR)

DESCRIPTION	NO.	DIMENSIONS				ESTIMATED QUANTITY	UNIT
PRIMER & 2 COATS EXTR OIL PAINT ON WOOD (as spec)							
Flush Extr Doors (extr face only)	1/	3-0	x	7-4	22 (÷9)	3	sy
Glazed Extr Doors (extr face only) wi single pane (about 14 sf) & measd over glass	4/	3-0	x	7-4	88 (÷9)	10	sy
Flush Panelled O/head Garage Door wi six panes (about 2 sf each) & measd over glass (b/sides measd)	2/	14-0	x	6-6	182 (÷9)	20	sy
PRIMER & 2 COATS EXTR OIL PAINT ON METAL (as spec)							
Parapet Flashings @ 18' above ground (as ROOFING p.2)	1/	120-0	x	1-9	210 (÷9)	24	sy
Cant Flashings x6"-9" girth @ 18' above ground (ditto)					63		
Counter Flashings x6"-9" girth @ 18' above ground (ditto)					7	70	LF
Wdw Head Flashings less than 3" wide @ up to 18' above ground (ditto)					50	50	LF
2 COATS CEMENT PAINT ON ROUGH STUCCO up to 18' above ground							
		(Area Stucco)			2926		
(narrow widths)	2/	18-0	x	0-3	9		
	5/	18-0	x	0-6	45		
					2980 (÷9)	330	sy
Scaffolding (incl with above items of work)							

FRANK R. WALKER CO., PUBLISHERS, CHICAGO

be measured in masonry units: bricks by the thousand and concrete blocks simply enumerated. Mortar is to be measured separately by the CY, and masonry reinforcement is to be measured in LF. But it is questionable whether this is the best method, and the simpler and more descriptive method of measuring masonry as *super* and *run* items is used in this example. Nevertheless, if measurement by enumeration is required or preferred, it is a simple step to convert the quantities of *super* and *run* items of masonry to numbers of masonry units. A block of nominal 8 in. × 16 in. face dims, including ⅜-in. joints, is ⁸⁄₉ths of a square foot. So to convert the area of block walls to number of blocks, add 12½ percent to the number of SF in the wall area, and the result is the equivalent number of 8 in. × 16 in. blocks.

The descriptions of items must be sufficiently detailed to permit accurate pricing; and insofar as the divisions of masonry required by the **MM-CIQS** are concerned, it is suggested that all *masonry work* should be identified by general location as well as by type and function. Thus, *work* in the small wall panels is separated from the warehouse walls, and the 4-in. blockwork at the roof level is so described.

Measuring masonry walls as *super* items is simplified if features of the walls such as bond beams and lintels (filled and reinforced) are measured as *extra over items*, as in the example. Alternatively, and especially if the masonry units are to be enumerated from the *super* and *run* quantities, such features as require special masonry units can be measured as regular *run* items (instead of *extra over items*), with appropriate deductions from the *super* item of masonry wall.

If blockwork dims are not multiples of the dims of the block units, the estimator should allow for the costs of cutting bocks by measuring *run* items of "straight cutting and *waste*." Alternatively, he should enumerate and describe cut blocks. In exposed *fair work*, the cutting may have to be done with an electric masonry saw, which is expensive, since one cut block costs about as much as two uncut blocks. Where block partitions are built, it is a common practice to conceal electrical conduits and boxes and plumbing pipes within the hollow blocks, and this practice may require a large amount of cutting. It is practically impossible to measure this cutting, and yet it may amount to a significant portion of the total *labor costs* of the *masonry work* and cannot be ignored by an estimator. In such cases, it is important that the estimator separately measure and properly describe the *masonry work* in which this cutting occurs, so that through *cost accounting* he can obtain cost data for different classes of masonry in different types of buildings. For example, the amount of cutting required for concrete block partitions (with concealed services) would be different in office buildings, schools, and hospitals; and there would probably be a difference between the cutting required in elementary schools and that required in high schools containing laboratories and workshops. However, the amount of cutting in the exterior walls of a warehouse would be insignificant, particularly if modular dims are used. Other job facts and conditions (besides concealed services) may significantly affect the costs of *masonry work*; for example, the building's size and shape. Consequently, cost data of masonry should always include references to these facts and conditions.

Warehouse Example

Metals: Structural

Measurement in this example is by weight obtained from the measured *super* and *run* items and from published unit weights. Alternatively, each piece could be enumerated and described in detail in terms of category, size, and weight. The **MM-CIQS** includes both of these methods of measurement; and although the latter is preferable for more accurate estimating, the weight method is more common, and it is much improved by stating the number of pieces in the total weight of each category of structural metal measured.

The steelwork dwgs, which are part of the *bidding documents* for a building with a structural steel frame, would not usually show the steel columns (and other structural steel components) to the same large scale as they are shown in the Warehouse Drawings, because these dwgs were prepared for the fabrication. Such large-scale dwgs are usually made by the *contractor* (or his *subcontractor*) as shop dwgs for fabrication and erection only after the contract has been awarded.[9]

An estimator usually has to measure structural steelwork for a steel framed building from a steel framing plan in which the beams are represented by single center lines and the columns by appropriate symbols. Connection details may be shown to a larger scale. Hence, the **MM-CIQS** calls for beams and purlins to be "measured from center to center line of columns," and columns to be "measured from top of base plate

[9] The Warehouse was both designed and built by the *contractor*, who also fabricated and erected the steelwork.

PRACTICAL "STANDARDIZED" FORMS FOR CONTRACTORS
Form 516 MFD IN U.S.A.

QUANTITY SHEET

PROJECT WAREHOUSE — ESTIMATOR KC — ESTIMATE NO. W-1
LOCATION — EXTENSIONS KC — SHEET NO. 1 of 3
ARCHITECT ENGINEER — CHECKED — DATE 1972
CLASSIFICATION MASONRY (CONCRETE BLOCKWORK)

DESCRIPTION	NO.	DIMENSIONS				ESTIMATED QUANTITY	UNIT
HOLLOW LOADBEARING STANDARD WEIGHT CONC BLKWK OF MODULAR UNITS REINFC'D EVERY 3RD COURSE & WITH 3/8" WIDE MORTAR JOINTS AS SPECS (p—)							
8" Extr W'house Walls (allow 2½% Waste)							
2/= 100-0 − 2-0 = 98-0 (W)	1/	98-0	x	22-0	2156		
33/8" = 22-0							
(22-0 − 18-0) = 4-0 (@ ends)	2/	1-0	x	4-0	8		
					2164		
DDT (door opening)	1/	3-4	x	7-4	−24 DDT		
					2140 (x 9/8)	2140	SF
						(2410	BLK)
8" Extr Office Walls in Small panels (x10'0" high) (allow 5% Waste)							
ELVTN 109-8 ″ −99-8 = 10-0 (W)	2/	4-0	(x	10-0)	8		
(W)	1/	7-4		″	7		
(E)	1/	22-8		″	23		
(SE)	1/	6-8		″	7		
					45 LF		
					x10-0 high		
					450		
DDT (door opening)	1/	3-4	x	7-4	−24 DDT		
					426 (x 9/8)	426	SF
						(480	BLK)
12" Extr W'house Walls (allow 2½% Waste)							
ELVTN 117-8 ″ −99-8 = 18-0 (N/S)	2/	200-0	x	18-0	7200		
27/8" = 18-0 (E)	1/	98-0	x	22-0	2156		
(22-0 − 18-0) = 4-0 (@ ends)	2/	1-0	x	4-0	8		
					9364		
DDT (openings)					DDT		
ELVTN 114-0 ″ −99-8	2/	16-0	x	16-0	− 512		
(opening) 14-4	2/	3-4	x	7-4	− 49		
(Lintel) +1-8 = 16-0	1/	4-0	x	7-4	− 29		
					−590		
					8774 (x 9/8)	8774	SF
						(9870	BLK)

FRANK R. WALKER CO., PUBLISHERS, CHICAGO

PRACTICAL "STANDARDIZED" FORMS FOR CONTRACTORS
Form 516 MFD IN U.S.A.

QUANTITY SHEET

PROJECT WAREHOUSE — ESTIMATOR KC — ESTIMATE NO. W-1
LOCATION — EXTENSIONS KC — SHEET NO. 2 of 3
ARCHITECT ENGINEER — CHECKED — DATE 1972
CLASSIFICATION MASONRY (CONCRETE BLOCKWORK)

DESCRIPTION	NO.	DIMENSIONS			ESTIMATED QUANTITY	UNIT
HOLLOW L/BEARING CONC BLKWK (CONT)						
4" Extr W'house Walls at eaves (allow 2½% Waste)						
(on 8" & 12" blk) (E/W)	2/	100-0 x 1-4	267	(x 9/8)	267 (300	SF BLK)
Extra Over for 12"/8" Corner Blks (allow 2½% Waste)						
	2/	22-0	44	(x 3/2)	44 * (66	LF BLK)
Extra Over for Lab & Mat in Bond Beams (x 8" high) in Blk Walls incl 2-#6 Rebars & conc filling all as specs (p–) (allow 2½% Waste)						
8" Bond Beams	1/	100-0	100			
(total office walls)	1/	45-0	45			
(over openings)	2/	6-0	12			
			157	(x 3/4)	157 * (118	LF BLK)
Closed Exposed Ends	1/	2	2			
	6/	1	6		8	No
12" Bond Beams	2/	200-0	400			
	1/	100-0	100			
(over openings)	2/	6-0	12			
"	1/	6-8	7			
4/20-0 = 80-0 (next -5-0 (office) 75-0	1/	75-0	75			
			594	(x 3/4)	594 * (446	LF BLK)
Closed Exposed Ends	1·3/	2	6		6	No
Extra Over for Lab & Mat in Filling Single Isolated 12" Blks Solid w/ Conc as Bearing under Joist Ends incl exp metal lath under as specs (p–)						
	26/	1			26	No

*When ordering, deduct quantities of Special Blks from Standard Blks.

FRANK R. WALKER CO., PUBLISHERS, CHICAGO

PRACTICAL "Standardized" Forms for Contractors
Form 516 MFD IN U.S.A.

QUANTITY SHEET

PROJECT WAREHOUSE — ESTIMATOR KC — ESTIMATE NO. W-1
LOCATION — EXTENSIONS KC — SHEET NO. 3 of 3
ARCHITECT ENGINEER — CHECKED — DATE 1972
CLASSIFICATION MASONRY (CONCRETE BLOCKWORK)

DESCRIPTION	NO.	DIMENSIONS		ESTIMATED QUANTITY	UNIT
HOLLOW L/BEARING CONC BLKWK (CONT)					
Extra Over for Lab & Mat in Filling wi Conc & 1-#4 Rebar Vertically in Single Cavity of Conc Blks at Jambs & at Free Ends of Walls, as specs (p-)					
(jambs)	5/2/	7-4	73		
"	2/2/	14-4	57		
(ends)	6/	10-0	60		
			190	190	LF
Extra Over for Lab & Mat in Forming Control Joints incl cutting blks up to steel cols b/s & Sealant, as specs (p-) & as detail (one joint measured at a steel column)					
(w'house N/S)	2/9/	18-0	324	324	LF

COLUMN
CUT CUT
12' BLOCK WALL
SEALANT

DESCRIPTION	NO.	DIMENSIONS		ESTIMATED QUANTITY	UNIT
8"x16" Blk Piers incl Filling wi Conc & 2-#5 Rebars, as specs					
(office) (N) (attached to 8" panel walls)	2·10/	10-0	120 (x3/2)	120 (180)	LF (BLK)
8"x24" Blk Pilasters (attached to 8" Blk Wall) incl Filling wi Conc & 2-#5 Rebars, as specs (p-)					
(w'house) (W)	7/	22-0	154 (x3/2)	154 (231)	LF (BLK)

[NB: Cast in place Conc Lintels to be meas'd in Div. 3- Concrete]

FRANK R. WALKER CO., PUBLISHERS, CHICAGO

PRACTICAL "STANDARDIZED" FORMS FOR CONTRACTORS
Form 516 MFD IN U.S.A.

QUANTITY SHEET

PROJECT WAREHOUSE — ESTIMATOR KC — ESTIMATE NO. W-1
LOCATION — EXTENSIONS KC — SHEET NO. 1 of 4
ARCHITECT ENGINEER — CHECKED — DATE 1972
CLASSIFICATION METALS: STRUCTURAL STEELWORK

DESCRIPTION	NO.	DIMENSIONS					ESTIMATED QUANTITY	UNIT
ASTM-A36 STRUCTURAL STEEL								
FABRICATED SHOP PAINTED & ERECTED AS SPECS (p-)								
Column Bases 3/4" thick Plate								
(MK A1)	9/	0-10	× 0-10	900	S In.			
(MK B1,C1,D1)	3/	0-7	× 0-8½	179				
(MK E1,F1)	5/	0-8	× 0-8	320				
(MK G1,H1)	2/	0-4	× 0-10	80				
(MK J1)	18/	0-10½	× 0-8	1512				
(TOTAL IN 37 NO B.PLATES)				2991	S In. × 0.2833 LB × 3/4" =		635	LB
Columns 8 WF × 31 LB								
(A1)	9/	20-2½		182				
(J1)	18/	20-8½		373				
(= 27 NO)				555	× 31 LB	=	17,205	LB
Columns 6 WF × 25 LB								
(B1,C1)	2/	16-2½		33				
(D1)	1/	16-4⅛		16				
(= 3 NO)				49	× 25 LB	=	1225	LB
Columns, Pipe								
3½" Φ (4 OD) (E1,F1)	2·3/	10-2½		51	× 10 LB =	510		
3" Φ (3½ OD) (G1,H1)	1·1/	10-2½		21	× 8.68 LB =	183	695	LB
(= 7 NO)								
(TOTAL: 37 NO COLS)								
Beam Seats, 3/4" thick Plate, in Connections to Glulam Beams as details (welding m/s)								
(A1)	9/	0-9¼	× 1-6	1499	S In.			
(B1,D1)	2/	0-5⅜	× 0-10½	113				
(C1)	1/	0-5⅜	× 1-3	81				
(E1)	3/	0-3½	× 0-8½	89				
(F1)	2/	0-5¼	× 0-8½	89				
(G1)	1/	0-3½	× 0-6	21				
(H1)	1/	0-5¼	× 0-6	32				
(J1)	18/	0-9¼	× 0-9	1499				
(TOTAL TO 37 NO COLS)				3423	S In. × 0.2833 LB × 3/4" =		730	LB

FRANK R. WALKER CO., PUBLISHERS, CHICAGO

PRACTICAL "STANDARDIZED" FORMS FOR CONTRACTORS
Form 516 MFD IN U.S.A.

QUANTITY SHEET

PROJECT WAREHOUSE	ESTIMATOR KC	ESTIMATE NO. W-1	
LOCATION	EXTENSIONS KC	SHEET NO. 2 of 4 R	
ARCHITECT ENGINEER	CHECKED	DATE 1972	
CLASSIFICATION METALS: STRUCTURAL STEELWORK			

DESCRIPTION	NO.	DIMENSIONS					ESTIMATED QUANTITY	UNIT
ASTM-A36 STRUCTURAL STEEL (CONT)								
Sides to Beam Seats, 1/4" thick Plate, in Connections to Glulam Beams, as details (welding m/s)								
(A1)	9/2/	1-6	x	1-0	27-0 SF			
(B1, D1)	2/2/	0-10½	x	1-6	5-3			
(C1)	1/2/	1-3	x	0-6	1-3			
(E1, F1)	2·3/2/	0-8½	x	0-6	3-6			
(G1, H1)	1·1/2/	0-6	x	0-6	1-0			
(J1)	18/2/	0-9	x	1-0	27-0			
(TOTAL IN 37/2 = 74 No. SIDES)					65-0 SF	x 144 x 0.2833 LB x 1/4" =	665	LB
Beam Connections, 1" Plate, (for Glulam to G. Beams) ditto (ditto)								
	10/2/	0-7	x	0-9¼	1295 S In.	x 0.2833 LB x 1" =	370	LB
Ditto 1/4" Plate (ditto) ditto (ditto)								
	10/2/	0-7	x	2-5¼	4095 S In.	x 0.2833 LB x 1/4" =	290	LB
1/4" Shop Fillet Welds (intermittent) btwn Bases & Cols, avg 4" long.								
(A1)	9/4/	0-3		(36 No)	108 L In.			
(B1, C1, D1)	3/4/	0-3		(12)	36			
2 +3 = 5 (Circ E1, F1)	5/3½/	0-4		(5)	63			
+1 = 2 (Circ G1, H1)	2/3½/	0-3½		(2)	22			
(J1)	18/4/	0-4		(72)	288			
(TOTAL TO 37 No BASES)				(127 No)	517 L In.	(127 No) =	517 L In. (43 LF)	
1/4" Ditto (ditto) btwn Cols & Beam Seats, avg 7½" long.								
(A1)	9/2/	0-7¼		(18 No)	131 L In.			
(B1, C1, D1)	3/2/	0-5¼		(6)	32			
(Circ E1, F1)	5/3½/	0-4		(5)	63			
(Circ G1, H1)	2/3½/	0-3½		(2)	22			
(J1)	18/2/	0-7¼		(36)	261			
(TOTAL TO 37 No SEATS)				(67 No)	509 L In.	(67 No) =	509 L In. (42½ LF)	
1/4" Ditto (continuous) to Beam Connection Plates, 7" long								
	10/2·2/	0-7		(40 No)	280 L In.	(40 No) =	280 L In. (23½ LF)	

FRANK R. WALKER CO., PUBLISHERS, CHICAGO

PRACTICAL "STANDARDIZED" FORMS FOR CONTRACTORS
Form 516 MFD IN U.S.A.

QUANTITY SHEET

PROJECT WAREHOUSE | ESTIMATOR KC | ESTIMATE NO. W-1
LOCATION | EXTENSIONS KC | SHEET NO. 3 of 4 R
ARCHITECT ENGINEER | CHECKED | DATE 1972
CLASSIFICATION METALS: STRUCTURAL STEELWORK

DESCRIPTION	NO.	DIMENSIONS				ESTIMATED QUANTITY	UNIT
3/16" Shop Fillet Welds (continuous) to Beam Connection Plates							
(A1)	9/2/1-6		(18 No)	27-0 LF			
(B1, D1)	2/2/0-10½		(4)	3-6			
(C1)	1/2/1-3		(2)	2-6			
(E1, F1)	5/2/0-8½		(10)	7-0			
(G1, H1)	2/2/0-6		(4)	2-0			
(J1)	18/2/0-9		(36)	27-0			
(TOTAL IN 37 No CONNCTNS)			(74 No)	69-0 LF		(74 No) = 69	LF
						(828 L In.)	
LABORS TO ASTM-A36 STRUCTURAL STEEL							
Machine Ends Cols True & Square:-							
8WF (637 LB/AVG)		27/2		54		54	No
6WF (408 LB/AVG)		3/2		6		6	No
3½ Φ (102 LB/AVG)		5/2		10		10	No
3 Φ (92 LB/AVG)		2/2		4		4	No
(TOTAL		37/2	=	TOTAL = 74 No)			
Drill Bolt Holes in Steel Plate:-							
1Φ in ¾" Bases	18·3·9/2			60		60	No
13/16"Φ in ¾" Bases		2·3/4		20			
		2/2		4			
(TOTAL IN 37 No BASES)				24		24	No
13/16"Φ in ¼" Connectn Sides							
(9+3+5) = 17		17/2/2		68			
(2+18) = 20		20/1/2		40			
(TOTAL IN 37 No CONNCTNS)				108		108	No
¾"Φ in ¼" Connectn Sides							
		10/2/2		40		40	No
(TOTAL IN 10 No CONNCTNS)							

FRANK R. WALKER CO., PUBLISHERS, CHICAGO

PRACTICAL "STANDARDIZED" FORMS FOR CONTRACTORS
Form 516 MFD IN U.S.A.

QUANTITY SHEET

PROJECT WAREHOUSE — ESTIMATOR KC — ESTIMATE NO. W-1
LOCATION — EXTENSIONS KC — SHEET NO. 4 of 4 R
ARCHITECT ENGINEER — CHECKED — DATE 1972
CLASSIFICATION METALS: STRUCTURAL STEELWORK

DESCRIPTION	NO.	DIMENSIONS		ESTIMATED QUANTITY	UNIT
ASTM-A325 BOLTS FOR STEELWORK, AS SPECS (p–)					
Anchor Bolts (to be set into conc) wi Nuts					
3/4"ϕ x 12" long		30/2	60	60	NO
5/8"ϕ x 15" long		5/4	20		
(30+5+2)=37		2/2	4	24	NO
(TOTAL	IN	84 NO HOLES)	(TOTAL 84 NO)		
Connection Bolts (for Glulam Beams) wi Heads, Nuts, & Washers					
5/8"ϕ x 12" long		9/2	18		
		18/1	18	36	NO
5/8"ϕ x 8" long		5/2	10		
		1/1	1	11	NO
5/8"ϕ x 6" long		3/2	6		
(9+18+5+1+3+1)=37		1/1	1	7	NO
(TOTAL	IN	108 NO HOLES)	(TOTAL 54 NO)		
1/2"ϕ x 12" long		10/2	20	20	NO
(TOTAL	IN	40 NO HOLES)			

STEEL WEIGHT SUMMARY

sht. #1	3/4"	Bases =	635 LB
	8	WF Col =	17,205
	6	WF Col =	1225
		Pipe Col =	695
	Connctn	Seats =	730
sht. #2	"	Sides =	665
(370+290) =	"	(ditto) =	660
(1306 LF/12 x 0.2 LB)	1/4"	Welds =	22
(69/12 x 0.15 LB)	3/16"	Welds =	1
158 NO		Bolts =	say 162
		TOTAL	22,000 LB = 11 TONS

DESCRIPTION	ESTIMATED QUANTITY	UNIT
SHOP-PAINTING STRUCTURAL STEEL, AS SPECS (p–)	11	TONS
ERECTING STRUCTURAL STEEL, AS DWGS & SPECS (p–)	11	TONS

FRANK R. WALKER CO., PUBLISHERS, CHICAGO

to elevation at 'top of steel'." In this example it is possible to measure the columns to exact lengths because of the large-scale dwgs provided.

Welds, like steel components, may be measured by weight, or by length, stating the unit weight as in the example. The number of intermittent welds is stated because this indicates their average length, which affects their cost. Labor for machining ends and drilling holes is described and enumerated, as are the anchor bolts and the connection bolts.

Erecting the fabricated components is part of the *work*; and, as stated in the **MM-CIQS**, each measured item is held to include "conveyance and delivery, unloading, hoisting, all labor setting, fitting and fixing in position..." in its description. Many estimators, however, will choose to price the erection of the steel as one separate and inclusive item based on the total weight, the number of pieces, and the time required for an erection crew and equipment. This is why the estimator should indicate the "number of pieces" when measuring steelwork by weight, because weight alone is not sufficient to accurately estimate erection time and costs. And this is why the better method of measuring structural steel is to enumerate and fully describe each piece.

Warehouse Example

Rough Carpentry: Decking and Beams

Wood decks made up of tongued and grooved planks can be measured as *super* items in SF, or the deck area and thickness can be converted to BF. This conversion requires an allowance to be made for the t & g joints, and for the loss in width caused by the lumber manufacturer's dressing the planks. Board foot measure is always based on nominal dims, and a t & g plank with a nominal width of 6 in. may have an actual overall-width of 5⅜ in. and, with a tongue ⅜ in. wide, an effective width of 5 in. Therefore, an allowance of 20 percent must be made for this loss in width from the nominal dim. (A plank 8 in. wide would require an allowance of about 15 percent.) A small allowance for *waste* caused by end cutting and loss should be made in the *unit price*. If lumber items are measured up to the nearest even dim, they can be described as "including *waste*."

Glued-laminated wood members are best measured by enumerating and fully describing them, the same as steel members. But the manufacturer of glulam products has to measure the lumber content as a basis for pricing, and this is done by calculating the number of BF from the number and nominal dims of the individual laminations (as explained in Chapter 8), or by calculating the beam's volume and weight and pricing it by means of a unit cost related to the units of measurement.

So, although from the *contractor's* viewpoint the easiest and best way to measure glulam products is by enumeration and description, that method does not lead to a full interpretation of costs. In this example the glulam beams are measured in LF, and the lumber content is calculated in BF. An understanding of prices and costs always requires their resolution into *unit prices* so that comparisons can be made.

Rough hardware for *carpentry work* is measured by description and enumeration except for common hardware such as nails and spikes, which can be allowed for in the *unit prices*.

Special labor required on carpentry items (such as "fire cut ends of joists") are usually described and enumerated as separate items so that the carpentry items themselves (the joists) can be priced as *basic items*. If a special kind of labor comprises a major part of a carpentry item's costs (such as with the "2 × 12 roof joists shaped to curve at top edge"), the two may be combined into one particular item for measuring and pricing; because the item is so unusual as to make it unlike any *basic item*, leaving no advantage to be gained by separating the labor item.

Descriptions of *carpentry items of work* must always include the grade of material and the method of fixing if other than the usual nailing. As explained before, if length of lumber is a cause of additional *material costs* (and possibly additional *labor costs* of handling long lengths), these lengths should be measured separately and so described.

Warehouse Exercises

All the *work* not measured in the examples can be measured in exercises; and as a major estimating exercise the entire project might be measured and priced at current, local *unit prices* and the total estimated costs compared with local costs of constructing similar warehouses. To provide more exercises and, at the same time, to develop a better understanding of costs, simple drawings of alternative methods of constructing the exterior walls, structural frame, and roof decks might be made, and the *work* measured and priced therefrom. A clearer understanding of the effects of differences in building dimensions on construction

PRACTICAL "STANDARDIZED" FORMS FOR CONTRACTORS
Form 516 MFD IN U.S.A.

QUANTITY SHEET

PROJECT WAREHOUSE | ESTIMATOR KC | ESTIMATE NO. W-1

LOCATION | EXTENSIONS KC | SHEET NO. 1 of 6

ARCHITECT ENGINEER | CHECKED | DATE 1972

CLASSIFICATION ROUGH CARPENTRY: DECKING & BEAMS

DESCRIPTION	NO.	DIMENSIONS		ESTIMATED QUANTITY	UNIT

DOUGLAS FIR "CONSTRUCTION" GRADE UNLESS OTHERWISE INDICATED

2" thick T&G Decking "STANDARD" grade laid in a "controlled random pattern" on wood joists @ 4'0" o/c (m/s) in Flat Roof Deck as specs (p-) (allow 3% Waste)

200-0 + 20-0 = 220-0 ✓ ; 20-0 × 11 Bays = 220-0 ✓

1/220-0 × 100-0 = 22000 SF
× 2 BF
44,000
(allow for T&G & Dressing) + 8800 (+20%)
52,800

53,000 BF
(22,000 SF)

3×14 Roof Joists in 20' lengths (incl Waste)

100-0 / 4' = 25 + 1 = 26 ; 10 Bays + 1 canopy = 11 No

26/11/20-0 = 5720 × 3½ BF

20,000 BF
(5720 LF)

Fire Cut Ends 3×14 Joists 26/1 = 26

26 NO

3×6 Runners (Ledgers) in 20' lengths bolted to Glulam Beams @ 4'0" o/c (bolts m/s) (incl Waste)

9/2/100-0 = 1800
(@ canopy) 1/1/100-0 = 100
1900 × 1½ BF

3000 BF
(1900 LF)

3×4 Joist Seats (Plates) in 20' lengths bolted to Conc Blk Bond Beam @ 4'0" o/c (bolts m/s) (incl Waste)

(W) 1/100-0 = 100 × 1 BF

100 BF
(100 LF)

3×6 Ditto (incl Waste)

(E) 1/100-0 = 100 × 1½ BF

150 BF
(100 LF)

FRANK R. WALKER CO., PUBLISHERS, CHICAGO

PRACTICAL "STANDARDIZED" FORMS FOR CONTRACTORS
Form 516 MFD IN U.S.A.

QUANTITY SHEET

PROJECT WAREHOUSE — ESTIMATOR KC — ESTIMATE NO. W-1
LOCATION — EXTENSIONS KC — SHEET NO. 2 of 6
ARCHITECT ENGINEER — CHECKED — DATE 1972
CLASSIFICATION ROUGH CARPENTRY: DECKING & BEAMS

DESCRIPTION	NO.	DIMENSIONS			ESTIMATED QUANTITY	UNIT
SUPPLY ONLY D. Fir Glulam Beams "Industrial" grade (for appearance) & "Interior" grade (for service) unless otherwise indicated (including 20% Waste in BF quantities for manufacturing)						
9"x27 5/8" Beams 50-0-(0-3+9-4)=40-5	9/	40-5	364	x34.0 BF/LF	12,376	BF
9"x29 1/4" Beams (50-0-0-3)+9-4 = 59-1	9/	59-1	532	x36.0 BF/LF	19,152	BF
5 1/4"x27 5/8" Exterior grade Beam (@ canopy)	1/	40-5	41	x20.4 BF/LF	837	BF
5 1/4"x29 1/4" Ditto (@ canopy)	1/	59-1	59	x21.6 BF/LF	1275	BF
ERECT ONLY Glulam Beams onto steel columns (generally about 20'0" high above floor) (connections & bolts m/s)						
9"x27 5/8" Beams	9/	40-5	364	x 52 LB/LF*	9 No = 18,928	LB
9"x29 1/4" Beams	9/	59-1	532	x 55 LB/LF	9 No = 29,260	LB
5 1/4"x27 5/8" Beam	1/	40-5	41	x 30 LB/LF	1 No = 1230	LB
5 1/4"x29 1/4" Beam	1/	59-1	59	x 32 LB/LF	1 No = 1888	LB

* At 30 LB/CF

[NB: Glulam Beams to Office Roof measured later.]

FRANK R. WALKER CO., PUBLISHERS, CHICAGO

PRACTICAL "STANDARDIZED" FORMS FOR CONTRACTORS
Form 516 MFD IN U.S.A.

QUANTITY SHEET

PROJECT WAREHOUSE — ESTIMATOR KC — ESTIMATE NO. W-1
LOCATION — EXTENSIONS KC — SHEET NO. 3 of 6
ARCHITECT ENGINEER — CHECKED — DATE 1972
CLASSIFICATION ROUGH CARPENTRY: DECKING & BEAMS

DESCRIPTION	NO.	DIMENSIONS			ESTIMATED QUANTITY	UNIT
GALV STEEL ROUGH HARDWARE (all Bolts incl Nuts & Washers)						
3/4"ϕ x18" Bolts wi two 2 5/8"ϕ Split Rings each & Drill 9" Glulam Beam & 2/3" Runners						
(3x6 Runners to Beam)	9/	24		216	216	No
3/4"ϕ x10" Bolts wi one 2 5/8"ϕ Split Ring each & Drill 5 1/4" Glulam Beam & 1/3" Runner						
(@ canopy)	1/	24		24	24	No
3"x3"x1/4" Angle Clips each 3/nailed to wood joists & to beams						
(@ outside joists)	9/2/	2		36		
(ditto, @ Canopy)	1/2/	1		2	38	No
1 1/2"x3/16" Flat (twisted) Joist Ties x18" long each 3/nailed to joist & set into Conc Blkwk						
(@ ends of joists	26/	2		52		
on Conc Blkwk)	26/	1		26	78	No
12" lengths Plumber's Strap each end 3/nailed to joists and beams as Ties						
(@ Canopy)	24/	1		24	24	No
(NB: Not @ outside joists wi angle clips)						
24" Lengths Ditto each end 3/nailed to ends of pairs of joists as Ties						
(@ W'house roof joists)	24/	9		216	216	No
(NB: Not @ outside joists wi angle clips)						
3/4"ϕ x8" Bolts & Drill 3" Joist Seats (Plates) & set into Conc Filling in Bond Beams						
	2/	26		52	52	No

[NB: Bolts for Glulam to Steel Connections Taken with Structural Steelwork.]

FRANK R. WALKER CO., PUBLISHERS, CHICAGO

PRACTICAL "STANDARDIZED" FORMS FOR CONTRACTORS
Form 516 MFD IN U.S.A.

QUANTITY SHEET

PROJECT WAREHOUSE ESTIMATOR KC ESTIMATE NO. W-1
LOCATION EXTENSIONS KC SHEET NO. 4 of 6
ARCHITECT ENGINEER CHECKED DATE 1972
CLASSIFICATION ROUGH CARPENTRY: DECKING & BEAMS — OFFICE

DESCRIPTION	NO.	DIMENSIONS		ESTIMATED QUANTITY	UNIT
DOUGLAS FIR "CONSTRUCTION" GRADE					
2x12 Roof Joists in 14' lengths (incl Waste)					
89-4 / 1-4 = 67, +1, +4 (extra) = 72	72/14-0	1008	x 2 BF	2000 (1000 LF)	BF
2x12 Ditto in 18' lengths (incl Waste)					
	72/18-0	1296			
(header)	1/90-0	90			
		1386	x 2 BF	2800 (1400 LF)	BF
2x12 Ditto in 10' lengths cut to 8'6" lengths & shaped wi Circ Top Edge (to large radius) as shown for vault-shaped office roof bays each notched one end for roof gutter & installed @ 90° across main roof joists, as detail					
31-4 −0-4 = 31-0; 31-0 / 1-4 = 24 +1 = 25	2/25			50	No
2x12 (D Dressed) Headers in 16' lengths installed against last joists (incl Waste)					
	2/32-0	64	x 2 BF	130 (64 LF)	BF
2x2 Blockings installed on top edges of joists (incl Waste)					
	2/7/32-0	448	x 1/3 BF	450 (150 BF)	LF
3x8 Runners (Ledgers) in 18' lengths bolted to Conc Blk Bond Beam @ 4' o/c (bolts m/s) (incl Waste)					
(on W'house wall)	1/90-0	90	x 2 BF	90 (180 BF)	LF
4x4 Cant Strips in 16' lengths planted on roof sheathing next headers (incl Waste)	2/32-0	64	x 2/3 BF	64 (45 BF)	LF

FRANK R. WALKER CO., PUBLISHERS, CHICAGO

PRACTICAL "STANDARDIZED" FORMS FOR CONTRACTORS
Form 516 MFD IN U.S.A.

QUANTITY SHEET

PROJECT WAREHOUSE — ESTIMATOR KC — ESTIMATE NO. W-1
LOCATION — EXTENSIONS KC — SHEET NO. 5 of 6
ARCHITECT ENGINEER — CHECKED — DATE 1972
CLASSIFICATION ROUGH CARPENTRY: DECKING & BEAMS — OFFICE

DESCRIPTION	NO.	DIMENSIONS			ESTIMATED QUANTITY	UNIT
1/2" thick D.Fir (Exterior) "Select" grade Sheathing Ply installed on joists in flat roof deck (allow 3% Waste)						
	1/89-4	× 31-4	2799		2800	SF
Extra Lab Nailing 1/2" Sheathing to Circ top Edges of roof joists						
	2/8-0	× 31-4	501		500	SF
1/2" Ply Gutter Lining (btm & side) x 8" total girth (use Waste above)						
	2/31-0		62		62	LF
SUPPLY ONLY D.Fir Glulam Beams "Industrial"/"Interior" grades a.b						
5" x 16 1/4" Beam (20-8 − 0-8) = 20-0	1/20-0		20	x 12.0 BF/LF	240	BF
3 1/4" x 16 1/4" Beams 2/14-0 = 28-0	2/28-0		56			
	1/12-0		12			
89-4; 1/2 − 1-4 = 88-0; −20-0; 68-0 ↑ 68-0; −56-0; 12-0			68	x 8.16 BF/LF	555	BF
ERECT ONLY Glulam Beams onto Steel columns (about 10'6" above floor) (connections & bolts m/s)						
5" x 16 1/4" Beam	1/20-0		20	x 17 LB/LF	1 No = 340	LB
3 1/4" x 16 1/4" Beams	2/28-0		56	x 11 LB/LF	2 No = 616	LB
3 1/4" x 16 1/4" Beam	1/12-0		12	x 11 LB/LF	1 No = 132	LB

FRANK R. WALKER CO., PUBLISHERS, CHICAGO

PRACTICAL "STANDARDIZED" FORMS FOR CONTRACTORS
Form 516 MFD IN U.S.A.

QUANTITY SHEET

PROJECT WAREHOUSE | ESTIMATOR KC | ESTIMATE NO. W-1
LOCATION | EXTENSIONS KC | SHEET NO. 6 of 6
ARCHITECT ENGINEER | CHECKED | DATE 1972
CLASSIFICATION ROUGH CARPENTRY: DECKING & BEAMS — OFFICE

DESCRIPTION	NO.	DIMENSIONS		ESTIMATED QUANTITY	UNIT
GALV STEEL ROUGH HARDWARE, as before					
3/4"φ x 12" Bolts & Drill 3" wood runner & side of Hollow Conc Blk Bond Beam & set into Conc Filling (m/s)					
90-0/4-0 = 23 +1 +5 (extra) = 29	29/1		29	29	NO
24" Lengths Plumber's Strap each end 3/nailed to ends of pairs of joists as Ties, as before					
	72/1			72	NO
Sheet Metal Joist Hangers 2x12 size					
(68-4) = 64	64/1			64	NO
Ditto 4x12 size					
	4/1			4	NO

[NB: Bolts for Glulam to Steel Connections taken with Structural Steelwork.]

FRANK R. WALKER CO., PUBLISHERS, CHICAGO

costs can be obtained if "change orders" are devised and written out, and if the costs of the ordered changes are estimated. For example, it could be vividly demonstrated in this way that a significant change in the height (and volume) of the warehouse does not make a proportionately significant change in total *costs of the work.*

Apartment Block Exercises

No examples are given for measuring *work* in the Apartment Block; because all parts of the building can be measured for exercises and a complete estimate of costs might be made as described for the Warehouse. Again, alternative construction methods can be devised and utilized and comparative cost analyses can be made.

As an approximate guide to accurate measurement, the approximate total quantities of some of the *major items of work* in the building are given below.

Concrete, all kinds	1520 CY
Steel rebar, all kinds	112 TON
Formwork, all kinds (contact area)	70,000 SF
Framing lumber, all kinds	170 MBF
Ply sheathing, walls & decks	47,000 SF
Exterior stucco, including soffits	1550 SY
Gypsum wallboard, walls and ceilings	118,000 SF

The sheathing, stucco and wallboard areas are net, with openings deducted. The gross quantity of wallboard (measured in standard sheet sizes) would be from 5 to 10 percent higher. Variations in total quantities will occur because of variations in measurement methods and in assumptions concerning details of construction. Nevertheless, general agreement should be possible. The concrete quantity is supported by actual delivery tickets and invoices, as well as by measurements. Two different persons measured the steel rebar and the results were the same within 3 tons. Three different persons measured the framing lumber and obtained total quantities within 5 percent of that shown. Likewise, the exterior stucco was measured and agreement was reached between two persons. The gypsum wallboard was measured twice by the same person.

Although the examples in this chapter involve *measurement of work,* both the examples and the exercises will prove to be more valuable to a student if they are priced. Pricing the quantities of *work* leads to a fuller understanding of measurement even if the *unit prices* are hypothetical (which they must be), because the purpose of measurement is pricing, and it is pricing that determines the requirements for measuring *construction work.* If you want to know how to measure any *item of work,* find out how it will be priced.

Most of the examples of measuring *work* in this chapter have been used at least once in estimating classes at the British Columbia Institute of Technology, and the earlier examples in this chapter have been used several times. Each time a group of students has done one of these examples, some new discovery has been made. Sometimes errors have been discovered, and even these have given rise to useful discussions that have helped someone to learn. Sometimes, new points of view have been expressed by students, which indicate that some widely accepted methods of measurement may need changing.

Estimating is not a pure science, for subjective judgments are part of it and it is not entirely a matter of deducing from self-evident truths. And whereas this is particularly true of the pricing part of estimating, it is also true of measurement to a lesser extent. Measurements of quantities are made primarily so that the *work* measured can be priced; and as an estimator views and understands the pricing of *work,* so will he measure it. Therefore, there is often room for more than one way of measuring any *item of work* and, therefore, there will always be different opinions about measurement methods.

However, it is a widely held opinion among quantity surveyors that better results are obtainable if it is *work* that is measured (rather than simply measuring the materials that form a part of it) and if the *work* is measured net in place, as required by the construction contract, with calculated allowances for such things as *waste, laps, swell* and *shrinkage* made in the *unit prices* that are calculated for the *items of work.*

Perhaps one of the most illustrative and illuminating *items of work* that can be studied in its economics and in its measurement and pricing is formwork for cast-in-place concrete, and also the reinforced concrete itself. This is why so many examples are based on this kind of *work.*

Reinforced concrete requires and contains within the measurement and pricing of its three primary parts —formwork, reinforcing steel, and concrete—most of the techniques and principles of construction estimating. Therefore, it is an ideal subject for most students of estimating, and the measurement and pricing of formwork vividly illustrates that it is *work* that is measured, not materials alone, and that the purpose of measurement is primarily pricing.

Measuring *work* is painful at the start, as many

students complain. It is painful because it demands a high level of sustained concentration and thought, and nothing hurts more. But like running and climbing, and other activities that cause pain when first begun, measurement soon becomes tolerable with practice and for some even a source of pleasure and satisfaction. An estimator who has measured the *work* in a project has built that project within his own head, wall by wall, and floor by floor. He probably knows more about the construction of the project and its details than the *designer,* and he is probably more aware of any errors and omissions in the drawings and specifications.

Having learned and applied the precepts and techniques of measurement, there is only one thing left to do—practice. Practice is the only way to overcome the initial pain, and it is the only way to become proficient in measuring *construction work.*

Questions and Topics for Discussion

1. In measuring wood studs, explain how the number of studs in, say, the framed exterior walls of a house should be measured, with particular reference to the requirements for extra studs for any purpose and reason.

2. Regarding slab screeds, (a) what is their purpose, (b) how are they installed, and (c) what is the best way to estimate the costs of slab screeds?

3. Formwork is generally measured by the "contact area" in square feet. Explain in detail why this is so, and why formwork is not measured as other work is generally measured. Explain also, by an example, why formwork is sometimes better measured as a linear item.

4. Explain precisely the differences between Light Bending and Heavy Bending of Rebar. Which is the more expensive, and why?

5. Explain all the differences between "framed" and "unframed" structural steelwork, and why they should be distinguished in an estimate. Also, explain the primary considerations in pricing both these kinds of structural steelwork.

6. Why should carpentry cross-bridging be measured as a linear item, or numerated in sets? Why not measure it in board feet in the same way most rough carpentry work is measured?

7. Explain fully the term "raking cutting and waste", and explain how it can be measured and priced in an estimate and verified by cost accounting.

8. Discuss the proposition: most errors in the measurement of rebar (apart from arithmetical errors) are errors of omission.

9. Compare the two methods of measuring structural steelwork described in Chapter 8 as applied to the structural steelwork in the Warehouse. Which is the better method in this case? Which is the better method generally?

10. Compose a standard method of measurement for one of the following sections of work: Formwork for concrete; Rough carpentry; Structural steelwork. Assume that in the future all estimating will be done with the aid of computers and that the estimating and cost accounting cycle (as described in Chapter 3) will be applicable.

10

Pricing Work: General

By pricing *work* we mean computing and applying *unit prices* to measured quantities to arrive at an estimate of the probable *costs of the work.* The different classifications of costs were examined in Chapter 5, and they are:

1. *Material costs*
2. *Labor costs*
3. *Equipment costs*
4. *Job overhead costs*
5. *Operating overhead costs*
6. *Profit.*

This chapter, and the next, deals primarily with the first three classifications of costs and their computation by means of *unit prices.* The last two classifications—*operating overhead costs* and *profit*—are not computed by means of *unit prices,* and their assessment goes beyond practical estimating into the areas of accounting, business, bidding strategy, and management policy. The first three classifications are called *direct costs* because they are by definition always attributable to the *items of work* of a specific construction job. The last two classifications likewise are called *indirect costs* because by definition they cannot be directly attributed to a specific job, as explained later.

The remaining classification, *job overhead costs,* by definition lies between the other two groups of costs, in that although these costs can be attributed to a specific job they cannot be attributed to specific *items of work.* Thus, they are among the *direct costs* of a construction job, but they cannot always be priced by *unit prices* because they do not always relate to specific *items of work,* but rather to the job as a whole.

Although we are primarily interested here in those *direct costs* that involve *unit prices,* a brief review of the *indirect costs* will be helpful in comprehending all construction costs and their places in estimates. *Profit* is included because it is a cost to the *owner.* Following a look at some general aspects of costs in this chapter, the next chapter looks at some specific aspects and examples of direct costs.

Profit and Overhead Costs

Profit is assessed in the light of risk and competition, the supply and demand for construction services, and the need for *work.* It is usually included in an estimate as a percentage of other costs to arrive at the total estimated *cost of the work.* It may be calculated as a percentage of the total of all other costs, or it may be based on one certain class of costs—usually *labor costs*—whereas the *profit* on materials may be looked for in volume discounts. The way in which costs are marked up to allow for *profit* really depends on the company and the *work* it does.

Operating overhead costs are the costs of being in business, whether *construction work* is being done or not. To include these costs in an estimate, a projection is made of future *operating costs* and the anticipated *volume of work* that will be done over, say, the next year. The calculated percentage of "*operating costs* to *volume of work*" is compared with similar percentages for previous years and, in this way, a percentage is established to be applied to the other *costs of work* to allow for *operating overhead costs.*

Profit and *operating overhead costs,* (or *indirect costs*), often simply referred to as "*profit and overhead,*" are usually the last costs to be included in an estimate, because they are related to the *total direct costs* and to the anticipated *total volume of work* to be done in the foreseeable future.

Job overhead costs are different from *operating overhead costs* because they can be attributed to a specific job. But they cannot be attributed to any specific *items of work* of that job. Because these costs are so numerous and various, they are computed in a variety of different ways: by lump sum; by salaries, wages, and rental rates on a time basis; and, in some instances, by *unit prices,* depending on the kind of overhead item. These costs are usually priced before the *profit* and *operating overhead costs* are added, and after the *measured work* has been priced. Figure 39 shows the common classifications of costs and the general order of pricing, ending with *operating overhead costs* and *profit.*

Before examining these classifications, it might be pointed out once again that, like all classifications, the common classifications of costs are somewhat arbitrary. They are not completely definitive, and not all costs can always be so precisely classified. Nevertheless, the classifications of costs in Fig. 39. are reasonably distinct except for *job overhead costs,* some of which might be classed as *direct costs* and others, at another time, as *indirect costs,* depending on the nature of the *work* and the contract and the organization of the construction company. But as the number of classifications are increased, the less useful the classifications become, so we use as few classifications as possible.

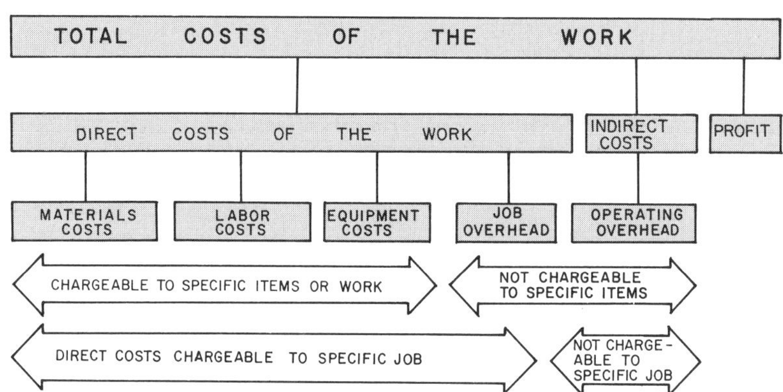

FIG. 39. The Classifications of Construction Costs.

A **unit price** has been defined as the costs of an *item of work* divided by the number of units of that item. As such, it is an **average price** that will vary depending on the number of units and any number of other conditions. Nevertheless, it is a convenient means whereby we can calculate and compare and analyze and price the *costs of work.*

It is necessary in defining a *unit price* to be specific about its subject—whether it is for *material costs, labor costs, plant and equipment costs, overheads and profit,* or for any combination of these. It is usually better to separate the *unit prices* and costs for each of these categories, but most standard estimating forms have separate columns for *material costs* and *labor costs* only. On the other hand, in *unit price contracts,* the *unit prices* are usually combined to include all the *direct costs* and sometimes they include all *direct and indirect costs,* thus providing one complete *unit price* for each *item of work.*

Pricing Plant and Equipment

Plant and equipment costs are dealt with in several ways. In some estimates they are entered against the appropriate *items of work,* and as there is often not a separate column on the estimate sheet for equipment, they are sometimes entered in the "*material costs*" column. Usually, the estimator wants the *total labor costs* separated and distinct; therefore, no other costs can be included with them, and the "materials" column is the only other place to enter *equipment costs.* Alternatively, an estimate form with a separate column for *equipment costs* can be used if they are a major part of the *work,* and if all types of costs are required to be shown separately.

In some estimates, *plant and equipment costs* are estimated separately at the end of the estimate, just before the final summary. One reason for this is that some plant and equipment is on the site for a long time and is used for many *items of work.* An overhead crane, for example, might be on site almost from job start to finish—or at least until the building structure is completed—and it might be used for lifting and placing concrete, steelwork, timbers, trusses, pieces of formwork, and loads of other materials. In such cases, the estimator estimates the period of time the crane is needed on site and computes the costs on a monthly rental basis, plus transportation, set-up, and removal costs. Many items, such as excavating equipment, table saws, and other plant, tools, and equipment, which have a number of different uses, are estimated in this way, with no direct reference to specific *items of work.* On the other hand, the costs of specialized equipment, such as that used in pumping concrete, can and should be attributed to one or several specific items and priced together with *material costs* and *labor costs* for those items and expressed as a separate *unit price,* if required.

In some other estimates, particularly in estimates for projects not requiring a large amount of heavy and expensive equipment, such as a shopping center project with no high buildings, estimators may allow for *plant and equipment costs* as a percentage of total costs or as a percentage of *labor costs.* The assumption here is that the *plant and equipment costs* are about the same proportion of total costs in similar jobs, and cost accounts may show that this proportion is fairly constant. Small tools are usually allowed for as a percentage of total *labor costs* in this way. Estimating any costs by proportion requires an estimator who can make a proper use of the cost data and a series of previous, similar jobs done under similar conditions, with cost accounts of those jobs showing reasonably constant relationships among the different costs. This method of estimating is contrary to the precepts covered in Chapter 6; and it is not recommended for estimates in which more accurate estimating methods can be used, or where they should be used because the *equipment costs* are a major part of the *work,* particularly if the estimate is for a *bid* for a stipulated sum contract.

Equipment costs have to be estimated the same as any other costs. If the equipment is leased, it means that the lessor has to calculate the rental rate he must charge. If the equipment is owned by a bidder making an estimate, it means that he must calculate a realistic rate per day (week, month) before he can estimate the total costs; and it is important that the rate be high enough to include all the *owning and operating costs* of the equipment, and low enough to be competitive in bidding. The elements of these costs were discussed in Chapter 5, and an example of pricing these costs will illustrate the basic method for pricing the costs of any tool, plant, or piece of equipment used in construction.

In Chapter 5 the *owning costs* were described as:

1. Depreciation (loss in value from any cause)
2. Maintenance (major repairs and replacement of parts)
3. Investment (costs arising from investment and ownership).

To price these costs we need three figures as a basis, and they are:

1. **Total investment** (total purchasing and delivery costs of the item to the purchaser less estimated salvage value)

2. **Estimated working life** (the number of years, and hours per year, the item is expected to be used and to remain economically usuable)

3. **Average annual investment** (the average amount of money the purchaser of the item will have invested in the item in each year of its life).

The total investment is made when the item is purchased; but as the item depreciates in value, the value of the investment gets progressively less. The older the item (tool, plant, or equipment), usually the less value it has. So over the total working life, the average annual investment is somewhere between 100 percent of the total investment and zero. Common sense tells us that the average annual investment should be about 50 percent of the total investment. Actually, it is never calculated at 50 percent because of the accountant's methods of working. For a five-year working life, the average annual investment is 60 percent of total investment; and for a ten-year working life, it is 55 percent. The average is calculated thus:

First day of 1st year	100 percent
First day of 2nd year	80
First day of 3rd year	60
First day of 4th year	40
First day of 5th year	20
First day of 6th year	0
5 years)	300 = 60 percent, average.

It depends on the years of working life, and the greater the number of years the closer the average comes to 50 percent when it is calculated this way. The average annual investment figure is required as the basis for calculating the investment costs per year for the item.

The estimated working life, as with all estimates, should be based on experience. If not on one's own, then it should be based on the experience of others, such as the average guide figures published by equipment associations. The number of hours per year of working life is usually taken at between 1000 and 2000 hours, and again it must be a realistic estimate based on experience. The working life figure is required to calculate both the average annual investment and the average annual depreciation.

The amount of the total investment made in purchasing the equipment is understandably the most important and basic figure in all estimates of *equipment costs*; because the investment, once made, should produce a return, so that at the end of the working life the total investment and a profit (return) on the investment will have been returned to the investor. This principle is fundamental to all investments. To achieve it, depreciation must be allowed for in the equipment rate used to price the use of equipment on a job, as explained in Chapter 5. The total depreciation will equal the amount of the total investment—the original costs less any salvage value obtained by selling the equipment when it is no longer economical to maintain and operate it. The average annual depreciation is calculated from the total investment and the working life, in years, by one of the three methods mentioned in Chapter 5; and for estimating purposes, it is usually calculated by the straight line method.

For example, a haulage truck for highway use might have a purchase price of $10,000 and an estimated working life of 5 years.

(*1*) *Total Investment*

Purchase price of truck	$10,000
Sales tax (5%), say, ...	500
Delivery costs, say, ...	200
	= $10,700
Less: estimated salvage value	− 700
	$10,000

(*2*) *Estimated Working Life*

5 years × 1800 hours .... = 9000 hours

(*3*) *Average Annual Investment*

60 percent of $10,000 .... = $ 6000

Estimate of Owning Costs (*per year/hour*)
Depreciation

Total Investment (above)	$10,000
Less: rubber tires and tubes[a]	− 1000
Total depreciation	$ 9000
Average annual depreciation (20%)	$ 1800

Note: *a* Because the truck will require more than one set of tires during its life they are dealt with separately. Alternatively, the other, say, two additional sets of tires could be included here in the total depreciation.

Maintenance

Taken as a percentage proportion of the depreciation (as the two are logically related), based on experience and published statistical data.

50 percent of $1800 average annual depreciation = $900

Investment

	(percent)
The annual interest rate for money borrowed or invested,[a] say, ...	10
Insurance and taxes, say, ...	2
Storage of truck, say, ...	3
	15 percent

15 percent of $6000 average annual investment = $900

Note: *a* The rate will vary with time, place, and person; but it should not be less than what the equipment owner pays for short-term financing for his business.

Depreciation per year	$1800
Maintenance per year	900
Investment per year	900
Estimated owning costs per year	$3600
(per hour)	$2.00

The operating costs of the truck might be calculated as follows. Theoretically, a gasoline engine will consume 0.065 gallon of gasoline per brake horsepower per hour when the equipment is operating fully loaded. Other running costs for oil, filters, and the like, are usually taken as a percentage proportion of the fuel costs based on experience. In some cases, with some equipment, an efficiency factor of less than 100 percent might be applied in calculating the fuel costs, assuming that it would not be operating all the time; but that assumption has not been made here in the case of the truck.

Estimate of Operating Costs (per year/hour)

Fuel 150 HP × 0.065 gal. gasoline @ $.30 gal. × 1800 hrs	= $5265
Lubricants and accessories, say, 20 percent of fuel costs	= 1055
Tires, 3 sets @ $1200 per set (average) = $3600[a] / 5 years	= 720
Tire repairs, say, 15 percent of tire costs	= 100
Running repairs[b], say,	= 60
Estimated operating costs (per year)	$7200
(per hour)	$ 4.00

Note: *a* Tires are sometimes included in *owning costs.*
b Includes only minor items such as fan belts; all other repairs are allowed for under *Owning Costs*—Maintenance.

Thus, the total charge-out rate for the truck, per hour, would be $2.00 + $4.00 = $6.00 per hour, plus the costs of a driver and *overhead and profit.*[1]

Whatever costs and *unit prices* are computed at the outset, they must be regarded as theoretical and tentative, to be checked and validated and revised as necessary by *cost accounting* in the light of actual experience with costs and productivity. Plant and equipment should be a subject of *cost accounting* the same as labor. In the past, *plant and equipment costs* have not received much attention, because they were often only a small part of total costs. Construction companies often do not charge adequate and proper costs for their own plant and equipment, particularly for equipment and plant used on stipulated sum contracts. This practice effectively reduces the *profit,* so that these construction companies are, in fact, gradually consuming their investment in plant and equipment. When the plant and equipment is worn out, either there is no money to replace it, and more credit is necessary to stay in business, or savings have to be used if any are available.

If equipment is to be hired for a contract, competitive *bids* should be obtained from rental firms as early in the bidding period as possible; and these *bids* should be dealt with the same as the other *bids* for the supply of materials and for *subcontract work,* as explained later. If owned equipment is to be used, rental rates quoted by rental firms may be a useful guide to an estimator in estimating the *owned equipment costs* for a project. However, the rental rates (or charge-out rates) are simply the costs per hour and, as with *labor costs,* the other factor to be estimated is the productivity, the *amount of work* that can be done. Productivity is discussed in the next chapter, under Site Work.

1 As an approximate guide to "bare rental rates" for equipment (excluding *operating costs*), it may be noted that the total bare rental rate per month is often between 4 and 5 percent of the total investment, usually closer to 5 percent. In the case of a truck this rate would include tires.

Pricing Materials

Materials costs are the easiest part of an estimate to price, if the materials are properly described and the *work* is accurately measured. The greatest possible care and attention should be given to estimating the costs of primary materials (those materials that make up the major part of the total *material costs*), and the estimator should obtain firm and competitive offers for their supply and delivery.

Requests for offers should be sent out as early as possible to ensure that they are obtained in good time, and the quotations received should be made a part of the estimate. Some estimators leave spaces in the estimate sheets for the immediate entry of all quoted prices and the suppliers' names next to the *items of work* as soon as they are received. Actual quotations and *sub-bids* (or photocopies) and *bids* received by telephone and immediately entered on a standard form for such *verbal bids* should be placed with the estimate. All data for an estimate should be put together in one place. This helps to ensure that the estimate is priced on the basis of the most competitive prices, and that an offer is not misplaced and overlooked. Time and confusion are great enemies of an estimator, and organization and systematic handling of data and information are essential. Errors are easily made in the rush to submit a *bid* by a deadline, and sheets of paper are easily misplaced.

Bidding construction firms should always send out requests for quotations and *sub-bids* and should not rely on suppliers and other firms to submit them automatically. Otherwise, a desirable offer may be missed. The requests can be in a standard letter or postcard bearing the company's name, with the title of the project entered, simply letting it be known that offers are required by the sender for the supply of materials, products, equipment, or for *subtrade work.*

Offers for the supply of materials should be checked on receipt to see that:

1. they are for the correct material or product, and for the correct project; that they refer to the proper specification items; and that descriptions are complete and accurate and conform to all specified requirements
2. they specify delivery to the required location (see reference to unloading and handling, below)
3. they include all sales taxes and all other charges required or sales taxes are specifically excluded
4. they specify the correct quantity, because quantity usually affects the offer and the price level
5. they contain the correct delivery date for late delivery of certain items may cause expensive delays (this may not be a guarantee of prompt delivery, but it is a start)
6. they are valid offers for a stated and adequate period of time, because it may be some weeks before the contract is awarded and some months later before the items are required on the site
7. they include proper discounts and proper credit periods and terms according to those terms that are expected by the contracting firm from its different *suppliers.*

The estimator must try to ensure that no probable costs are omitted from the estimate. And he should ensure that, either in the offers received or in his own estimate, there are allowances for all costs of unloading and handling deliveries, and that any hoisting and storage costs are included.

Allowances for *waste,* and for increases and decreases in bulk of excavated and fill materials, and the like, should be stated in the descriptions of the items and included in the *unit prices.* If an allowance of, say, 25 percent is included in an item's description, the quotations for the supply of the material should be based on the gross quantity; that is, the net quantity measured in the estimate plus 25 percent. But in pricing the item in the estimate, the *quoted unit price,* plus 25 percent, should be entered in the *unit price column* of the estimate. For example:

Code	*Item*	*Material Unit Price*
0204	River sand fill in trenches (allow 25% for *shrinkage* by consolidation)	$2.50 per CY (*i.e., $2.00 per CY quoted plus 25%*)
	Quotations	
	1. Amazon Co. $2.00 per CY	
	2. Mississippi Co. $2.20 per CY	

In this way, the estimate explicitly records all the necessary data. The quantities in the estimate are net quantities—the amounts required by the *contract documents*—and the estimated factors (for *swell, shrinkage, waste, etc.*) are clearly shown and are, therefore, available for validation or revision as required and as determined through *cost accounting.*

Material costs will usually include sales taxes, and it is often necessary to keep these costs separate so that sales taxes can be readily added or deducted. In some government contracts sales taxes may have to be dealt with in special ways. This is one more reason to separate all the *different costs of work* and to make every estimate as analytical as possible.

Pricing Labor

Labor costs are the most difficult part of an estimate to price, because labor productivity is the most difficult thing to predict. The *unit prices* for *labor costs* are derived from:

1. *wage rates* and *labor rates*
2. productivity factors.

These terms were explained in Chapter 5.

Some estimators price estimates by using *wage rates* in *unit prices* for labor, and they allow for all fringe benefits and other related costs in the summary of the estimate, usually as a percentage added to the estimated total of wages. No other costs are added into the "labor column" of the estimate so that an accurate total of wages can be obtained as a basis for calculating the *indirect labor costs.*

Other estimators compute and use *labor rates* in *unit prices* for labor. These rates are computed by calculating a tradesman's wages for a standard week and adding all other costs to the employer in respect to that employee (as explained in Chapter 5), and dividing the total of the *direct* and *indirect labor costs* by the number of hours of *work*. If the *work* is for an out-of-town job, or if other special conditions exist, the costs of those conditions are also taken into account in calculating the *labor rates*. Which is the better method depends on the construction company and the type of *work* in the estimate. There are arguments in favor of both methods.

If *wage rates* are used in *unit prices*, the total estimated direct costs of wages can be readily compared with payroll totals periodically and during the course of the *work*. The *indirect costs* of fringe benefits and statutory payments are then added as a percentage of the *direct costs*; and this percentage varies depending on the mixture of trades, each trade having different fringe benefits. In 1970 *indirect labor costs* were about 20 percent to 23 percent of the *direct labor costs*.

If *labor rates* are used in *unit prices*, greater accuracy may be obtained by pricing the *work* of different trades done under different conditions, especially if the conditions are unusual and require the consideration of overtime rates, costs of subsistence, and the like. Also, the full costs of specific items can be seen more readily if this method of pricing *labor costs* is used.

Productivity factors are the other ingredient of *unit prices* for *labor costs*, and they are one of the primary reasons for *cost accounting*. In pricing the *labor costs* of an *item of work*, an estimator's first source of productivity data should be the analyzed cost accounts of his own past jobs. But there are times when suitable data are not available. Estimating and *cost accounting* have been described as a cyclical process, whereby estimating gives rise to *cost accounting*, which, in turn, feeds back data for future estimates. It is, therefore, reasonable to ask, How do you start? How is an item priced when no historical cost data are available?

This is a common situation, because many firms do no *cost accounting* at all, and many of the firms that do some *cost accounting* do not derive much benefit from it. Besides, everyone has to start somewhere. As a result, many estimates are priced by intuition (particularly the *labor costs*), by guesswork, by references to estimating handbooks, and by a varying mixture of these means.

If it is possible to make an estimate of the probable total costs of a construction project by a similar process of analysis and synthesis, it should be possible to make an estimate of any part of the whole, such as the *labor costs* of an *item of work*. The work processes in any *item of work* can be analyzed step by step, and estimates can be made of the time required to carry out each activity until the total time for the complete item has been estimated. This requires the estimator to have a good knowledge of construction, in order to analyze the work processes, and it also requires that he have enough time to apply this "method study" to those items of an estimate that require this technique.[2]

In competitive bidding, an estimator is usually short of time, and it is not suggested that he can analyze the work processes of all *items of work* for which he does not have historical data to arrive at a *unit price*. But in most jobs there is usually a relatively small number of important items that account for a large part of the total costs, and it is these items that may require and warrant careful method study and analysis if cost data are not available. For example, in a job costing $2 million, the costs of the reinforced concrete structure might be accounted for as follows:

COST OF A REINFORCED CONCRETE STRUCTURE (1969)

Items	*Materials and Other Costs*	*Labor Costs*[a]	*Total Costs*
1. Concrete (placed)	$100,000	21,500	121,500
2. Equipment Rental	12,500	—	12,500
3. Test and Inspection	3000	—	3000
4. Patching and Rubbing	1000	6000	7000
Concrete (placed)	$116,500	27,500	144,000
5. **Formwork** (erected and removed)	24,000	180,000	204,000
6. **Reinf Steel** (placed)	72,000	29,000	101,000
Total costs (including profit and overhead)	$212,500	236,500	449,000

Note: *a* Including all fringe benefits and other charges. In the above costs (from an actual job), the *labor costs of formwork* are just over 40 percent of the total costs of the structure, whereas the *labor costs of placing concrete* are less than 5 percent. Obviously, the only place where significant savings might be made by the *contractor* is in the formwork labor; and it is here that the estimator, and others of the *contractor's* staff, should apply the greatest amount of time and attention both before and after bidding.

[2] This aspect of an estimator's work may involve "work simplification" and "methods engineering" that at times might necessitate the aid of an expert, although all estimators and superintendents should be familiar with the principles so that they can be applied when necessary and where possible before a *bid* is made. For further reading and study see Deatherage, in the Bibliography.

An estimator's time and effort should be applied where it is most effective. In most jobs, as in the one above, most of the costs are more or less fixed and minor variations will have little effect on the total costs. One-half of $.01 per pound is a major variation in the *labor costs* for reinforcing steel, but it makes a difference of about two-thirds of 1 percent in the total costs of the above structure. Similarly, the *labor costs* for placing concrete cannot be reduced so as to make a significant difference in the total costs.

The *quantity of an item of work* has a significant effect on productivity and on the *labor costs* for two reasons:

1. the start up and finishing time required
2. the learning process at the beginning.

All *items of work* require a certain time at the beginning for setting out and organizing the *work*, and at the end a certain time is necessary for dismantling and cleaning up. This start up and finishing time is not proportionate to the total amount of *work* done, and it tends to remain more or less constant despite the quantity of *work*. This means that smaller quantities require proportionately more time than larger quantities of *work*.

Each *item of work* in each job has certain characteristics that are the same in all jobs and other characteristics that are unique to a particular job. No *item of work* is absolutely and completely the same, or different, on different jobs. To the extent that an *item of work* varies from one job to another, the tradesmen doing the *work* have to adapt to the new conditions and to the unique characteristics of the job. This process of adaptation and learning may be so brief as to be practically nonexistent, especially if the workmen have much experience with the particular *item of work* and are soon working at optimum productivity. Or the learning process may be much longer because there are new problems to be solved and new conditions to which to adapt, and it may be some hours, or days, before all the snags are unraveled and the *work* is proceeding well.

Ideally, the effect of *quantity of work* on *labor costs* should be apparent from historical cost data, and with sufficient data it should be possible to establish guidelines for pricing. However, until *cost accounting* is well established and practiced, there will probably not be enough data to indicate such modes and trends; and in the meantime the estimator has to interpolate and adjust prices according to quantity through his own judgment. The important thing is to do it, and to do it consciously and deliberately, and then to try and check the accuracy of what has been done by *cost accounting*. This should be an estimator's constant attitude—to get to the facts about costs and to pursue them with a positive desire to know them. The one with the opposite attitude can always find reasons and excuses why costs cannot be estimated, and he will justify poor practices, such as careless measurement and pricing, by pointing to the unknown factors.

Pricing *labor costs* in *alteration work* and *work* in small jobs, such as minor additions to existing buildings, requires special considerations, because the conditions under which this kind of *work* is usually done usually entail low productivity and high *labor costs*. There is a big difference between wood framing and gypsum wallboard in extensive partitioning in a new building, and the same kind of *work* done in small quantities to fill existing openings in an alteration job. In fact, despite their superficial similarities, they are entirely different items from the point of view of *labor costs*. Yet some estimators will relate the *labor costs* of one to the other simply because the materials are the same in both cases.

Work in alterations and in small jobs for which there is no precedent can be estimated by the methods study approach mentioned earlier. The only alternatives are guesswork or intuition. An experienced estimator's intuition is sometimes a wonder to behold, but it is not guesswork. Rather, it is the rapid comprehension and working of a mind programmed by experience, and perhaps the estimating effort could be less if we understood intuition more.

If the alterations include an *item of work* to fill in an existing doorway with framing and gypsum wallboard, the *labor costs* of the item might be analyzed and estimated as follows:

LABOR COSTS: FILLING EXISTING DOORWAY

Carpenter	*Minutes*
1. Take out existing door (3′ 0″ × 7′0″)	5
(removal by helper; see below)	–
2. Take out existing wood frame and trim	30
(removal by helper; see below)	–
3. Cut and install 2 × 4 floor plate in 3′3″ opening with two ramset fasteners	10
4. Cut and install two 2 × 4 studs at jambs	10
5. Ditto one 2 × 4 at head	10
6. Ditto two 2 × 4 studs between jambs	10
7. Ditto ½″ wallboard sheets (4′ × 8′) to both sides of opening	30
8. Collect tools and clean up	5
9. Move to next location	10
	120
Allow for 75 percent efficiency (+⅓)	+40
	60)160

Carpenter: total time = 2 2/3 hours

Helper (with Carpenter)	*Minutes*
1. Deliver materials from store to location of work	15
2. Remove existing door to stores	10
3. Remove frame for disposal	10
4. Clean up	5
	40
Allow for 67 percent efficiency (+½)	+20
	60)60
Helper: total time =	1 hour

LABOR COSTS*: FILLING EXISTING DOORWAY

Carpenter, 2⅔ hours @ $10.00 =	$26.67	
Helper, 1 hour @ $ 7.50 =	$ 7.50	
	$34.17	
Total Costs	$35.00	(per doorway)

* National (U.S.A.) *average labor rates* in 1971 were about $8.00 per hour for tradesmen and about $6.00 per hour for helpers. Rates vary by more than 30 percent above and below the national average, according to locality. Consequently, no attempt has been made in examples to use any particular *wage rate* or *labor rate* applicable to any particular time or place, and readers should rework all examples using current *local labor rates.* Productivity also varies from place to place, and from time to time.

An efficiency factor is always necessary, because 100 percent efficiency is never achieved; and according to some studies the widely used factor of 75 percent efficiency might be too high for some jobs. Of course, the factors could be included in the estimates of time required for each step of the operation, but it is easier to apply them as in the example.

Subcontract Work

A major part of the general estimates made by prime bidders are the *sub-bids* submitted to them by subcontract firms; and a major part of a general estimator's work is to compare and select *sub-bids* and incorporate the most competitive and favorable (usually the lowest) into the general estimate. Remember, a *bid,* or a *sub-bid,* is an offer, the acceptance of which will lead to a contract, or a sub-contract, if all the other necessary ingredients are present.

Each *sub-bid* should be in writing, stating precisely and without qualification the scope of the *work* included in the offer, and the amount of the offer. Frequently, a *written bid* is dispensed with, at least in the first instance and between parties who have had previous dealings; and a *sub-bid* is often made by telephone and confirmed later in writing. There is, obviously, some risk attached to this practice, and written confirmation at least is essential. Risk probably arises more from possible misunderstandings of the scope and terms of a verbal offer than from any intended dishonesty; but, either way, care should be taken to avoid trouble.

The scope and terms of *sub-bids* are a frequent source of misunderstanding, often leading to incomplete *general bids* or *bids* with parts of the *work* included twice (in more than one *sub-bid*), thereby causing an inflated *general bid.* If the *general bid* is incomplete because of *work* excluded in error from *sub-bids,* the *general bid* will be lower than it should be, and may, thereby, be accepted as the *lowest bid.* Later, there may be a dispute about who bears the loss because of the exclusion. If the *general bid* is erroneously inflated by the duplication of *work* included in *sub-bids,* it is not likely to be accepted.

One of the common causes of such misunderstanding and error are the *bidding documents,* particularly the specifications. Specifications for stipulated sum contracts are properly directed to prime bidders, one of whom usually becomes the *contractor.* The specifications form part of the *contract documents* of the contract between the *owner* and the *contractor,* and they must, therefore, be directed to the *contractor,* who is responsible to the *owner* for all the *work* required by the *contract documents.* The distribution of the greater part of this *work* among *subcontractors* (who are responsible for its performance to the *contractor*) is the responsibility of the *contractor* unless the *designer* has nominated certain *work* to be done by particular *subcontractors* who, nevertheless, still have a subcontract with the *contractor.* The *owner* looks only to the *contractor* for performance of a stipulated sum contract.

The scope of each subcontract is mainly determined by the *sections* of the specifications and by trade practices, and if the specification *sections* and trade practices are compatible no misunderstanding is likely to occur. But the contents of the *sections* depend on the *work,* and upon the specification writer. Some writers know and follow local trade practices closely because they know the *sections* will be the scope and basis of *sub-bids.* Other writers do not always know or care and, in the final view, the specification writer is contractually correct when he says that the specifications are directed to the *contractor* who is responsible to the *owner* for all of the *work,* and its division among *subcontractors* is the *contractor's* concern.

It is apparent to many in the industry that the stipulated sum contract for *work* that must be done by a large number of specialized *subcontractors* is an anachronism, particularly in the matter of contractual responsibility. Hence, the appearance of the management type of contract, in which those who do *work* on a site have a direct contractual responsibility to the *owner.* Nevertheless, stipulated sum contracts will probably be in use for some time, and the prime bidder must analyze and compare the *sub-bids* he receives, not only their amount, but also their scope.

For large *work*, the *sub-bids* are best tabulated on a chart as they are received, checked, and if necessary confirmed with the sub-bidders. In some projects, there are *sub-bids* for separate and for combined *work* such as plumbing and heating and ventilation *work*. Some firms may bid for both, and some may bid for only one; and a *sub-bid* for plumbing may be less when the plumbing is combined with heating and ventilating than when it is to be done alone. (Usually, this is because of the fixed *job overhead costs*, which do not increase as the amount of the other *direct costs* increase.) This can make the selection of *sub-bids* difficult, but clear *bidding documents* and tabulation will reduce the difficulty.

A general bidder may have all the *work* to be done by his own forces estimated well in advance of the "closing time for bids," and yet still be unable to avoid a rush in the last few hours. Bid depositories have helped to eliminate this problem by extending the time available after receiving *sub-bids* and before submitting *bids* to the *owner* to a day or more. Without a bid depository, a general bidder might receive *sub-bids* up until the last hour, or less, before the *general bid* has to be submitted to the *owner*.

To ensure that a proper *general bid* is submitted complete with a bid bond and the other prerequisites, it is a common practice, where permitted, to submit the *bid* and the bond a day or more in advance, and to send a telegram up to within a few hours of the closing time for bids, including amendments that reflect lower *sub-bids* or other information. This procedure is quite confidential, of course, because a telegram that states, "Reduce our *bid* for the Temple Gates Project by $9500.00 (nine thousand five hundred dollars); signed. . . ." is quite meaningless to anyone but the *owner* who has received the *general bid*. It is also safe for the bidder because the *first bid* to be made is a relatively high figure that is later reduced. Should something happen to prevent the *amending bid* from being made, the bidder is not likely to find himself with an unwanted contract for a low amount.

Unit Prices and Calculations

Delusions about accuracy are probably more common in pricing than in measuring *work*. Somehow, $.01 on a *unit price* of $2.50 per cubic yard seems much more significant than "allow 25 percent for *shrinkage* in consolidation of fill material." The fill is visualized as a muddy morass, and a 25 percent allowance sounds close enough. The single cent on an invoiced price of $2.51 apparently conflicts with the estimator's price of $2.50 and appears as a definite shortage and mistake.

If the *labor rate* is $6.00 an hour (which is $.10 a minute), may not the estimator round off a price of $2.48 to $2.50 per cubic yard, if there are only a few hundred cubic yards? The multi-million dollar *bid* that ends with $.50 and the *unit price* given to three decimal places are rarely justifiable, even if they do no harm. What is more important is the attitude that these things may indicate—perhaps, a preoccupation with a numerical accuracy that may not be a true reflection of the facts, and may be, or may become, a substitute for common sense.

Common sense is the first requirement in all pricing and measurement, and it can save time and trouble for an estimator. Common sense says that the simple is preferable to the complex, and that a hammer is not the best means to kill a fly. Many *unit prices* in an estimate can be rounded off to the nearest $.05 without any loss in accuracy, because they are applicable to relatively small quantities. Other *unit prices* that are applicable to large quantities may sometimes need to be calculated to a fraction of a cent; and in most cases, a simple fraction (1/4, 1/3, 1/2, etc.) will suffice and will avoid the risk of being misread, which is always a risk with the decimal fractions.

One way of avoiding the need for fractions in *unit prices* is to use a larger unit of measurement. The square yard is not so widely used as the square foot, and yet the larger unit is often much more appropriate to many *super items of work*, such as earthwork, shoring, dampproofing, insulation, drywall, and the like. Many trades customarily do measure their *work* in units larger than a square foot, usually either in square yards or Squares (of 100 square feet); and there appears to be no good reason for the customary use of the "square foot" unit in many trades. The Square is perhaps too large for some items; and one minor disadvantage with the yard is its basis of 3 feet (*super*) and 9 feet (*cube*), which does not fit into a decimal monetary system. When the change to "metric" comes to North America, presumably the meter will be the basic unit of measurement as it is elsewhere.

Finally, a word about those items in an estimate for which no cost data are available, and on which an estimator must put a price so as to complete his estimate—the kind of item that cannot be found in any data or publication when it is needed. How does an estimator find a reasonable *unit price* for such an item when time is short? The first thing to do is to assess the relative importance of the item to the estimate as a whole. It doesn't make sense to agonize over an item whose cost is very small when compared with the tolerances of accuracy that are inevitable and acceptable for the major *items of work* in a job. At the same time, a good estimator does not want to throw away the chance of obtaining the job, or having obtained it, to throw away money because of under-priced items.

Having assessed the importance of the item in question, the estimator should make an appropriate effort to estimate its probable costs; and if the costs are relatively small, he should not spend too much time and effort on it. If the costs are likely to be significant, the estimator should gather whatever relevant data he can, and make the best possible effort to estimate the costs. This should be done by stating briefly in the estimate the data used and the basis for the estimate of the costs of that item. There should be some rational basis and some usable data available, no matter how tenuous, to which the estimator can refer. Above all, he should not simply guess and show no rationale, because at some later date he may not be able to remember why he included the costs he did. On the other hand, if he states the data and the basis of his rationale in the estimate, he can refer to it later and say, if necessary, "There is the data that were available at the time, and on that basis, that is what I decided. Can you tell me a better way?" The chances are that most critics could not. However, there may be times when the risk is great and when the estimator should make it known to his directors and obtain their advice and instructions.

Estimating publications, such as those by Walker and by Means, can be of great help to an estimator when he is in need of cost data, if he uses the published data intelligently. And one of the more reliable ways of using published cost data is to relate them to personal experience and personal data, as follows.

The word "data" is the plural of "datum," and a datum is often used in construction as a physical reference point, or level, from which other points and levels are established. A similar method can be used to establish probable costs in an estimate from a cost datum.

An estimator may often need outside cost information about an *item of work* in an estimate. Let us suppose that the *item of work* involves placing steel rebar in sizes #10 and #11, for which he has no personal cost data. He finds that a publication indicates a *unit rate* of, say, 13½ man-hours per ton for placing bars of sizes #10 and #11; and for placing smaller bars of sizes #5 and #6 (for which he does have his own data) he finds that the publication indicates 18 man-hours per ton. His own cost records indicate a *unit rate* for placing bars of sizes #5 and #6 of 20 man-hours per ton.

Applying his own costs to the published data, he finds that there is a difference of 11 percent, or a factor (F) of 1.11 applicable to the published data. Applying the same factor to the other published data, he obtains an estimated *unit rate* of 15 man-hours (13.50×1.11) for the larger bars to use in his estimate. Of course, the factor can only be used in this way if the estimator can foresee that conditions of productivity similar to those that have applied to the *rebar work* he has done in the past will also apply to future *rebar work* in which larger bars are placed. Nevertheless, relating personal cost information to published cost data in this way will generally enhance the value of both sets of data. The method can be represented thus:

$$A^e = A^d \times F$$
$$B^e = B^d \times F$$

wherein

A^e is the estimator's own information about the *item of work, A*
A^d is the published data for item A
B^d is the published data for item B
F is the calculated factor
B^e is the data required by the estimator about item B

Do not be impressed with mathematical symbolism; a great deal of common sense is dressed up in this manner nowadays. But it can be helpful in remembering such things as long as we do not come to believe that we are dealing with scientific laws and then forget to apply common sense.

Common sense may tell us that the published data used above were probably derived from jobs in which the workmen were experienced in placing steel bars of larger sizes, whereas we may know that our own workmen have not yet had that experience. So we should apply yet another factor for learning a new kind of *work* and allow for a *unit rate* of, say, 18 man-hours per ton (15.00×1.20) for this job, and hope that we shall have an opportunity later to test our estimate by *cost accounting*.

Guessing is not estimating, and often what an experienced estimator may refer to as a guess (or a "guestimate") is a shrewd judgment, modestly clothed. To make such judgments requires experience and an instinct for the costs and prices of *construction work* that can be deliberately acquired and cultivated. First, an estimator should read a wide selection of publications on the subject, particularly those issued periodically, from which he should make comparisons with his own cost data. Next, he should be inquisitive and curious about costs, so that whenever he comes across some cost information he should try to convert it into a *unit price* so that he can compare and remember it. He should have a general knowledge of costs beyond his own field such as the costs of basic materials that are also used in related industries, the rates of different kinds of transportation, and the rental rates for many things, from helicopters to hacksaws. At the same time, he should be aware of the many human and business factors that sometimes make a mockery of the scientific approach and that sometimes produce a spread of 100 percent between *bids.*

Perhaps some of the most valuable experience an estimator can have, if he chooses to acquire and make use of it, is to work as an estimator in another country, preferably in one which is industrially under-developed and in which the job conditions are very different from those in North America. If not as an estimator, he might work in any other part of the local construction industry. In a totally different environment, and lacking all the aids usually taken for granted, there is an opportunity to enlarge the imagination and the capacity for ingenuity, and to see the economics of construction from another point of view.

Questions and Topics for Discussion

1. In pricing *equipment costs* for the use of a bidder's own equipment in an estimate, describe briefly two ways in which an estimator can establish rates for the equipment.

2. Describe how an estimator should deal with quotations for the supply of materials received by telephone, and explain why.

3. State those things that an estimator should look for in checking offers for the supply of materials to a construction project.

4. Explain how and why the quantity of an *item of work* often has an effect on labor productivity.

5. Discuss briefly the major factors affecting the scope of *work* done by *subcontractors.*

6. Explain why a sub-bidder might offer to do both *plumbing work* and *heating and ventilation work* for a total amount that is less than the total amount of his two separate and simultaneous *sub-bids* for the plumbing and for the heating and ventilation.

7. Show in two ways how a *unit price* for steel rebar (supplied and installed) calculated at $0.1575 per pound might be better written in an estimate.

8. Explain what an estimator should do to estimate the *labor costs of an item of alteration work* with which he has little experience, and for which he has no cost data.

9. Explain briefly the precepts an estimator should follow in estimating the *costs of an item of work* that appear to be relatively important to the estimate but for which he has little data and limited time to estimate the costs.

10. Explain how a bidder can often leave the submission of the actual amount of his *bid* to the very last but, at the same time, ensure that all the bidding requirements are properly met.

11

Pricing Work: Particular

Examples of pricing are available in many estimating texts and reference books,[1] and as this book is primarily concerned with fundamentals, precepts, and methods, it does not attempt to be a reference book of factual data as well. Nevertheless, some examples will be useful to illustrate the practical application of the estimating precepts and methods explained in previous chapters and to point up particular points in pricing *work* in the various *divisions*.

The *standard divisions* of the "Uniform Construction Index" that were followed in Chapter 8, Measuring Work: Particular, are also followed in this chapter, which, in addition, refers to Division 1, General Requirements, previously mentioned only briefly because General Requirements contains items creating *overhead costs,* which require no special explanation in terms of measurement.

Readers are urged to obtain current *wage rates* and *labor rates* in their own localities and to recalculate all examples using local rates and prices.[2]

[1] See Bibliography—*Estimating and Cost Accounting.*

[2] As pointed out before, no attempt has been made to use any particular *wage rates* or *labor rates.* In fact, different rates have been deliberately used in the text to avoid emphasizing any particular rate and to indicate that rates do continually change from time to time, and from place to place, the same as productivity.

In this way, the examples will not be used as if they were of valid, actual *unit prices,* and students will obtain some idea of local *unit prices.* It must be emphasized that the examples in this book illustrate only methods and do not reflect actual cost data, although the examples have been made as realistic as possible.

Pricing General Requirements (Division 1)

These General Requirements are referred to in Chapter 5, Construction Costs, under *Job Overhead Costs,* where some of the costs that arise out of the articles and conditions of the contract are listed. Most of these *overhead costs* are easily priced, if not always so easily recognized.

Supervision and staff salaries and expenses are usually straightforward; but the provision of construction camp facilities for staff and workmen in remote areas is sometimes more difficult to estimate, because costs per man, per day, depend partly on the maximum number of men provided for by the camp facilities, the degree of occupancy, and the duration of the project.

Construction camp facilities and living standards for workmen at remote sites are covered by union agreements in most places. Typical requirements are: 112 square feet of bedroom space for 2 men; room fully furnished, lighted, and heated, with individual thermostats; clean bed linen weekly, and blankets to be laundered every 3 months; 1 shower fixture for every 10 men; 1 washbasin for every 5 men; and toilets provided according to specified requirements related to population. Recreation rooms and canteens are to be provided separately. Food standards generally have to be high, and catering is usually subcontracted to a catering firm. The menus and variety of choice are often superior to many second-class hotels, and they have to be, under the circumtances. Also, higher standards generally seem to follow higher *wage rates.*

Camp and subsistence costs in British Columbia, Canada, in 1967, were calculated at:[3]

[3] By Concosts Services, Ltd., publishers of *Concosts, Construction Costs Reports* (Vancouver, B.C. Canada), who have provided much of the cost information in this chapter. In 1970, the upper range of costs shown were still representative, and subsistence costs based on $8.00 per man, per day, were still valid. Since then, some standards and costs have risen.

CAMP AND SUBSISTENCE COSTS PER MAN-HOUR (40-HOUR WEEK)

Maximum Number of Occupants	*Six-Month Contract Period*	*Twenty-Four Month Contract Period*
25	$1.61 to 2.22	$1.54 to 2.08
50	1.37 to 1.95	1.32 to 1.86
75	1.31 to 1.88	1.27 to 1.80
100	1.29 to 1.86	1.25 to 1.79
200	1.22 to 1.78	1.20 to 1.72

These costs were based on contracted catering costs of from $5.00 to $8.00 per man, per day, depending on quantity, quality, and location. They were also based on 100 percent occupancy of the construction camp. At 80 percent occupancy the costs increased by about 10 percent for 25 occupants, and by 7½ percent for 200 occupants. At that time, the local carpenter's basic *wage rate* was $4.14 per hour, and the local laborer's basic *wage rate* was $3.24 per hour. So it can be seen that on small jobs the camp and subsistence costs could be as high as 50 percent of the basic wages.

Premiums for construction bonds depend on the location, size, and duration of the project, and on the percentage of the value of the project to be bonded. Typical Canadian rates in 1970 are given below to show how they are arranged. Rates in the U.S.A. were reported to be from 15 to 20 percent higher.

CONSTRUCTION BOND RATES

BID BONDS. A premium of $10.00 is charged regardless of size for both Bid Bond or Letter Consenting to Surety. This premium will be refunded if the *contractor* is successful and a Performance Bond is issued.

PERFORMANCE BONDS

(*a*) *Short Term Projects:—*

(i) Up to twelve months:

*Percentage**	*Rate per Annum*	*Basis of Premium*
1– 10%	$17.50 per $1,000	Bond Amount
11– 50%	3.50 per 1,000	Full Contract Price
51–100%	5.25 per 1,000	Full Contract Price

(ii) Balance of Project Time. If term of project exceeds 12 months a renewal premium at the original rate is charged based on the uncompleted portion of the contract price.

(*b*) *Long Term Projects:—***

(i) Up to 24 Months. Basis of Premium: Full Contract Price.

Percentage	*Amount*	*Rate per Annum*
1–50%	1st $2,500,000	$5.70 per 1,000
	Next 2,500,000	4.50 per 1,000
	Next 2,500,000	4.00 per 1,000
	Over 7,500,000	4.00 per 1,000

Percentage	*Amount*	*Rate per Annum*
51–100%	1st $2,500,000	7.50 per 1,000
	Next 2,500,000	5.75 per 1,000
	Next 2,500,000	5.50 per 1,000
	Over 7,500,000	5.00 per 1,000

* Percentage of contract price. Probably the most common bonds are performance bonds for an amount equal to 50 percent of the contract price (amount), on the assumption that the need for a performance bond, should it arise, will most likely occur in the latter stages of the contract when the *work* is at least 50 percent complete.

** Usually only economical if project cost exceeds $10,000,000 or time exceeds 24 months.

(ii) Balance of Project Time. If term of project exceeds 24 months, a renewal premium charge of 1% of the premium per month is made.

(*c*) *Over-runs and Under-runs:—*

Additional adjusting premiums will be charged or credited at the same rate as the original premium in the event the contract is increased or decreased. This adjustment is usually ignored unless the adjustment exceeds $50.00.

PAYMENT BONDS

(*a*) *Short Term Projects:—*

(i) Up to 12 Months. $2.00 per $1,000 of the full contract price regardless of bond percentage.
(ii) Balance of Project Time. As per Performance Bond.

(*b*) *Long Term Projects:—*

Included in Performance Bond rate.

SUPPLY BONDS

Percentage	*Rate per Annum*	*Basis of Premium*
1– 50%	$1.75 per $1,000	Full Contract Price
51–100%	2.65 per 1,000	Full Contract Price

MAINTENANCE BONDS

Percentage	*Rate per Annum*	*Basis of Premium*
1– 50%	$1.00 per $1,000	Full Contract Price
51–100%	1.60 per 1,000	Full Contract Price

As a rough guide, performance bond premium costs are from ½ to 1 percent of project's total costs. The requirement for payment bonds, supply bonds, and maintenance bonds is not nearly so common as the requirement for a performance bond; and it is becoming common practice to require *major subcontractors* to provide performance bonds.

Premiums for insurances vary considerably according to job location, type of construction, value of contract, and the types of coverage required. An estimator should always obtain quotations for a project's builder's risk and public liability insurance and include the premium amounts as a *job overhead cost.* Costs of workmen's compensation and employer's liability insurance are better treated as *indirect labor costs,* although some may treat them as a job overhead. Some other insurance premiums may have to be treated as an *operating overhead* because they provide for general coverage. Builder's risk and public liability insurance premiums may be as low as ½ of 1 percent of the contract amount for a large multi-million dollar contract, and as high as 1 percent on smaller contracts.

Photographs to show *work* in progress cost about $20 per shot at downtown sites by a professional photographer, and naturally they cost more if traveling costs are involved. Other General Requirements are numerous and varied, but once identified and classified they are not usually difficult to price.

Any problem in this *division* is more likely to involve identifying the need for an overhead item and in estimating its extent or quantity, rather than in pricing it. Toward this end, many estimators use a check list of overhead items and general requirements; and some use a standard estimate summary sheet, designed for their company's needs, on which are listed those overhead items that experience has shown are most likely to be required in the majority of their jobs. Some estimating offices go further and publish a manual for the use of their estimators that contains check lists for most *divisions of work* and guides and data for estimating the costs of the more common items.

The majority of general requirements can be priced in the same way that similar items in other *divisions* are priced. For example, fences, screens, and other such *temporary work* require measuring and pricing in the same way as framing and sheathing; and temporary offices, and the like, are charged for like plant and equipment. The majority of *job overhead costs* are related to either the site, or to the cost or duration of the *work*, and knowledge of the site's location and characteristics, and of the total costs and time for the *work* are essential.

Pricing Site Work (Division 2)

Demolition generally must be priced either at the site or from extensive notes and measurements and sketches made at the site. However, massive demolitions of complete structures are beyond the scope of this text.

Alterations generally involve labor, tools, and equipment; and each *item of work* should be priced according to the crew required, together with the requisite tools and equipment, by the day or by the hour. Remembering the well-known principle that "*work* expands to fill the time available," the estimator should consider estimating most minor items in quarter-day units. (The work day is divided into four quarters by two coffeebreaks and lunch.) Major items should be estimated up to the nearest full day.

A contingency allowance for *unforseen work* and contingencies should be included in an estimate of *alteration work*. The best method is to price the items as they are seen, and to show a contingency as a separate and specific item in the estimate rather than to allow for contingencies by loosely increasing the crew-times and costs for the *items of work*. The first method is more rational and permits better *cost accounting* and checking of the estimate. An estimator should always try to facilitate subsequent *cost accounting* through his estimates by indicating his reasons for including costs and allowances, which in turn will help to produce more cost data to help in making other estimates.

Excavating is done mainly by equipment such as dozers, tractor-shovels, scrapers, trenchers, and the like; and with most kinds of excavating equipment there are certain economic considerations to be taken into account in estimating the *equipment costs* of doing *work*. In Chapter 10, the method of estimating rates (per year, hour) for plant and equipment was discussed and illustrated. The other primary consideration is productivity—the amount of time, and therefore the costs, required for a machine to do a certain amount of measured *work*.

In estimating the productivity of equipment, the basic unit of time is the cycle time: the number of minutes required for a machine to complete a work cycle. In the case of a shovel, for example, cycle time might include loading the shovel with earth, hauling it to a truck (for dumping) or directly to a dump site close at hand, dumping the earth, and returning to load the shovel again and to begin another cycle. Most earthwork, such as excavating, is made up of a series of such repetitive cycles that can be observed and timed in operation and that, on the basis of collected data, can be calculated for an estimate and subsequently be validated or corrected by *cost accounting*.

A cycle time consists of (1) fixed time, and (2) variable time. Variable time depends on speed and distance, the distance that the machine has to travel in hauling and returning. Fixed time is the balance of a cycle time, and is so-called because it is more or less fixed by the type of machine and its method of operation. Therefore, fixed times can be established by calculation, observation, and experience for different machines of different types, if the times are periodically checked and revised as necessary. Variable times must be calculated for each project according to the job and site conditions, hauling distances, and machine speeds.

Machine manufacturers publish specification data for their machines, including machine speeds in the several gears, both forward and reverse. Variable times are therefore calculated as follows:

$$\text{Variable time} = \left(\frac{\text{Haul Distance (ft)}}{\text{Haul Speed (mph)} \times 88}\right) + \left(\frac{\text{Return Distance (ft)}}{\text{Return Speed (mph)} \times 88}\right)$$

Different gears and speeds are usual for hauling (loaded) and returning (unloaded). The factor 88 is to convert from miles per hour to feet per minute.

Having calculated a variable time, the fixed time is added to produce cycle time for the particular job; and not much can be done to minimize this cycle time other than to use a suitable machine and to plan the job so that the variable time can be kept to the minimum. Since the basis of *equipment costs* is time, the next part of the estimate is to calculate the number of trips per hour by dividing the cycle time into 60 minutes and finally to estimate the production by dividing the quantity of earth to be excavated by the amount in each load. However, there are two other factors to be considered as well.

Excavations are properly measured by "bank measure"; that is, the quantity of earth in place (in bank) before it is excavated. But we know that when earth is excavated *swell* occurs, that the earth increases in bulk. Therefore, a *swell* factor must be applied in calculating the production of an excavating machine. This factor will depend on the type of ground and the ground conditions, as previously explained.

The other factor to be applied to an estimate of any machine's productivity and production is commonly referred to as an "efficiency factor." It is a factor that is applied to allow for inefficiency (and should more precisely be referred to as an "inefficiency factor," but that sounds negative), because no machine can continuously operate for 60 minutes in every hour. Commonly used efficiency factors are (1) for machines on wheels, 0.75 (45 min/hr); and (2) for machines on tracks, 0.83 (50 min/hr).

As an example of the application of the above factors and of cycle time to estimating *equipment costs*, consider a basement excavation in loamy soil, size 50 ft × 30 ft × 6 ft deep, a total of 388 CY, in which a tractor shovel (1½ CY capacity) is used to bulldoze earth to both ends of the excavation, as in Fig. 40. By this typical method, more earth is moved in each cycle as the bucket of the machine is filled and an additional amount of earth builds up in front of the bucket.

FIG. 40. Excavating a Basement.

If the machine's travel begins at the center and extends beyond the end of the excavation, the distance forward may be taken as about 35 ft. Then, if the average speed forward is 1.5 mph and 2.0 mph in high reverse:

$$\text{Variable time} = \frac{35 \text{ ft}}{1.5 \times 88} + \frac{35 \text{ ft}}{2.0 \times 88}$$
$$= 0.265 + 0.200 = 0.465 \text{ min}$$
$$= (\text{say}),\ 0.5 \text{ min per cycle.}$$

The fixed time for shifting the machine is, say, 0.25 min per cycle. Cycle Time = (0.5 + 0.25) = 0.75 min. With a 1½ CY shovel used as a dozer to push the earth, and assuming that another ½ CY would build up in front of the machine, 2 CY would be moved in each cycle. The number of cycles per hour would be a maximum of 80 cycles (60 min divided by 0.75 min per cycle), thus moving a maximum of 160 CY per hour (80 cycles × 2 CY each). But the factors for *swell* and inefficiency have yet to be applied.

Allowing a factor for *swell* of 0.80, and allowing an efficiency factor for a tracked machine of 0.83, the estimated maximum amount of earth excavated per hour would be 160 CY × 0.80 × 0.83 = 106 CY. The time required to excavate 388 CY would therefore be 3.66 hr (388 CY divided by 106 CY per hr); and allowing for cleaning up and squaring the excavation, the estimated minimum total time required would be 4 hours. Time for mobilization and demobilization, transporting to and from the site, and any idle time would be extra. If the tractor shovel could not be used to push the earth as described, the excavation might take twice as long to complete. The example (based on data from a machine manufacturer) is intended only to illustrate the method of estimating a cycle time and the time required to do a particular job from data obtained from machine manufacturers, equipment associations and, above all, from observation and *cost accounting* of actual jobs.

Having estimated the total time required for the equipment, including that required for its transportation to and from the job site, the *equipment costs* can be estimated by applying the rate for *owning and operating costs,* calculated as explained in Chapter 10, and the *labor rate* for the operator. In addition, the costs of a flat-bed truck (and driver) to transport a machine on tracks will be required. Finally, any required contingency allowances and *overheads and profit* will be added to arrive at the total costs.

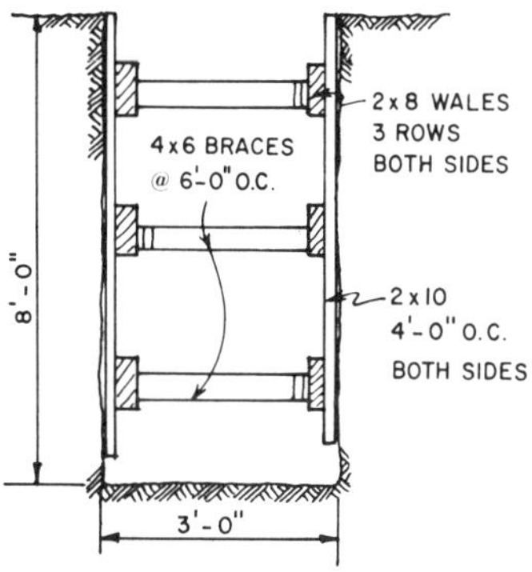

FIG. 41. Temporary Shoring of Trench Excavation.

Temporary shorings for excavations involve mostly *labor costs* if the lumber is re-used several times. Shoring the sides of a trench 12 ft long × 3 ft wide × 8 ft deep, as shown in Fig. 41, might require the lumber as shown here, depending on the soil conditions.

Materials

2/ 3/8 ft 2 × 10 boards = 80 fbm (feet board measure)
6/12 ft 2 × 8 wales = 96 fbm
6/3 ft 4 × 6 braces = 36 fbm

Total: 212 fbm @ \$0.10 = \$21.20
Divided by number of uses (say) = ÷5
= 4.24
Divided by area supported (2/12 × 8 ft) = ÷192 SF
Material costs (say) = \$0.02 per SF

Labor

Labor installing: 2 men × 2 hrs = 4 hrs
Labor removing: 2 men × 1½ hrs = 3 hrs

Total: 7 hrs @ \$6.00 = \$42.00
Divided by area supported ÷192 SF
Labor costs (say) = \$0.22 per SF

This illustrates one method of pricing *temporary work* in which the materials are re-used, as in concrete formwork. The measurements in the estimate are in "contact feet," and the materials are not measured until the *unit prices* are analyzed. This simplifies measurement of *work* in which the number of uses is a major factor in *material costs.*

Gravel and sand are used for many things, including fill, road bases, and for aggregates. Often the cartage from source to site is as much (or more) than the cost of the materials at the source, so source and site locations are important. Remember that moisture content affects the weight and volume of these materials, and that allowances must be made for *shrinkage* of loose materials that are placed and compacted as fill. These allowances can be approximately calculated from average figures in reference books, but actual figures for local materials should be established from experience. Usually, the costs of different types of gravel and sand from the same source do not vary greatly. The biggest difference is between unwashed materials used for fill and washed and screened materials used for aggregates and other special purposes. For example, washed sand used in mortar, concrete, or plaster, may cost several times as much as sand used as fill.

Pricing Concrete (Division 3)

In a reinforced concrete building, the complete costs of reinforced concrete per cubic yard in place is about four to six times the cost of ready-mixed concrete, and the following analysis illustrates the approximate distribution of costs:

REINFORCED CONCRETE BUILDING FRAME

	Percentage of Total Costs		
	M & E +	L[(a)] =	Total
Concrete, in place[(b)]	25 +	5 =	30%
Formwork, erect and strip	10 +	35 =	45%
Rebar, in place	17½ +	7½ =	25%
	52½ +	47½ =	100%

a M & E = Materials and Equipment; L = Labor, including all benefits and charges.
b Including inspection and testing.

These figures are approximate, but indicative of the relative costs of the different parts. The total *labor costs* (including *indirect labor costs*) are about 50 percent, and the other 50 percent is in the *material costs* and *equipment costs.* Current, national average figures can be obtained from annual publications on construction costs in the United States and Canada.[4]

At a total cost of $110.00 per CY (equals 100 percent) the costs would be:

REINFORCED CONCRETE BUILDING FRAME

	Costs per CY in Place		
	M & E +	L[(a)] =	Total
Concrete, in place[(b)]	$27.50 +	$ 5.50 =	$ 33.00
Formwork, erect and strip	11.00 +	38.50 =	49.50
Rebar, in place	19.25 +	8.25 =	27.50
	$57.75 +	$52.25 =	$110.00

a M & E = Materials and Equipment; L = Labor, including all fringe benefits and charges.
b Including inspection and testing.

An estimator should have a general knowledge of such cost figures in order to make an approximate check of an estimate and to appreciate the relative importance of the various parts of the *work* he is measuring and pricing so that he make reasonably accurate decisions.

Ready-mixed concrete prices depend primarily on the cement content, because the cement is by far the most expensive ingredient, making up about one-third of the total cost. Concrete mixes with aggregate sizes of ¾″ and 1½″ maximum may have the same prices, but concrete with ⅜″ aggregate is often more expensive. Lightweight concretes and semi-lightweight concretes (with natural sand) are also more expensive. Quotations for specific projects and quantities are usually obtained, because prices are affected by location and quantity as well as by specifications.

Some estimators allow up to 5 percent for *waste* in placing cast-in-place concrete.[5] It appears that actual quantities placed are often higher than the net quantities in the estimate, probably because of oversized forms and some spillage. Reinforcing steel represents about 1 to 2 percent of the total volume of reinforced concrete, which is not usually deducted in measuring

[4] *Building Construction Cost Data,* published each year by Robert Snow Means Company, Inc., Engineers and Estimators, P.O. Box G., Duxbury, Massachusetts 02332. This reasonably priced publication is a useful source of costs and cost analyses. See Bibliography—*Estimating and Cost Accounting*—for other publications.

[5] Some say divide the number of cubic feet in the total measured quantity by 25 (instead of 27) to obtain the number of cubic yards, thereby making a constant allowance of 8 percent for *waste.* This is a typical example of the kind of thinking and gimmickry that is practiced by some people who call themselves estimators.

(120 pounds of steel equals about ¼ cubic foot, which equals about 1 percent of a cubic yard). Five percent *waste* appears to be excessive, as a general rule, but it depends on the job and the *contractor*. Actual *waste* should be accounted for and the data recorded for future estimates.

Considerable interest is always expressed in the costs of placing concrete, probably because the actual costs are easily segregated and compared with the estimated costs. But the placing costs do not amount to a large and important part of the total costs of *concrete work*. If the concrete is placed by a chute directly from the mixer-truck—the most economical means—the cost is about ½ man-hour per cubic yard. To place concrete above ground by crane requires about 1 man-hour per cubic yard plus the crane costs.

Formwork costs are a major part of the costs of cast-in-place concrete, and the variations and probabilities of this *temporary work* are numerous. With the form materials, the type and the number of uses involved are the most important items; and before pricing an estimate a decision about these items must be made, depending on the number of stories, the size of the building, and the variations in sizes of concrete members. *Material costs* include the wood sheathing (or rental of steel forms); the support lumber; the hardware, including wedges and bolts; and the consumable items such as form-ties, oil, and nails. Plywood and steel are the most common materials used at the contact surfaces. Lumber boards are not so widely used now as in the past, except to produce an architectural concrete finish. Plywood is usually ⅝ in. and ¾ in. thick and comes in sheets usually 4 feet × 8 feet. Special formply, 11/16 in. thick, is available with coatings of resin and other plastics to protect the form's contact surface and to eliminate or reduce the need for oiling prior to each use. Typical numbers of possible uses of sheathing are:

Type	Location	No. of Uses	Avg No.
Uncoated formply	Slabs	4– 6	5
	Walls	6– 8	7
	Columns	8–12	10
Coated formply	Slabs	8–12	10
	Walls	12–20	16
	Columns	20–30	25
Fir boards	Slabs	1– 3	2
	Walls	1– 3	2
	Columns	1– 3	2

Excessive cutting and fitting, damage and repairs, and the design of the building may curtail the number of uses. Employing well-made, modular-sized, form panels and careful stripping and handling may increase the number of uses. Each use usually causes some *waste* through cutting and overlapping, and an allowance of 10 percent is commonly made for this contingency. The number of uses of fiberglass and steel forms is usually much higher than that for wood, but the higher initial costs may tend to balance out with the number of uses.

Support lumber is usually Construction or Standard grade (No. 1 or 2) S4S fir, or the equivalent, for studs, walers, and light braces. Secondary supports may be Construction grade rough sawn fir, or patented steel shores and scaffolding, which must be allowed for at proper rental rates. Apart from forms for footings, the costs of using the form sheathing and the support lumber are about equal. Form panels that have a variety of uses should be designed to withstand handling and moving. As a result, the sizes and quantities of support lumber used may be larger than those required solely for the support of filled forms.

Formwork labor costs vary according to design and site conditions, and actual experience is the only true means of making an accurate estimate. As a guide, the following figures are reproduced:[6]

LABOR HOURS PER 100 SF OF CONTACT AREA OF FORMWORK

	Fabricate	Erect	Strip	Clean
Continuous footings	2–3	5– 6	1½–2	1½–2½
Isolated footings	6–7	7– 9	2–3	2–3
Pedestals and piers	4–6	6– 8	2–3	2–3
Grade beams and foundation walls	4–6	4– 6	2–2½	1–2
Walls	4–6	2– 4	1½–2	1–2
Columns	5–7	4– 6	3–4	1½–2½
Flat slab soffit	2–4	2– 4	1½–2½	1–2
Beam sides	4–6	3– 5	2–3	2–3
Beam soffits	2–4	8–10[a]	2–3	2–3

a Includes reshoring time at soffits.

As a guide to pricing formwork two examples are

[6] Concosts, *Construction Costs Reports* (Vancouver, B.C.: Concosts Services, Ltd., 1970), Division 3—Concrete.

given, and following the same procedure as in the examples it is possible to price any formwork once the basic decisions about design and re-use have been made.

FORMWORK TO FOUNDATION WALLS × 8 FT HIGH, PER 100 SF CONTACT AREA

Materials		*First Use*	*Re-uses*
100 SF 5/8″ formply @ $250.00 MSF		$25.00	—
10% *waste* allowance		2.50	$2.50
150 BF lumber @ $150.00 MBF		22.50	—
10% *waste* allowance		2.25	2.25
Hardware (ties, nails, use of bolts, wedges)		4.25	4.25
		$56.50	$9.00
Labor			
Fabrication: carpenter (first use)	4 hr @ $9.00	36.00	—
Fabrication: carpenter (re-use)	1 hr @ $9.00	—	9.00
Erection: carpenter	4 hr @ $9.00	36.00	36.00
Stripping: carpenter	2 hr @ $9.00	18.00	18.00
Stripping: helper	1 hr @ $7.00	7.00	7.00
Cleaning and moving: helper	$1\frac{1}{2}$ hr @ $7.00	10.50	10.50
	Total per 100 SFCA	$164.00	$89.50

With these two figures for "first use" and "re-uses," the average cost of any number of uses can be easily found. For example, assuming seven uses:

$$\frac{\$164.00 + (6 \times \$89.50)}{7} = \frac{\$164.00 + \$537.00}{7}$$
$$= \$100.00 \text{ (average) per 100 SFCA}$$
$$= \$1.00 \text{ (average) per SFCA}$$

In the above example, notice that 10 percent is allowed for *waste* due from cutting in the first use, and for recutting in subsequent re-uses. Hardware is not measured in detail because it is only about 4 percent of the total costs and practical measurement is not possible. But the allowance for hardware can and should be periodically validated or revised by *cost accounting*. Initial fabrication assumes that panels are prefabricated so that only adjustment and fitting is required in the re-uses, but this depends on type of *work*.

FORMWORK TO PAN JOIST SLAB, PER 100 SF CONTACT AREA (2′6″ × 40′0″)

Materials		
150 BF lumber to joist soffits (2 × 6) and beam support under (3 × 10) @ $150.00 MBF ÷ 5 uses		$ 4.50
10% *waste* allowance for lumber		0.45
Nails (say)		1.05
		$ 6.00
Labor		
Erection of lumber and shores:	carpenter 4 hr @ $9.00	$ 36.00
	helper 2 hr @ $7.00	14.00
Removal of ditto:	helper $1\frac{1}{2}$ hr @ $7.00	10.50
Reshoring joists:	helper $\frac{1}{2}$ hr @ $7.00	3.50
Cleaning and moving lumber:	helper $\frac{1}{2}$ hr @ $7.00	3.50
		$ 73.50
Equipment		
Rental of 3 adjustable shores to 3 × 10 beam (under joists) @ 70 cents each (for 1 month)		$ 2.10
Rental of steel pans, 24 in. wide × 40 ft long, including installation and removal by *subcontractor*		$ 30.00
		$105.60
Alternatively:		= $1.06 per SFCA
Cost of 2 × 6 joist forms per lin. ft		$1.90
Cost of 24 in. wide steel pans per lin. ft		$0.75

In the last example, the lumber quantity depends on the width and the spacing of the joists and the need for intermediate support and support at the slab's perimeter. There is no prefabrication required, so each use is essentially the same, unlike in the previous example. The lumber cost is divided by the expected number of uses. In some jobs, some of the lumber intended for rough carpentry can be pre-used for formwork, if this is not prohibited by the contract, in which case the only lumber cost for forms might be an allowance for *waste*.

Steel pans are often rented, and installed, and removed by the rental company at a *unit price* per square foot of slab area (excluding slab beams) or per linear foot of pans. If the *contractor* owns the steel pans, his costs will be similar and should be charged in the same way. Extra costs for the use of any special end-pans should be allowed.

Other steel forms, such as for walls, are either owned or rented and are usually handled by the *contractor*. Either way, the estimator must allow for their costs; largely, depreciation and repairs. Rental rates often include services such as erection drawings and some site supervision. Special forms to fill in to job dimensions can be provided at extra cost if required. Forms for beams (including slab beams in pan joist slabs) and columns are more likely to be specially made, because standard panels are less easily utilized for these members. But the pricing method is the same, with beam soffits requiring shores and reshoring costs.

One of the best ways to understand formwork costs is to view formwork as plant, or equipment. This is particularly easy to visualize if the forms are hired in prefabricated panels. The formwork costs then can be identified and classified as *owning costs* and *operating costs* (or using costs) containing all of those cost elements previously described.

Reinforcing steel bars vary in cost for many reasons, as explained in Chapter 8, and they are usually "shop fabricated" (cut to length and bent) and delivered to the site marked and ready for installation. Site bending is usually more expensive, but it is sometimes necessary. Some contracting companies buy bars in large quantities and fabricate them in their own shops. Light bending costs about 100 percent more than heavy bending per unit weight. Small quantities of bars can cost up to 50 percent more than large quantities when installed. Placing costs vary considerably, but specialists can usually place steel much more economically than a *general contractor.*

The Mean's cost data book gives a national average price (1970) for reinforcing steel of $180 per ton, fabricated and delivered, and $105 per ton for accessories, handling, and placing at site, with an extra $10 per ton in the West. The same book provides data on other variations for quantity, sizes, lengths, and grades. Accessories for steel bars, such as spacers, tie wire, and the like, are included and priced at about $10.00 per ton of steel, which is about 3½ to 4 percent of the total installed price. It is not always practical nor necessary to the accuracy of an estimate to measure these accessories, unless the design requires the use of chairs and bolsters.

The estimator should check his total costs of reinforced concrete by comparing the average price per cubic yard, including formwork and reinforcing steel, with the prices for past jobs. He should also check the estimated quantity of reinforced concrete by dividing the total volume (in cubic feet) by the total floor area of the building and comparing this ratio with that of past buildings. He might also make further comparisons among the quantities of concrete, formwork, and steel rebar in the several parts of different buildings, such as the slabs, columns, beams and walls, as an accuracy check and to obtain useful data.

Pricing Masonry (Division 4)

Masonry units come in many kinds, both clay and concrete, with a few being composed of less common materials such as glass and gypsum. *Waste* from breakage depends on the type of masonry unit, and how it is delivered. Packaged units on pallets (wooden platforms) usually cost more, but they are usually more economical because of fewer breakages, perhaps about 1 percent. Breakages may be 5 percent, or higher, for unpackaged units. Allowances for *waste* from breakage should be made with all kinds of units, according to experience.

Concrete blocks that are not packaged and protected and that get wet from rain are difficult to lay properly, and higher *labor costs* and lower quality *work* may be the result. In some areas, particularly those in which clay bricks are not an indigenous material, concrete blocks are produced to high standards of quality, often approaching the quality of ashlar masonry. In some areas where clay bricks are widely used, concrete blocks are a second-class product used only in cheap *work,* and these local circumstances are reflected in the local costs of different kinds of masonry.

Masonry mortar may be delivered ready-mixed in barrels or it may be mixed at the site. The costs are usually about the same, but a better controlled mix is possible with ready-mixed mortar. *Waste* of mortar is high. Mortar joints in hollow concrete blocks are of two kinds: "face shell" joints, and "full" joints. In the more common "face shell" joints, mortar for the bed-joint is placed only along the two faces of the block and not on the cross-webs. In this type of joint, therefore, the amount of mortar does not vary significantly with the thickness of the concrete block.

The costs of scaffolding and hoists for masonry are often priced as a "lump sum" in the *unit prices,* and about $5.00 per 100 SF appears to be general. But on large masonry jobs the scaffolding should be designed and priced in detail and charged for like other plant and equipment according to the time required. *Isolated work,* such as on chimneys and *work* starting above ground, may require more scaffolding.

There are two major considerations associated with *masonry labor costs*: the weather, and the number of helpers required. The weather affects productivity, the quality of the *work,* and the need to protect both bricklayers and masonry. If wet weather is likely, it may be possible and desirable to allow for temporary shelters. It also may be necessary to allow for lost time because of weather. *Work* that has been soaked when laid or when still "green" (fresh) may require more finishing and cleaning because of smeared joints and mortar stains.

From several observations, the usual number of bricklayers to helpers on concrete blockwork such as in warehouse walls appears to be about four or five bricklayers to two helpers. For brickwork, more helpers may be required because more mortar is used. Other factors affecting the number of helpers are the amount of scaffolding, the type and the height of the *work*, the handling of masonry units, the use of ready-mixed mortar or mortar mixed on site, and union agreements. The relative figures above were obtained from jobs on which ready-mixed mortar was used, and the blocks were delivered packaged on pallets and placed close to the bricklayers. In some areas the number of helpers to bricklayers is reported to be as high as one to one.

The degree and quality of finish of *masonry work* affects the *labor costs,* and apprentices are sometimes employed to tool joints and to clean masonry; but it is probably better to price the *work* as though done by fully trained tradesmen. The physical characteristics of the masonry units may affect *labor costs,* and units of the same type that vary excessively in size may cause higher costs because of problems in laying out the *work* and in bonding the masonry units.

Another consideration in *masonry labor costs* is the amount of cutting required for conduits, electrical outlets, plumbing, and the like. Extensive cutting up to soffits and other similar items should be measured and priced separately; but detailed cutting cannot be measured practically and thus must be estimated. This allowance can be accurately established only by *cost accounting.*

Some examples are given here to illustrate the analysis of *masonry unit prices.*

6″ LIGHTWEIGHT CONCRETE BLOCK PARTITION WALLS, PER 100 BLOCKS/PER BLOCK/PER SF

Materials		
100 – 6″ × 8″ × 16″ Blocks @ $0.35 each	=	$ 35.00
2% waste allowance (breakage and cutting)	=	0.70
1/4 CY mortar @ $20.00 per CY	=	5.00
Scaffolding and hoist (allow)	=	5.00
		$ 45.70
Labor		
Bricklayer: 6 hr @ $9.00	=	54.00
Helper: 3 hr @ $7.00	=	21.00
Total per 100 blocks		$120.70
Total per block		$ 1.21
Total per SF[a]		$ 1.36

[a] An 8″ × 16″ nominal block face (with 3/8-in. joints) has an area of 128 sq. in., which is 8/9 SF.

In the example above, the labor time assumes some detailed cutting of blocks, as this is common in partition walls; but again it must be emphasized that these examples are to illustrate precepts and methods, not to provide factual data about productivity and costs.

EXTRA COST OVER 6″ LIGHT WEIGHT CONCRETE BLOCK PARTITION WALL FOR LINTELS, PER 100 LF/PER LF

Materials		
75 – 6″ × 8″ × 16″ Lintel Blocks @ $0.50 each	=	$ 37.50
300 BF Lumber for supports @ $100 MBF ÷ 5 uses	=	6.00
Concrete fill, 1 CY @ 20.00 per CY	=	20.00
Reinf steel 2—#4 bars @ $0.08 LF	=	16.00
		$ 79.50
Labor (extra)		
Bricklayer: 8 hr @ $9.00	=	72.00
Helper: 4 hr @ $7.00	=	28.00
		$179.50
Deduct (to obtain "extra costs")		
75 standard blocks @ $0.35 each		−26.25
Total per 100 LF (*Extra*)		$153.25
Total per LF (*Extra*)		$ 1.53

In the above example of pricing concrete block lintels, the *extra cost* per LF over the cost of a plain wall has been estimated. The *extra cost* of the lintel blocks is obtained by deducting the cost of standard blocks. The additional cost of concrete fill and steel bars is included. The *extra labor costs* of placing steel, filling lintels with concrete, and installing and removing temporary supports under the lintels' soffits are included. The small cost of the use and *waste* of support lumber has been allowed for in the same way as for formwork items. The total cost is an *extra over cost,* which simplifies measurement. But the lintels could be measured and priced as a *complete item of work* in the same way as the wall is priced. Equipment may be necessary for concrete filling.

The next example is of clay unit masonry:

CLAY FACE BRICK VENEER, 4″ THICK, COMMON BOND, PER M/PER SF

Materials		
1000 face bricks @ $100.00 per M	=	$100.00
$2\frac{1}{2}$% *waste* allowance	=	2.50
$\frac{1}{2}$ CY mortar @ $24.00 per CY	=	12.00
Scaffolding and hoist (allow)	=	8.00
		$122.50

Labor		
Bricklayer: $16\frac{1}{2}$ hr @ \$9.00	=	148.50
Helper: $12\frac{1}{2}$ hr @ \$7.00	=	87.50
Total per 1000 Bricks		\$358.50
Total per SF[a]		\$ 2.27

[a] \$358.50 × 0.00633 bricks.

In this example, the mortar is priced higher, assuming that the brick veneer requires an exterior grade cement mortar instead of a cement-lime mortar (1:1:6) such as is used in partition walls. The allowance for scaffolding is as before, at about \$.05 per SF. The number of bricklayers to helpers is greater. The *unit price* may be calculated per thousand bricks or per SF based on 6.33 standard bricks per SF, with ½-inch wide joints.

It should be pointed out that brick prices vary greatly across the continent–by as much as 100 percent for the same item according to one publication[7]–and in some locations common bricks cannot be purchased for the price of face bricks in other locations. Also, there is a great variety of different types of face bricks available, with corresponding variations in face brick prices and in physical characteristics.

Developments in design and changes in building economics over the last fifty years have sparked a return to clay bricks as a major structural material. Reinforced concrete design has reached a high degree of efficiency; but formwork costs are high, and the logic of erecting and removing formwork for cast-in-place concrete is not always very sound. The advantages of clay bricks over many other materials include lower initial costs, lower maintenance costs, and a better appearance.

Structural clay masonry, which combines clay brickwork and steel reinforcing bars with concrete to unite them, has inherited the advanced design techniques of reinforced concrete; but it does not require expensive formwork. Structural masonry in lintels, beams, and floor systems usually requires some temporary supports, but nothing like the formwork required by cast-in-place concrete. Instead, clay brickwork becomes the permanent form for the concrete and steel in structural clay masonry, and it does not need painting.

Pricing structural masonry requires the same considerations given to other masonry, with special attention to the costs of hoists and scaffolding, especially on high buildings. At the same time, the costs of steel rebar, concrete, and temporary supports must be taken into account in much the same way as in reinforced concrete. Recent developments in the design, fabrication, and erection of prefabricated, structural masonry wall panels indicate that some estimating practices for structural clay masonry will follow those for precast concrete. This is representative of a general trend in which more building components are produced under the controlled conditions of mass production, and the major construction costs involve erection and installation.

[7] *Building Construction Cost Data—1970* (Duxbury, Massachusetts: Robert Snow Means Company, Inc.).

Pricing Metals (Division 5)

Most *work* in this *division* is done by specialist *subcontractors*, and the general estimator may have only handling and erection or installation costs to estimate.

Structural steelwork is fabricated in a shop, and this detailed and precise *work* is sometimes highly automated and is priced by a steelwork estimator. The pricing of steelwork is deceptively simple, because *unit prices* are usually quoted per ton for supply, fabrication, and erection, as well as for painting the steel. To give an idea of the distribution of costs; a breakdown of the costs of steelwork for a one-story warehouse building with open web joists supported on trusses on WF columns appears as follows:

STRUCTURAL STEEL FOR WAREHOUSE (1970)

Steel delivered to shop	\$180	40%
Detailing costs	18	4%
Shop fabrication, priming, and shipping	87	20%
Cost delivered to site	\$285	64%
Unload and erect 10 hr @ \$7.00	70	$15\frac{1}{2}$%
Rental of equipment	15	$3\frac{1}{2}$%
Direct costs: erected at site	\$370	83%
Indirect costs: Overhead and profit	80	17%
Total cost per ton	\$450	100%

Site painting, if required, would be about \$15 per ton extra. Steel costs in 1970 averaged \$165 delivered to shop.[8] The warehouse costs listed above were for the West only, but they do show the typical distribution of steelwork costs.

Unframed steelwork, such as a steel beam over an

[8] For the South, East, and Midwest. Add \$15 per ton for the West. *Building Construction Cost Data—1970* (Duxbury, Massachusetts: Robert Snow Means Company, Inc.).

opening in a masonry wall, would be priced differently. For example:

ONE 10″ × 25.4 LB I-BEAM, 20 FT LONG OVER OPENING

Material

Beam 20 ft × 25.4 lb	=	508 lb
Cost per lb delivered	=	×11 cents
		$55.88
Cost for one cut	=	1.12
Cost of beam delivered	=	$57.00

Erection

Crane ½ hr @ $15.00	=	7.50
Labor 2 hr @ $7.00	=	14.00
Total cost each	=	$78.50
(per ton	=	$309.00)

This unframed steelwork is better priced as a *number* item, which is in accordance with one of the methods in **MM-CIQS**, described in Chapter 8. The weight measure and the grouping of items are not conducive to accurate pricing of the *labor and equipment costs* of this kind of *work.*

Pricing Carpentry (Division 6)

Rough framing is not difficult to price if it is properly measured according to the proper classifications and *items of work.* Some *residential framing work* is done by "labor only" contracts at a *unit price* per SF of floor area, with the *owner* supplying the lumber; it is possible to price the framing of houses and apartment blocks in this way if the estimator has sufficient productivity and cost data available from other similar jobs. However, a detailed estimate is always better, even if several items are eventually priced at the same *unit price.* The following examples illustrate the methods of pricing framing items in detail.

SILL ON FOUNDATION WALLS, PER MBF/PER LF

Materials

2 × 4 standard grade fir, per MBF	$140.00
5% *waste* allowance (cutting to lengths)	7.00
250 anchor bolts (½ in.) @ $0.15 each	37.50
(Note: bolts are often measured separately)	$184.50

Labor

Carpenter: 28 hr @ $9.00	$252.00
Helper: 8 hr @ $7.00	56.00
Total per MBF (2 × 4)	$492.50
Total per LF (2 × 4)	$ 0.33

Sills and plates are often measured with the general framing, in which case the additional labor in laying-out and setting can be measured and priced as an *extra over* item.

FRAMED 2 × 4 STUD PARTITIONS, PER MBF

Materials

2 × 4 standard grade fir, per MBF	$135.00
2% *waste* allowance (cutting)	2.70
Nails, 15 lb @ $0.15 per lb	2.25
	$139.95

Labor

Carpenter: 20 hr @ $9.00	$180.00
Helper: 2 hr @ $7.00	14.00
Total per MBF	$333.95

Studs 8 feet long usually cost less than other dimension lumber. Exterior framed walls will require less labor if they are framed on the deck and "tilted up" into position.

FRAMED 2 × 10 FLOOR JOISTS, PER MBF

Materials

2 × 10 standard grade fir, per MBF	$145.00
10% *waste* allowance (cutting)	14.50
Nails, 8 lb @ $0.15 per lb	1.20
	$160.70

Labor

Carpenter: 15 hr @ $9.00	$135.00
Helper: 3 hr @ $7.00	21.00
Total MBF	$316.70

Smaller joists would require more labor time per MBF and larger joists slightly less. *Waste* allowances are not always necessary.

Fir dimension lumber prices vary as much as 100 percent according to location. The lowest prices are in the Northwest where the fir lumber originates. Prices for carload lots of 20 to 35 MBF are about 10 to 20 percent less than for lots of 10 MBF and less. Prices also vary according to size and length. To illustrate the variation in prices, a synopsis of a typical dimension lumber price list is given below. The prices were about average across the continent in 1969–70. Lumber prices have fluctuated greatly in the last few years, and special quotations are necessary for each project of any size.

DOUGLAS FIR DIMENSION LUMBER (1970)

Standard grade (No. 2 Common) in random lengths of 8 to 20 ft; in carload lots; delivered to site.

Size	*Price per MBF*	*Extras*
2 × 2	$141	1. For specified lengths,
2 × 3	137	10 to 20 ft add $5
2 × 4[a]	130	2. Ditto, 22 to 24 ft
2 × 4	138	add $10–$20
2 × 6	138	3. For 10 MBF lots
2 × 8	138	add 10 percent
2 × 10	143	4. For smaller lots add
2 × 12	143	20 percent

a 8-foot studs.

Notice that 2 × 4, 2 × 6, and 2 × 8 are the same price; 8-foot studs are less; 2 × 10 and 2 × 12 are more, as are specified lengths. Fir planks and small timbers cost more; for example, 3 × 4 to 6 × 8 timbers cost $152 to $187, and larger timbers (up to 14 × 14) cost about $205 MBF. Larger orders of specified sizes and lengths can be purchased at more competitive prices than list prices. Prices for different grades and sizes fluctuate considerably according to supply and demand.

Plywood has generally replaced lumber boarding for sheathing because of the wide selection of types and qualities and lower installation costs. Typical prices in 1970 for unsanded sheathing (standard grade) were: ⅜″ thick, $100.00; ½″ thick, $150.00; ⅝″ thick, $180.00; ¾″, $215.00 MSF. Typical *unit prices* are analyzed below. If there is not much handling involved, all the *work* may be done by carpenters. *Waste* varies considerably with the job.

PLYWOOD WALL SHEATHING, PER MSF/PER SF

Materials

3/8″ ply sheathing, per MSF	$100.00
10% *waste* allowance	10.00
Nails, 5 lb @ $0.15 per lb	0.75
	$110.75

Labor

Carpenter: 20 hr @ $9.00	$180.00
Total per MSF	$290.75
Total per SF	$ 0.29

PLYWOOD FLAT ROOF SHEATHING, PER MSF/PER SF

Materials

5/8″ ply sheathing, per MSF	$180.00
5% *waste* allowance	9.00
Nails, 10 lb @ $0.15 per lb	1.50
	$190.50

Labor

Carpenter: 10 hr @ $9.00	90.00
Helper: 2½ hr @ $7.00	17.50
Total per MSF	$298.00
Total per SF	$ 0.30

This example assumes that intermediate edge supports at right angles to the joists are not required. Edge supports can be provided by inserting short lengths of 2 × 4 between joists; using H-clips on the plywood edges; and using tongued and grooved plywood sheets, all of which increase the costs. Large deck areas can mean a big reduction in installation costs, and the amount of *waste* is also less.

Furrings, and the like, where strips of lumber or plywood are fixed to stud walls or masonry to receive and to plumb drywall finishes, should be measured and priced in LF, not in BF. The *unit price* per LF can be converted to a price per SF for a specified spacing of the material.

FURRING ON FRAMING, PER MLF/PER LF

Materials

1 × 4 standard grade fir @ $100.00 MBF/@ $33.33 MLF	$ 33.33
10% *waste* allowance	3.33
Nails, 10 lb. @ $0.15 per lb	1.50
	$ 38.16

Labor

Carpenter: 15 hr @ $9.00	$135.00
Total per MLF	$173.16
Total per LF	$ 0.17½

The *unit price* for 1 × 3 furring would be about $0.16½ per LF, the only difference being in the cost of the lumber.

Wood posts and columns are usually more easily measured and priced as *run* or *number* items rather than in board feet. The cost of installing a 6″ × 6″ fir post, 8 feet long, cannot be much different from the cost of installing an 8″ × 8″ post 8 feet long; but the number of board feet in one and the other differs by about 80 percent. If rough carpentry items are priced on a board foot basis, there must be a different *unit price* for each size of lumber.

Finish carpentry is generally more difficult to price than rough carpentry, because of the higher proportion of *labor costs* in *finish work* and the greater variation in standards of workmanship. Also, the materials and components in finish carpentry are generally more complex or require special treatment.

For example, consider comparatively simple items like exterior wood siding or interior paneling. Wood siding installation costs depend in part on the treatment at corners–whether rough and covered by trim; or fitted up to a vertical corner trim; or mitred, which is the most expensive method. Similarly, interior paneling costs depend on the treatment at perimeters and corners, and covering them with trim is much cheaper than exposed edges, either set in or scribed, and mitred corners.[9]

With siding and paneling, the *laps* (between boards) should be measured in BF quantities, and *waste* from cutting and fitting should be allowed for in the *unit price* because it is an estimated variable. Instead of measuring this *work* in BF, it can be better measured by the installed net *super* area (in SF), in which case the *laps* as well as the *waste* have to be allowed for in the *unit price*. Each such allowance should be included in the item's description.

9 Exterior wood siding and eaves soffits are sometimes classified as *rough carpentry work;* but if rough carpentry is properly defined as "concealed, and generally structural in nature," siding, soffits, and other *exposed work* in which appearance is important, and in which edges and joints should be neatly cut and tightly fitted, are more accurately described as finish carpentry.

RED CEDAR PANELING, PER MBF/PER MSF

Description (in estimate):
1″ × 8″ Western Red Cedar Paneling, "A" Grade, BCLMA pattern No. 101; secret-nailed vertically to framing with no end joints (estimated waste—15%)

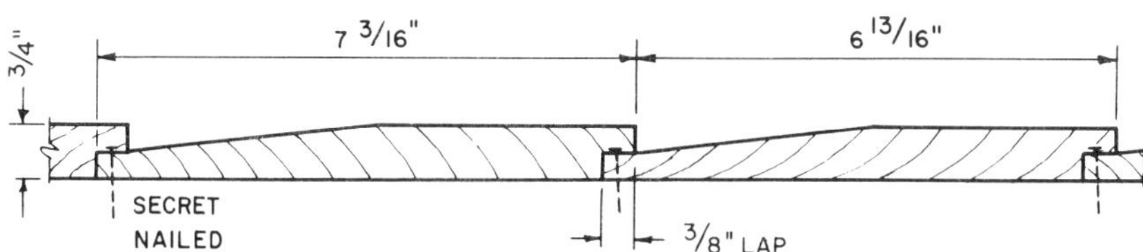

FIG. 42. Cedar Paneling Material.

To calculate the *unit price* of the paneling materials, per MBF (excluding *laps,* already allowed for in measured quantity).

CEDAR PANELING MATERIALS, PER MBF

Quoted price, delivered, per MBF	$185.50
15% *waste* allowance	27.83
Nails, 15 lb @ $0.20½ per lb	3.07
Cedar shingles (for shimming), say,	1.10
Total per MBF	$217.50

To calculate the *unit price* of materials, per MSF (including *laps*), the *laps* are ⅜ in. wide, out of a board 7$\frac{3}{16}$ in. in actual width, out of 8 in. in nominal width. This gives an effective covering width of 6$\frac{13}{16}$ in., or a loss of about 15 percent from the 8-in. nominal width on which the BF quantity is based. Which means that 1 BF that in theory covers 1 SF (with a nominal 1-in.-thick board), in practice, and in this case, covers only 85 percent of a square foot (with an actual ¾-in.-thick board).

CEDAR PANELING MATERIALS, PER MSF

Quoted price, delivered, MBF	$185.50
17½% *lap* allowance[a]	32.46
	217.96
15% *waste* allowance	32.70
Nails, 18 lb @ $0.20½ per lb	3.69
Cedar shingles (for shimming), say,	1.28
Total per MSF	$255.63

a 850 SF + 17½% = 1000 SF, approx.

The other materials (nails, cedar shingles) are increased by 17½ percent, because they are now required for 1175 BF, not 1000 BF, as before. A paneled wall 12½ ft long × 8 ft high contains 100 SF net area of paneling and requires 117½ BF of materials, plus *waste*. Either way:

117½ BF @ \$217.50 MBF = \$25.56
100 SF @ \$255.63 MSF = \$25.56

A slight but common discrepancy is often made by calculating the loss between 8 in. and 6$^{13}/_{16}$ in. as 15 percent, and adding 15 percent instead of 17½ percent for *laps*. But 1000 − 15% = 850; 850 + 17½% = 1000.

The *labor costs* of installing the paneling can be calculated either way. If cost accounts show that it requires 24 hours carpenter's time to install 1000 BF; then, it will require

$$\frac{24 \text{ hr}}{1} \times \frac{1000}{850} = 28.23 \text{ hr to install 1000 SF net area.}$$

Either way:

Labor

Carpenter: 24 hr @ \$9.00	=	
Total per **MBF**		\$216.00
Carpenter: 28.23 hr @ \$9.00	=	
Total per **MSF**		\$254.03
And the *labor costs* for paneling the wall are:		
117½ BF @ \$216.00 **MBF**	=	\$ 25.38
or 100 SF @ \$254.00 **MSF**	=	\$ 25.40
(The small difference results from rounding off.)		

Many carpentry items, including shiplap sheathing, siding, interior matching (tongued and grooved), and tongued and grooved heavy wood decking can be measured and priced in the same ways. In some cases, helpers or laborers are required for handling, and in other cases, such as with horizontal siding, two carpenters may have to work together. These things and their costs cannot be determined without the guidance of historical *labor costs* and experience with the *work*. One reference book for estimators considers such aspects of *labor costs* by describing typical work crews for *items of work* and by stating the approximate daily output of the crews. It is much more helpful and realistic to read the following description:[10]

Placing Wood Chair or Dado Rail. In large rooms, long straight corridors, etc., a carpenter should fit and place 275 to 300 lin ft of wood chair rails per 8 hr day, at the following labor costs per 100 lin ft:

	Hours	
Carpenter	2.8	.....
Labor	0.5	.....
Cost per 100 lin. ft.	3.3	.....

In small kitchens, pantries, closets, bath rooms, etc., a carpenter will place only 160 to 180 lin ft of chair rail per 8 hr day, at the following labor costs per 100 lin ft:

Carpenter	4.7	.....
Labor	0.5	.....
Cost per 100 lin ft	5.2	.....

than to read that it requires 4.25 man-hours (average) to install 100 LF of rail, without any indication of job conditions. Cost account records should contain qualified cost data, indicating job features and conditions so that the data can be properly used in future estimates.

Finish carpentry and millwork specifications are frequently based on one (or more) of the three standards of woodwork described in standard specifications:[11]

1. **Economy:** wherein price outweighs quality of *work*
2. **Custom:** for good quality regular *work*
3. **Premium:** for the finest *work*, such as in monumental buildings.

The *labor costs* should be priced accordingly, and recorded cost data should be identified with one of these three standards of woodwork.

It is not uncommon for the *labor costs* of installing millwork to be priced as a percentage of the total costs of the millwork supplied to the job. The maxim here is: Do what you like, but know what you do. If an estimator has historical cost data that are pertinent and show that the relationship between *labor costs* and *material costs* for a particular class and standard of *work* has been consistent within a certain limited deviation, he may be in a position to use a percentage of the *material costs* to estimate the *labor costs.* Of course, inaccuracies and discrepancies easily arise if an estimator has nothing better than a rough rule of thumb; and a detailed estimate of *labor costs* is generally preferable, especially if each of the *items of work* can be priced separately on the basis of data obtained through *cost accounting.* But better a general rule of proportion supported by experience than a detailed estimate without any such support.

10 *The Building Estimator's Reference Book* (Chicago: Frank R. Walker Company, 1967), p. 962, 963.

11 Specifications are published by the Architectural Woodwork Institute, 1808 West End Building, Nashville, Tennessee, 37203, U.S.A., and the Millwork Contractors' Association of British Columbia, 2675 Oak Street, Vancouver, 9, B.C., Canada. These publications contain a detailed explanation of the three standards of woodwork.

Pricing Moisture Protection (Division 7)

This *division* embraces a great variety of *items of work,* the most common of which items include bituminous materials and membranes such as waterproofing and built-up roofing.

Built-up roofing prices are simplified by the fact that most of the membrane materials (felts and papers) are sold in rolls containing enough material to cover 4 Squares (400 SF). Some of the heavier materials, such as smooth surface roofing, are sold in rolls to cover one Square, and some others to cover two Squares. Each roll contains 8 percent additional material for edge and end *laps,* so that in a roll of No. 15 asphalt felt there are 432 SF, and in a roll of the heavier, smooth surface roofing there are 108 SF.

Roofing prices should be analyzed according to job conditions such as location; building height; accessibility to roof; roof area; degree of slope, or pitch, of the roof; and the number of interruptions in the roof areas caused by changes in slope and level, stacks, mechanical buildings, fixtures on the roof, and the like. In other words, the ideal roofing job is, perhaps, an unobstructed flat roof of over 200 Squares, over a single-story warehouse that is located about a mile from the roofer's yard and that at the time of roofing has no exterior walls yet erected, and consequently, no problems with spillage on wall surfaces. In addition, and ideally, the deck will be bone-dry, and the *roofing work* is to be done during a fine and sunny week in May.

Such roofing jobs may be too infrequent, but a good method of pricing any *work* is to price an ideal job and to use it as a datum for pricing other jobs. This method of pricing from a datum (or base price) established as a basis for pricing specific jobs has been used successfully in estimating and *cost accounting* in manufacturing industries, and there is no reason why these techniques cannot be adopted by the construction industry. This point is discussed again later in this chapter.

Two other factors to be considered in pricing roofing are the optimum crew size for a job, and the fixed costs of setting up and clearing away at completion. A certain minimum crew is necessary to tend the kettle (in which the asphalt is heated); to transport and hoist, or pump, the hot asphalt to the place of application; to lay roofing felts; and to apply the asphalt to the felts. Depending on the roofing specifications and the job conditions, a crew of 5 might apply from 15 to 25 Squares daily, of from 3- to 5-ply roofing on flat decks. That represents an average of about 2 man-hours per Square, which is worthwhile remembering because it appears to be an approximate figure for many built-up roofing applications.

Some examples of the analysis of *unit prices* for built-up roofing illustrate the pricing method:

5-PLY FELT, ASPHALT, AND GRAVEL ROOFING (UP TO 100 SQUARES) ON FLAT WOOD DECKS, PER SQ.

Materials

No. 2 sheathing paper, 1/4 roll @ \$1.60	=	\$0.40
No. 15 asphalt felt, 1 1/4 rolls[a] @ \$3.60	=	4.50
120 lb asphalt @ \$3.00 per 100 lb	=	3.60
400 lb gravel, 1/7 CY @ \$4.20	=	0.60
Nails, fuel, mops, etc., allow	=	0.50
		\$9.60

[a] 1¼ rolls × 4 Sq. = 5 Sq. × 1 ply; or 1 Sq. × 5 ply.

Labor

Roofer: 2 hr @ \$8.50	17.00
Foreman: ½ hr @ \$0.50 (extra)	0.25
Total: per Sq.	\$26.85

In the above example, one ply of sheathing paper is nailed down and covered by two plies of felt over which is mopped 20 lb of asphalt per Square. Three additional plies are laid and mopped with 20 lb of asphalt per Square under each of them and are covered with a flood coat of 60 lb of asphalt, which is then covered with 400 lb of gravel per Square. It is assumed above that one of the crew is a working foreman. For over 100 Squares the *labor costs* might be 10 to 20 percent less. *Overhead and profit* might be an additional 20 to 30 percent.

INSULATION BOARD (INSTALLATION ONLY) ON 2-PLY FELT AND ASPHALT VAPOR BARRIER (UP TO 100 SQUARES) ON FLAT WOOD DECKS, PER SQ.

Materials (excluding insulation)

No. 2 sheathing paper, 1/4 roll @ \$1.60	=	\$ 0.40
No. 15 asphalt felt, 1/2 roll @ \$3.60	=	1.80
40 lb asphalt @ \$3.00 per 100 lb	=	1.20
Nails, fuel, mops, etc. allow	=	0.20
Insulation[a]		\$ 3.60

Labor (incl. laying insulation)

Roofer: 1 hr @ \$8.50	=	\$ 8.50
Foreman: 1/4 hr @ \$0.50 (extra)	=	0.12½
Total per Sq.	=	\$12.22½

[a] Insulation material varies in type, thickness, and price, and so has not been included above.

Prices for different classes of insulation vary greatly; but if prices are related to the "R factor" of the insulation material, it will be seen that prices increase as the R factor increases and that different insulation materials of different thicknesses but with the same R factor cost about the same.

The following set of prices for insulation according to R factors indicates a constant ratio of about 1 to 4 (for this set) between "R factors" and prices per Square.

R Factor	*Price per Sq.* ($)	*Factor* (*Price*/*R*)
3.13	12.40	3.96
3.85	14.20	3.69
4.38	18.40	4.20
5.00	19.50	3.90
5.65	24.70	4.37
6.25	26.30	4.20
7.52	30.50	4.06
10.00	37.80	3.78

That is, fiberboard 2″ thick and blue polystyrene 1⅛″ thick both have an "R factor" of about 5.65, and both cost about the same. Other sets of prices will show different factors (Price/R).

On concrete decks omit sheathing paper and add asphalt primer.

4-PLY FELT, ASPHALT, AND GRAVEL ROOFING (UP TO 100 SQUARES) ON FLAT CONCRETE DECKS, PER SQ.

Materials

No. 15 asphalt felt, 1 roll @ $3.60	=	$ 3.60
140 lb asphalt @ $3.00 per 100 lb	=	4.20
400 lb gravel, 1/7 CY @ $4.20	=	0.60
Fuel, mops, etc., allow	=	0.40
		$ 8.80

Labor

Roofer: 1 7/8 hr @ $8.50	=	15.94
Foreman: 3/8 hr @ $0.50 (extra)	=	0.19
Total per Sq.	=	$24.93

ASPHALT PRIMER TO CONCRETE DECKS, PER SQ.

Material

1 gallon asphalt primer @ $1.50	=	$ 1.50

Labor

Roofer: 1/10 hr @ $8.50	=	0.85
Foreman: 1/40 hr @ $0.50 (extra)	=	0.01
Total per Sq.		2.36

On concrete decks there is no need for sheathing paper as there is on wood decks (to stop hot asphalt from dripping through) and no nailing to concrete decks is required. But concrete decks do need priming to ensure adhesion of the asphalt.

5-PLY FELT ASPHALT AND MINERAL SURFACED ROOFING (UP TO 100 SQ.) ON SLOPING WOOD ROOFS, PER SQ.

Materials

No. 2 sheathing paper, 1/4 roll @ $1.60	=	$ 0.40
No. 15 asphalt felt, 3/4 roll @ $3.60	=	2.70
80 lb asphalt @ $3.00 per 100 lb	=	2.40
M. surfaced roofing, 2 rolls (90 lb) @ $5.00	=	10.00
Nails, fuel, mops, etc., allow	=	0.50
		$16.00

Labor

Roofer: 2½ hr @ $8.50	=	21.25
Foreman: 1/2 hr @ $0.50 (extra)	=	0.25
Total per Sq.	=	$37.50

This specification may be applied to slopes on which surface asphalt and gravel cannot be retained on the roofing. *Labor costs* are higher because of the heavy sheets and the slope.

Warranties stipulating that the roofing be inspected while it is being installed require from $1.50 to $2.50 per Square to be added to the cost of the roof.

Sheet metal flashings in conjunction with built-up roofing on, say, a 200 Square warehouse roof may not amount to more than 5 percent of the total roofing costs. On an institutional building wherein some of the sheet metal flashings may be a visible architectural feature, the flashings may cost several thousand dollars and amount to a much larger proportion of the roofing costs.

Sheet metal materials vary considerably in cost, whereas the *labor costs* of fabricating and installing them tend to vary more with the complexity of the *work* than with the material. The materials in usual order of comparitive cost, with the cheapest first, are:

1. **Galvanized steel** (1¼-oz zinc coating), 26 gauge (GSG) 0.0217 in. thick–*$0.15 per SF*
2. **Galvanized copper-bearing steel,** (2-oz zinc coating) 26 gauge (GSG) 0.0217 in. thick–*$0.20 per SF*
3. **Aluminum sheet,** 24 gauge, 0.020 in. thick–*$0.20 per SF*
4. **Zinc alloy sheet,** 0.027 in. thick–*$0.45 per SF*
5. **Stainless steel sheet,** easy formed, 28 gauge, 0.16 in. thick–*$0.55 per SF*
6. **Copper sheet,** cold rolled, 16 oz per square foot–*$1.10 per SF*

7. **Lead sheet,** 4 lb per square foot—*$1.30 per SF*

All of these products are sold by weight, and prices (per 100 lb) must be converted to prices per SF, or *super* quantities in the estimate must be converted to the equivalent weights. The above prices are approximate for large quantities, and prices for smaller quantities may be 20 percent higher, or more. The prices shown indicate only the general and relative cost levels; some sheet metal costs fluctuate greatly, especially the price for copper. It is possible to obtain one metal cheaper than another by buying in large quantities, even though it is shown above to be more expensive. Prices also vary not only according to quantity, but according to locality and availability.

Stainless steel of the "Ezeform" type, which is very soft and ductile and easy to bend, may require less labor than galvanized steel. Similarly, sheet lead is easier to form than most other metals. There are many other flashing materials, such as plastics, metal alloys, and coated metals such as stainless steel coated with copper, and aluminum with plastic and enamel coatings, which are cheaper than the traditional copper and lead and have a pleasing appearance.

Waste will vary according to the required widths and girths of the *sheet metal work* and the standard sheet sizes. Most sheet metals are available in a variety of sizes, although the most common sheet dimensions are 30- and 36-in. widths and 96- and 120-in. lengths. In galvanized steel, 36-in. widths are standard and 30- and 48-in. widths are slightly more expensive. A supplier's catalogue is necessary to obtain all the information on sizes and prices.

Unit prices for labor on sheet metal should be obtained through *cost accounting,* in which it is simpler to relate productivity to the *super* area of metalwork (describing the kind and thickness) and to convert to material prices per square foot in order to record data and to price an estimate rather than relating labor to the total weight of the metal. *Labor costs* depend largely on the amount and complexity of the *work* and whether or not forming can be done in the shop. Large quantities of flashings of a simple profile and with only two or three "breaks," which can be made on a bench and delivered to the site ready for installation, are the most economical in terms of *labor costs.*

GALVANIZED (COPPER BEARING) STEEL 26-GAUGE SHEET EAVES FLASHING (IN LENGTHS NOT EXCEEDING 10 FT) × 18 IN. GIRTH, WITH 4 BREAKS AND TWO HEMMED EDGES, PER LF

Materials (for one 10-ft length)

15 SF galv sheet @ $0.20 per SF	= $ 3.00
(15 lb galv sheet @ $0.20 per lb)	
Clips, wedges, screws, etc., allow	0.30
	$ 3.30

Labor (for one 10-ft length)

Shop labor, 3/4 hr (fabricating)[a]	
Site labor, 3/4 hr (installing)	
1 1/2 hr @ $9.00	$13.50
Total 10-ft length	$16.80
Total per LF	$ 1.68
Total per SF	$ 1.12

Note: *a* Includes two men (2 × ⅜ hr) cutting, forming, and back painting.

In the above example, 26-gauge galvanized steel sheet weighs 100 lb per 100 SF, so the conversion is easy. Flashings are often specified with an underlay of roofing felts for which a *unit price* can be analyzed, as for built-up roofing. Sheet metal roofing is priced in a similar manner to flashings, although more of the labor may be performed at the site.

Many other analyses and explanations of *unit prices* for *work* in this *division,* including wood, asbestos, and asphalt shingles; slate roofing; sheet metal roofing; and corrugated aluminum, steel, and asbestos cement roofing and siding are to be found in Walker's *The Building Estimator's Reference Book* (Footnote 10).

Pricing Doors, Windows, and Glass (Division 8)

Doors are usually standard products, and custom made wood doors are a millwork item. In either case, they are delivered to the site ready for installation. Consequently, their pricing is relatively simple for the *general contractor.*

Door installations vary immensely in cost, from about $30 for a single interior door in a light steel frame with a latchset up to $1500 or more for double doors to a hospital's operating room, with stainless steel frame and hardware, including closers.

Door assemblies, complete with frame and hardware from one supplier, are replacing the traditional system of various doorway parts assembled at the site by a carpenter. Quotations can be obtained for complete door assemblies, and the general estimator has to add only for installation time at the site. For example:

RESIDENTIAL INTERIOR DOOR ASSEMBLY

Comprising:

One 1 3/8" thick hollow, rotary mahogany (stain grade) prefinished flush door, with cedar styles
One 18-gauge pressed steel frame with 4 5/8-in.-wide throat (for wood frame and drywall finished walls) and matching snap-on trim
One latchset-Beverley design-26/D finish, 2 3/4-in. backset with hinges (attached to frame) and door stop (spring type)

Total price of assembly, delivered	$21.00
Labor handling and installing assembly:	
Carpenter: 1/3 hr @ $9.00	$ 3.00
Helper: 1/3 hr @ $7.00	$ 2.34
Total each unit	$26.34

Wood and metal windows, like doors, are easy to price once quotations have been obtained for their supply and delivery. Aluminum windows are an important building component and are widely used. Residential windows are often standard products, delivered from stock; but the heavier and more expensive commercial and institutional windows are usually custom made because of the variations required in their designs.

As an illustration of estimating and pricing this class of *custom work,* in which most of the *work* is done in a factory, an example of a price analysis for custom made aluminum windows follows. The prices and rates used in this example include *profit and overhead* and were valid for a particular location in 1970. Obviously, they will not apply elsewhere at different times, and the prices shown should be taken for what they are—a means of illustrating an estimating method and an analysis of the *costs of the work.*

CUSTOM MADE COMMERCIAL AND INSTITUTIONAL ALUMINUM WINDOWS

Materials (including profit and overheads)

Aluminum windows solid sections, per LF	
Main frame, universal type, $1\frac{1}{2}$" deep	$0.45
Vent frame, ditto (Ixx = 0.069 in.^{4})	$0.45
Vent frame, ditto (Ixx = 0.167 in.^{4})	$0.55
Tee bar, ditto (Ixx = 0.076 in.^{4})	$0.40
Tee bar, ditto (Ixx = 0.133 in.^{4})	$0.50
Tee bar, ditto (Ixx = 0.596 in.^{4})	$0.70
Glazing bead (extruded section)	$0.10
Drip cap (ditto)	$0.10
5 1/2" sill (ditto) incl. chairs	$1.35
8" sill (ditto) ditto	$2.20
Aluminum window tubular sections, per LF	
Vent frame, universal type, (Ixx = 0.121 in.^{4})	$0.75
Tee bar, ditto (Ixx = 0.498 in.^{4})	$1.05
Tee bar, ditto (Ixx = 0.807 in.^{4})	$1.40

In these *unit prices* sections are cut to length, and a 10 percent *waste* allowance is included. Prices vary according to weight of section, and whether solid or tubular, and usually amount to about $0.50 to $0.60 per pound from the extrusion manufacturer.

Aluminum Finishes to Sections, per LF	
Clear anodizing	$0.12
Bronze anodizing	$0.12
Gold anodizing	$0.13
Colored enamel	$0.12

Vent Hardware, including stainless steel Anderburg hinges, white bronze fastener, and weatherstripping, per set $8.50

Sheet glass (B-Quality), including risk, cutting, and waste, per SF

Single strength	
(3/32 in. thick; 18 oz. per SF) (0–10SF)	$0.30
Double strength	
(1/8 in. thick; 24 oz. per SF) (0–10SF)	$0.40
(5/32 in. thick; 32 oz. per SF) (10–15SF)	$0.65
Heavy sheet	
(3/16 in. thick; 40 oz. per SF) (15–20SF)	$0.70
Heavy sheet	
(7/32 in. thick; 45 oz. per SF) (15–20SF)	$0.75
Heavy sheet	
(1/4 in. thick; 50 oz. per SF) (25–50SF)	$1.05

Polished Plate, or Float, including risk, cutting, and waste, per SF

(1/4 in. thick; 50 oz. per SF) (25–50SF) $3.00

Glass prices vary greatly according to sizes and the quantity purchased, and they may vary as much as 100 percent between prices for bulk purchases by the carload and the case and prices for panes ex-stock cut to size. The above prices are representative of those charged by a window manufacturer who purchases glass by the case or pallet and who establishes an in-shop price list for use by his own estimator, but who buys glass at continually fluctuating prices at the lowest price possible.

Labor (including profit and overheads)

Factory labor wage rate	= $4.00 per hr
Factory labor overhead (50%)	= 2.00
Factory labor rate	= $6.00 per hr
Profit and office overhead (33 1/3%)	$2.00
Factory labor charge-out rate	= $8.00 per hr

Fabrication (including profit and overheads)

1. Fabrication of main frame, with four arc-welded corners including glazing beads:
   20 min @ $8.00 per hr = $2.67 each
2. Fabrication of vent frame, with four arc-welded corners, including glazing beads and hardware:
   40 min @ $8.00 per hr = $5.34 each
3. Installation of tee bar mullion, with two mechanical joints at ends, including beads:
   3 min @ $8.00 per hr = $0.40 each

Glazing (including profit and overheads)

Labor glazing in factory, per light:
15 min @ $8.00 per hr = $2.00 each

Site glazing is usually more expensive than shop glazing but it is necessary if the windows have to be transported far and if there is a risk of damaging the glass and glazing in the window after leaving the factory and before installation.

Window installation (including profit and overheads)

Materials, per LF (of frame)

Elastomeric calking compound $\frac{\$2.00}{25LF}$ per tube	=	\$0.08
Plastic foam backing	=	0.03
Plugs and screws, say,	=	0.10
		\$0.21

Labor, per LF (of frame)

Installing: 20 LF per hr @ \$7.00	=	0.35
Calking: 50 LF per hr @ \$7.00	=	0.14
		\$0.70
Profit and office overhead (33 1/3%)	=	0.23
Total per LF (of frame)		\$0.93

This example assumes that scaffolding does not have to be provided by the *window subcontractor* and that it is available for use at no charge by the *general contractor.*

Following is an example using the *unit prices* above.

COMMERCIAL TYPE ALUMINUM WINDOW, 5 FT WIDE × 4 FT HIGH, WITH SIDE-HUNG CASEMENT VENT, DRIP CAP AND 5½-in. WIDE SILL, AND WITH CLEAR ANODIZED FINISH

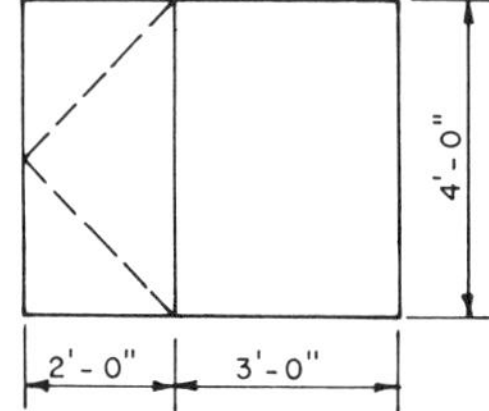

Materials

Main frame, universal type, 1½" deep, 18 LF @ \$0.45	=	\$ 8.10
Tee bar mullion (Ixx = 0.133 in.^{4}), 4 LF @ \$0.50	=	2.00
Vent frame, universal tubular type (Ixx = 0.121 in.^{4}), 12 LF @ \$0.75	=	9.00
Vent hardware, 1 set @ \$8.50 each	=	8.50
Glazing beads (12 + 14 LF) = 26 LF @ \$0.10	=	2.60
5½" sill, including chairs, 5 LF @ \$1.35	=	6.75
Drip cap 5 LF @ \$0.10	=	0.50
Clear Anodizing 70 LF @ \$0.12	=	8.40
(18 + 4 + 12 + 26 + 5 + 5 = 70 LF)		\$45.85

Labor (as above)

Fabricate main frame, 1 frame @ \$2.67	=	2.67
Fabricate vent frame and install hardware, 1 vent @ \$5.34	=	5.34
Install mullion, 1 @ \$0.40	=	0.40
Total		\$54.26

Glass and Glazing to Aluminum Window, 5 ft × 4 ft

Materials

2 lights, 32-oz clear sheet (2.0 × 4.0) + (3.0 × 4.0) = 20 SF @ \$0.65 =		\$13.00
Glazing tape	26 LF @ \$0.02 =	0.52
		\$13.52

Labor

2 lights @ \$2.00 each =	4.00
Total	\$17.52

Window and Installation (Summary)

Fabricated window, as above	=	\$54.26
Glass and glazing as above	=	\$17.52
Delivery to site, allow	=	\$ 1.25
		73.03

Installation

Window frame, 18 LF @ \$0.93	=	16.74
Sill, 5 LF @ \$1.30[a]	=	6.50
Drip cap, 5 LF @ \$0.93[b]	=	4.65
		(\$90.92)
Total Cost: Window Installed (including *profit* and *overheads*), say,		\$91.00[c]

ab The analysis of these *unit prices* is not shown. *c* This represents a *unit price* of \$4.55 per SF of window installed, which is a typical price.

The same estimating approach may be used for glass and metal storefronts and entrances to stores, offices, and institutional buildings. This class of *work* is estimated and done by specialist *subcontractors,* whose estimators are sometimes required to prepare shop drawings and to do a certain amount of site supervision and general office business unless the company is quite large. As a result, these estimators become very knowledgeable about the *specialty work* with which they are dealing, and most *construction work* is estimated by such estimators as these.

Finish hardware was described in Chapter 8 as a significant part of the *work* in a building and which is often included in a contract under a cash allowance prescribed by the *designer.* Many other *sections of work* may be dealt with in the same way; and whatever can be said here with regard to finish hardware can also be applied to other *work,* particularly that covered by cash allowances and also some *work* in *divisions 10, 11, 12, 13,* and *14.*

Cash allowances are used by *designers* usually because they have to delay making decisions about certain parts of a project, such as selecting specific items of finish hardware. At the same time, the *designer* wishes to include that part of the *work* in the contract so that its costs are part of the contract amount. This may be because the items are required to be purchased through and installed by the *general contractor*; or it may be for another reason, such as the financing from a mortgage company based on the contract amount of a stipulated sum contract.

It follows, therefore, that if a cash allowance is included in a contract by a *designer,* there will probably be a lack of information about the *work* it includes. If the *designer* had all the information required, there would be no need for a cash allowance. Nevertheless, the *contractor* is, at least, usually required to allow for coordinating the *work* covered by the cash allowance with the other *work* in the project; and at times he may be required to allow in his bid for installing items such as finish hardware and equipment supplied and paid for out of a cash allowance.

A careful study of the wording of the *bidding documents* is first necessary to establish precisely what is covered by a cash allowance and what is not. Whatever is not must be included elsewhere in the estimate. For example, the cash allowance may be for the supply of finish hardware (or equipment) by a firm to be selected later by the *designer*. But does the allowance also include getting the items into the building that is under construction?[12] If the items are refrigerators and stoves for an apartment block of 100 suites, the handling costs may be quite high. Even the exact nature of the items supplied may be significant. For example, finish hardware may be described as including door closers. However, there is a difference between installing surface-mounted closers and concealed closers. Also, $100 worth of door hardware may take only one hour of installation time, but $100 worth of plastic door numerals may require many hours to install. The *contractor* should know what he is required to do so that his estimate of the costs can be fair and reasonable.

In addition to allowing for any handling and installation costs not included in the cash allowance, the *contractor's bid* must include any other costs he may require for *overhead and profit,* and any other expenses. It is a common practice to allow a lower percentage for *overhead and profit* on cash allowances than on the *costs of other work,* but it is probably unrealistic to include no markup on cash allowances on the supposition that there will be no *overhead costs.*

Pricing Finishes (Division 9)

This *division* includes a great variety of different *work* that, for pricing purposes, can be grouped as follows:

1. **Drywall work:** in which manufactured boards and panels such as plywood, composition, and gypsum wallboards are installed with linear trim at edges and corners (and sometimes at joints) as a finish material, sometimes with special fire protective and acoustical qualities
2. **Tile work:** in which ceramic and other tiles, blocks, and slabs of various compositions are adhered to a base surface
3. **Sheet work:** in which flexible sheet goods are applied to base surfaces as resilient, acoustic, and decorative coverings
4. **Wet finish work:** in which plastic materials are mixed and applied to a base surface in such thicknesses as are required to give sufficient strength and density for finishes with various qualities, including durability, imperviousness, and acoustical and decorative qualities
5. **Coating work:** in which plastic materials are applied to a base surface in one or more thin coats to provide a protective and decorative finish
6. **Beds and wet backing work:** similar to *wet finish work,* but preparatory to, and a backing for, *finish work*
7. **Lathing and dry backing work:** in which supports and backings for *finish work* are constructed with dry materials, usually either metal lath or gypsum lath board for *wet work* and backing boards for drywall and tile finishes
8. **Supporting work:** such as suspended systems for ceilings and vertical stud systems for partitions, usually of light metal members, designed to be self-supporting and to carry finishes, lathing, or backings.

As before, these classifications are not entirely definite, and some *finish work* might be placed in more than one classification. But by classifying the *work* in this way it is possible to see common facts and similarities and to discuss the pricing of a great number of different kinds of *work* without referring to each one separately.

Material costs and *labor costs* of finishes are affected by the *quantity of work*; the sizes and shapes of rooms and areas; the amount of cutting required; and the amount of trimming or other labor done at edges, joints, and corners; consequently, changes in plane increase the costs. The type and quality of the surfaces or *supporting work* affect costs, because shimming or other treatment or adjustment of the base surface may be necessary. Hoisting and handling materials create additional *labor costs,* and restricted working space may reduce productivity.

Drywall work is usually done by carpenters, who install the boards and panels to the *supporting work.* Some *work,* such as acoustic ceilings, is often done by other specialist trades. The *material costs* include the boards or panels and the fasteners. *Waste* may be a significant cost, depending on sizes of boards or panels and room dimensions. Trim should be measured and

[12] The American Institute of Architects' *standard form of contract* (AIA Document A201), *"General Conditions of the Contract for Construction," Article 4, para. 4.8, Cash Allowances,* states that "these allowances shall cover net costs of the materials and equipment delivered and unloaded at the site, and all applicable taxes. The Contractor's handling costs on the site, labor, installation costs, overhead, profits and other expenses contemplated for the original allowance shall be included in the Contract Sum and not in the allowance."

priced separately. *Labor costs* involve cutting to size and installing the drywall material and installing trim at joints, edges, and corners. Sometimes, trim may not be used and labors such as fair cutting, scribing, and mitering may be required. These labors are expensive and should be measured and priced separately. In suspended acoustic ceilings, the panels are sometimes simply laid into the supporting system, and only cutting at perimeters is required.

Tile work is done by several trades, according to the type of material, whether it be ceramic tiles, precast slabs and blocks, wood and composition blocks, resilient tiles, acoustic tiles, or others. The *material costs* include the units and the adhesive, or mortar. *Waste* is not usually a major cost unless areas are small or irregular or unless the units are easily broken. Waterproof adhesives are expensive (about $8.00 per gallon), but the cost of ordinary adhesives represents only a few cents a square foot.

Labor costs of tiling are increased by patterns and mixed colors and sizes. Exposed edges requiring fair cutting are expensive, especially with hard materials. Tiling installed with traditional mortar is usually more expensive than that installed with thinner applications of fast-setting adhesives because of the longer time required and because preparatory beds and backings are often required for "mud" (mortar) setting. With thin-set mortars and latex adhesives, the base surfaces must be true and free from irregularities, otherwise some additional *preparatory work* may be necessary. Beds and backings for tiling should be measured and priced separately.

Sheet work is most common in floor coverings, such as linoleum and vinyl plastic sheet goods; but, there are pricing similarities with sheet wall coverings such as vinyl fabrics and wallpapers. Cutting and *waste* occurs at edges, and at joints and seams, particularly with patterned goods for which extra cutting is necessary to match patterns at seams. The specific locations of seams may also create additional cutting and *waste*, particularly with a floor covering in which seams should not be placed across door openings but parallel to lines of traffic to avoid transverse wear at seams. Good trade practices usually prohibit the use of narrow pieces of trade goods, and this restriction creates more *waste* of material and additional labor in setting out and cutting.

Wallpapers may be put in this classification for pricing purposes. Papers are usually sold in rolls, or double rolls, containing 36 and 72 square feet, respectively. The *waste* is usually quite high, from 15 to 20 percent or more with some special types and patterns. *Labor costs* generally increase as the price and quality of the material increases.

Wet finish work is fundamentally different from the preceding three classifications. The materials are mixed with water to produce a wet and plastic mixture (like mortar), which is applied to the base surface in one or more coats. Usually, more than one coat is necessary for soffits and walls, because the finish coat is different from the others, and because of the difficulty of applying and curing an adequate thickness of plastic material in one coat. Cement toppings, mastic flooring, and terrazzo are often laid on floors in one application.

Material costs for *wet finish work* include the dry ingredients and the mixing agent plus any special additives for color, hardness, and other qualities. *Waste* is not usually high for applications to floors, but it is high with other applications, in which a large amount of material usually is dropped or spilled during application.

Equipment costs for mixers, sprays, and scaffolding of various kinds are usually necessary for this *work*, and they should be included.

Labor costs for *wet finish work* include the costs of erecting and removing scaffolding, mixing the materials, applying the plastic mixture to the base surfaces, curing the applied *work*, and cleaning up. Patching after other trades is also usually necessary. Applications made in several coats with curing periods between applications increase *labor costs*, especially if the *work* is of such quantity or arrangement that it is necessary for workmen to leave the site for periods of several days between applications. Quantity is particularly important to costs, and the most economical job is one containing enough *work* to enable it to be scheduled in a continous sequence such as in a multistory building.

Coating work is typified by painting, but also includes other thin, wet, plastic coatings applied by sprays and other means. The materials are usually pre-mixed and delivered in cans or drums ready for application, although some site mixing may be required. Different materials are often used for undercoats and finish coats. The *material costs*, including consumable items such as brushes and rollers and mixing and cleaning agents, are usually much lower than the *labor costs*. A painter may apply 2 to 3 gallons of paint in a day with, say, $20.00 worth of materials and $60.00 for labor.

Labor costs of *coating work* vary according to:

1. **the material applied:** heavy materials like enamel require more time than thin materials such as stain and varnish

2. **the type of surface being coated:** rough and absorbent surfaces require more time than smooth, impervious surfaces

3. **the means of application:** spray equipment covers an area faster than brush or roller

4. **the number of coats:** the later coats go on faster than the first coat

5. **the area and planes to be coated:** large flat areas require less time than small broken areas, multi-plane surfaces, and narrow widths: and walls are easier to coat than ceilings

6. **the amount of "cutting in" required:** between contiguous coatings of different colors or types and around openings, and the like

7. **the number of colors and types of coating applied:** as in the preceding item, and, in addition, the resultant interruptions and additional cleaning of brushes and equipment required.

Labor costs are incurred by handling scaffolding and other equipment to provide working access, and *equipment costs* must also be included.

Beds and wet backing work can be compared to *wet finish work* for purposes of price analysis. It includes items such as cement plaster and toppings to receive other *finish work,* such as *tile and sheet work.*

Lathing and dry backing work can be compared to *drywall work* when using gypsum board lath and gypsum backing boards to receive *finish work,* except that both the *material costs* and the *labor costs* of lathing and backings are less than for drywall, because there is no need for any special joint treatment and because gypsum lath and backing board does not cost as much as gypsum wallboard. Also, there is less *waste,* because the location and number of joints is less important in *concealed work.* In addition to metal beads and other trim, 2″ × 2″ cornerite lath is installed at internal corners, and strip lath, 3 and 4 inches wide, is installed at heads of openings and at certain joints and changes in base surfaces. It is better to measure and price all such *work* separately unless the amount involved is very small.

Supporting work done by carpenters, lathers, and other finishing trades to carry finishes is similar to framing carpentry in that it is made up of light members (such as metal tee bars and steel studs), erected in supporting systems. The linear members, together with wire hangers, tie wire, fasteners, and other accessories, may be priced piecemeal as *run* and *number* items. Or the *material costs* of a specific supporting system (with members at specified spacings) may be analyzed and priced as a *super* item. Similarly, *labor costs* may be related to the area of the specific system or to the amount of material erected. For simplicity in measurement and pricing, the former is preferable. Scaffolding and other *equipment costs* should also be included.

Other costs of *finish work* may arise from inspection and testing requirements, special warranties, provision of samples, extra materials for *owner's* maintenance, and patching and touching up after other trades have finished their *work.*

In summary, the pricing of finishes, although materials and methods vary widely, does permit some generalization about the precepts and methods of measuring and pricing. *Unit prices* and costs of finishes are affected by:

1. *Work* to floors
2. *Work* to walls
3. *Work* to soffits
4. Area sizes and shapes
5. Repetition of similar areas
6. Perimeter/area ratios
7. Total quantities
8. Changes in plane
9. Number of corners
10. Number of openings
11. Quality of finish
12. Sizes of boards, panels, sheets
13. Installation methods
14. Number of joints
15. Location of joints
16. Treatment of joints
17. Treatment of edges and perimeters
18. Treatment of corners
19. Use and type of trim
20. Type of base surface and adhesive.

These factors, when applicable, should be reflected in the descriptions of the *items of work* and in the quantities and methods of measurement employed in the estimate so that they can be taken into account in pricing the *work.*

Pricing Specialties and Similar Work (Divisions 10, 11, 12, 13, and 14)

These *divisions of work* were mentioned in Chapter 8, Measuring Work: Particular, with regard to their measurement and description in estimates so that *items of work* can be priced.

Because *work* in these *divisions* is, generally, highly specialized and performed by specialist firms, pricing the *items of work* is not usually a general problem. It is usually done by a specialist firm whose estimator is experienced in that kind of *work*. However, there are two areas in which problems do frequently arise for the *general contractor* and his estimator. These areas include (1) items from suppliers installed by the *general contractor* (for example, metal storage shelving), and (2) co-ordinating *subcontractor's work* in these *divisions* with the other *work* on the job site (for example, special construction in rooms such as studios, radiation centers, and laboratories).

Pricing the installation of items supplied to the job site can be a problem, as was explained before with respect to finish hardware. So often the estimator does not have enough information, because the specific items have not been selected by the *designer* and they are to be supplied under a cash allowance in the contract, or the estimator has no experience in dealing with the items because they are uncommon. It is simply a matter of a risk to be calculated on the basis of the information available.

Similarly, in the matter of co-ordinating *work*, it is a question of information and risk in dealing with uncommon items. All major buildings contain elevators, and all *contractors* will have experience in dealing with this *subcontractor's work*, so the problems should be few. However, food service equipment and radiation protection may be a different matter. First, the general estimator should obtain all the information available, and he should try to learn as much about the *special work* as he can. This will possibly mean pressing the *designer* to obtain and pass on all the information he can get from the specialist firm. (The *designer* has not necessarily included all the information he has in the specifications.) Second, the estimator must carefully examine the related specifications and the *sub-bids* submitted for the *work* to ensure that all the *work* required by the contract is included and properly and clearly allocated to the *contractor* and his *subcontractors*. Scheduling is particularly important; and if the *contractor* has a contract containing an article on liquidated damages, he should try to minimize his risk by extending a similar contractual obligation to his *subcontractors*.

Above all, it is a question of information and communication—and scheduling. Which may lead us to consider the possibilities of reducing risks attributable to a lack of information by doing the *work* within some contractual arrangement other than a stipulated sum contract.

Pricing Mechanical (Division 15)

In some respects, estimating the *costs of mechanical work* appears to be more straightforward than other estimating, whereas in other respects it appears to be more difficult. It is generally easier to price if certain precepts and methods are followed, and often more difficult to measure. There are several publications written solely for the mechanical estimator, probably because of the distinctive nature of *mechanical work*, and of mechanical systems in buildings. Nevertheless, because of the nature of *mechanical work*, mechanical estimating (if the term can be used without causing confusion) can be done more systematically. For although each mechanical system in a building is unique in its own arrangement and scope, it is still made up of numerous standard manufactured components such as pipes, fittings, valves, fixtures, and mechanical equipment.

The *labor costs* of installing mechanical systems are always more or less subject to site and job conditions, the same as all *labor costs*. But, at the same time, they are generally more directly related to the materials installed and somewhat less to site and job conditions than *labor costs* in many other kinds of *construction work*. Joining one pipe length to another by means of a pipe fitting is a distinct operation whichever mechanical system it is in, and it is this relationship between materials and labor in *mechanical work* that makes it easier to rationalize and systematize the estimating of the *costs of mechanical work*. Each pipe fitting requires two or three or four joints, depending on its type. And each type of pipe requires a certain type of joint and fitting. Consequently, mechanical estimating largely depends on a detailed take-off of all the materials: the pipes, the fittings, the valves, the fixtures, and the items of mechanical equipment; and the *labor costs* are directly related to the material items, particularly to the pipe fittings (and their joints) and to the fixtures and equipment (and the connections to them).

Published "labor units" are available to mechanical estimators,[13] based on a systematic approach to

[13] The *Labor Calculator* published by the National Association of Plumbing, Heating, Cooling Contractors, Washington, D.C. 20036. This useful publication uses the term **"labor units"** in reference to the data it contains concerning man-hours required for a wide variety of items of *mechanical work*. Some other publications use other terms, such as "labor factors" and "labor rates." There is no uniform terminology in this field.

mechanical estimating, together with guidelines for establishing efficiency factors related to specific jobs and specific mechanical firms through *cost accounting.* In other words, there is a published set of data for labor productivity with an explanation of how to adjust and utilize them with efficiency factors for *mechanical work.*

Earlier in this chapter, reference was made to the use of a datum in pricing. The use of published *unit prices* as a datum and as a basis for bidding and contracting is not something new. Published *schedules of unit prices for construction work* are used by some government departments and private companies,[14] and bidders can make offers by quoting the percentage additions or deductions required by the bidder to be made to or from the *unit prices* in a prescribed schedule, leading to *unit price contracts* in which the *work* done is measured and paid for at the adjusted *unit prices.* To make an estimate and a *bid* for *work* to be done on this basis, an estimator must be able to compare the *published unit prices* (the datum) with *actual unit prices* (established for his company by *cost accounting*) to arrive at a factor for different items, or, more practically, for different sections or systems of the *work.* A labor unit is also a datum; but whereas a price datum applies to the *unit prices* of a group of related *items of work,* a labor unit applies to the labor time (in man-hours) required for a specific *item of work.* Adjustments to labor units (the data) are made, as required, by "efficiency factors" calculated for specific projects, or for specific parts of projects such as mechanical systems or sub-systems.

A series of labor units for screwed steel water piping with malleable iron fittings might be as follows:

LABOR UNITS X EFFICIENCY FACTORS FOR SCREWED STEEL PIPING

	Efficiency Factor										
Pipe size	0.5	0.6	0.7	0.8	0.9	1.0	1.1	1.2	1.3	1.4	1.5
in.	*Labor Unit (man-hr per Ell-fitting)*										
1/2	0.30	0.36	0.42	0.48	0.54	0.60	0.66	0.72	0.78	0.84	0.90
3/4	0.35	0.42	0.49	0.56	0.63	0.70	0.77	0.84	0.91	0.98	1.05
1	0.40	0.48	0.56	0.64	0.72	0.80	0.88	0.96	1.04	1.12	1.20

14 Such as those published by Richardson Engineering Services, Inc., 10021 Tecum—P.O. Box 726, Downey, California 90241. See the Bibliography—*Estimating and Cost Accounting.*

For ease of calculating, the above table (shown in part) shows a practical range of efficiency factors and the products of the labor units and the efficiency factors.

The *item of work* in the table includes cutting and threading two pipe ends and fixing the pipe fitting. The tabulated labor units should be based on wide, actual experience so that average productivity (or the mean) is about equal to that indicated by the labor unit times the "efficiency factor" of 1. That is to say, according to the table, a 1-in. diameter pipe fitting with two pipe joints, on the average should normally require 0.80 man-hour to install. Under certain circumstances it might conceivably require up to 1.20 man-hours, or as little 0.40 man-hour. The immediate, usual reaction to many published labor units is one of surprise at their magnitude, and in some instances the published figures are undoubtedly higher than some practical experience would indicate. On the other hand, most forget (or do not know) that a large part of a workman's time is spent in doing things other than actually working at his trade, as explained in Chapter 5.

Establishing an efficiency factor is critical in using a labor unit, and it is suggested that this can only be done, in the first instance, by considering those things that mostly affect labor productivity; namely,

1. **Labor availability and skill**
2. **Supervision availability and skill**
3. **Type of mechanical work**
4. **Design and inspection of work**
5. **Job organization**
6. **Site and work conditions**
7. **Delays, deliveries, and details**
8. **Season and weather**
9. **Market supply and demand**
10. **Estimating data.**

These are examined in more detail, since each item has several aspects. In fact, it would be possible to further subdivide each of the above ten items, and so on, ad infinitum; but it would not be practically possible for an estimator to effectively rate the influence on labor productivity of more finely divided factors without the help of a computer and a specially designed program and much more information.

Labor availability and skill depend, in part, on the amount of *work* at hand and the project's location. If the project is far from town, it may be necessary to hire local labor. If the mechanical contracting company is busy, all its regular men may be employed on other jobs and new and unknown workmen may have to be hired, with a possibly lower level of productivity, and certainly with a greater risk.

Supervision availability and skill are affected in the same way as labor availability and skill. The success of a job is often decided by the quality of the site supervision, and individual supervisors might be rated for this purpose.

Different types of mechanical work may have different labor efficiency factors, and an assessment should be made for each major system in a large project in terms of the probable labor productivity in installing that system. The size of the contract may also be considered here, since small jobs are often less productive that larger jobs.

Design and inspection of work (by the *designer* or his engineering consultant) often has an effect on labor productivity. The design includes not only the drawings and their effectiveness and practicability, but also the specifications and the extent to which these may or may not be observed. The effect of bylaws and regulations and their enforcement by officials should also be considered.

Site and work conditions include many factors that affect productivity. The organization of the job as a whole, and the mechanical part in particular, should be considered. The *mechanical contractor's work* is affected by the *general contractor*, by the other *subcontractors,* and by the way in which the various parts of the job are put together. Access, storage and working spaces, hoisting facilities, and facilities for workmen are some of the things to be considered under site conditions and their effect on labor productivity.

The deliveries of materials and equipment, and the timely receipt of design details and additional instructions from the *designer,* are critical to the good progress of the *work,* and delays may result in lower productivity.

The season and the weather affect some types of *mechanical work* more than others, and for this reason it is necessary to review each mechanical system separately. *Underground work* outside the building may be seriously affected, whereas interior plumbing may remain unaffected by the weather.

The market's supply and demand for construction in an area will have a direct effect on labor productivity and on some of the preceding factors, such as availability of labor and supervision.

Estimating data varies in availability and in reliability, and for the more standard types of mechanical systems more valid data is usually available. An estimate is as good as the data on which it is based, and this last item enables the estimator to take some account of the quality of the data he is using.

The estimator's assessment for each item is listed and the average factor is calculated; for example,

EFFICIENCY FACTOR RATING—PLUMBING SYSTEM

	Item	Factor
1.	Labor availability	1.1
2.	Supervision availability	0.8
3.	Type of *work*	1.1
4.	Design and inspection	1.2
5.	Job organization	0.9
6.	Site and *work* conditions	1.0
7.	Delays, deliveries, details	0.9
8.	Season and weather	1.2
9.	Market supply and demand	1.1
10.	Estimating data	1.0
	Total	10.3
	Average factor	1.03

If the estimator has no opinion on any item above he rates that item 1.0 (average).

Once labor units are established and used with efficiency factors, established as above, it is essential that the validity of the factors be confirmed or revised by continually comparing *estimated labor costs* with *actual labor costs* through *cost accounting.*

Pricing Electrical (Division 16)

Almost everything that has been said about Mechanical can be said about Electrical, because similar conditions exist in *mechanical* and *electrical work* and estimating, and the same measurement and pricing methods are applicable. Like mechanical, *electrical work* can be broken down into systems; and the basis of an estimate is a schedule of materials to which labor units can be related and for which efficiency factors can be assessed, as explained before, for the several *systems of work* in the project.

In both mechanical and electrical estimating, the measurement (take-off) must be done by an estimator experienced in the design and installation of *systems of work.* Then the material items can be priced from published lists and the labor times for items can be calculated from published labor units by an assistant. When this has been done, the senior estimator can make ready the estimate for completion by applying efficiency factors to the total labor times (in man-hours) for each system, or sub-system, and by applying *labor rates* to the totals of man-hours adjusted by factors to obtain *labor costs.* Similarly, the senior estimator can prepare the *material costs* for calculation by applying quoted discounts to the totals of the list prices for the systems measured in the estimate.

An Estimating Method

This method of estimating, which uses published price lists, quoted discounts, scheduled data, and labor units and efficiency factors to adjust varying price and productivity levels, has several advantages for bidders and their estimators.

First, it requires that data obtained by *cost accounting* be systematically scheduled, which produces several benefits. Next, it enables junior estimators to do useful and instructive work under the supervision of a senior estimator. At the same time, it relieves the senior estimator of much of the routine labor of estimating, which enables him to handle more estimates and to apply his time and experience where it can be most effective. Finally, the adoption of such a method of estimating encourages the standardization of terminology and estimating methods through the publication of schedules of data on labor productivity by trade associations. And this development, in turn, encourages the general rationalization of estimating and the collection of data.

Such an estimating method could be adopted by the members of any trade association for their mutual benefit through better estimating and through the pooling of data without any detrimental disclosures or loss of autonomy. It can be adopted by individual firms and operated independently, but there is much more advantage to be gained in pooling certain information.

Questions & Topics for Discussion

1. Identify and explain the application of two different factors (stated as percentages) required in analyzing the time for excavating a basement with a piece of equipment.

2. In considering economies to be made in the construction of a reinforced concrete framed building, which specific *costs of the work* offer the greatest possibilities for reductions and why?

3. If a cast-in-place concrete structure cost an average of $100.00 per cubic yard of reinforced concrete, formed and placed, indicate the approximate and proportionate costs (per cubic yard) of the concrete, the forms, and the steel rebar, showing the *labor costs* of each separate from the other costs.

4. Show by an example the analysis of a *unit price* for a 6-in.-thick conc blk ptn wall, assuming the conc blks cost $0.31½ each; the cmt mtr, $18.00 per CY; and the mason, $8.10 per hour.

5. Describe briefly two ways of measuring and pricing structural steelwork, and explain why one should be more accurate than the other.

6. Analyze *unit prices* for framed floor joists, sizes 2 × 6 and 2 × 8, for comparison, and explain if and why one should cost more than the other.

7. In accounting for and recording the costs of built-up roofing for future estimating, state the major job conditions you would want to record to explain and qualify your cost data.

8. Sketch a two-light aluminum window with an opening vent, showing the several dimensions (of your choice), and measure the various items that you as an estimator with a window manufacturer would have to price in detail.

9. Describe the major considerations in estimating the *costs of finish work* generally classified herein as *tile work.*

10. Describe in detail how the costs of *mechanical work* can be estimated by use of published data on prices and productivity.

12

Cost Accounting Practices

The purposes and precepts of *cost accounting* (CA) were explained in Chapter 6 together with the purposes and precepts of estimating, because we believe that *cost accounting and estimating* cannot be properly understood unless they are seen as two parts of one function—construction management.

Both information and experience are essential to all management, and reports indicate that a lack of these factors is a major factor in the inefficiencies and business failures in the construction industry.[1] The estimator must have information and the experience of past *construction work* in order to estimate the probabilities and the costs of future *work*. The construction superintendent must be able to oversee the entire project to bring it to successful completion, and at the same time, he must know the details of the *work* and its costs as they were estimated and as they are actually incurred.

Construction is essentially a management function, rather than one of investment by the *contractor*, whose investment in the *work* is usually

[1] In the study on manpower utilization, referred to in Chapters 3 and 5, and in the annual reports of business failures by Dun and Bradstreet, Ltd., referred to in Chapter 5.

relatively small and temporary. The so-called management contracts, discussed earlier, have come about as a result of the recognition of general construction as primarily a management function. This function has become more apparent as building construction has become more complex and as proper management has become more necessary. The demand and the need for good management has been the deathly angel for many *general contractors.* Many could not provide it, because they did not know how to provide it. So many construction jobs in the past have not gone well because they lacked good management.

A *designer* is not usually hired by an *owner* to continuously supervise the *work,* and the *general contractor* often saw himself simply as the *contractor* for the site works and the structure; that is, simply, as the first one of a group of *specialist contractors* who were each responsible for doing parts of the *work.* He often did not see himself primarily and properly as the general manager, but only as *a contractor* who, in doing the part of the *work* that he did, was therefore required to see that the other parts of the *work* fitted together. Of course, this was not true of all *general contractors*; but it was true of many, and it still is true of some.

Now, there are management companies that were *general contractors,* that no longer have plant and equipment and a labor force but have retained an expert and experienced staff of foremen, supervisors, estimators, managers, and engineers. Many no longer use the title *contractor* because it does not describe their new function as contract managers. Their assets are management expertise, experience, and information rather than capital, plant, and equipment.

Data are essential as a basis for inference, for estimating probabilities and for making decisions. Information is the stuff of decision-making and of management, and it must be systematically obtained and analyzed and put into usable forms. *Contractors* often have at hand large amounts of information that is in a raw and unusable form because they have no systematic process for handling and analyzing it, and much of it is acquired incidentally and without deliberation through other systems such as the payroll.

It is not necessary to use a computerized estimating —*cost accounting*—management system, as described in Chapter 3, to obtain and analyze data. But it is necessary to have a systematic process; and then a subsequent change to electronic data processing (EDP), when it becomes economical and desirable, should not be too difficult. The greatest difficulties with CA are usually at the outset and not with the CA system but rather with people.

CA is not usually a popular subject, and for everyone in a company who realizes its value there will probably be several others who will resist it. Some, because they are suspicious that it is a waste of time and money, and that it will only lead to more paper work and larger overhead. Others, because they do not want the additional responsibility and still others who would prefer not to have someone know as much as they do about the costs of the jobs that they are supervising.

Some of these attitudes are justified, at least at the outset. Some CA is not worth the expense, because data is obtained and is never used; and no doubt CA has been made an end in itself, in some companies. As to the matter of responsibility, it should be understood that CA is an integral part of construction management and supervision, and it is, therefore, part of the normal duties of supervisory staff. Finally, no competent foreman or supervisor need fear anything from CA, because it is an objective extension of the estimating process, primarily to report on an estimate's validity and to provide data for estimating and management. Perhaps the main reasons for the opposition to CA come from a lack of understanding and knowledge, which is understandable when one sees the dearth of courses and training for foreman and supervisors, particularly in estimating and CA.

Codes for CA

First, it is necessary to have a code for CA whereby *items of work* are represented by a series of symbols; and at this point it would be worthwhile to review the meaning of the term *item of work.* A code is necessary for several reasons, but primarily, for brevity, accuracy, and for better communication.

Brevity is obtained by using, say, up to six numerals or letters to represent the description of an *item of work* that would otherwise require several or many words. Accuracy and better communication are both obtained by standardizing and codifying the descriptions of items. Coding makes it necessary to first carefully define and describe the items if a code is to have any value. Once coded, the descriptions are crystallized and fixed within the code so that future misinterpretations and misunderstandings are less likely than with original and loose verbal descriptions.

For example, if the code for "labor placing concrete in footings" is 0303, that numerical symbol will always mean that same *item of work* in the same CA system. A verbal description, on the other hand, would be much longer and could vary slightly with each writer, so that it might not always be clear whether the item included "labor," or "material," or both. Or, the location "in footings" might be omitted, or "footings" and "foundation walls" might be combined in one item.

Many agree that six numerals or letters are a practical maximum to represent an *item of work,* and some say that four are enough. It depends on the structure of the code. Four numerals provide for almost 10,000 items (0001 to 9999), which, theoretically, is more than enough. But the *divisions* and *sections of work* and the need for flexibility may preclude the use of a large portion of the consecutive numbers from, say, 1 to 9999.

Several books containing cost codes have been published. For example, one uses letters and numbers:

C-40. All labor costs for handling materials, placing and removing runways, mixing, hoisting, placing and consolidating concrete for reinforced concrete construction, or as follows:

C-41. Placing concrete for columns

and so on.[2]

Another published code uses only numbers. For example,

5-12. Column Concrete: together with a list of third-number classifications to be added:

1. Freight	38. Placing
2. Cartage	39. Scaffolding
3. Unloading	40. Laying out[3]. . .etc.

In this last code, "labor placing concrete for columns" would be "**5–12–38.**" This entry requires seven numerals and signs, which are too many in the opinion of some. Not only is the number of numerals or letters important to the person writing them down at the site, but it is also important in the use of the computer hardware, which has a limited number of spaces in the width of the paper used for its printed reports.

Another cost code is published by the Construction Specification Institute within the Uniform Construction Index, referred to previously. This code is based on the *sixteen standard divisions of construction work.* Each item requires two digits (from **01** to **16**) to identify the *standard division,* such as (**01**) General Requirements, (**02**) Site Works, (**03**) Concrete,. . . through to (**16**) Electrical. A further breakdown of items requires at least two or three more digits, and to obtain proper identification, a total of seven or more digits is probably required, which again may be too many. The Uniform Construction Index endeavors to embrace all kinds of *construction work,* and consequently it has to use many numbers initially in identifying the *divisions* and broad-scope classifications, such as "cast in place concrete," and "concrete unit masonry." Most contracting firms need to codify a few hundred items in a limited scope of operations, not hundreds and thousands of items over a very wide scope of *work.* Most *contractors* need a code, for example, that will enable a distinction to be made between "placing concrete in foundations," and "placing concrete in columns;" not between "cast in place concrete" and "library equipment." They require a code of limited scope, and one of some detail. Or, more precisely, they require a code that can develop the amount of detail that at times may be required.

All construction jobs are in some way unique, and there are many differences among most jobs. One job may contain 100 cubic yards of cast-in-place concrete below ground level, and none above ground. Another job may have 3,000 cubic yards of concrete between the footings and the fifteenth floor. The "concrete placing" costs on the one job may be less than $400, and on the other job more than $12,000. Can the same cost code be used on both jobs? The answer is, yes, if the code has been designed for such flexibility.

There are several ways in which flexibility can be built into a cost code. Numbers that are multiples of 10 can be used to codify the headings of groups of similar items, as in the first example above, in which C-40 represents "labor handling and placing reinforced concrete"; and "labor handling and placing reinforced concrete in columns" (slabs and beams, and so on), is represented by C-41, (C-42, etc.). In this way, the ten-multiple code number (C-40) can be used for placing all reinforced concrete on small jobs, whereas some or all of the subsidary numbers (C-41 to C-49) can be used on larger jobs to segregate the costs of placing the various kinds of reinforced concrete.

Another aspect of flexibility in cost codes is the difference between what might be called the *basic* and the *particular items* of a code. The *basic items* are those that form the framework of the code, including the group headings and subheadings, and the items that are to be found in almost all jobs, such as "labor handling and placing concrete in footings" A *particular item* in the *concrete division of work* might be "labor in heating concrete aggregates for cold weather concrete work."

Each cost code should be specially formulated for a construction company to reflect its work activities and *basic items of work,* none of which will be redundant. Therefore, the original code should contain a minimum number of items. As estimates are made, the estimators will come across items that are not among the *basic items* of the code. Each of these *particular*

[2] *The "Practical" Construction Cost Schedule* (Chicago: Frank R. Walker Co.)

[3] George E. Deatherage, P. E., *Construction Office Administration* (New York: McGraw-Hill Book Company, 1964), pp. 209, 225.

items is then given an appropriate code number, according to its type and the heading under which it should be placed in the code; and that will be the code number for that item in that job. If a particular item continually appears, it should be made a *basic item* and given a permanent number in the code. Similarly, *basic items* that prove to be rarely used should be dropped from the code to become *particular items* if and when they are needed.

No cost code can be completely comprehensive, and attempts to produce and use a standard, rigid, and very detailed code are usually doomed. Such a code will probably be too long and too inflexible, and it will eventually prove to be incomplete. Better to create an organic code with small beginnings, and with a framework that has the potential for growth, so that with each job the code can grow to suit the *work* for which it is intended. This means that the estimator must not only use the cost code when estimating, but that he must also work with it in creating code numbers for *particular items,* which means that the code cannot be a petrified thing but must be something living and malleable. It also means that the site personnel, who are allocating workmen's time to *items of work* each day, must refer to a copy of the estimate to discover the code numbers for the *particular items of work* of that job. This procedure, it follows, helps to overcome the general inclination to ignore the estimate and to codify items according to individual memory and opinion.

Reporting for CA

There is only one person who can do the original reporting for the distribution of *labor costs* to *items of work,* and that is the foreman directing the labor. On a small job, it may all be done by the general foreman, or by the superintendent; but on larger jobs it must be done by the foremen, because it is they who know best what the workmen are doing. If the information is recorded by the timekeeper, it is still the foremen's responsibility to distribute the workmen's time to *items of work,* because they direct the workmen to their various tasks in the first place. This reporting should require not more than 15 to 20 minutes a day of a foreman's time, for what should be one of his primary responsibilities.

There are many printed forms designed for reporting labor distribution to *items of work,* and samples of the headings and layouts of several forms are shown and explained on page 242.

Daily reports have the value of immediacy, and probably greater accuracy, in that the foreman does not encounter a delay in completing the distribution of time if reports have to be submitted each day. Daily reports with a separate form for each *item of work* are probably most effective on large jobs wherein each foreman should not have more than two or three to complete, containing up to a maximum total of, say, twenty individual entries daily. If the foreman is in charge of ten men, and if each man works on two different items in one day, that will require twenty entries, which is not too many; and even forty entries should not take more than about 20 minutes for the foreman to enter.

Any number of variations on the daily report form is possible, according to the company's requirements and the kind of *work* done. There are two fundamental parts:

1. the daily checking and recording of each man's total time, and the entry of the *wage rate* and amount for payroll
2. the daily distribution of each man's total time to *items of work* (represented by code numbers) for CA.

These two items may be completed together on the same form or entered separately. In larger firms and on larger jobs, they are probably better done separately, if the total time is checked from one form to the other.

Information for both purposes may be entered on a Daily Time Card, designed especially for use with EDP, which is sent each day to the head office for data processing with the information entered as shown on page 243.

Many standard forms are published and sold for estimating, cost accounting, and bookkeeping by construction firms; and rather than reproduce them here it is suggested that publishers' catalogues containing illustrations of the forms and explanations of their uses be consulted. One of these publishers is the Frank R. Walker Company of Chicago,[4] which also publishes *Practical Accounting and Cost Keeping for Contractors,* an inexpensive book that explains these subjects and the use of the Walker's forms in some detail.

Equipment time and costs should be distributed to *items of work* when possible, and this can be done by the daily reporting of equipment hours and their distribution in the same way as for labor. "Idle time" should be distributed to *items of work* pro rata to the distributed "working time." Equipment and plant such as hoists and tower cranes, which have many uses,

[4] Publishers of standard forms and complete bookkeeping systems for contractors, architects, and engineers, and of several books on estimating. See the Bibliography.

HEADINGS AND LAYOUTS OF SEVERAL FORMS USED IN REPORTING FOR COST ACCOUNTING

DAILY LABOR DISTRIBUTION REPORT

Job name/number: *Date:* *Sheet number:*

Employees Name/No.	Time In	Time Out	Hours	O/T Hours	Rate $	Amount $	LABOR DISTRIBUTION			
								(*code numbers*)		
(*1*)	(*2*)	(*3*)	(*4*)	(*5*)	(*6*)	(*7*)	(*8*)	(*9*)	(*10*)	(Etc.)

(*1*) Some use only the name, or only a number; but a combination of both is better. (*2*) Starting time. (*3*) Finishing time. (*4*) Hours worked, up to maximum of a standard day, depending on the labor agreement. (*5*) Overtime hours worked, entered on a separate line to permit entry of a different rate, and its extension. (*6*) and (*7*) These entries may be by a timekeeper, or clerk. (*8*) (*9*) (*10*), etc., are for entries to distribute total time in (*4*) and (*5*) to different *items of work,* represented by code numbers entered above the columns.

DAILY LABOR DISTRIBUTION REPORT

Job name/number: *Date:* *Sheet number:*

Employees Name/No.	Rate $	Hours									Total Hours
		1	2	3	4	5	6	7	8	O/T	
(*1*)	(*2*)	(*3*)								(*4*)	(*5*)

(*1*) Some use only the name, or only a number; but a combination of both is better. (*2*) May be entered by timekeeper, or clerk. (*3*) Each hour (up to 8 hours) of each day is distributed to *items of work* by entering code numbers in columns, in spaces below. (*4*) Overtime is entered here. (*5*) Total hours are entered as a check. This form is separate from any payroll functions.

LABOR DISTRIBUTION REPORT

Job name/number: *Date: from to* *Report/Sheet number*

Work Description	Code No	Hours	O/T Hours	Rate $	Amount $	Total Cost this period	Total Cost to date	Estimated Total Cost	Difference (+ or −)
(*1*)	(*2*)	(*3*)	(*4*)	(*5*)	(*6*)	(*7*)	(*8*)	(*9*)	(*10*)

(*1*) Brief verbal description. (*2*) Matching code number. (*3*) Hours worked, up to maximum for a standard day. This may show "total hours" for the indicated period; or, it could contain, say, five columns for daily entries in a week. (*4*) Overtime, entered on a separate line as before. (*5*) to (*9*) These entries may be made by a timekeeper or payroll clerk. (*10*) Shows difference between (*8*) and (*9*) to indicate state of job costs.

DAILY LABOR REPORT

Job name/number: *Date:* *Code number:*

Employees Name/No	Work Description	Hours	O/T Hours	Rate $	Amount $
(*1*)	(*2*)	(*3*)	(*4*)	(*5*)	(*6*)

(*1*) Some use only the name, or only a number; but a combination of both is better. (*2*) Description should match the code number at top. **A separate Daily Report** is made for ***each item of work*** by the foremen. (*3*) Hours worked, up to maximum of standard day. (*4*) Overtime, entered on a separate line, as before. (*5*) and (*6*) These entries may be made by timekeeper, or clerk.

DAILY TIME CARD

Man No. and Name *Date*

Job No.	Job Cost Code	Hours	Rate $	Class
3004	1008	4	6.00	*
3004	1107	4	6.00	*
3004	9089	1	6.00	*TT*[(a)]
3004	1107	1	9.00	*OT*[(b)]
(*1*)	(*2*)	(*3*)	(*4*)	(*5*)

(*1*) Identifies the job, by number. (*2*) Identifies the *item of work*, and other costs such as "travel time". (*3*) Indicates hours distributed. (*4*) Hourly *wage rates*. (*5*) Indicates the times and rates for (*a*) travel time, (*b*) overtime, etc., in (*3*) and (*4*). This example shows only the information that is entered at the site. The extensions and labor distributions done by EDP to produce the printed payroll and cost reports are not shown.

and which, therefore, cannot always have their time distributed to different items, should nevertheless have their total idle time and their working time reported daily. An effort should be made to distribute the time to *types of work*, if not to specific *items of work*; and when a crane is used for such items as placing forms, steel rebar, and concrete the time can and should be segregated and identified. If any information can be obtained without too much effort, the effort should be made and the information obtained.

Materials should be charged to *items of work* by means of the cost code, and every order, invoice, and delivery ticket should bear both a job name and number, and a code number, for the *item of work*. With many distinctive materials and components this is easily done. The problems arise with the ordinary materials, such as ready-mixed concrete and construction lumber, which are used in many *items of work* and are sometimes also used in *job overhead items*, such as temporary fences and barriers.

Ready-mixed concrete can usually be accounted for and its costs distributed to *items of work* because of the delivery tickets and the entries in Daily Reports regarding concrete pours. Delivery tickets should be marked with a job number and a code number, and Daily Reports should indicate the ticket numbers and the amounts and locations of concrete placed. If both are not done daily, it may be that extra yards of concrete required for *waste* or for *extra work* cannot be accurately accounted for.

Construction lumber is more difficult to accurately distribute to *items of work*. It may be on the site some time before it is used, and once used and installed, it is often soon concealed by other *work*. With this and similar materials, an estimator should try to include in the estimate all quantities of the material required, not forgetting the temporary uses of lumber for forms and supports, which should be made specific items in the estimate, identifiable by description and by code number. Further identification for distribution purposes can sometimes be made in terms of the species and grades of lumber; but it is not unusual for, say, Standard grade (No. 2 common) lumber, in 2×4 and 2×6 sizes, to be used for several different items of *permanent* and *temporary work*. It is also not unusual for 2000 or 3000 board feet of lumber to be unaccounted for on a job, because (1) it was not allowed for in the estimate; (2) it was used on the job without any attempt to record its use; (3) some was stolen, or borrowed for another job. These reasons should be obviated by (1) specific and measured quantities in the estimate for all types of *work*, including *temporary work* and *job overhead items*; (2) daily records on the site; (3) the use of proper records for all materials moving on and off site, and to and from stores and suppliers.

If there is no other provision made to account for the use of such materials, notes should be made in the Daily Report about the amount and use, particularly if the material is used for a *job overhead item* for which there may be costs allowed in the estimate but no specific quantities of materials indicated. If no effort is made to record the use of these basic materials and to allocate them to specific items, the looseness and lack of knowledge concerning them will be perpetuated, and the misplaced costs of several thousands of board feet of lumber will make attempts at accuracy elsewhere in the estimating and CA rather futile.

It is true that CA, like estimating, cannot be absolutely accurate, but this is not a valid argument against its practice and use. Practical CA is as feasible as practical estimating, if the precepts in Chapter 6 are followed.

Measuring for CA

Of those construction companies that do CA, many are not able to make full use of the cost information they obtain because it is incomplete, and above all, because they lack *interim measured quantities of work.* It is of little value to know the *labor costs* expended on an uncompleted *item of work* if the quantities of *work* done, and yet to be done, are not known. Without these quantities, the costs have little meaning.

It is a common practice simply to look at the *items of work* and to mentally estimate the amount, proportion, or percentage of *work* completed. But this procedure is unsatisfactory and often inaccurate for most types of *work,* and actual measurement is usually necessary to obtain accurate quantities.

The precepts of measurement in estimating and in CA are given in Chapter 6, and some practical advice on measurement in CA is also offered. Here, it is sufficient to stress the fact that CA without the *interim measurement of work* is, for the most part, valueless. Interim *labor costs* may be a by-product of the payroll. Similarly, interim *material costs* may be a by-product of other normal accounting routines. But a practical CA system must entail *interim measurements of work,* and it is at this point that some firms fail by not periodically recording the quantities of *work* done. Part of the reason for this failure is a reluctance to employ a person to do the measuring. On many jobs, the measurement can and should be done by foremen; but on other jobs, because of their size or complexity, the job may require a measurer or a surveyor.

As an illustration of practicability, we can refer to a multi-million dollar community complex consisting of many kinds of *work,* including reinforced concrete high-rise apartment buildings with underground parking and stores, for which all measurement and all segregation and allocation of *labor costs* and *material costs* for the job were done at the site by a young woman who had trained as a building technologist, and who reported directly to the project engineer. *Cost accounting* is possible, but a practical and realistic approach is essential.

Cost accounting undoubtedly creates more work, costs money, and must be done by technical staff. It is easy to cheat a CA system; it should objectively report bad news as well as good news, and it does not always produce immediate results. Site staff are, by nature and training, practical and physically active, and paper work does not sit well with them. Therefore, the amount of paperwork for CA must be kept to a minimum at the site. *Cost accounting* is itself the cause of an *overhead cost,* and it must be economical and simple. If it is expensive it cannot be justified, and if it complicated it will fail from lack of support by the site staff.

If necessary, foremen and supervisors who are unfamiliar with CA and its relationship to estimating and construction management should be educated in these subjects. Unless they appreciate the value of CA, some may believe that the allocation of costs to the wrong *item of work* is not serious, because (it may be argued) it is all part of the total job costs. A foreman may lend a hand at a particularly difficult *item of work* without charging any of his time to the item, instead charging all of his time to supervision. As a result, it may appear that the costs of the item involved have been over-estimated.

Ideally, most CA should be done by estimating staff, and the best way to train junior estimators is to have them measuring and allocating costs on the site, and calculating and recording *unit prices* from completed jobs in the office. Sometimes the title, *cost accounting,* misleads the executive and CA is wrongly made a responsibility of the accounts department. Because most accounting and clerical staff lack the necessary technical knowledge for *construction cost accounting*—**which is different from cost accounting in other businesses and industries**—the CA system may be a failure because the costs are not consistently related to measured quantities of the *items of work* done.

The need for estimating is immediate and obvious most of the time, whereas, the need for CA is not immediate and obvious and it may not produce immediate and obvious results. An estimate, and a subsequent *bid,* may produce a contract within a few weeks of starting the estimate, whereas CA at its best produces only information. Yet because estimating and CA are really parts of the whole construction management process, the difference between them is more apparent than real; and estimating without CA is not truly estimating, because "calculating the *costs of work* on the basis of probabilities" requires historical data and recorded experience.

Questions & Topics for Discussion

1. State three possible arguments against the institution of *cost accounting* in a construction company, and state three counter-arguments for its institution.

2. Present an argument for the use of *cost accounting* by a construction company that obtains most of the *work* it does by negotiating rather than by competitive bidding.

3. Who should allocate workmen's time to the *items of work* done for *cost accounting,* and why?

4. Compose a suitable cost code for *concrete work* by a *general contractor,* briefly indicating and explaining the main features of the code.

5. Explain briefly the differences between *basic items of work* and *particular items of work* as they are included in and affect construction specifications and cost codes.

6. Show the arrangement and column headings of a "daily labor distribution report" suitable for a construction firm doing a great variety of *work.*

7. Explain briefly the problems of accounting for framing lumber on a construction job and what should be done to make the accounting as accurate as possible.

8. What makes *cost accounting* in the construction industry unique, and what site activities are usually essential to it?

9. Discuss the main features and practices of *cost accounting* for *plant and equipment costs.*

10. In measuring conc ftgs for *cost accounting* it is found that, although the drawings show conc ftgs to be 24 in. wide, the actual widths are from 25 to 26 in. State which widths you would record to calculate the amounts of concrete placed, and explain why. Explain how you would determine and deal with the actual quantities of concrete placed.

13

Quantity Surveying

The title "quantity surveyor" requires some explanation because, although it is becoming more familiar, it is still not widely known and understood in North America. The two words in the title are common enough, but the word "surveyor" in this particular context causes some confusion. In North America, the title "surveyor" is generally understood to be that of a land surveyor who determines the boundaries and measurements of land, and plots them on plans and maps. But this is a limited meaning of the historical title.

"The name of a Surveiour is a French name and is as moche to saye in Englysche as an Overseer," it was explained in 1523 in a book of surveying.[1] The same book states that the surveyor should know *"what the walles, tymber, stone, lead, sclate, type or other of coverynges is worth"* and should have a knowledge of *"Castels and other buyldinges."*[2]

The office of Surveyor of the King's Works was responsible to the Crown for the construction and maintenance of royal castles and other buildings. In the eleventh century, when the office was first created, the

1 John Fitzherbert, *Book of Husbandry and Book of Surveying* (1523) (1539 ed.), f. 38 v.

2 *Ibid.*, f. 4.

great Domesday Survey[3] of the English realm was made, somewhat like a national census and assessment for taxation purposes, as well as a registration of property rights and titles. This great survey of all the realm, which was made at the command of King William the Conqueror, established and recorded by written description and sworn testimony all the existing property rights and titles, together with their extent and value, and their liabilities, in the great and famous *Domesday Book.*

Surveys of landed estates were made later and from time to time to reafirm and to re-establish rights and titles and to find out and to rectify any abuse, damage, or waste. These occasional surveys and the day-to-day management of the estates, including the erection and maintenance of buildings, appears to have been the earliest work and occupation of surveyors in England, and from this beginning there developed the various kinds of work now done by surveyors.

In the sixteenth century, the social upheaval and religious reformation caused radical changes in the traditional ownership of land and real property. More accurate land records became necessary, and the precise delineation of boundaries by surveyors' plans gradually replaced the verbal "metes and bounds" descriptions and sworn testimonies of earlier days. About the same time, chains and other measuring instruments were invented and developed, and land measurement again became a science—an ancient science which had been lost and that had to be rediscovered. The sixteenth century saw social changes that required new skills, and the seventeenth century saw those skills develop. By the end of the seventeenth century there existed in England the embryo of a "profession of the land" to deal with its management and development, and by the eighteenth century the several specialties of the surveying profession were well established if not united in one body.

In the eighteenth century the distinctions among architects, surveyors, and builders were not very clear. Designing buildings was the sole occupation and livelihood of only a few architects, as it was also the gentle art and pastime of many educated gentlemen who had visited Venice and Rome. In 1792 several architect-surveyors who managed urban property and buildings for the large mercantile City Companies of London established the Surveyors' Club within a few months of the establishment of the Architects' Club. The next 100 years witnessed the founding of several new institutions. The Land Surveyors' Club was formed in 1834, and in the same year the Royal Institute of British Architects was established with a membership exclusively concerned with building design, and specifically excluding any architect-builder or architect-surveyor in order to identify and distinguish the architect-designer. In 1868 the Institution of Surveyors was created by a small group of surveyors of several kinds, including several members of the Land Surveyors' Club and two quantity surveyors. The Institution rapidly became the representative organization for the surveying profession, and in 1881 it was granted a royal charter. Today, the Royal Institution of Chartered Surveyors has a worldwide membership of some 30,000 professional members, of which about one-quarter are chartered quantity surveyors.[4]

In addition to the chartered quantity surveyors, who are for the most part in private practice or public service, there are in Britain and elsewhere the members of the Institute of Quantity Surveyors, most of whom are employed by construction companies. In other countries there are local societies of quantity surveyors, including the Institute of Quantity Surveyors of Australia, the South African Chapter of Quantity Surveyors, and the Canadian Institute of Quantity Surveyors. There are also branches of the Royal Institution of Chartered Surveyors in several other countries, particularly in the Middle East, Asia, and Africa; and there are organized groups of chartered quantity surveyors elsewhere, such as those in eastern and western Canada. Some British firms of chartered quantity surveyors have recently opened offices in several European countries; and quantity surveyors are now working in France, Holland, Austria, Germany, Spain, and Italy. In 1970 delegations from France, Switzerland, and Japan visited the headquarters of the RICS in Westminster, London, seeking information about the quantity surveying profession; and similar inquiries have been received from other countries.[5]

It is said that "the roots of quantity surveying as a distinct activity with its own literature go back to the

[3] Domesday (or "dooms day") refers to the final day of judgment. The survey was so-called because "it spared no man, but judged all men indifferently, as the Lord in that great day will do."

[4] In 1966 the total membership of the RICS was 30,238 of which 18,290 were professional members, 11,917 were students and probationers, and 31 were ordinary members. The membership includes surveyors in agriculture and land agency, valuation (appraisal), and property management, as well as building and quantity surveyors, land surveyors, and mining surveyors.

In 1971, following unification with two related bodies in 1970, total membership rose to over 42,000 members, probationers, and students.

[5] Reported by the RICS in 1970.

seventeenth century and the Great Fire of London,"[6] which was in 1666. The building boom that followed the Great Fire appears to have been the reason why measuring and valuing *construction work* for payment became a separate occupation, when previously it had been performed by the architect-surveyor and by the craftsman. With such an abundance of buildings to be designed and built and measured for payment, greater specialization and developments in these techniques were a natural result.

In medieval times, the *owner* often bought or supplied the building materials, and he paid the laborers and craftsmen for *work* by the day. Later, it was the practice to pay for *work* at agreed rates, or *unit prices,* and the *work* was measured and valued as it was completed. At first, two measurers were required—one for each party—with the craftsman sometimes measuring for himself. But this duplication became unnecessary with the appearance of independent measurers, who developed better methods of measurement and a greater skill and reliability.[7]

By the nineteenth century, the construction of buildings by independent craftsmen was giving way to general contracts with *contractors* who were responsible to the *owners* for the *work* of all trades. Many large and famous buildings, such as the Houses of Parliament at Westminster, London, were built in this way; and the costs of these buildings were often estimated in the early stages of the design by quantity surveyors, who also prepared the bills of quantities that were priced by the bidders and were the basis of the building contracts.[8]

Bills of quantities contain lists, or schedules, of all the *items of work* required in a building, or in other construction. They are not simply bills of materials. The modern bill of quantities is usually part of the *contract documents;* and like the contract specifications in North America, it is usually the written instrument of the contract, complementary to the contract drawings, and containing the articles of the agreement and conditions of the contract, often by reference to a standard form of construction contract. Many bills of quantities also incorporate the trade specifications for materials and workmanship in the form of preambles to the scheduled quantities and in the descriptions of the *items of work* for which the quantities are given. Some sample pages from recent bills of quantities are shown in Fig. 43, (A), (B), and (C).[9] By North American standards, bills of quantities are generally concise; and despite the fact that they contain all the *items of work* and their quantities, as well as the cash columns for pricing of the items by the bidders, bills of quantities are not usually the large volumes that we have come to expect for contract specifications.

Other formats for bills of quantities are in use in which the quantities and other specific information about the *work* have been analyzed and arranged in different ways; for example, according to the *elements* of a building, or according to the operations and stages in executing the *work* instead of solely according to trades. Many documents are now produced by EDP (computers) so that the information about a project can readily be utilized for other purposes besides that of bidding.

The work of the quantity surveyor in Britain and in those other countries where the "quantities method" of contracting exists does not consist only in producing bills of quantities, and many quantity surveyors believe that their title is no longer adequately descriptive. In some universities and institutes of technology, the courses are now called "building economics" rather than "quantity surveying" because of the evolution of the quantity surveyor's work, which now includes economic feasibility studies of proposed developments and the cost planning of projects while the design is being developed and translated into the *contract documents.* The quantity surveyor also negotiates contracts and controls the costs of jobs in progress. He negotiates changes in the contract amount and makes interim valuations of completed work for certificates and reports. If construction contracts are based on bills of quantities, the quantity surveyor is named in the contract along with the architect as an agent of the *owner.* Sometimes, the architect and the quantity surveyor are professional partners, and one firm may then provide a complete service to the *owner,* including cost planning and cost control.

[6] F.M.L. Thompson, *Chartered Surveyors, the Growth of a Profession* (London: Routledge and Kegan Paul, Ltd., 1968), p. 66. In this chapter I am indebted to F.M.L. Thompson and his book, which should be read by anyone interested in the beginnings of the design and construction industry and its related professions.

[7] Many were "sworn measurers," or "ordained measurers" as they were known in Scotland, who were examined for their competence and given legal authority to practice by the Sheriff. Only their reports were acceptable in legal proceedings. The first standard methods of measurement were established by these measurers, who were the immediate ancestors of the modern quantity surveyor.

[8] For an account of the construction contracts for the new Houses of Parliament and the estimating methods employed see F.M.L. Thompson, *Chartered Surveyors: the Growth of a Profession* (London: Routledge and Kegan Paul, Ltd., 1968), pp. 87–90.

[9] By permission of C. John Mann & Son, and Yeoman and Edwards, Chartered Quantity Surveyors, 11 Charlotte Street, London, W. 1., England.

CJM. 602 - Bill No. 1. Preliminaries.

Item £ s d

Contract Particulars.

1 B015 The Conditions of Contract shall be those contained in the "Agreement and Schedule of Conditions of Building Contract" issued under the sanction of the Royal Institute of British Architects and others PRIVATE EDITION (WITH QUANTITIES) 1966 Edition, as amended by these Bills of Quantities and the revision to Clause 11(4)(C) issued in January, 1967.

1 B016 The following is a Schedule of the Clause headings of the Conditions of Contract and the Contractor shall allow for any expenses which he may incur in observing them.

1. Contractor's Obligations.
2. Architect's Instructions.
3. Contract Documents.
4. Statutory obligations, notices, fees and charges.

For rates on temporary buildings see the Provisional Sum included in Bill No. 2. (Item C 310).

Allow for paying all other fees or charges legally demandable.

5. Levels and setting out of the Works.
6. Materials, goods and workmanship to conform to description, testing and inspection.
7. Royalties and patent rights.
8. Foreman-in-charge.
9. Access for Architect to the Works.
10. Clerk of Works.
11. Variations, provisional and prime cost sums.

Daywork Sheets shall have entered on them the names of each individual workman engaged on daywork. See also Bill No. 6.

To Collection £

1/5.

FIG. 43 (a)

CJM.602 - Bill No. 4. Concrete Work.

Item				£ s d
4 F184 (Contd.)	(Contd.) foregoing shall be carried out for the purpose of obtaining information in connection with the structure. The cost of such tests shall be borne by the Contractor if the results of the test are unsatisfactory, but otherwise by the Employer.			
4 F185	The Contractor will be held responsible for the stability of the structure during the progress of the works and is to shore up or otherwise support any portion of the building as may be necessary whether or not specifically described.			
	PLAIN CONCRETE (1:2:4 - $\frac{3}{4}$" aggregate) 3000 lbs. as described.			
4 F105	Locker plinth.	1	Cub Yd	
4 F120	Filling in base to boiler flue.	1	Cub Yd	
4 F130	Padstone size $13\frac{1}{2}$" x $6\frac{1}{2}$" x 5" high cast on top of brick wall including formwork to sides and part soffit and casting in and including 2 No. $\frac{3}{4}$" diameter ragbolts 7" long with nut and washer.	24	No.	
	FORMWORK TO PLAIN CONCRETE.			
4 F150	Edge of locker plinth over 6" and not exceeding 9" high.	10	Lin Yds	
	REINFORCED CONCRETE.			
	REINFORCED CONCRETE (1:2:4 - $\frac{3}{4}$" aggregate) 3000 lbs. as described.			
4 F190	4" Suspended canopy slab.	8	Sq Yds	
4 F195	5" Suspended landing.	4	Sq Yds	
4 F196	6" Suspended roof slab.	8	Sq Yds	
	To Collection £			

4/10.

FIG. 43 (b)

CJM.602 - Bill No. 4. Roofing.

Item	Description	Quantity	£	s	d
	SHEET METAL FLASHINGS.				
	Lead.				
4 M351	4 lb. Lead flashing 6" girth lapped 6" intermediately, the top edge turned into groove in brickwork and lead wedged and dressed down over upstand of felt roofing (measured net - no allowance made for laps).	8 Lin Yds			
	Aluminium.				
4 M450	Aluminium shall be plain mill finish. Preformed flashings shall be pressed to an approved profile at the works before delivery to site. They shall be accurately and neatly formed with clean sharp arrises and edges and shall not be subsequently dressed or hammered on site. Screws for fixing shall be aluminium and where exposed to the weather shall be sealed with approved plastic washers and caps.				
4 M451	1" x $\frac{3}{4}$" x $\frac{1}{8}$" Channel and fixing as capping to top of fascia and box gutter including drilling holes for $\frac{1}{4}$" diameter bolts through both flanges at 2'0" centres (bolts and holes through fascia and gutter measured separately).	51 Lin Yds			
4 M461	22 S.W.G. preformed capping 10" girth, four times bent and with interlocking seamed joints and plugging and screwing with aluminium screws, plastic washers and caps to weathered top of concrete kerb.	8 Lin Yds			
4 M462	Irregular angle.	3 No			
4 M463	Notched end.	2 No			
4 M464	Ditto but 13" girth.	28 Lin Yds			
4 M465	Angle.	2 No			
4 M466	Returned end.	2 No			
	To Collection £				

4/45.

FIG. 43 (c)

The British chartered quantity surveyors maintain a Research and Information Group, and there are several research fellowships at universities. A cost information service is also operated by chartered quantity surveyors in London. *The Standard Method of Measurement of Building Works* (refered to in earlier chapters) is published jointly by the RICS and the National Federation of Building Trades Employers; and the RICS has been a prime mover in the conversion to the metric system in Britain.

The preparation of bills of quantities is still the daily bread and basis of the quantity surveyor's work in private practice; and since 1965 many bills have been produced with the aid of computers and EDP. Dictionaries and thesauri of standard phrases for bills of quantities have been written and published for computer use, and several specially designed computer services are available to quantity surveyors.

One question invariably asked in any discussion of the quantities method, is, What happens if the quantity surveyor makes a mistake and omits part of the *work* from his measurements in the bills of quantities? The quantity surveyor carries the same professional responsibility for his errors as the architect or the engineer. The *contractor,* in a contract based on a bill of quantities, can question the correctness of the *quantities of work* at any time; and he can require measurement of the *actual quantities of work* done, and is paid accordingly. This is set down in the Conditions of the Standard Form of Building Contract,[10] which state:

(1) The quality and quantity of the work included in the Contract Sum shall be deemed to be that which is set out in the Contract Bills which Bills unless otherwise expressly stated in respect of any specified item or items shall be deemed to have been prepared in accordance with the principles of the Standard Method of Measurement of Building Works 5th edition Imperial revised 1964/5th edition Metric by the Royal Institution of Chartered Surveyors and the National Federation of Building Trades Employers, but save as aforesaid nothing contained in the Contract Bills shall override, modify, or affect in any way whatsoever the application or interpretation of that which is contained in these Conditions.

(2) Any error in description or in quantity in or omission of items from the Contract Bills shall not vitiate this Contract but shall be corrected and deemed to be a variation required by the Architect.

[10] Article 12, Private Edition with Quantities; issued by the Joint Contracts Tribunal, copyright of the Royal Institute of British Architects.

This means that the *owner* pays for the *work* that he receives through the contract; which is equitable. If the quantity surveyor has been negligent, the *owner* may sue him for any damages he has suffered. But the quantity surveyor's professional reputation is always the best safeguard of the *owner's* interests. However, many quantity surveyors in private practice carry professional indemnity insurance against errors and omissions.

In North America, there are only a few quantity surveyors in the United States; but there is a much larger number in Canada, despite the fact that Canada is the only major country in the British Commonwealth that does not use the quantities method. Most of them are members of the Canadian Institute of Quantity Surveyors, which was founded in 1959 with the following objectives:

1. To provide through its members, professional advice to Contractors, Architects, Consulting Engineers, Public Authorities and Building Owners on all matters relating to quantity surveying in construction costs, management and administration of projects.
2. To collaborate with other professions and organizations in the interests of the Construction Industry.
3. To promote and advance the professional status and gainful employment of Quantity Surveyors.
4. To establish and maintain a high standard of professional competence and integrity by limiting membership to persons who either:
   a) have passed the examinations prescribed by, and acceptable to, the Institute;
   b) have satisfied the requirements of training, practical knowledge, experience and integrity prescribed by the Institute.
5. To promote fellowship and to provide a medium for the interchange of current construction knowledge and other topics of interest to members.
6. To take appropriate action when possible infringements of the aims and objectives of the Institute arise from other sources anywhere in Canada.

In the CIQS publication, **"Quantity Surveying-Career Information and Admission Requirement"**[11] the quantity surveyor and his work are defined as follows:

1. The Rules and Regulations of the Institute define a Quantity Surveyor as follows:

[11] Obtainable from the Executive Secretary, Canadian Institute of Quantity Surveyors, 2872 Kingston Road, Scarborough, Ontario, Canada. This extract is published by permission of the CIQS. There are also several publications available explaining the work of chartered surveyors, obtainable from their headquarters in London.

A Quantity Surveyor is a person who, by virtue of his training, experience and qualifications, is capable of performing the following functions:

a) The preparation of construction tenders[12] from information provided by Architects and Engineers.

b) The management, administration and coordination of all types of construction projects, including subcontracts; preparation of construction progress schedules; setting up and operation of cost control systems; valuation of changes and finalization of accounts.

c) The giving of advice on construction cost planning to prospective Owners, Architects, Engineers and Public Authorities.

d) The preparation and/or interpretation of tender documents, specifications, general conditions and forms of contract.

e) The checking and analyzing of tenders.

f) The conducting of arbitration in contract disputes and acting as expert witness.

2. The education and training of a Quantity Surveyor is designed to provide him with a detailed and comprehensive knowledge of construction and construction methods and the ability to analyze costs and prepare estimates for the work of all trades. A separate division of the Institute comprises Members specializing in Mechanical and Electrical installations but all Quantity Surveyors should have a working knowledge of the practical aspects and design features of such installations. The Quantity Surveyor should also possess a thorough knowledge of the laws relating to construction projects and the accounting and administration procedures essential to the successful management of construction contracts.

3. As might be expected in a profession covering such a wide field, Quantity Surveyors tend to specialize. The more clearly defined groups are:

a) Estimators—The majority of the Institute's Members are employed as estimators by the major building, civil engineering, mechanical and electrical contractors.[13] In this capacity they are responsible for preparing bids for all types of construction work and in the case of many firms, estimators may also be responsible for the management of construction contracts.

The successful estimator will be well informed about current market conditions affecting the cost of construction; he must study the methods and idiosyncrasies of the architects and engineers whose work he bids and he will keep up to date records of proposed construction work so that he may select for bidding those projects containing features favorable to any special skills or equipment his company may possess.

An ever-increasing part of any construction project involves the work of specialist subcontractors. The contractor's estimator must compile lists of all such firms operating in his locality and be aware of their competency and financial capacity. He must build up and maintain a good working partnership with the best of these firms. Only then will he be successful in bidding and avoid involving his firm in the disastrous consequences resulting from the employment of incompetent and financially unstable sub-contractors.

b) Private Practice—Members in private practice fall into two main groups:

i) Firms engaged in the preparation of Schedules of Quantities for contractors who may wish to bid on a project but are unable to prepare their own quantity "take-off". Such firms often provide a complete estimating service and members engaged in this type of practice must possess all the attributes of Estimators employed directly by Contractors, be exceptionally fast and accurate in their work and be good administrators.

ii) Firms specializing in Preliminary Estimates and Cost Planning. The preparation of preliminary estimates often necessitates working from single line sketch drawings and the Quantity Surveyor engaged in such work must possess so intimate a knowledge of buildings and structures that he is able to visualize the project and assess quantities and cost from minimal information. Preliminary estimates provide data for decisions governing the scope of a project. Cost planning is an extension of preliminary estimating whereby the Quantity Surveyor revises the estimates through the various stages of design and preparation of working drawings and provides comparative estimates for alternate materials and construction methods. The money available for a project is seldom sufficient to satisfy the ambitions of the Owner, the Architect and the Engineer; cost planning

12 *"Construction bids"* (The terms *tender* and *bid* are both used in Canada; although there are no equivalent alternatives to "bid bond" and "bid depository" in current use. A general use of the term *bid* would be more clear and consistent.)

13 This is said here of the members of the Canadian Institute of Quantity Surveyors, and the same could be said of the members of the Institute of Quantity Surveyors in Britain; but Chartered Quantity Surveyors (members of the Royal Institution of Chartered Surveyors) are for the most part in private practice or public service. As with the accounting profession, the quantity surveying profession provides a diversity of services through several professional bodies.

permits the establishment of priorities and refinement of design to ensure that the final cost will be within the budget.

Quantity Surveyors engaged in this type of practice work in close cooperation with Owner, Architect and Engineer and often are required to give advice on methods of calling Tenders and the selection of appropriate forms of contract.

c) Public Service and Commerce—An increasing number of Quantity Surveyors are employed in the Construction Branch of Public Service (Federal, Provincial and Municipal) and by commercial organizations such as Trust Companies. The duties of such Quantity Surveyors generally follow those of the Private Practitioner described in 3 (b) (ii).

d) Architectural Administration—With the increasing complexity of modern buildings and engineering and the necessity for stringent control over the financial aspect of Construction Contracts, many of the large firms of Architects and Engineers employ Quantity Surveyors on their staff to take over the financial administration of work in progress and the processing of Change Orders. Quantity Surveyors so employed also prepare Preliminary Estimates and often draft those parts of the specifications relating to strictly contractual matters.

It is indicative of the training and experience of quantity surveyors that many of them are able to do the variety of work described above, and that so many quantity surveyors have been able to transplant their skills from one country to another, despite the differences in contracting methods. The quantity surveying approach is essentially practical and analytical, because estimating is its basic discipline. But estimating is not quantity surveying, and estimators are not always quantity surveyors.

Quantity surveying began with measurement and estimating and has now become a profession that continuously draws on and adds to a body of knowledge and techniques based on soundly established principles and abstract concepts. Its practice, in the fullest sense, requires the wide technical knowledge of a generalist in the design and construction industry, together with the specialized application of scientific analysis and synthesis in construction economics and technology. Such professional practice requires the research, collection, and dissemination of theories, data, and new techniques, and their publication and discussion in books, papers, and theses, and through courses and seminars. It also requires an involvement in tertiary education, so that the accumulated knowledge and experience of quantity surveyors can be passed on to practitioners, probationers, and students. All of this is part of, and only possible within, a society of qualified practitioners with an ethical regard for the part they play in society.

A review of quantity surveying over the past 25 years shows a large and steady growth in the demand for quantity surveyors' services wherever the profession is established, particularly in Britain and Australia. It shows the establishment of quantity surveyors in many of the developing nations that have Commonwealth ties with Britain; and it shows that in the last few years quantity surveyors have gone outside and beyond the English-speaking and Commonwealth nations, particularly into Europe. They anticipated entry into the European Common Market by becoming one of the prime movers in the adoption of the Metric System in Britain.

Everything points to similar developments in North America within the next decade. But although during the last decade several hundreds of practicing quantity surveyors have begun work in Canada, the quantities method of contracting has not become established. In its present form, it may not be suitable for North America. Any method that has evolved over centuries inevitably carries with it elements that are redundant, and a transplantation is the proper time to prune. Certainly, the quantity surveyors who have moved to North America have shown that their training and methodology is adaptable, and this fact is now being increasingly realized.[14]

The separation of architect-designers from architect-builders and architect-surveyors when the Royal Institute of British Architects was created in 1834 was undoubtedly one of the reasons for the growth of the surveying profession and its membership of building and quantity surveyors. But that same period saw the beginnings of several other professions, which eventually found their way to those countries with historical and linguistic connections with Britain. As to why quantity surveying has not developed in North America to the extent that it has developed in some other countries, we can point to several likely reasons, including an earlier political separation and a greater independence and vigor that resulted in a booming rate of growth and construction that has, until now, precluded by its momentum the time and the need to develop other professions in the design and con-

14 Firms in the United States have advertised for staff with training in quantity surveying by advertising in Canadian and British periodicals during 1971. In that same year, a group of British quantity surveyors was invited to tour the United States of America to represent the professional practice of quantity surveying, and a report on this tour is included in the Appendices.

struction industry. But what about the future, which starts today?

Building costs and economics are increasingly important to everyone, and we cannot afford to be inefficient and wasteful in the use of our land and constructed resources. We have to find more efficient and economical ways and means to provide an immense amount of construction in the next quarter-century. Many of the current practices of designing, constructing, and using buildings need to be changed; and the functions of designing and constructing should be related in ways that ensure the best results for the *owner* and for society.

Building economics should be a discipline of the *designer* and a basic subject in a common program of education for both design and construction. At the same time, there is a need in this industry for generalists with ability and experience based on training in building technology and management, with a working knowledge of contracts and a specialized knowledge of building economics, which starts with *estimating* and *cost accounting*.

Questions & Topics for Discussion

1. In current North American practice, what replaces the "bills of quantities" used in some other countries to bring about a stipulated sum contract?

2. Present an argument for, and an argument against, the use of bills of quantities (prepared by a consultant quantity surveyor) as part of the *bidding documents* and of the *contract documents* for a construction project.

3. Explain the fact that in the so-called "quantities method" of contracting the *contractor* need not suffer from an error of omission in the bills of quantities.

4. What are some of the functions performed by quantity surveyors in the construction industry in the United States and Canada?

5. What is a profession, and specifically what is required of a profession that deals with the *costs of construction work* to advance knowledge in this field?

6. Which professions in North America deal with and often have responsibility for the *costs of construction work*?

7. Assume the position of an *owner* about to embark on a design and construction project to build an office block on his own property. List and briefly explain the steps you would take to get the *work* done within a budget (not yet established) for tenancies (not yet leased).

8. Which contracting method in North America is similar to the quantities method, and in what ways are they different?

9. Compose a syllabus for a course of studies for a quantity surveyor, and explain briefly the need for each subject in the syllabus.

10. Illustrate and explain the scope and details of construction contracts and construction economics by means of a fictitious, historical account of the life of one of the following construction components:
    a. a clay brick
    b. a 2 × 4 stud
    c. a cubic yard of concrete

    from its manufacture until its destruction as part of a demolished building. Each event in the life of the selected item should be recorded along with descriptions or reproductions of all the necessary illustrations of, and references to, the item that might be made by all the different persons involved with the item's selection and utilization.

Glossary

Certain terms and phrases in the text are printed in *italics* to indicate that they are used in the text with a specific meaning, which is given in this Glossary. Some of these terms have a specific contractual or economic meaning when used in a construction context that might otherwise be overlooked, because many of the terms have other more general and more common meanings or simply because they are too often loosely used. Other terms have been coined and defined in this Glossary to represent and convey particular concepts, such as the term *item of work*. This term is made up of three commonplace words, and the term itself is not uncommon and is in common use, in a general sense. But, beyond that general sense, the term has been given a special meaning in this text; that is, to represent an aspect of *construction work* that is fundamental to *cost accounting* and, therefore, to estimating.

In a few instances, adjectives and other words have been italicized with the words from the Glossary; and, at times phrases in the Glossary have been slightly rearranged in the text to ease the eye over them. But the special meaning and identity of such a phrase should remain apparent to the reader because the words are in italics.

The Glossary incidentally provides a useful way of reviewing important aspects of the subject through the meanings of the terms it contains, and teachers can use the Glossary as a basis for tests and discussions.

One or two pages of the text are unavoidably filled with italicized words, and this goes against the general principle and recommended use of italics. Nevertheless, and despite this fact, it was decided to persist because of the general advantage gained.

Alteration work: *new work* done to and in conjunction with *existing work* (both which see); other than complete demolition.

Basic items (of work): those *items of work* described and included in specifications and cost codes and that occur in most of the construction jobs done in any particular field of construction activity, and that can, therefore, constitute part of a basic specification or a basic cost code in that field, as opposed to *particular items of work* (which see). For example, in most building construction contracts, "concrete footings" are a *basic item of work* and as such can be included in a basic specification by a *designer* and in a basic cost code by a *contractor.* The term *basic item* is not completely definitive; but the distinction between *basic items* and *particular items* is an expedient one in preparing specifications and cost codes, and in the analysis and collection of cost data through the isolation of *basic items* from *extra over items* (which see). (*Note*: a basic specification contains only *basic items,* whereas a master specification usually contains a much wider selection).

Bid: an offer to do *construction work* for payment, the acceptance of which constitutes a contract between an *owner* (who has accepted the *bid*) and a *contractor* (who has made the *bid* to the *owner*). Sometimes referred to as a *general bid* (made by a *general contractor*) to distinguish it from a *sub-bid* (which see). (*Note*: the term tender (instead of *bid*) is often used in Canada, although "sub-bid" and "bid bond" appear to have no equivalent alternatives, and the consistent use of the term *bid* would be better).

Bidding documents: they should be the same as the *contract documents* (which see), plus the Instructions to Bidders, and the Bid Form, and any other informative documents related to the bidding. Instructions to Bidders should contain only information and instructions regarding bidding. In some instances, the Bid Form is identified as one of the *contract documents* because of what it contains, such as a list of proposed *subcontractors,* or a list of *unit prices,* submitted by the bidder and required to be made part of the contract.

Construction work: see *Work.*

Contract amount (sum): the amount of money paid by the *owner* to the *contractor* for the *work,* according to the terms of the contract. In stipulated sum contracts it is the amount stipulated by the bidder and later stated in the articles of the contract. In some other types of contract (e.g., Cost Plus Fee) the *contract amount* is implicit in the articles and conditions of the contract; but it is not explicit until it is determined according to the amount of *work* done and the terms of the contract.

Contract documents: those drawings, specifications, and other documents prepared by the *designer* (first as *bidding documents*) that illustrate and describe the *work,* the terms and conditions under which the *work* is to be done for the *owner,* and the terms and conditions of payment for the *work* by the *owner* to the *contractor.* The *contract documents* should contain nothing that is not in the *bidding documents* (which see).

Contract manager: the party to a contract to provide construction managerial services to an *owner* (the other party to that contract), particularly services for organizing and managing a construction project in which the *work* is done for the *owner* by a number of *contractors.* The *contract manager* and the *designer* are both agents of the *owner* in a management contract. (The term has also been used for an employee charged with the management of *construction work* for a *contractor.*)

Contractor: the party to a construction contract who does *construction work* for the *owner* (the other party). The *contractor* may have *subcontracts* with *subcontractors* to do parts of the *work* that he has undertaken to perform for the *owner.* (See also *general contractor*).
(*Note*: the term *contractor* is sometimes used in a comprehensive sense to include by implication the *subcontractors* (and their *sub-subcontractors*) of the *contractor,* because that which applies to the *contractor* usually applies to some or all of the *subcontractors* and *sub-subcontractors.*)

Cost accounting: that part of construction management by which actual *costs of work* are segregated and attributed to specific *items of work* and to specific construction jobs, after which the cost data is analyzed and formulated for use in job planning and cost control and in estimating the costs of other jobs.

Costs in use: all the costs incurred by an *owner* through his ownership of a building, apart from and subsequent to the *initial costs of the work* incurred by constructing the building, including depreciation, maintenance, fianancing, taxes, insurance, vacancies, and building operating costs.

Costs of (the) (construction) work: all the *direct costs* and *indirect costs of work* (which see), generally classified as *labor costs, material costs, plant and equipment costs, job overhead costs, operating overhead costs,* and *profit* (all of which see). These are the costs to the *owner* for the *work* done, and in place.

Cost planning: estimating the *costs of work* during the design stage (usually by *elements,* or other such means) and including, if necessary, selecting materials and construction methods and adjusting the design so as to complete the *work* within a budgeted cost. Designing the *work* so that it can be done for a specified cost.

Cube: an item in an estimate with its quantity expressed as a volume, usually in cubic feet, or cubic yards (or cubic meters). With a *cube* item, no dimensions need appear in the item's description, although in some cases it is more informative if some dimensions are included. (See also *super, run,* and *number* items.)

Designer: a party to a contract to provide professional design and other services to an *owner* (the other party to that contract). Usually, the *designer* is an architect or a professional engineer. In certain cases, a *designer's* services may be part of more comprehensive design and construction services provided to an *owner,* which may also include land acquisition as well as *construction work,* as in so-called "package deals" or "turn key" projects.

Designer's consultant: a party to a contract to provide professional design or other services to a *designer* who, in turn, has undertaken to provide design and other services to an *owner,* of which the services of the *designer's consultant* are a part. The usual fields of *designer's consultants* are listed in Chapter 1.

Developer: a person who develops land through *construction work,* and who becomes an *owner* in order to have the *work* done.

Direct costs (of work): those *costs of work* generally classified as *labor costs, material costs, plant and equipment costs,* and *job overhead costs,* all of which are directly attributable to a specific construction job. (See also *indirect costs.*)

Direct labor costs: those *labor costs* paid by a *contractor* directly to an employee; usually based on a wage agreement. (See also *indirect labor costs*).

Division (of a specification): one of the sixteen *standard divisions of construction work* in the "Uniform Construction Index" (see Bibliography–*Construction Specifications*). Standard *divisions* are subdivided into non-standard *sections* (which see) by the specification writer according to the nature and extent of the *work* specified to facilitate the production of the specifications and to facilitate their use in bidding and construction.

Duodecimals (from Latin "duodecim," meaning twelve): a number system based on twelfths, used in calculating quantities from dimensions given in feet and inches (twelfths). The product of multiplying feet and inches as duodecimal fractions is in units and twelfths–e.g., 2.4 × 3.9 = 8.9 (2 ft, 4 in. times 3 ft, 9 in. = eight and nine-twelfths of a square foot, or eight and three-quarters of a square foot).

Element (of a building): has been defined as "that part of a building which always performs the same functions irrespective of building type." A building may be made up of any number of *elements* within practical maxima and minima. A minimum of *elements* might include:

1. Substructure, including lowest floor and finishes
2. External walls complete with finishes
3. Internal walls complete with finishes
4. Upper floors, each with part of the supporting structure, complete with finishes
5. Roof, with part of the supporting structure, complete with finishes.
6. Services (inside the building)
7. Site work and services (outside the building).

Most *elements* consist of several *items of work,* and always at least one; e.g. an *external wall element* of 8-inch-thick concrete blockwork (with no finishes inside or outside). The practical maximum number of *elements* is usually about thirty, or even less; and lists of *elements* are published by several authorities such as the RICS and the CIQS. The first list of *elements,* and the elemental estimating technique, were developed by chartered quantity surveyors in 1955. See Douglas J. Ferry's, *Cost Planning of Buildings* (London: Crosby Lockwood and Son, Limited, 1964) for a description and a brief history.

Equipment costs: see *plant and equipment costs.*

Existing work: *work* that already exists when a contract is made for other *work,* and which by definition has some physical relationship or connection with the *work* of the contract. (See also *new work* and *alteration work.*)

Extra over (item): a feature or part of an *item of work* measured and priced separately from the main item to facilitate and simplify the measurement, pricing, and subsequent costing of the *work;* e.g., pipe fittings such as bends and tees are often measured as *extra over items* (enumerated) whereas the *run* of pipe is measured over the fittings, the lengths of which are not deducted. The fittings are priced at the *extra cost* over the cost of the pipe that they displace. Measuring *extra over items* eliminates the need for many minor deductions and adjustments. It also enables the theoretical isolation of *basic items* by separating them from their minor and variable features, thus effectively increasing the available and useful cost data of *basic items of work.*

General contractor: a *contractor* who has *subcontractors* to do some (or all) parts of the *work* that he has undertaken to do for an *owner.* (Previously, the term referred to a *contractor* who employed workmen of several different trades, and who undertook to do most (or all) parts of *work,* as distinct from a *specialist (trade) contractor,* who normally undertook the *work* of only one trade).

General requirements: temporary services and other requirements for *work* provided by a *contractor,* which, because of their general nature, are related to the *work* as a whole rather than to specific *items of work.* They are properly to be found in *Division 1* (of the Uniform Construction Index), in job specifications. Certain general conditions of some contracts might be more accurately called *general requirements* in that they more directly relate to the *work* rather than to the terms of the contract.

Indirect costs (of work): those *costs of work* generally classified as *operating overhead costs* and *profit* as opposed to *direct costs* (which see).

Indirect labor costs: those *labor costs* paid by an employer on an employee's behalf for such things as insurance, social security (pension plan), etc. (See also *direct labor costs*).

Item of work: a part of the *work* that, by its nature, can be observed, identified, and distinguished from other parts of *work* for purposes of estimating, *cost accounting,* and construction management. Costs can be segregated and allocated to specific *items of work.* For example, a masonry wall is an *item of work,* whereas the masonry units and the mortar joints separately are not, because they do not fit the above definitions, and because all related costs cannot be separately allocated to each of them, since they are both inseparable parts of the wall. *Items of work* usually involve only one trade.

Job overhead costs: those *direct costs of work* that because of their general nature, cannot be allocated to specific *items of work* but can be allocated to a specific job. As distinct from *operating overhead costs* (which see). For example, temporary facilities and services on the job site and the cost of the building permit.

Labor costs: that part of the total *costs of work* expended on labor, dependent on the *labor rates* paid for workmen and their productivity. They include both *direct* and *indirect labor costs* (which see). The *costs of work* other than *material costs, plant and equipment costs, overhead costs,* and *profit* (which see).

Labor rate: the total costs per hour for labor paid by the employer, including all *direct* and *indirect labor costs* for a specific period of time (and place) divided by the number of hours worked during that period. (See also *wage rate.*)

Laps: additional material required by and incorporated into the *work* because of the dimensions of the material product and of the *work* and the resultant need for joining the material by overlapping. Consequently, the actual amount of additional material required for *laps* usually depends on the dimensions of the material product and the dimensions of the *work.* Generally, however, the smaller the critical dimension of the material product (relative to the critical dimensions of the *work*), the more the additional amount of material required for *laps* becomes a constant. For example, with shiplap boards the additional material in *laps* is practically a constant function of the boards' nominal width/actual width and the actual width of the *lap.* In the case of splices (*laps*) in steel reinforcing bars, the amount of material required in *laps* is a function of bar lengths and *work* dimensions and *laps* must be individually identified and measured. *Laps* are sometimes measured as *work;* but, in many cases, it is better to allow for them in the *unit price* instead, particularly if the amount required for *laps* is practically a constant. (See also *waste.*)

Material costs: the costs of all materials, products, building components, fixtures, and building equipment required to be incorporated and installed in the *work,* including delivery to the site and taxes. The *costs of work* other than *labor costs, plant and equipment costs, overhead costs,* and *profit* (which see).

New work: *work* (as defined) as opposed to *existing work* (which see).

Number: an item in an estimate with its quantity expressed by enumeration. With a *number* item, all three dimensions must appear in the item's description. (See also *cube, super,* and *run* items.)

Operating costs: see *owning and operating costs.*

Operating overhead costs: with *profit,* the indirect *costs of work;* those costs of operating a construction business that, because of their general nature, cannot be allocated to specific jobs; as distinct from *job overhead costs, plant and equipment costs, material costs,* and *labor costs* (which see). Often misleadingly referred to as "head office overheads."

Overhead costs: the *job overhead costs* and *operating overhead costs.* Usually, the more specific terms should be used (which see).

Overhead and profit: usually refers to *operating overhead costs* and *profit,* which are *indirect costs* (which see).

Owner: the party to a construction contract who pays the *contractor* (the other party) for *construction work;* the party who owns rights to the land on which the *work* is done and, therfore, owns the completed *work* itself. Also, the client of the *designer.*

Owning and operating costs: are total *plant and equipment costs.* The *owning costs* are those incurred by ownership of plant and equipment and consist primarily of the costs of investment, maintenance, and depreciation. The *operating costs* are those incurred by operating and using plant and equipment (over and above the *owning costs*) and consist mainly of running repairs, fuel, lubricants, and the *operator's labor costs.* Mobilization and demobilization costs must also be included in an estimate.

Particular items of work: those *items of work* that are not *basic items of work* (which see) and that therefore require an original description in the specifications or cost code for the jobs in which they occur. (In construction specifications, *basic items* are sometimes prescribed in so-called "master specifications;" which may be extant in a printed form or stored in a computer.) *Particular items* are originated as required for particular projects and are interleaved among the *basic items* to create the project specifications. Similarly, *particular items* are originated in an estimate and given a code number specially for that project so that the item can be identified in *cost accounting* and other construction management procedures. Also, "rogue" or "maverick" items.

Plant and equipment costs: all the *owning* and *operating costs* of plant and equipment (which see). The *costs of work* other than *material costs, labor costs, overhead costs,* and *profit* (which see). They also include *costs of tools* other than the costs of those provided by the workmen for their own use.

Profit: the excess of income from doing *work* over total expenditure. For the *owner,* the *profit* is a *cost of the work.* For the *contractor,* it is a motive to do the *work.* The *contractor's* total expenditures are *material costs, labor costs, plant and equipment costs,* and *overhead costs.* These must be accounted for before the amount of *profit* can be ascertained, which is one reason for proper estimating and *cost accounting.*

Profitability: the return on investment; often represented as a percentage ratio (obtained by dividing the *profit* ($\times 100$) by the "tangible net worth" of a business).

Run: an item in an estimate with its quantity expressed as a linear measurement, usually in linear feet, or linear yards (or meters). With a *run* item, two dimensions must appear in the item's description. (See also *cube, super,* and *number* items.)

Section (of a specification): a distinct part of a specification *division* (which see) with its own title and reference. *Section* references are numerical–alphabetical, comprising a number (1 to 16) of one of the sixteen standard *divisions,* followed by an alphabetical letter (which usually varies according to the job). The *work* specified in a *section* depends on the job's requirements as interpreted by the specification writer. Usually, the *work* specified in any *section* should be:

1. recognizable as a distinct entity or component; i.e., as consisting of one or more related *items of work*
2. done by one construction trade
3. the subject of not more than one trade's *sub-bids,* so that all *sub-bids* have one (or more) *sections* as their scope, and so that no *sub-bid* includes only part of the *work* of a *section.*

This breakdown is not always possible, because a specification writer cannot always correctly anticipate how *work* will be divided between subtrades. Therefore, specification writers should create as many valid *sections* as possible in the specifications they write to facilitate bidding by subtrade firms.

Standard forms (of contracts): are prescribed forms containing articles of agreements and general conditions for different types of construction contracts, usually written (in conjunction with construction associations), published, and sold by architectural institutes and engineering institutes. (A list is included in the Bibliography.)

Sub-bid: an offer made by a subtrade firm to a *contractor* to do *construction work* for payment, the acceptance of which constitutes a subcontract between the *subcontractor* (who made the *sub-bid*) and the *contractor* for part of the *work* that the *contractor* has undertaken to perform for the *owner.* (See also *bid.*)

Sub-sub-bid: an offer made by a subtrade firm to a *subcontractor* to do *construction work* for payment, the acceptance of which constitutes a sub-subcontract between the *sub-subcontractor* (who made the *sub-sub-bid*) and the *subcontractor* for part of the *work* that the *subcontractor* has undertaken to perform for the *contractor.* (See also *sub-bid* and *bid.*)

Subcontractor: a party to a construction subcontract who does *construction work* for the *contractor* (the other party to the subcontract) that is part of the *work* that the *contractor* has undertaken to perform for the *owner.* (See also *sub-bid.*)

Sub-subcontractor: a party to a construction sub-sub-contract who does *construction work* for a *subcontractor* (the other party to the sub-subcontract) that is part of the *work* that the *subcontractor* has undertaken to perform for the *contractor,* who, in turn, has undertaken to do all the *work* for the *owner.* (See also, *sub-sub-bid.*)

Super: an item in an estimate with its quantity expressed as a superficial area, usually in square feet, or square yards (or meters). With a *super* item, one dimension must appear in the item's description. (See also *cube, run,* and *number* items.)

Supplier: one who supplies stock materials, products, or building components for *construction work* to a *contractor, subcontractor, sub-subcontractor,* or *owner.* A *supplier* does not usually do any of the *work,* at the site or elsewhere, as a *subcontractor;* but the distinction is not always clear.

Swell and shrinkage: the increase (*swell*) and decrease (*shrinkage*) in bulk of excavated materials and imported fill. Soils increase in bulk as they are excavated, and excavated materials and imported fills usually decrease in bulk when placed and consolidated. Amounts depend on the nature and moisture content of the materials. Typical amounts are:

	Swell (%)	*Shrinkage* (%)
Sand and gravel	15–20	10–15
Loamy soil	15–25	15–20
Ordinary soil	20–30	20–25
Heavy clay	25–40	25–30
Solid rock	50–75	–

E.g., 1 CY ordinary soil "in bank measure" becomes, say, 1.25 CY when excavated and 0.80 CY when compacted.

But such figures are only generally indicative, and actual amounts should be established for specific materials with different moisture contents in different localities.

Tender: see bid.

Unit Price: the *unit price* of an *item of work* is an **average price of a unit of the work** calculated thus:

$$\textit{unit price} = \frac{\text{total costs of } \textit{item of work}}{\text{total quantity (no. of units) of item}}$$

"Total costs" and *unit prices* depend on the nature of the item (whether it includes *material costs, labor costs,* etc.), on the job conditions, and on the "total quantity." "Total quantity" depends on the contract. Therefore, in theory, *unit prices* for similar *items of work* in different jobs will never be exactly the same. However, *unit prices* for similar items in different jobs may be practically the same because of the limitations of practical accuracy. The primary function of *unit prices* is to enable costs to be compared.

Unit rate: the *unit rate* of an *item of work* is the average productivity rate for a unit of the *work* calculated thus:

$$\textit{unit rate} = \frac{\textit{unit price} \text{ for } \textit{labor costs} \text{ of } \textit{item of work}}{\textit{wage/labor rate.}^{1}}$$

Note:[1] The *labor costs* may be based on one or more *wage rates,* or one or more *labor rates,* depending on the method of estimating and the composition of the labor crew. If the crew is paid at different rates, an *average wage rate* (or an *average labor rate*) for the crew should be calculated.
A *unit rate* is expressed in man-hours per unit to indicate the average rate of labor productivity for an *item of work;* and since the *unit rate* is independent of the *wage rate* it can be compared with other *unit rates* for *work* done at other times and for different *wage rates.*
The term "labor unit" is used in Chapter 11, Pricing Mechanical (and Electrical), instead of *unit rate,* because "labor unit" is the term used in the published Labor Calculator. The two terms are synonymous.

Void: a deduction made for an opening (door, window, etc.) or for a minor area within a major area, as illustrated in Chapter 7. (See also *want.*)

Wage rate: the *direct labor costs* per hour exclusive of *indirect labor costs* (which see).

Want: a deduction made for a deliberate over-measurement of *work* in an estimate, usually in the case of an irregular area with *want(s)* adjoining the area's perimeter, as illustrated in Chapter 7. (See also *void.*)

Waste: construction material that is additional to the actual quantity required in the *work,* as indicated in the *contract documents,* but that is, nevertheless, required by or used in performing the *work,* and that therefore contributes to the *material costs of the work.* Some *waste* may be inevitable and necessary, as in cutting a standard sheet of material to a smaller size required by the job, thus rendering the offcut unsuitable for any other use. Other *waste* may be attributable to poor design or construction practices, and it may be avoidable with better design, purchasing, or planning. So, to some extent, *waste* is subject to individual efficiency. It is, therefore, essentially variable; and usually it is more easily accounted for (by *cost accounting*) if it is stated and kept separate from the net, measured quantity of *work* by allowing for it in the synthesis of the *unit price.*

Work: labor, materials, and the use of tools, plant, and equipment, and all other things and services required of the *contractor* in a contract, for which the *owner* pays.

(See also *alteration work, costs of work, existing work, extra work, item of work,* and *new work*).

It is appropriate that the Glossary should end with the term *work,* for that is what this book is all about. The estimator and the quantity surveyor both measure the *work* to be done for the *owner* so that its probable costs can be calculated, with the intention that the *owner* should pay for what he receives from the *contractor.*

Bibliography

The amount of published material related to the subject of this text and available in North America and Europe is immense; and this bibliography is limited to a few publications that are particularly relevant, and that enlarge on some of the subject matter. Some sources of other publications, catalogs, and bibliographies are also included.

The following divisions are made for convenience, even though they are not always completely indicative of the listed publications' full contents:

1. Construction Contracts
2. Architectural and Engineering Practices
3. Construction Specifications
4. Estimating and Cost Accounting
5. Quantity Surveying and Construction Economics
6. Construction Management
7. General and Miscellaneous.

Construction Contracts

Standard Forms in the U.S.A.

These are published by the American Institute of Architects, 1735 New York Avenue, N.W., Washington, D.C. 20006, from whom a catalog of publications can be obtained, which includes:

A101–Owner-Contractor Agreement Form–Stipulated Sum

A107–Short Form for Small Construction Contracts–Stipulated Sum

A111–Owner-Contractor Agreement Form–Cost Plus Fee

A201–General Conditions of the Contract for Construction

A401–Contractor-Subcontractor Agreement Form

A501–Recommended Guide for Bidding Procedure & Contract Awards

B131–Owner-Architect Agreement–Percentage of Construction Cost

B231–Owner-Architect Agreement–Multiple of Direct Personnel Expense

B331–Owner-Architect Agreement–Fee plus Expense.

Standard Forms in Canada

These are published by the Royal Architectural Institute of Canada, 151 Slater Street, Ottawa 4, Ontario, Canada, from whom a catalog of publications can be obtained, which includes:

No. 6–Client and Architect Agreement Form

No. 10–Construction Tender Form

No. 12–Standard Form of Construction Contract–Stipulated Sum

No. 13–Standard Form of Construction Contract–Cost Plus Fee

–A Guide to Construction Tendering Procedures.

Similar standard forms of construction contracts are published by the Association of Consulting Engineers of Canada, 176 St. George Street, Toronto 5, Ontario, Canada, for contracts in which an engineer is the *designer* and the agent of the *owner*; and, in addition,

No. 4–Standard Form of Construction Contract–Stipulated Unit Prices.

All of the above Canadian standard forms are approved by the Canadian Construction Association, 151 O'Connor Street, Ottawa 4, Ontario, Canada, which represents the Canadian construction industry and which also has a catalog of publications available.

Standard Forms in Britain

These are published by the Joint Contracts Tribunal, under the sanction of several bodies, including the Royal Institute of British Architects, 66 Portland Place, London, WIN4AD, England (which holds the copyrights), the National Federation of Building Trades Employers, and the Royal Institution of Chartered Surveyors, among others. All of these bodies have available lists of publications that include the following standard forms of contracts:

Private Edition with quantities
Private Edition without quantities
Local Authorities Edition without quantities.

The private editions are for private contracts, and the others are for contracts in which the *owner* is an office of local government, such as in contracts for public housing and schools. Contracts in which bills of quantities are provided to the bidders include the bills of quantities among the *contract documents.*

Architectural and Engineering Practices

The publications listed are included because they have some specific significance to this text. They are not particularly included as general references for architectural and engineering practices.

The primary sources of information on this subject, apart from texts by individual authors, are the professional associations and institutes from whom catalogs and lists of publications can be obtained.

Architect's Accounting Documents and *Architect's Office and Project Documents* (F Series and G Series), Washington, D.C. 20006, U.S.A., The American Institute of Architects, Documents Division. A wide selection of documents for use in architects' offices for their own businesses and for the administration of construction contracts.

Architects' Handbook of Professional Practice, Washington, D.C. 20006, U.S.A., The American Institute of Architects.

Canadian Government Specifications Board, *Standard on Architectural Drawing Practices,* C.G.S.B. 38–GP–7, Ottawa, Canada, National Research Council, 1961. This standard applies to practices recommended for the preparation of architectural working drawings, and it embraces modular dimensioning.

Ferry, Douglas J., ARICS, *Cost Planning of Buildings,* London, England, Crosby Lockwood and Son, Ltd., 1964. An introduction to cost planning describing the theory and practice of elemental cost planning, elemental cost analysis, preliminary estimates, cost criteria, and cost research.

Hunt, William Dudley, Jr., Editor, *Comprehensive Architectural Services* (prepared by The American Institute of Architects), New York, Toronto, London, McGraw-Hill Book Company, 1965. Each chapter is written by a specialist in at least one aspect of development or architectural services to explain the analysis, promotion, design and planning, construction, and supporting and related services required to provide comprehensive services to *owners* and *developers.*

Construction Specifications

The primary sources of information about construction specifications in North America are the Construction Specifications Institute, Suite 300, 1150 Seventeenth Street, N.W., Washington, D.C. 20036, U.S.A.; and the Specification Writers Association of Canada, 1250 Bay Street, Toronto 5, Ontario, Canada.

Many manufacturers of construction materials and components also publish literature useful to specification writers, and several source lists are available from the above organizations and from government publications.

Architect's Specification Documents (K Series), Washington, D.C. 20006, U.S.A., The American Institute of Architects, Documents Division. A set of informative documents on various kinds of construction trade work for the specification writer.

CSI Manual of Practice and Specification Series, Washington, D.C. 20036, U.S.A., Construction Specifications Institute.

Rosen, Harold J., *Principles of Specification Writing,* New York, London, Reinhold Publishing Corporation, 1967. Probably the most current introduction to the subject in North America by a well-known expert.

Uniform Construction Index, A System of Formats for Specifications, Data Filing, Cost Analysis and Project Filing, obtainable from the Construction Specifications Institute and the Specification Writers Association of Canada. This published system is based on sixteen *standard divisions of work* and is endorsed by the architectural institutes and the specifications institutes of both the U.S.A. and Canada.

Estimating and Cost Accounting

The texts and manuals listed below provide factual information for estimators about the physical properties of *construction work* and also data as a guide for estimators who have to make estimates from the information and experience of others because they do not yet have enough of their own. They also provide useful information for the preparation of approximate and preliminary estimates.

Building Construction Cost Data, published annually by Robert Snow Means Company, Inc., P. O. Box G, Duxbury, Massachusetts 02332, U.S.A. Provides average *unit prices* and a relative construction cost index and labor index for many major U.S. and Canadian cities. It is inexpensive and useful.

Labor Calculator (man-hour units), Washington, D.C., U.S.A., National Association of Plumbing Heating Cooling Contractors, revised 1971. Excellent data and estimating guide for mechanical estimators. The concept and arrangement of the publication should be of interest to estimators in all kinds of construction firms because it illustrates methods that can be applied to *construction work* of all kinds.

Walker, Frank R., *The Building Estimator's Reference Book,* ed. by McClurg-Shoemaker, Chicago, Illinois, U.S.A., Frank R. Walker Company. Since 1915 there have been more than sixteen editions of this widely used book, which is full of information and data for the building estimator. Probably the most comprehensive, single-volume reference book for estimators in North America.

Walker, Frank R., *Practical Accounting and Cost Keeping for Contractors,* Chicago, Illinois, U.S.A., Frank R. Walker Company. It is described as "illustrating and describing in easy, understandable language, Bookkeeping and Accounting Systems for Contractors, giving complete instructions and examples of the proper methods of keeping time and compiling costs on all classes of construction work."

Richardson Engineering Services, Inc., *Manual of Commercial-Industrial Estimating and Engineering Standards (volume 1); Field Manual of Commercial-Industrial Construction Costs (volume 2); Manual of Residential—Light Construction Estimating and Engineering Standards (volume 3).* A very detailed annual publication of construction cost information with revised and additional pages sent periodically to annual subscribers. Published annually by Richardson Engineering Services, Inc., 10021 Tecum—P.O. Box 726, Downey, California 90241.

Quantity Surveying and Construction Economics

The primary sources of information on quantity surveying are the Canadian Institute of Quantity Surveyors, Suite 401, 8 Colborne Street, Toronto, 1, Ontario, Canada; and the Royal Institution of Chartered Surveyors, 12 Great George Street, Parliament Square, London, S.W.1, England. Both publish a monthly periodical and a standard method of measurement, and the RICS publishes a large number of papers on quantity surveying, construction economics, and other related subjects. Lists of publications are available from both these bodies. The CIQS has chapters in many of the major Canadian cities. The RICS is represented in Canada by a few regional committees, and its members are found in many other countries, particularly those in the Commonwealth.

Chartered Surveyor, the journal of the Royal Institution of Chartered Surveyors, published monthly in London, England. Available to non-members by subscription, the journal is published for all classes of chartered surveyors, not only for chartered quantity surveyors. Consequently, it also contains papers on such subjects as appraisal (valuation) and land surveying.

Cipher, the official publication of the Canadian Institute of Quantity Surveyors, published in Toronto, Canada.

Method of Measurement of Construction Works, Toronto, Ontario, Canada, The Canadian Institute of Quantity Surveyors. The third edition is current at this time, and additions and revisions are issued periodically to holders. Its chief object is to promote a "method of measurement of building works that may be used for teaching quantity surveying." There are similar publications in other countries, such as Australia, wherein the quantities method is practiced.

Standard Method of Measurement of Building Works, authorized by agreement between the Royal Institution of Chartered Surveyors and the National Federation of Building Trades Employers, London, England. The fifth edition, and the fifth edition metric are the latest editions at this time for use in Britain. Both publications are of value and interest in North America for their principles and precepts as well as for their technical contents. Revision of the document is under discussion.

Thompson, F.M.L., *Chartered Surveyors–the Growth of a Profession,* London, England, Routledge and Kegan Paul, 1968. A history of the profession, which includes chapters entitled "The Sixteenth Century Surveyor" and "The Origins of Quantity Surveying" and a view of the development of the design and construction industry in Britain as it relates to the surveying professions. Apart from its special interest to quantity surveyors, estimators, and designers, it is a generally interesting view of the industrial revolution that began in Britain.

Construction Management

Sources of information are the Associated General Contractors of America and the Canadian Construction Association. The development of management techniques in the last two decades has resulted in many publications on this subject, but not all have completely succeeded in applying the general techniques to the construction industry.

Benson, Ben, *Critical Path Methods in Building Construction,* Englewood Cliffs, N.J., U.S.A., Prentice-Hall, Inc., 1970. A well-illustrated introduction to practical CPM for the student and the superintendent.

Coombs, William E., *Construction Accounting and Financial Management,* New York, Toronto, and London, McGraw-Hill Book Company, 1958. Comprehensive and useful to both student and practitioner.

Deatherage, George E., P.E., *Books on Practical Construction Management, (1) Construction Company Organization and Management, (2) Construction Office Administration, (3) Construction Estimating and Job Preplanning, (4) Construction Scheduling and Control,* New York, Toronto, London, McGraw-Hill Book Company, 1964. This series gives a broad view of the subject and introduces some techniques not yet generally used by *contractors,* including "work simplification" and "methods engineering."

Geary, R., *Work Study Applied to Buildings,* London, England, George Godwin, Ltd., 1962. A short but good introduction with simple illustrated examples.

Lockyer, K.G., *An Introduction to Critical Path Analysis,* London, England, Sir Isaac Pitman and Sons, Ltd., 1964. A lucid explanation of the fundamentals of the technique.

O'Brien, James, *CPM in Construction Management,* New York, Toronto, and London, McGraw-Hill Company, 1965. Contains many figures and construction examples.

Suggested Guide for Field Cost Accounting, AGC Form No. 18, The Associated General Contractors of America, 1957 E Street, N.W., Washington 6, D.C., U.S.A., 1961. A well-illustrated guide to the fundamentals.

General and Miscellaneous

The following publications are listed because they enlarge on matters only briefly mentioned in the text.

Hollingdale, S.H., and G.C. Tootill, *Electronic Computers,* Pelican Book A524, Penguin Books, 1965.

Robinson, Joan, *Economic Philosophy,* Pelican Book A653, Penguin Books, 1964. A book that provides a better insight into economics than many of the larger and weightier books.

Saxon, James A., and Wesley W. Steyer, *Basic Principles of Data Processing,* Englewood Cliffs, N.J., U.S.A., Prentice-Hall, Inc., 1967. An easily understood introduction to the basics of electro-mechanical and computerized data processing systems.

Appendix A

Data and Probabilities of Construction Costs

The greatest need in applied construction economics is historical cost data and its analysis at an international level. It is obvious that most of the things that affect the costs of construction are essentially variable; but, because of the fundamental rationale underlying the design of most buildings, it should be possible to relate the costs of almost any construction project to the costs of previous projects that are similar in construction, function, and purpose. Likewise, at a deeper level of analysis, it should be possible to relate the costs of most building *elements* to the costs of similar *elements* in previous building construction projects.

Certainly, as the world shrinks to a village in which we all shall soon know more of one another and in which our life styles will merge, there will be fewer differences in our buildings and more similarities in our architecture and methods of construction. Already it is practically possible to collect and establish construction cost data at a continental level from which the probable costs of many types of buildings can be calculated—taking into consideration the particular factors of time and space (location)—on the basis of producing a building designed for a specific function and under the normal constraints that have applied to the earlier examples.[1] It appears that there are two major reasons why there is not a much wider utilization of historical cost data in design and construction that, in turn, could lead to more rational designs, more economical procurement of construction services, more effective management of construction, more equitable distributions of risk, and greater efficiency in all phases of design and construction. The two major reasons are:

1. a lack of recognition of the use and value of applied historical cost data.
2. an absence of means of collecting, analyzing, and integrating useful data.

These two reasons are related like the chicken and the egg.

Recognition of the value of utilizing historical cost data in design and construction may not come until data can be applied and its use demonstrated; and it may not be possible to effectively demonstrate its use until there exists the means of obtaining and utilizing the data.

Many who have tried to obtain construction cost data through *cost accounting* have found that the results have often been disappointing because of the sparseness of the data. And those who have tried to supplement their own data with that from technical publications have often been frustrated by the apparent absence of the relationships among data that they expected to be present. Obviously it is not enough to recognize that cost data (in adequate quantities) have a value. It is also necessary to have the means whereby data can be analyzed and factored and integrated into usable forms; and this requires a standard method and format that can be applied and used as widely as possible—that is, internationally.

First, a standard method of measurement of construction works is necessary at an international level, based on rational measurement of *work* (in place) for the purpose of costing. This would mean replacing some traditional methods with more rational and probably more simple measurement methods. The metric system would have to be used to achieve uniformity of units.

Next, it is necessary to have a uniform format for cost analysis on the basis of standard *elements* of buildings, and the application of this format and the standard method of measurement should be taught in all educational programs for design and construction. *Designers* must learn about economic constraints at the same time that they learn about the constraints of statics and building regulations; and *contractors* must account for all the *costs of construction work* that they originate.

Above all, *designers and contractors* must work together to produce the best and the most economical construction for society. Now that we have the mathematical methods and the computers that are able to handle and utilize large amounts of data, the probabilities of costs and risks can be computed so that risks can be minimized; and the residual risks can then be shared equitably between those who hope to gain from the construction contract. Competition need not be removed, only inordinate and unnecessary risk; and the inevitable remaining risk can be measured and its probability can be calculated.

A technique for risk analysis used by a major corporation has already been explained elsewhere,[2] and the simple theory of probability can be applied in making most judgements and decisions. It can be applied either through statistical probability based on historical data, or through inductive probability based on subjective judgements of the probabilities, or through a combination of both statistical and inductive probability, since there are invariably many differences between a subject currently under consideration and past subjects from which data have been derived.

Simple probability is easy to comprehend. If you toss a coin, the probability of a "head" is 0.5, and the probability of a "tail" is also 0.5. The probability of either a head or a tail is (0.5 + 0.5) = 1.0, which is a certainty. This assumes that the probability of the coin landing on edge is zero, or an impossibility.

The probability that there will be sunshine tomorrow (in March) might be as follows:

24 hours of sunshine in one day (no chance)	0.000
12 hours of sunshine in one day (most unlikely), say	0.001
6 hours of sunshine in one day (possible), say, ...	0.333
3 hours of sunshine in one day (most likely), say, ...	0.500
1/4 hour of sunshine in one day (unlikely), say,..........	0.166
That the sun will rise tomorrow (a certainty)	1.000

Of course, it depends on what we mean by sunshine in the example, but it does simply illustrate the theory. That there will be no night and twenty-four hours of sunshine tomorrow may not be an absolutely impos-

[1] For example, the building cost data contained in the volumes of the *Boeckh Building Valuation Manual* and the *Building Cost Modifiers* that are issued several times a year to provide subscribers with "Time—Location Modifiers" to adjust the basic cost data for most cities in North America.

[2] Douglas W. Campbell, Refinery and Chemical Division, Bechtel Corporation, San Francisco, California, "*Risk Analysis*" (a paper presented at the 14th Annual Meeting of the American Association of Cost Engineers, June 1970, San Francisco, California). *AACE Bulletin*, Morgantown, West Virginia, August/October 1971, Vol. 13, Nos. 4 and 5, pp. 8–11.

sible phenomenon,[3] but most of us would bet all we have against it. That is to say, if the proposition were made that the sun would shine for twenty-four hours tomorrow, we should have to say that statistically and inductively the probability is zero. Probability relates to a proposition.

Estimators use the inductive method of judging probabilities all the time, often without realizing it. Also, they do not usually record their inductive processes; and even bearing in mind the pressure under which many estimates are prepared, there are nevertheless many times when a brief record would make subsequent checking of an estimate against the eventual facts both possible and valuable. This procedure would then enable an estimator to refine his estimates and to improve his judgment of probabilities; and also it would help to produce statistics from which data statistical probabilities could be calculated to assist in the inductive method.

As an example of the inductive method, let us assume that one of many *unit rates* to be estimated is for the *labor costs* of a particular *item of work*. With historical cost data we might apply the statistical method, and this might give us this result:

2400 CY @ 0.5 man-hours per CY = 1200 man-hours

However, because the *unit rate* (0.5 man-hours per CY) has been taken from jobs that were not identical to the subject project, and because the job conditions are different, it is necessary to make a subjective judgment. The probabilities might be induced as follows:

Manhours per CY	*(P)*	
0.40	0.00	(an impossibility)
0.50	0.20	
0.60	0.70	
0.70	0.10	
0.80	0.00	(an impossibility)

The probability (*P*) that the *work* in question will take less than 0.5 man-hours per CY is counted (in the example) as an impossibility. Likewise, that it will go as high as 0.8 man-hours per CY is also counted as an impossibility. The other probabilities are given the values shown, based on as much information as is available, and finally on subjective judgment. The probabilities listed add up to 1.0. The final probability is simply calculated as follows:

$$
\begin{aligned}
0.40 \times 0.00 &= 0.00\\
0.50 \times 0.20 &= 0.10\\
0.60 \times 0.70 &= 0.42\\
0.70 \times 0.10 &= 0.07\\
0.80 \times 0.00 &= \underline{0.00}\\
& 0.59
\end{aligned}
$$

This is the *unit rate* (0.59 man-hours) to be used in the estimate. When the *work* is done and cost accounted, the *unit rate* can be checked and the new bit of cost information added to the data for future estimates. Subjective judgments may be aided substantially by statistical data and by the use of histograms compiled for the most important *items of work*. Minor items can be judged subjectively without the aid of statistics illustrated in histograms, and without seriously impairing the whole estimate.

With sufficient data it would be possible to show the most probable costs of all *items of work* in an estimate; and by weighting the costs of each item against the total costs, it would be possible (with a computer) to arrive at the most probable *costs of the work* for the project. By random sampling techniques applied to the many variations of costs possible within the estimate, the probable effect of the variables on the total estimated costs could be shown, thus indicating mathematically the probabilities of the actual costs being more or less than the estimated costs. With this information, executive management would then be in a better position to make decisions about the risk involved in doing *work* for a specified amount, and about the amount of contingency and *profit* that should be added to the estimated costs to arrive at the sum to be bid or the limits for negotiation.

Probabilities of construction costs can be predicted much more accurately than they frequently are; and if all the means available for getting the most for the least are not utilized, one of the parties to the contract will have to bear a heavy risk. Risk should be borne by either party according to their ability to control it, and that risk that cannot be controlled should be borne according to the amount of ultimate benefit to be gained.

[3] See Joshua 10:12–13. Also, Immanuel Velikovsky, *Worlds In Collision* (New York: Dell Publishing Co., Inc., 1965), Chapter 1.

Appendix B

A Preliminary Report and Commentary on the RICS Chartered Quantity Surveyors' Mission to the United States of America in 1971

In the fall of 1971, five British chartered quantity surveyors visited the United States at the invitation of the American Institute of Architects to sell the idea of quantity surveying as a profession in the North American construction industry. Their hopes no doubt were warmed by the fact that so many other people had visited Britain in the past few years to find out about quantity surveying and had gone away expressing at least interest, and in some cases a desire, for its adoption.

The five members of the group were chosen for their past and present positions as chartered quantity surveyors and as representatives of both the private and public sectors of the profession. But the absence from the group of a "landed immigrant," if not indigenous quantity surveyor—someone who knew better the lie of the land and the language—might have been a little rash. It is a common mistake to believe that going from Britain to North America is not the same as, say, going from Britain to Asia when, in fact, because of the false sense of familiarity with language and custom, the first assumption can be just as hazardous. However, the first reports show that the group did well, and that they had no illusions about anything. They talked about cost budgets, cost planning, and cost controls; about independent cost consultants and cost advice; and about bidding and contracts and the procurement of cost data. They talked to hundreds of persons at almost forty meetings and lectures, and they tried to sell an idea.

The group apparently recognized that the main point of misunderstanding about the quantity surveying profession and the quantities method undoubtedly lay in the general North American view of construction costs from the contractor's position. The quantity surveyor's view is essentially that of the *owner*, who is concerned primarily with the completed project and its costs. The professional quantity surveyor is, after all, like the architect, usually an agent of the *owner*; and with the architect he is primarily concerned with his client's interests—the interests of the *owner*. While in North America the study of construction costs is usually by those primarily interested in production rather than in the final product itself.

The quantity surveyor sees, for example, the exterior walls of a building as an essential *element* with specific functions that must be provided for a cost; and he knows that the costs of that *element* and its functions should be properly related to the costs of all the other *elements* in the building and to their functions. He sees the costs, including the *contractor's overhead and profit*, as a charge against the *owner* that is determined by the economics of construction, and that can be varied, within certain limits, by changing the design and specifications of the *work* and the method of procuring the *work*. The *contractor*, on the other hand, primarily sees the building as a designed requirement consisting of materials, labor, plant and equipment, and the *work* of *subcontractors*, all to be integrated at whatever costs the current market produces and to bear a mark-up for his *overhead and profit*. He is primarily interested in production rather than the product, because the decision to utilize a particular product (*element*) in the building has already been made, and he is required only to produce it.

More than one of the visiting surveyors saw that generally the North American view is different in this respect from their own and that, perhaps in part because of the traditional reference to quantities in their title, they were often regarded, if not as makers of materials lists, then as glorified estimators. One North American journal put it well when it suggested with light pleasantry that quantity surveyors might be described as contract managers with British accents. At least, the emphasis in that journal's description was on the quantity surveyor's services to his client, the *owner*.

Perhaps not unrealistically, the first published report[1] nowhere indicated any reference by any one of the group of five to the existence of the Canadian Institute of Quantity Surveyors or its published method of measurement and standard form of cost analysis, although they did refer to the need for such things in North America. It is true that in the past the CIQS has rather tended to give backward looks toward Britain and other Commonwealth countries rather than toward a new concept for the New World, even to perpetuating the obsolescent title of "quantity surveyors," which is understandably mistaken by many even in its country of origin.

Given the rising tempo of change in the design and construction industry, as in everything else, it may be that the quip about contract managers is prophetic. After all, the birth and development of the profession of quantity surveying owed a lot to the withdrawal of British architects and their counterparts from a real working involvement in construction economics. If there is a space to be filled in the construction industry in North America, perhaps it is already being filled by that much hailed and latest arrival the *contract manager*. It is true that until lately he was in most instances a contractor-builder (although in Canada he may have been a quantity surveyor). But if he recognizes the need for standards of practice and the advantages of sharing and accumulating information, and if he and others get together to promote these ends, we may find that the North American variety may prove to be what is needed in other continents as well.

[1] Published in the British design and construction journal, *BUILDING*, Friday, 26 November 1971 (No. 6706, Vol. CCXXI), pp. 72–75.

Appendix C

Mensuration of Areas and Volumes

Areas of Regular Figures

Area = length of one side × ½ altitude (altitude is perpendicular distance to corner opposite the side measured); i.e., $A = z \times \frac{1}{2}a$

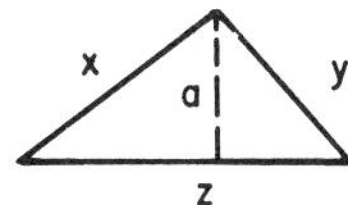

TRIANGLE

Area = length of one side × altitude (altitude is perpendicular distance to opposite side); i.e., $A = z \times a$

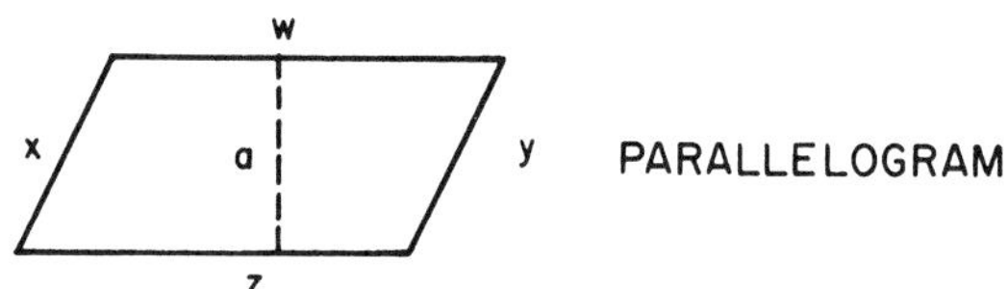

Area = half the sum of the parallel sides × altitude (altitude is perpendicular distance between the parallel sides); i.e., $A = \frac{1}{2}(w + z) \times a$

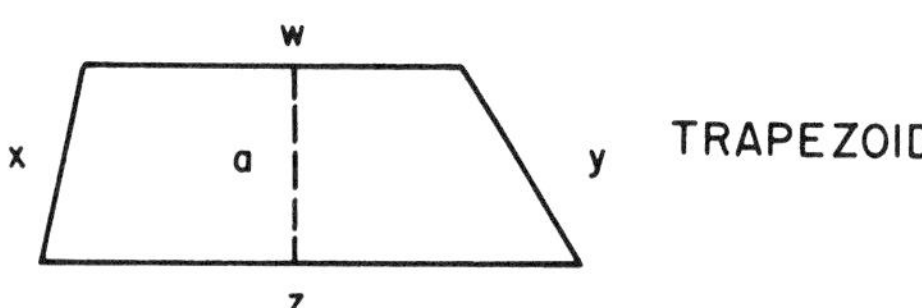

Area = ½ sum of all sides × inside radius

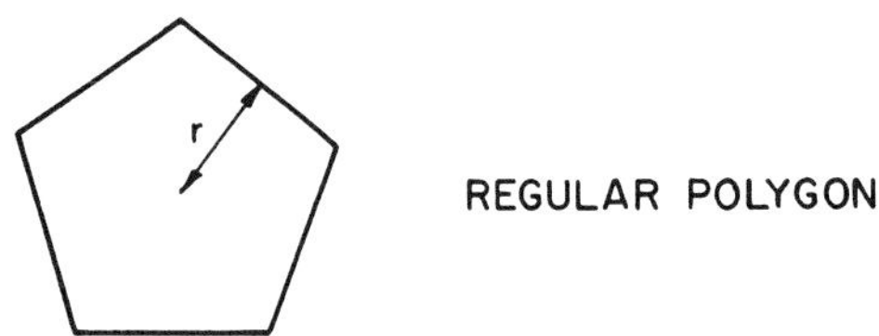

Area $= \pi r^2$
$= 0.7854 \times \text{diameter}^2$
$= 0.0796 \times \text{circumference}^2$

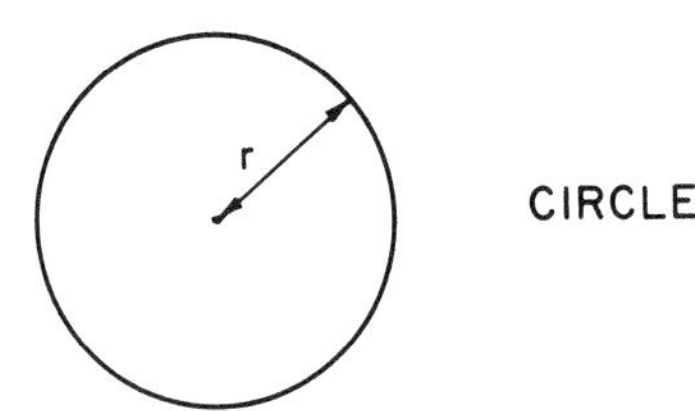

Area = length of arch × ½ radius
$= \frac{\alpha^\circ}{360^\circ} \times \pi r^2$

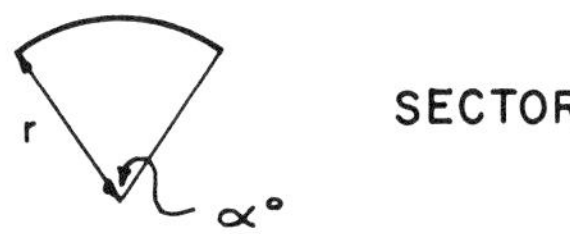

Area = major axis × minor axis × 0.7854; i.e., $A = a \times b \times 0.7854$

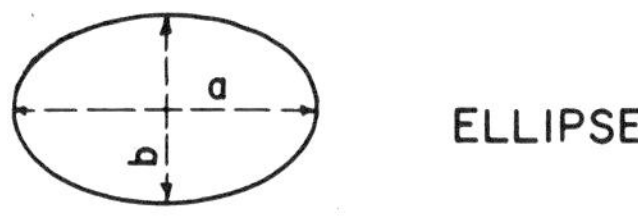

Area = base × ⅔ altitude; i.e., $A = b \times \frac{2}{3}a$

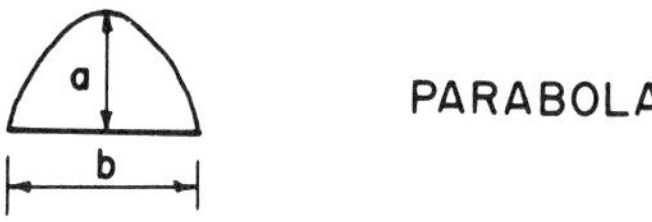

Area of an Irregular Figure

Procedure. Divide the irregular figure into strips of any equal width (d) by equally spaced parallel lines. Measure the length of each of the parallel lines. Apply one of the following rules to calculate the area:

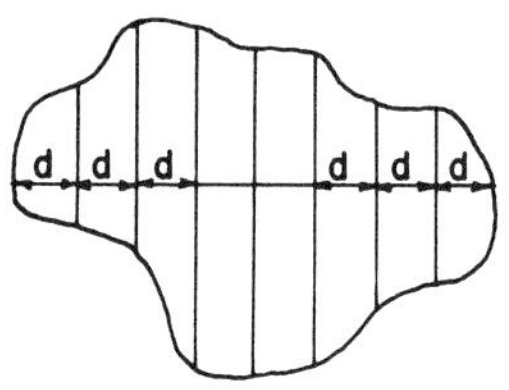

1. **Trapezoid Rule.** Add together the lengths of all the parallel lines and deduct ½ the length of the first line and ½ the length of the last line. Multiply the total by the standard distance (d) between the lines. (Usually accurate enough for estimating site work and similar items).

2. **Simpson's Rule.** This rule requires an **even number of strips.** Add together the lengths of the parallel lines, taking the first and last lines at actual length (1 × value); the second, fourth, six, etc., **from each end** at 4 × value; the third, fifth, etc., **from each end** at 2 × value. Multiply the total by one third of the standard distance ($1/3d$). (Very accurate for figures bounded by smooth curves).

3. **Durand's Rule.** Add together the lengths of the parallel lines, taking the first and last lines at $\frac{5}{12}$ value; the second **from each end at** $\frac{13}{12}$ × value; and

all the others at full value (1 × value). Multiply the total by the standard distance (*d*) between the lines.

Notes: The smaller the width of the strips—the standard distance (*d*)—the more accurate are the results. The areas also can be measured with a planimeter.

Surface—Areas and Volumes of Regular Solids

Volume $= \frac{4}{3} \pi r^3$
$= 0.5236\, d^3$
S.Area $= 4 \pi r^2$
$= 3.14159265\ d^2$

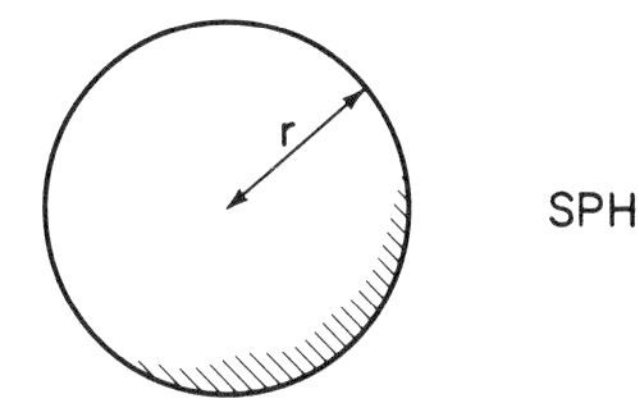

SPHERE

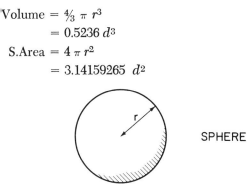

Volume $= \frac{1}{6} \pi abc$
S.Area (no simple rule)

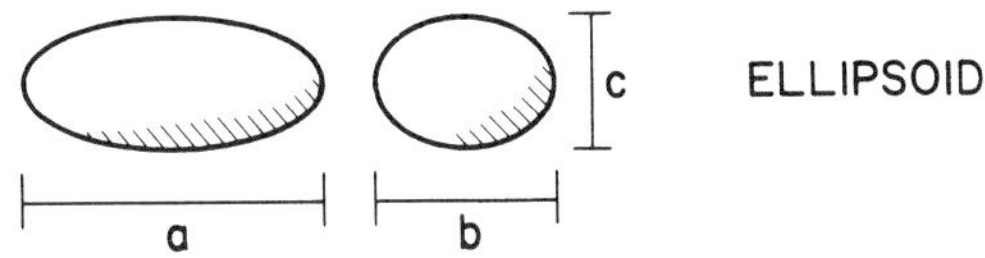

ELLIPSOID

Volume = Area circular base × ½ height; i.e., $V = \pi r^2 \times \frac{1}{2} h$

S.Area (no simple rule)

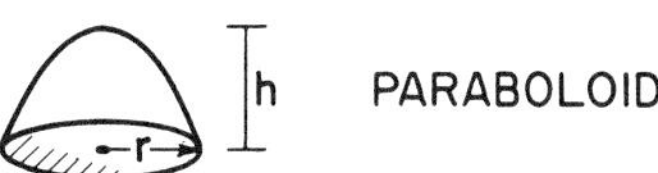

Volume of an Irregular Solid

The Prismoidal (Prismatoid) Formula. This formula is reasonably accurate for measuring volumes contained by two parallel planes at the ends connected by planes containing straight lines or smooth curves, as shown.

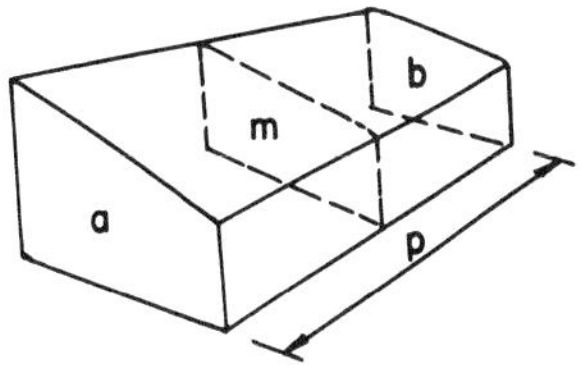

Volume = sum of the areas of the two parallel planes plus 4 × area of the parallel mid-section multiplied by ⅙ of the perpendicular distance between the two parallel ends; i.e., $V = (a + b + 4m) \times \frac{1}{6} p$

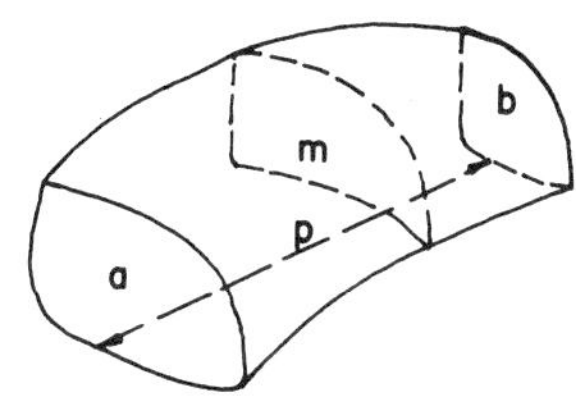

Index

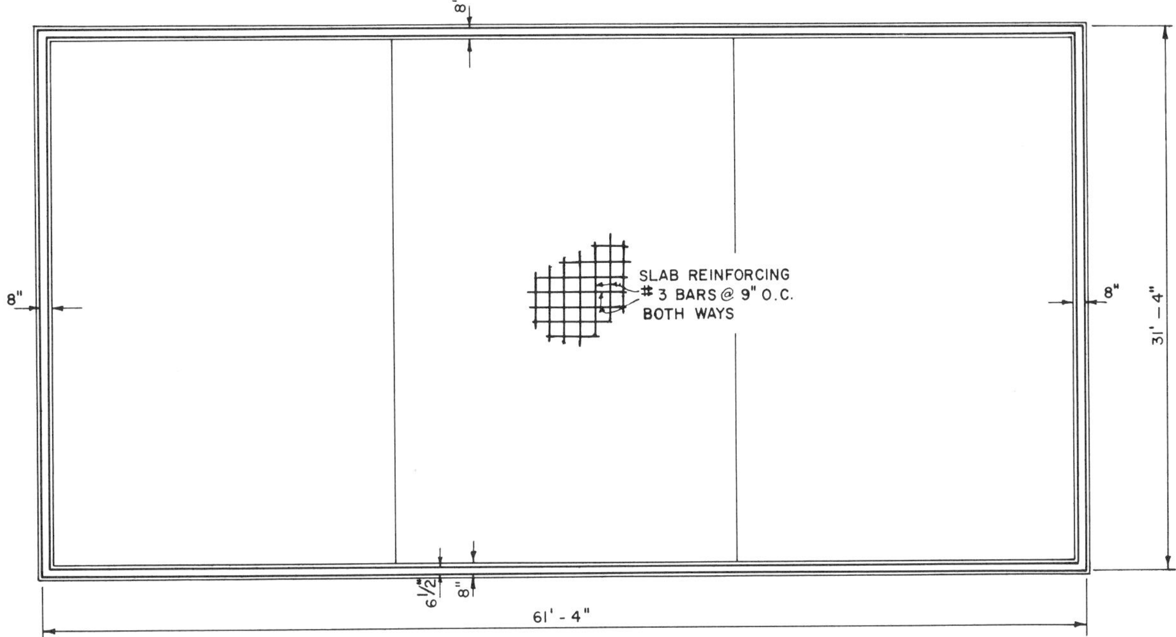

PLAN

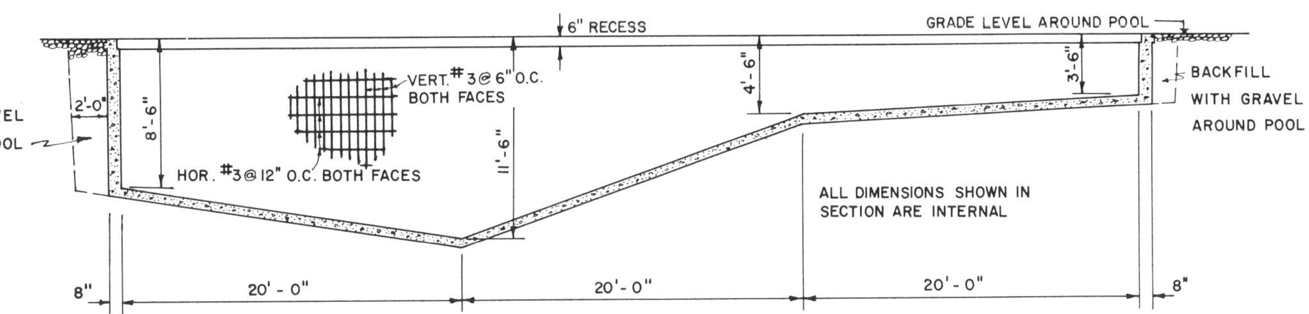

LONGITUDINAL SECTION (NOTE: SHAPE OF POOL SECTION AS SHOWN FOR PURPOSE OF EXAMPLE ONLY)

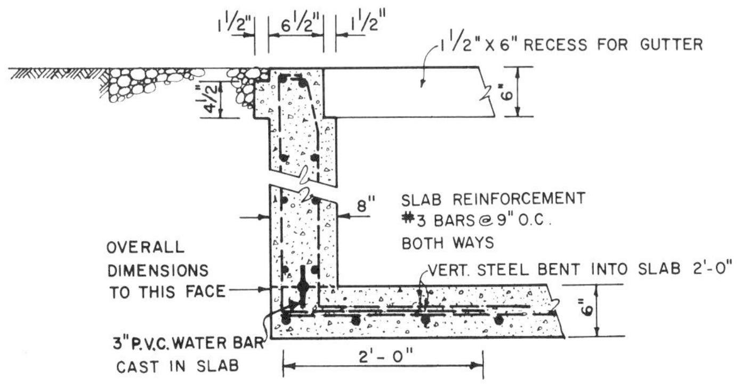

TYPICAL SECTION AT SIDES

(END WALLS SIMILAR)

NOTES

1. 3500 P.S.I. - 3/4" MAX. AGGREGATE; UNFINISHED CONCRETE.
2. #3 BARS : (0.376 LB. PER LIN. FT.) GRADE 60
3. VERT. BARS BENT 90° , 2'-0" INTO SLAB.
4. CONC. SLAB (ZERO SLUMP) WITH FLOATED FINISH; LAID ON NATURAL GROUND

GROUND CONDITIONS & FILL

1. COMPACTED GRAVEL & CLAY
2. NO TOPSOIL AT SURFACE; SITE LEVEL
3. FILL WITH IMPORTED PIT-RUN GRAVEL AGAINST WALLS

P-3
SWIMMING POOL

4" CONC. PAVING SURROUND NON-SLIP FINISH
ON 6" GRAVEL UNDER BED

CONTROL JOINT AT WALL &
IN PAVING MAX. 20'-0" APART.

10'-0"
42'-0" (INSIDE CONCRETE)
10'-0"

4" CONC PAVING SURROUND NON-SLIP FINISH
ON 6" GRAVEL BED

10'-0" 82'-6" (INSIDE CONCRETE) 10'-0"

PLAN

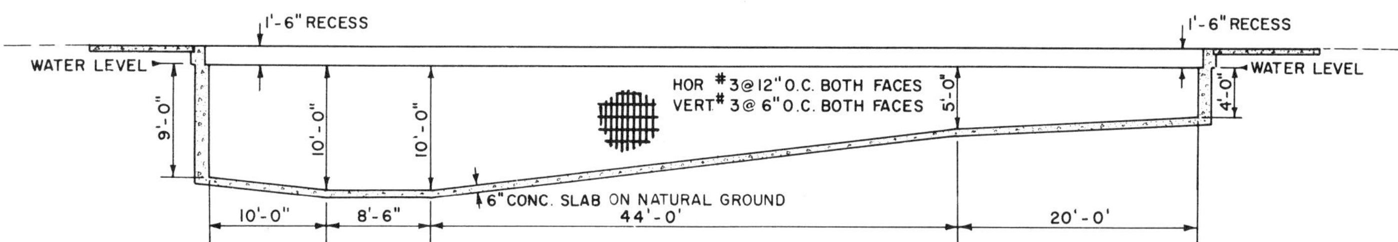

TYPICAL SECTION

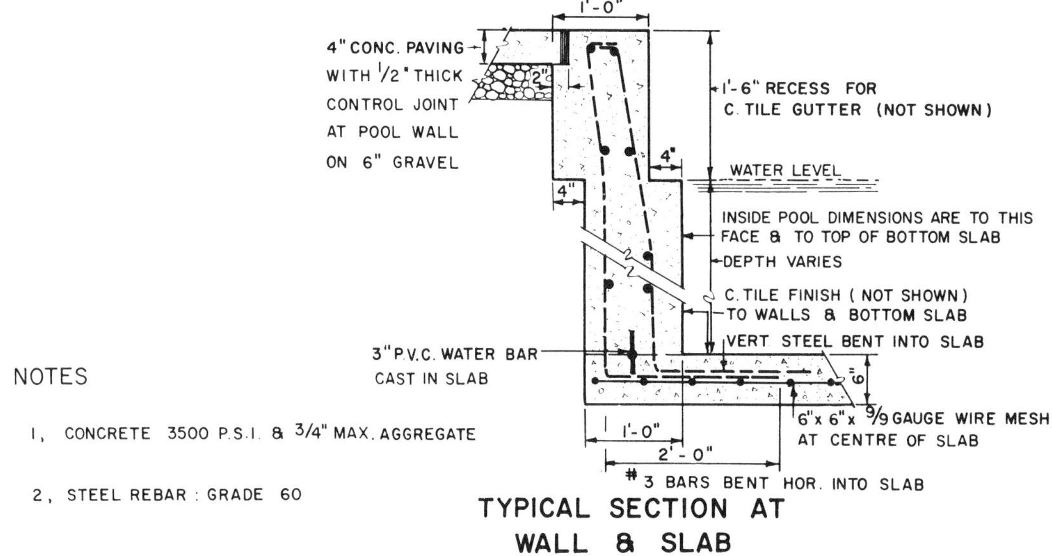

TYPICAL SECTION AT WALL & SLAB

NOTES

1, CONCRETE 3500 P.S.I. & 3/4" MAX. AGGREGATE

2, STEEL REBAR : GRADE 60

P-3A
SWIMMING POOL
(for estimating exercise)

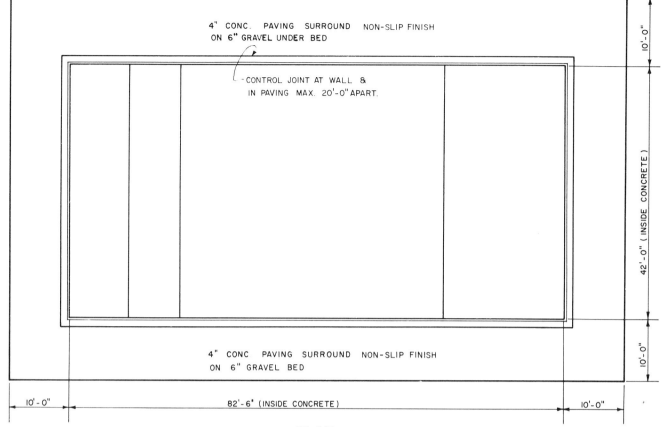

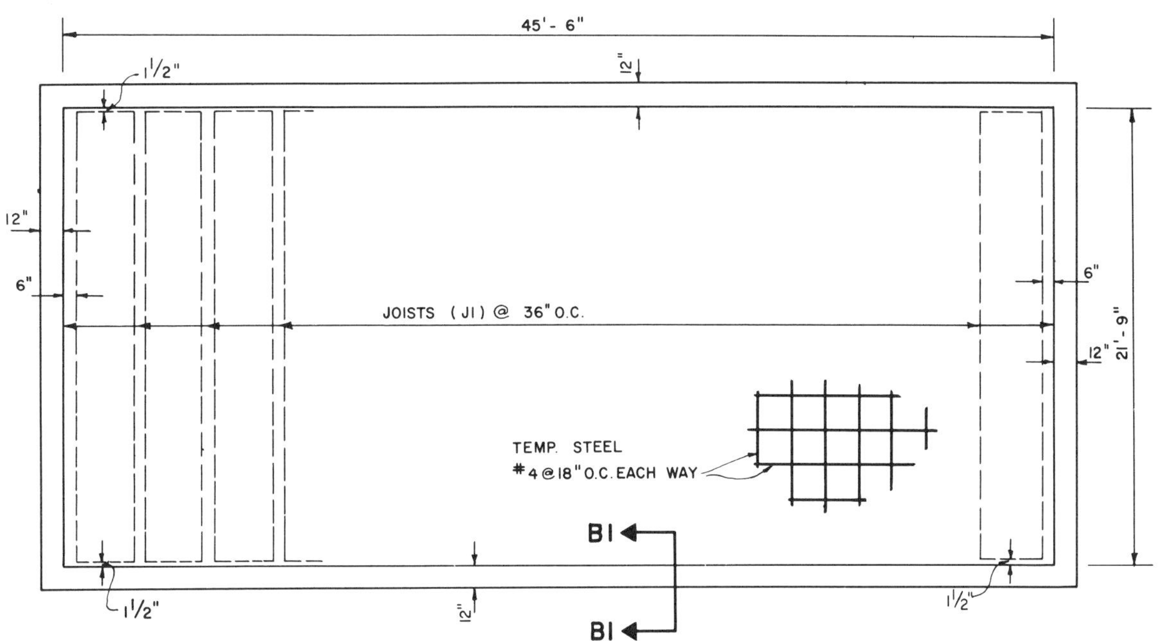

PLAN OF ROOF SLAB

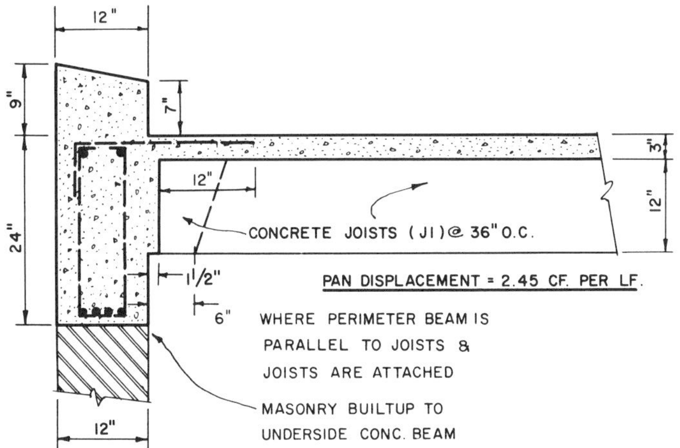

SECTION (B1) TYPICAL PERIMETER BEAM - B1

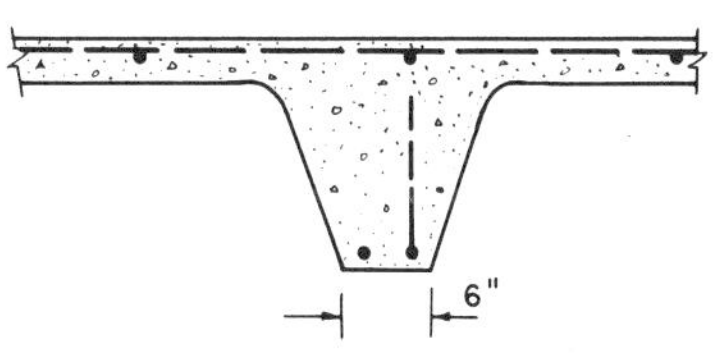

TYPICAL JOIST - J1

SPEC. NOTES:

1. CONCRETE 3500 P.S.I. AT 28 DAYS, 3/4" MAX. SIZE AGGREGATE
2. STEEL GRADE 60, MAX. BAR LENGTH 30 FT, SPLICED LAPS 36 x DIAMETER

REBAR SCHEDULE - JOISTS AND BEAMS

MK.	SECTION	NO	SIZE	LENGTH	NO	SIZE	LENGTH	BENDS	SPACING
		STRAIGHT			BENT				
J1	3, 12, 6 (joist section) NOTE: J1- JOISTS ATTACHED TO PERIM. BEAM ARE NOT REINFORCED	1	#7	22-9	1	#8	25-6	4-0, 1-4 1/2, 4-0; 1-0, 12-9, 1-0	36" O.C.
B1	9, 24, 12 (beam section)	4	#8	CONT.	4x4	#8	6-0	3 3	SPLICES AT CORNERS
		2	#5	CONT.	4x2	#5	4-0	2 2	
					-	#4	5-0	90° STANDARD STIRRUP HOOKS	18" O.C.
					-	#4	2-9	9, 2-0	18" O.C.

WEIGHTS OF REBAR: (PER LIN. FOOT)

#4 - 0.668 LB. #7 - 2.044 LB.

#5 - 1.043 LB. #8 - 2.670 LB.

P-4
CONCRETE PAN JOIST SLAB

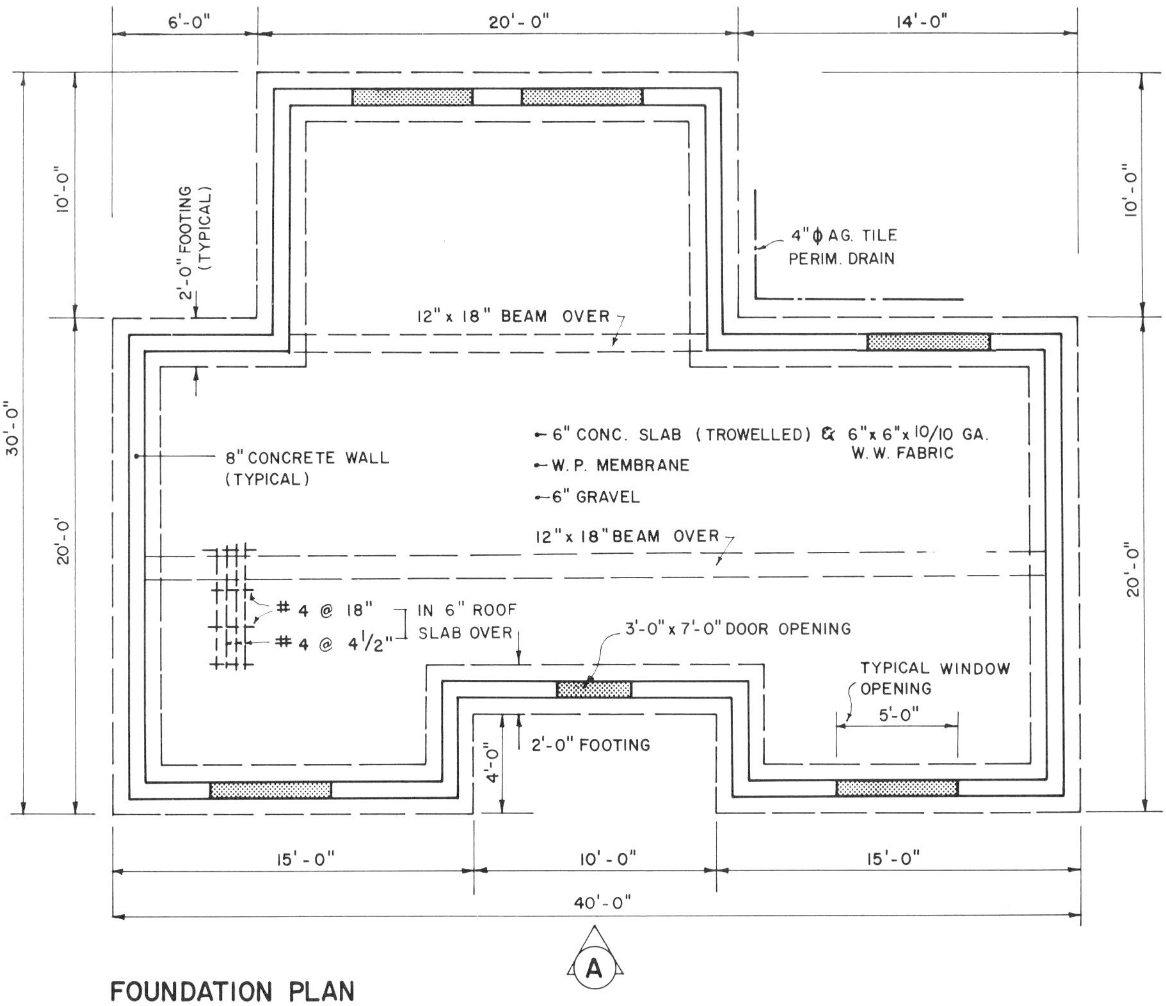

6'-0"
20'-0"
14'-0"
10'-0"
30'-0"
20'-0'
10'-0"
20'-0"
2'-0" FOOTING (TYPICAL)
4" φ AG. TILE PERIM. DRAIN
12" x 18" BEAM OVER
8" CONCRETE WALL (TYPICAL)
6" CONC. SLAB (TROWELLED) & 6" x 6" x 10/10 GA. W. W. FABRIC
W. P. MEMBRANE
6" GRAVEL
12" x 18" BEAM OVER
4 @ 18"
4 @ 4 1/2"
IN 6" ROOF SLAB OVER
3'-0" x 7'-0" DOOR OPENING
TYPICAL WINDOW OPENING
5'-0"
4'-0"
2'-0" FOOTING
15'-0"
10'-0"
15'-0"
40'-0"
A

FOUNDATION PLAN

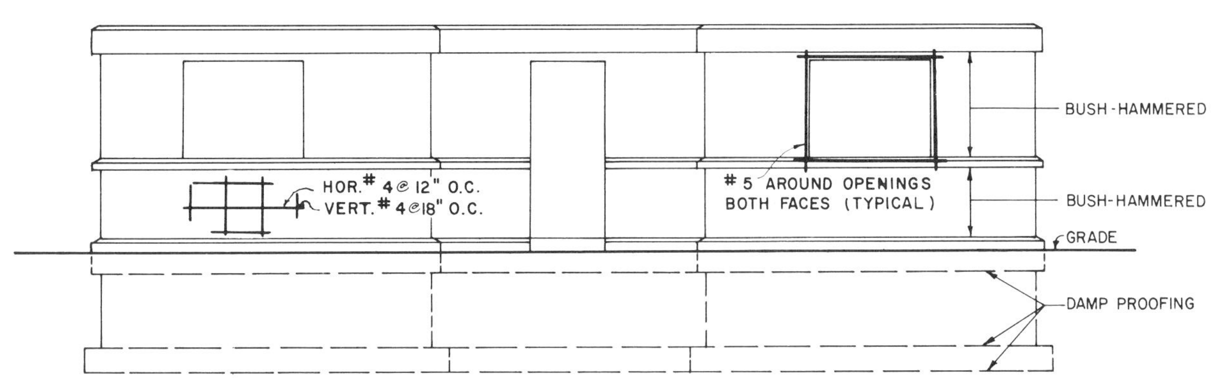

(A) FRONT ELEVATION (OTHERS SIMILAR)

SHEET 1 OF 2

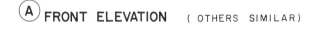

P-5
CONCRETE BUILDING —PLAN & ELEVATION

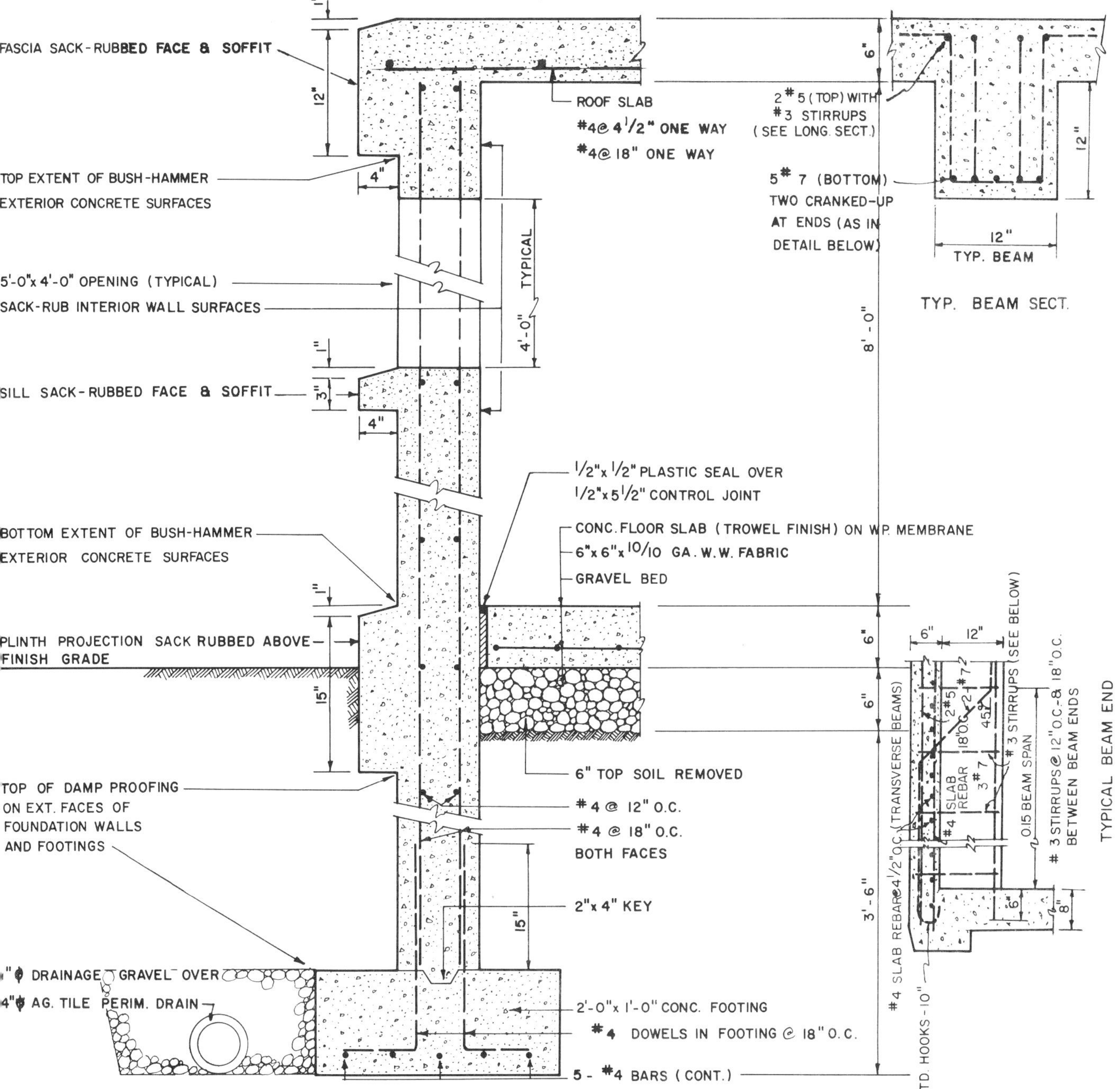

TYPICAL SECTION

SPECIFICATION NOTES

1, ALL CONCRETE 3000 P.S.I. (1 1/2" MAX. AGG. SIZE IN FOOTINGS, 3/4" MAX. ELSEWHERE)

2, BUSH-HAMMER ALL EXTERIOR VERTICAL SURFACES, AS INDICATED.

3, SACK RUB REMAINING VERTICAL SURFACES WHERE EXPOSED & SOFFITS OF SILL & FASCIA (NOT SLAB)

4, STEEL TROWEL FLOOR SLAB. (TO RECEIVE FLOOR TILING) AND SLOPING TOPS OF PROJECTIONS

5, WOOD FLOAT ROOF SLAB (TO RECEIVE INSULATION & ROOFING)

6, DAMP PROOF EXTERIOR SURFACES CONC. WALLS AND FOOTINGS BELOW PLINTH.

7, SITE ASSUMED TO BE LEVEL.

SHEET 2 OF 2

P-5

CONCRETE BUILDING -TYPICAL WALL SECTION

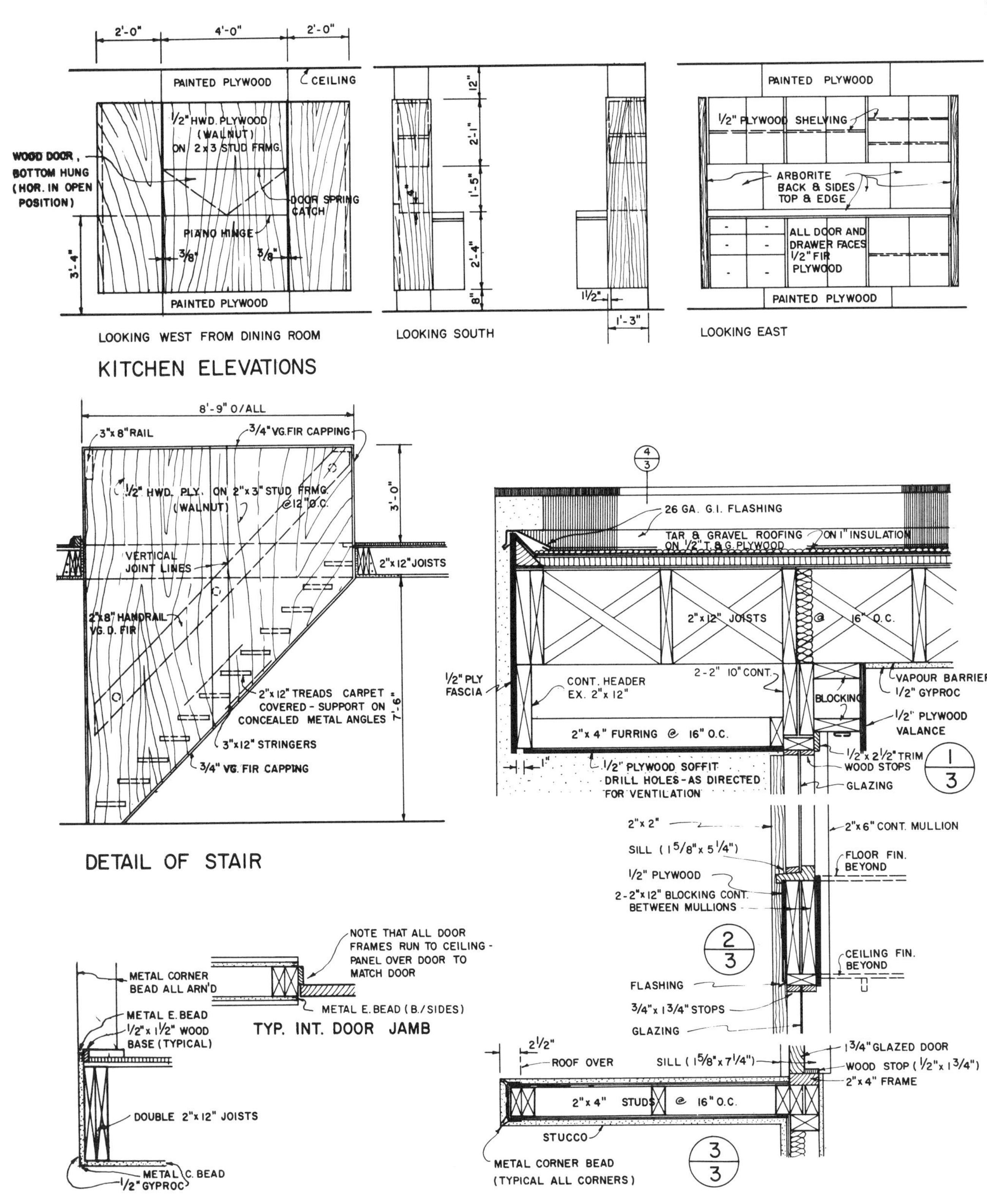

2'-0"
4'-0"
2'-0"
PAINTED PLYWOOD
CEILING
1/2" HWD. PLYWOOD (WALNUT) ON 2 x 3 STUD FRMG.
WOOD DOOR, BOTTOM HUNG (HOR. IN OPEN POSITION)
DOOR SPRING CATCH
PIANO HINGE
3/8"
3/8"
3'-4"
PAINTED PLYWOOD
LOOKING WEST FROM DINING ROOM
12"
2'-1"
1'-5"
2'-4"
8"
1 1/2"
1'-3"
LOOKING SOUTH
PAINTED PLYWOOD
1/2" PLYWOOD SHELVING
ARBORITE BACK & SIDES TOP & EDGE
ALL DOOR AND DRAWER FACES 1/2" FIR PLYWOOD
PAINTED PLYWOOD
LOOKING EAST
KITCHEN ELEVATIONS
8'-9" O/ALL
3"x 8" RAIL
3/4" VG. FIR CAPPING
1/2" HWD. PLY. ON 2"x 3" STUD FRMG. @ 12" O.C.
(WALNUT)
3'-0"
VERTICAL JOINT LINES
2"x 12" JOISTS
2"x 8" HANDRAIL VG. D. FIR
2"x 12" TREADS CARPET COVERED - SUPPORT ON CONCEALED METAL ANGLES
3"x 12" STRINGERS
3/4" VG. FIR CAPPING
7'-6"
DETAIL OF STAIR
4/3
26 GA. G.I. FLASHING
TAR & GRAVEL ROOFING ON 1/2" T. & G. PLYWOOD
ON 1" INSULATION
2"x 12" JOISTS @ 16" O.C.
2-2" 10" CONT.
VAPOUR BARRIER
1/2" GYPROC
1/2" PLY FASCIA
CONT. HEADER EX. 2"x 12"
BLOCKING
1/2" PLYWOOD VALANCE
2"x 4" FURRING @ 16" O.C.
1/2 x 2 1/2" TRIM
WOOD STOPS
1/3
1"
1/2" PLYWOOD SOFFIT DRILL HOLES - AS DIRECTED FOR VENTILATION
GLAZING
2"x 2"
2"x 6" CONT. MULLION
SILL (1 5/8" x 5 1/4")
FLOOR FIN. BEYOND
1/2" PLYWOOD
2-2"x 12" BLOCKING CONT. BETWEEN MULLIONS
2/3
CEILING FIN. BEYOND
FLASHING
3/4" x 1 3/4" STOPS
GLAZING
NOTE THAT ALL DOOR FRAMES RUN TO CEILING - PANEL OVER DOOR TO MATCH DOOR
METAL CORNER BEAD ALL ARN'D
METAL E. BEAD (B./SIDES)
TYP. INT. DOOR JAMB
METAL E. BEAD
1/2" x 1 1/2" WOOD BASE (TYPICAL)
DOUBLE 2"x 12" JOISTS
METAL C. BEAD
1/2" GYPROC
DETAIL OF UPPER FLOOR EDGE
2 1/2"
ROOF OVER
SILL (1 5/8" x 7 1/4")
1 3/4" GLAZED DOOR
WOOD STOP (1/2" x 1 3/4")
2"x 4" FRAME
2"x 4" STUDS @ 16" O.C.
STUCCO
3/3
METAL CORNER BEAD (TYPICAL ALL CORNERS)
TYPICAL WALL & ROOF DETAILS

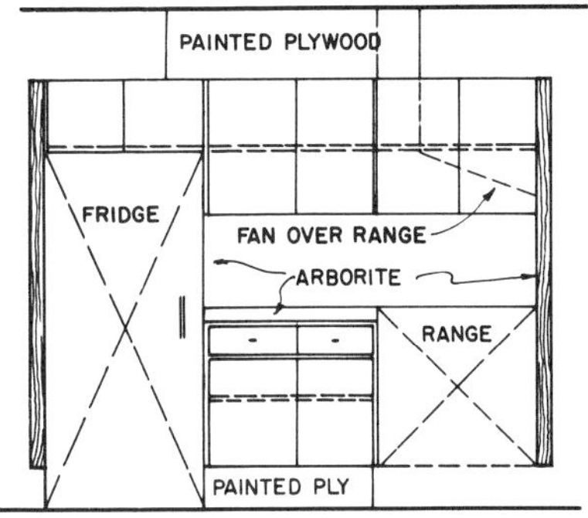

LOOKING WEST

NOTES

1. ALL WOOD & PLYWOOD D. FIR UNLESS OTHERWISE INDICATED
2. ALL EXPOSED SURFACES PAINTED TWO COATS UNLESS OTHERWISE INDICATED
3. HWD. PLYWOOD SURFACES, OILED & RUBBED FINISH

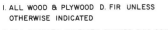

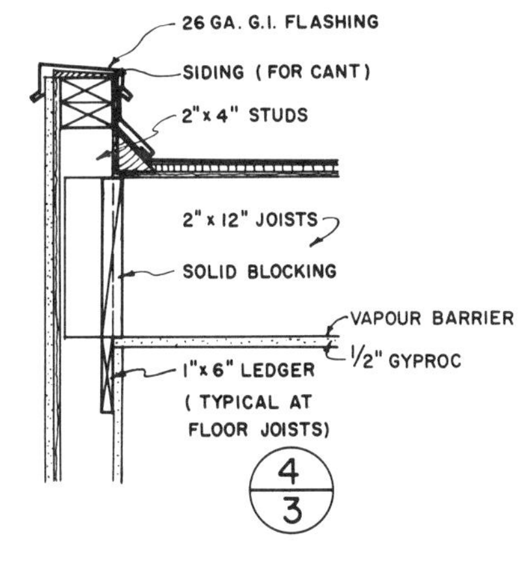

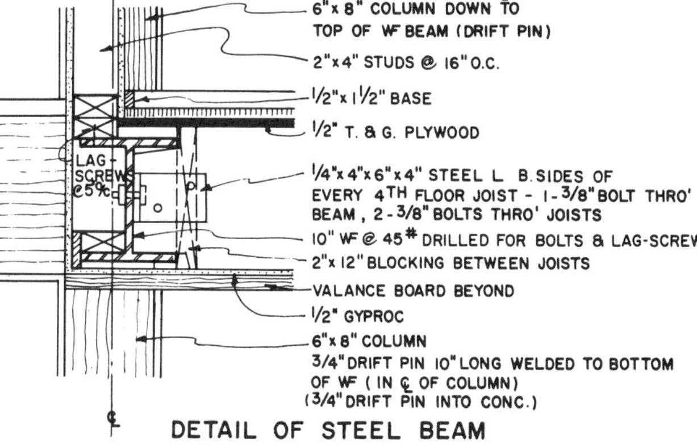

DETAIL OF STEEL BEAM

TO COLUMN CONNECTION

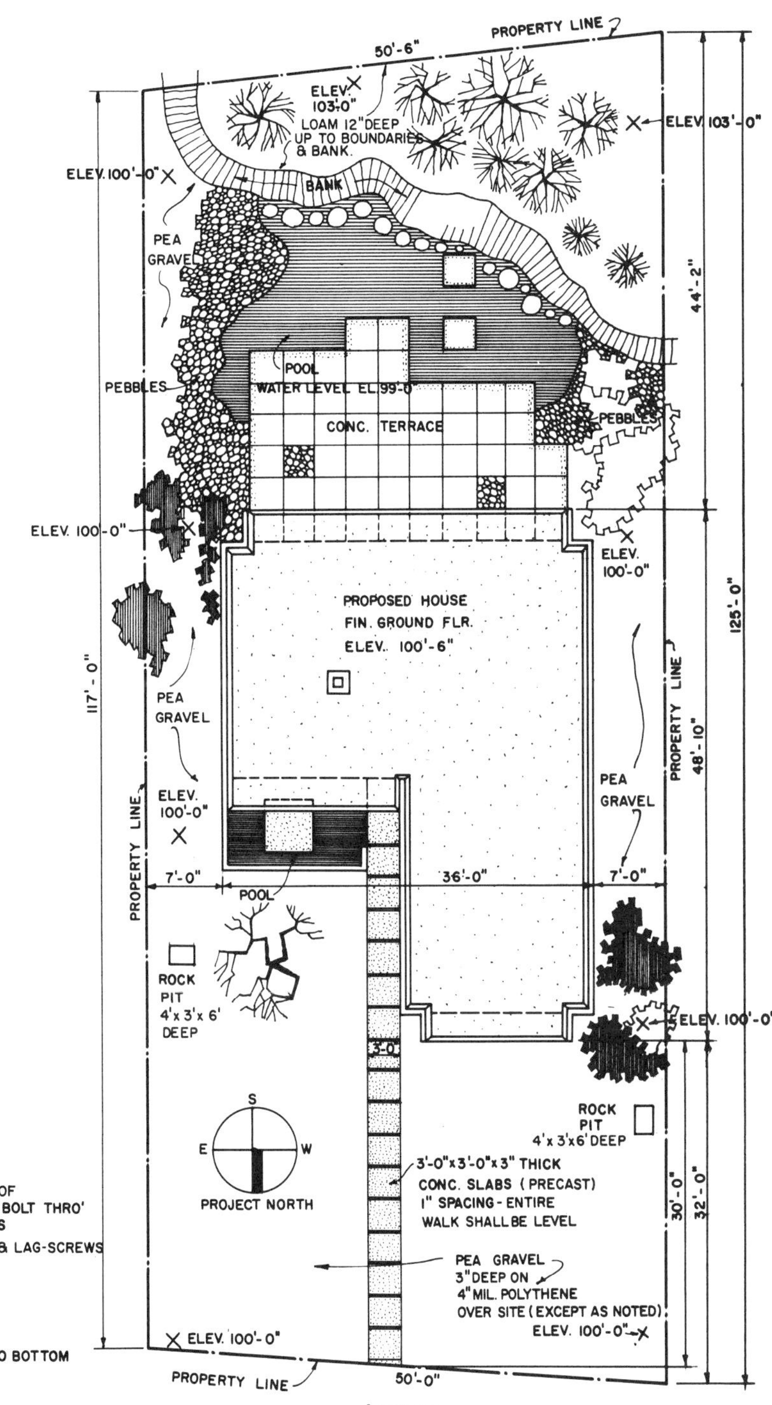

SITE PLAN

RESIDENCE 

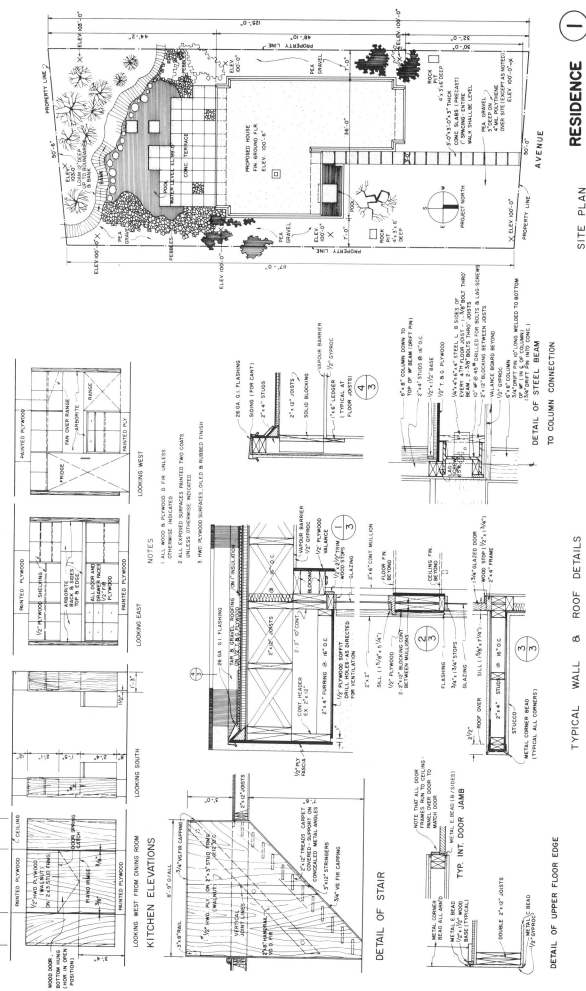

KITCHEN ELEVATIONS

LOOKING WEST FROM DINING ROOM

2'-0" | 4'-0" | 2'-0"
PAINTED PLYWOOD
CEILING
1/2" HWD. PLYWOOD (WALNUT) ON 2 x 3 STUD FRMG.
WOOD DOOR, BOTTOM HUNG (HOR. IN OPEN POSITION)
DOOR SPRING CATCH
PIANO HINGE
3/8" | 3/8"
3'-4"
PAINTED PLYWOOD

LOOKING SOUTH

12" | 2'-1" | 1'-5" | 2'-4" | 8"
1 1/2" | 1'-3"

LOOKING EAST

PAINTED PLYWOOD
1/2" PLYWOOD SHELVING
ARBORITE BACK & SIDES TOP & EDGE
ALL DOOR AND DRAWER FACES 1/2" FIR PLYWOOD
PAINTED PLYWOOD

LOOKING WEST

PAINTED PLYWOOD
FRIDGE
FAN OVER RANGE
ARBORITE
RANGE
PAINTED PLY

NOTES

1. ALL WOOD & PLYWOOD D. FIR UNLESS OTHERWISE INDICATED
2. ALL EXPOSED SURFACES PAINTED TWO COATS UNLESS OTHERWISE INDICATED
3. HWD. PLYWOOD SURFACES, OILED & RUBBED FINISH

DETAIL OF STAIR

8'-9" O/ALL
3"x 8" RAIL
3/4" VG. FIR CAPPING
1/2" HWD. PLY. ON 2"x 3" STUD FRMG. (WALNUT) @ 12" O.C.
3'-0"
VERTICAL JOINT LINES
2"x 12" JOISTS
2"x 8" HANDRAIL VG. D. FIR
2"x 12" TREADS CARPET COVERED - SUPPORT ON CONCEALED METAL ANGLES
7'-6"
3"x 12" STRINGERS
3/4" VG. FIR CAPPING

DETAIL OF UPPER FLOOR EDGE

METAL CORNER BEAD ALL ARN'D
METAL E. BEAD
1/2"x 1 1/2" WOOD BASE (TYPICAL)
DOUBLE 2"x 12" JOISTS
METAL C. BEAD
1/2" GYPROC

TYP. INT. DOOR JAMB

NOTE THAT ALL DOOR FRAMES RUN TO CEILING - PANEL OVER DOOR TO MATCH DOOR
METAL E. BEAD (B/SIDES)

TYPICAL WALL & ROOF DETAILS

4/3
26 GA. G.I. FLASHING
TAR & GRAVEL ROOFING ON 1/2" T. & G. PLYWOOD
ON 1" INSULATION
2"x 12" JOISTS @ 16" O.C.
1/2" PLY FASCIA
CONT. HEADER EX. 2"x 12"
2-2" 10" CONT.
VAPOUR BARRIER
1/2" GYPROC
BLOCKING
2"x 4" FURRING @ 16" O.C.
1/2" PLYWOOD VALANCE
1"
1/2" PLYWOOD SOFFIT DRILL HOLES - AS DIRECTED FOR VENTILATION
1/2"x 2 1/2" TRIM WOOD STOPS
GLAZING
1/3

2"x 2"
SILL (1 5/8"x 5 1/4")
1/2" PLYWOOD
2-2"x 12" BLOCKING CONT. BETWEEN MULLIONS
2"x 6" CONT. MULLION
FLOOR FIN. BEYOND
CEILING FIN. BEYOND
FLASHING
3/4"x 1 3/4" STOPS
GLAZING
2/3

2 1/2"
ROOF OVER
SILL (1 5/8"x 7 1/4")
1 3/4" GLAZED DOOR
WOOD STOP (1/2"x 1 3/4")
2"x 4" FRAME
2"x 4" STUDS @ 16" O.C.
STUCCO
METAL CORNER BEAD (TYPICAL ALL CORNERS)
3/3

26 GA. G.I. FLASHING
SIDING (FOR CANT)
2"x 4" STUDS
2"x 12" JOISTS
SOLID BLOCKING
VAPOUR BARRIER
1/2" GYPROC
1"x 6" LEDGER (TYPICAL AT FLOOR JOISTS)
4/3

DETAIL OF STEEL BEAM TO COLUMN CONNECTION

6"x 8" COLUMN DOWN TO TOP OF WF BEAM (DRIFT PIN)
2"x 4" STUDS @ 16" O.C.
1/2"x 1 1/2" BASE
1/2" T. & G. PLYWOOD
1/4"x 4"x 6"x 4" STEEL L. B SIDES OF EVERY 4TH FLOOR JOIST - 1-3/8" BOLT THRO' BEAM, 2-3/8" BOLTS THRO' JOISTS
10" WF @ 45# DRILLED FOR BOLTS & LAG-SCREWS
2"x 12" BLOCKING BETWEEN JOISTS
VALANCE BOARD BEYOND
1/2" GYPROC
6"x 8" COLUMN
3/4" DRIFT PIN 10" LONG WELDED TO BOTTOM OF WF (IN ℄ OF COLUMN)
(3/4" DRIFT PIN INTO CONC.)
LAG-SCREWS @ 5"

SITE PLAN

PROPERTY LINE
50'-6"
ELEV. 103'-0"
LOAM 12" DEEP UP TO BOUNDARIES & BANK
ELEV. 103'-0"
ELEV. 100'-0"
BANK
PEA GRAVEL
44'-2"
PEBBLES
POOL WATER LEVEL EL. 99'-0"
CONC. TERRACE
PEBBLES
ELEV. 100'-0"
ELEV. 100'-0"
125'-0"
PROPOSED HOUSE FIN. GROUND FLR. ELEV. 100'-6"
117'-0"
PEA GRAVEL
PROPERTY LINE
48'-10"
ELEV. 100'-0"
PEA GRAVEL
7'-0" | 36'-0" | 7'-0"
POOL
ROCK PIT 4'x 3'x 6' DEEP
3'-0"
ELEV. 100'-0"
ROCK PIT 4'x 3'x 6' DEEP
PROJECT NORTH
S | E | W
3'-0"x 3'-0"x 3" THICK CONC. SLABS (PRECAST) 1" SPACING - ENTIRE WALK SHALL BE LEVEL
30'-0"
32'-0"
PEA GRAVEL 3" DEEP ON 4" MIL. POLYTHENE OVER SITE (EXCEPT AS NOTED) ELEV. 100'-0"
ELEV. 100'-0"
PROPERTY LINE
50'-0"
AVENUE

RESIDENCE

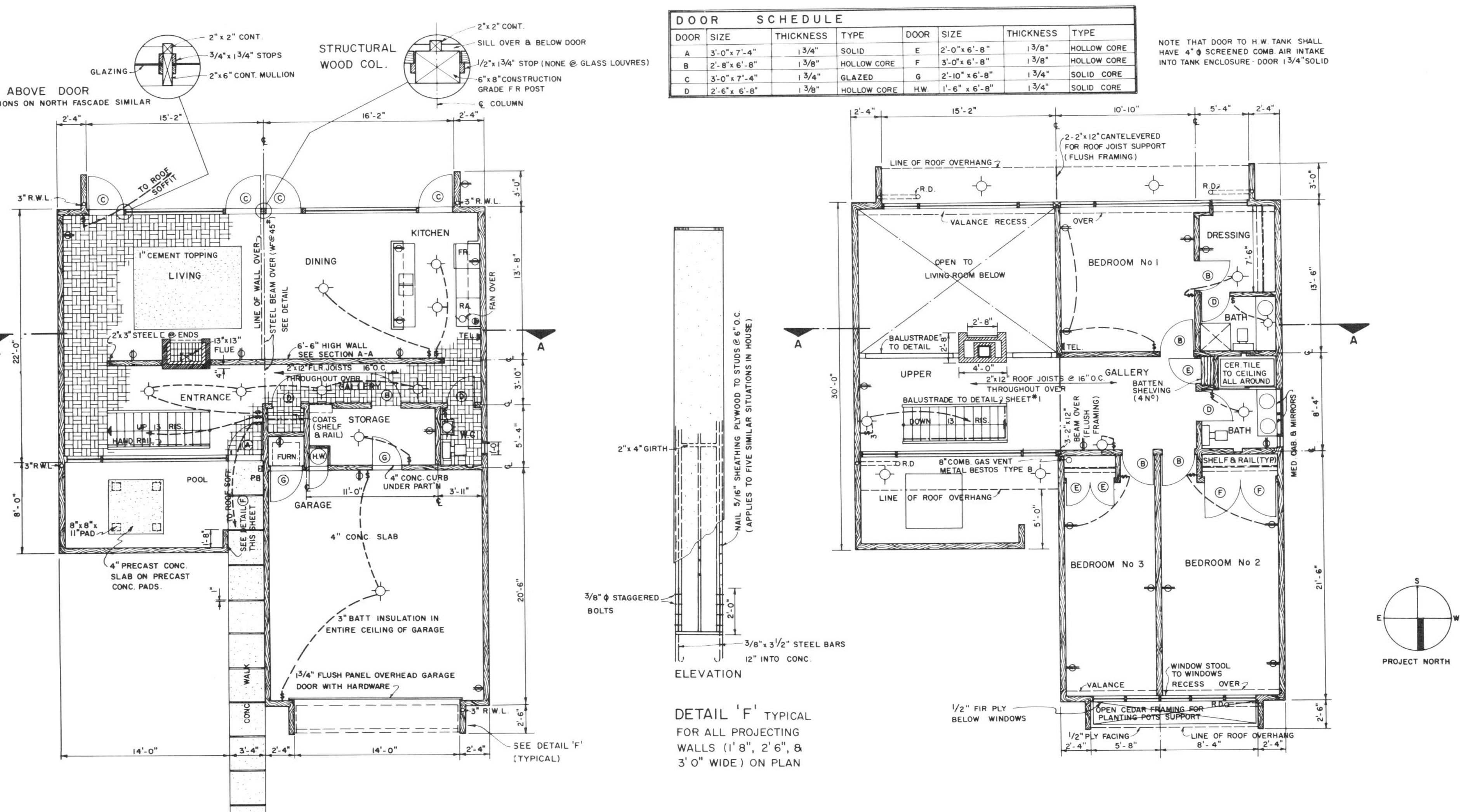

DOOR SCHEDULE

DOOR	SIZE	THICKNESS	TYPE	DOOR	SIZE	THICKNESS	TYPE
A	3'-0" x 7'-4"	1 3/4"	SOLID	E	2'-0" x 6'-8"	1 3/8"	HOLLOW CORE
B	2'-8" x 6'-8"	1 3/8"	HOLLOW CORE	F	3'-0" x 6'-8"	1 3/8"	HOLLOW CORE
C	3'-0" x 7'-4"	1 3/4"	GLAZED	G	2'-10" x 6'-8"	1 3/4"	SOLID CORE
D	2'-6" x 6'-8"	1 3/8"	HOLLOW CORE	H.W.	1'-6" x 6'-8"	1 3/4"	SOLID CORE

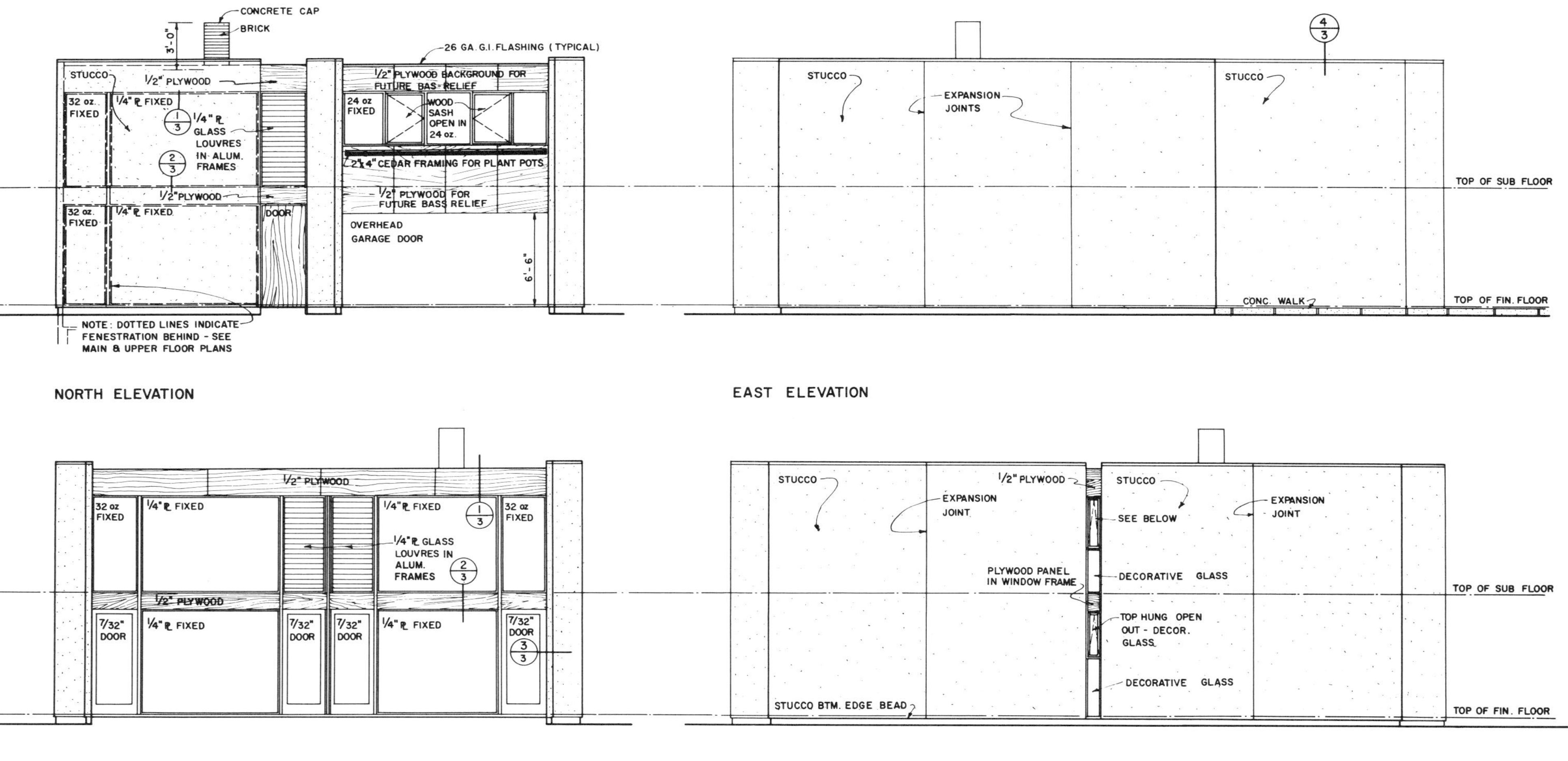

CONCRETE CAP
BRICK
3'-0"
26 GA. G.I. FLASHING (TYPICAL)
STUCCO
1/2" PLYWOOD
32 oz. FIXED
1/4" PL FIXED
1/4" PL GLASS LOUVRES IN ALUM. FRAMES
1/3
2/3
1/2" PLYWOOD BACKGROUND FOR FUTURE BAS-RELIEF
24 oz FIXED
WOOD SASH OPEN IN 24 oz.
2" x 4" CEDAR FRAMING FOR PLANT POTS
1/2" PLYWOOD
1/2" PLYWOOD FOR FUTURE BASS RELIEF
32 oz. FIXED
1/4" PL FIXED
DOOR
OVERHEAD GARAGE DOOR
6'-6"
NOTE: DOTTED LINES INDICATE FENESTRATION BEHIND - SEE MAIN & UPPER FLOOR PLANS
NORTH ELEVATION
4/3
STUCCO
EXPANSION JOINTS
STUCCO
TOP OF SUB FLOOR
CONC. WALK
TOP OF FIN. FLOOR
EAST ELEVATION
1/2" PLYWOOD
32 oz FIXED
1/4" PL FIXED
1/4" PL FIXED
1/3
32 oz FIXED
1/4" PL GLASS LOUVRES IN ALUM. FRAMES
2/3
1/2" PLYWOOD
7/32" DOOR
1/4" PL FIXED
7/32" DOOR
7/32" DOOR
1/4" PL FIXED
7/32" DOOR
3/3
SOUTH ELEVATION
STUCCO
EXPANSION JOINT
1/2" PLYWOOD
STUCCO
SEE BELOW
EXPANSION JOINT
PLYWOOD PANEL IN WINDOW FRAME
DECORATIVE GLASS
TOP OF SUB FLOOR
TOP HUNG OPEN OUT - DECOR. GLASS
DECORATIVE GLASS
STUCCO BTM. EDGE BEAD
TOP OF FIN. FLOOR
WEST ELEVATION
RESIDENCE 3

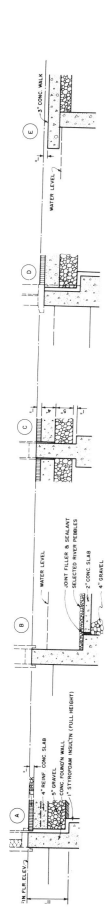

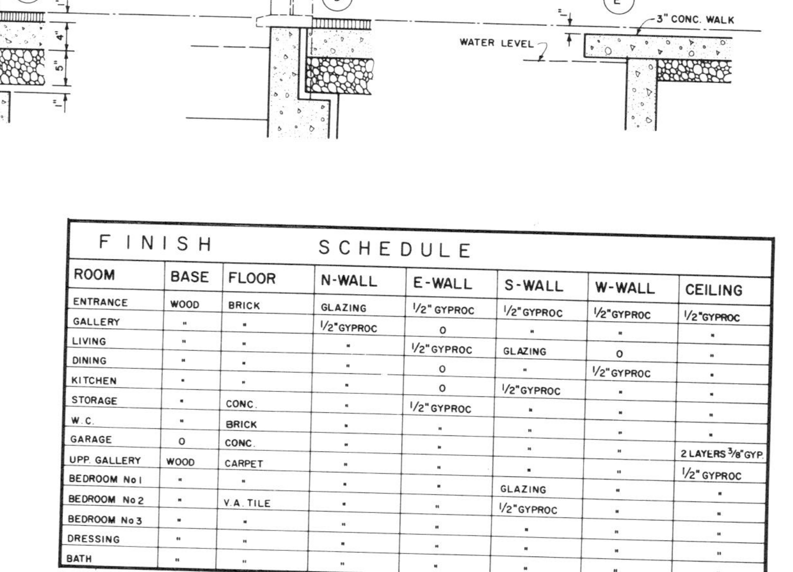

FINISH SCHEDULE							
ROOM	BASE	FLOOR	N-WALL	E-WALL	S-WALL	W-WALL	CEILING
ENTRANCE	WOOD	BRICK	GLAZING	1/2" GYPROC	1/2" GYPROC	1/2" GYPROC	1/2" GYPROC
GALLERY	"	"	1/2" GYPROC	O	"	"	"
LIVING	"	"	"	1/2" GYPROC	GLAZING	O	"
DINING	"	"	"	O	"	1/2" GYPROC	"
KITCHEN	"	"	"	O	1/2" GYPROC	"	"
STORAGE	"	CONC.	"	1/2" GYPROC	"	"	"
W.C.	"	BRICK	"	"	"	"	"
GARAGE	O	CONC.	"	"	"	"	"
UPP. GALLERY	WOOD	CARPET	"	"	"	"	2 LAYERS 3/8" GYP.
BEDROOM No 1	"	"	"	"	GLAZING	"	1/2" GYPROC
BEDROOM No 2	"	V.A. TILE	"	"	1/2" GYPROC	"	"
BEDROOM No 3	"	"	"	"	"	"	"
DRESSING	"	"	"	"	"	"	"
BATH	"	"	"	"	"	"	"

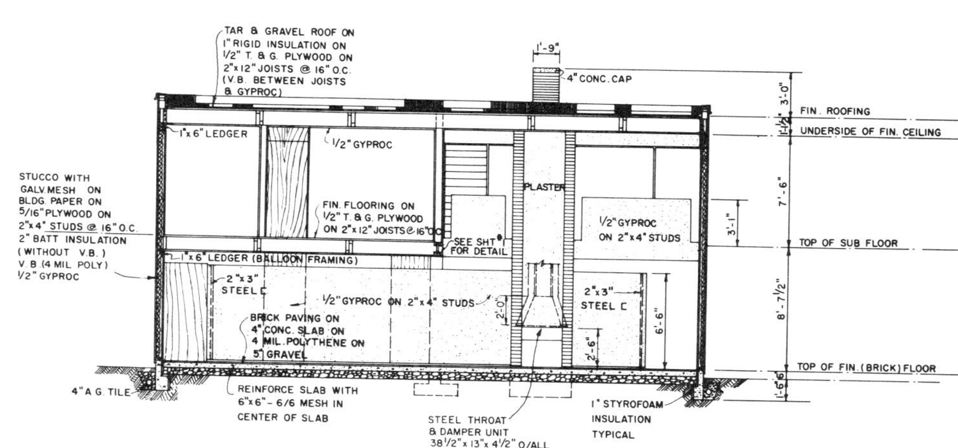

SECTION A - A

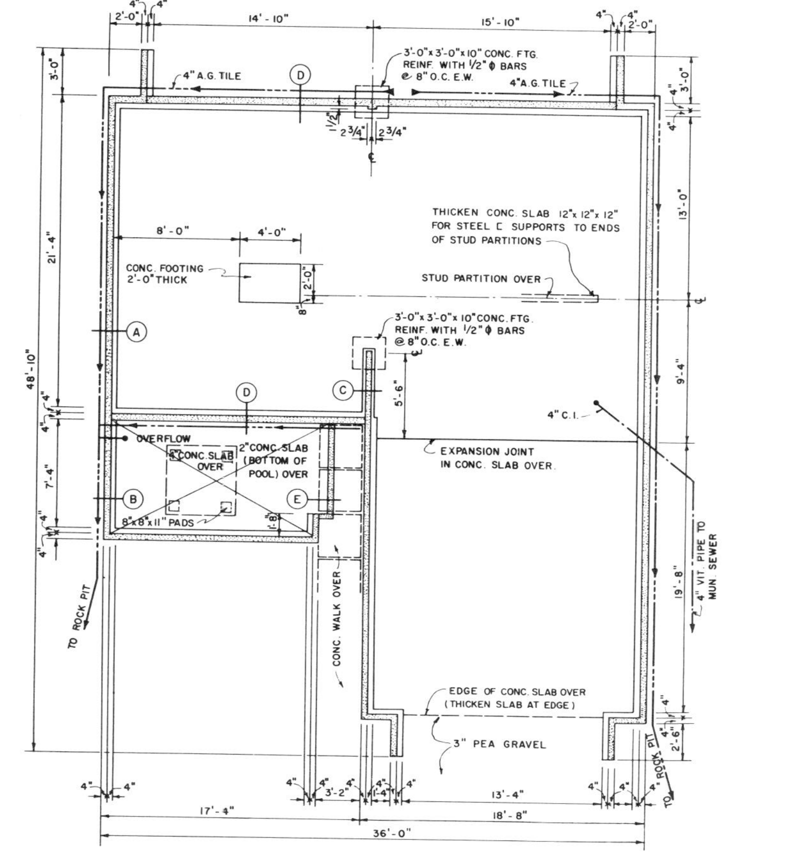

FOUNDATION PLAN

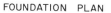

RESIDENCE 4

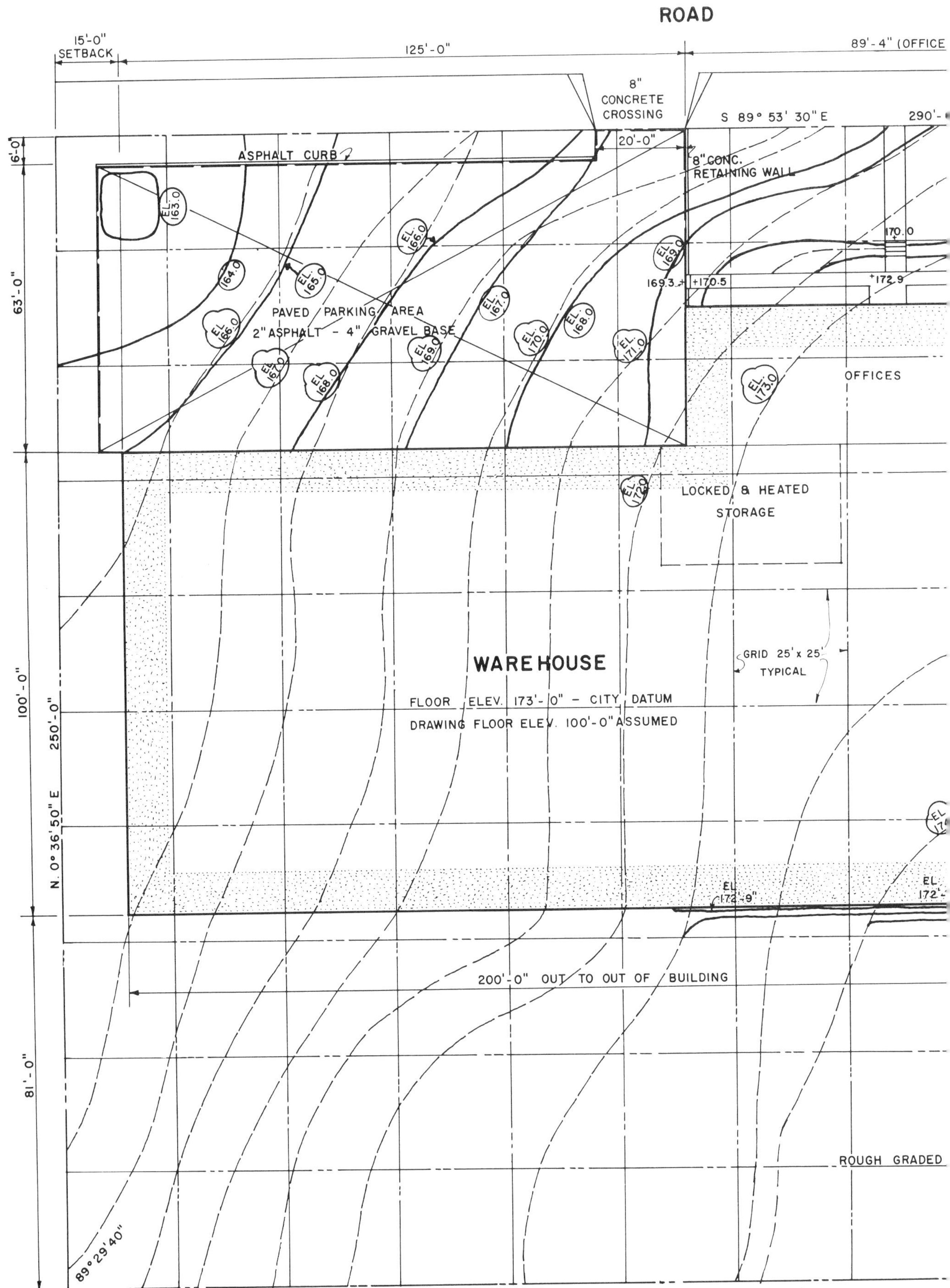
ROAD
15'-0" SETBACK
125'-0"
89'-4" (OFFICE
8" CONCRETE CROSSING
S 89° 53' 30" E
290'-
20'-0"
ASPHALT CURB
8" CONC. RETAINING WALL
6'-0"
63'-0"
EL. 163.0
EL. 166.0
EL. 169.0
170.0
169.3
+170.5
+172.9
164.0
EL. 165.0
EL. 166.0
PAVED PARKING AREA
2" ASPHALT - 4" GRAVEL BASE
EL. 167.0
EL. 168.0
EL. 169.0
EL. 167.0
EL. 170.0
EL. 168.0
EL. 171.0
EL. 173.0
OFFICES
EL. 172.0
LOCKED & HEATED STORAGE
WAREHOUSE
GRID 25' x 25' TYPICAL
FLOOR ELEV. 173'-0" - CITY DATUM
DRAWING FLOOR ELEV. 100'-0" ASSUMED
100'-0"
250'-0"
N. 0° 36' 50" E
EL. 172'-9"
200'-0" OUT TO OUT OF BUILDING
81'-0"
ROUGH GRADED
89° 29' 40"
S 89° 53' 30" E 305'-0"

SITE PLAN

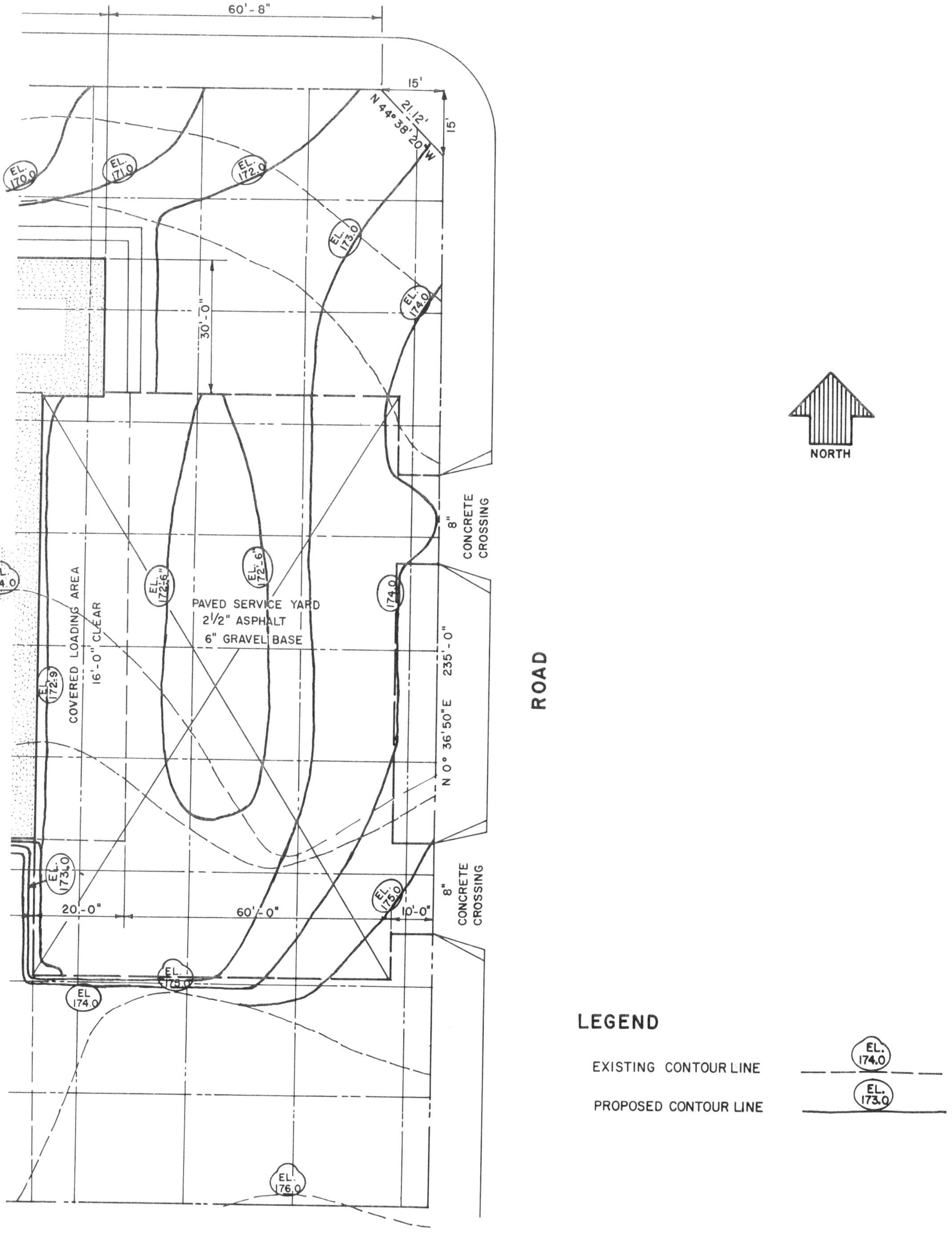

60'- 8"

15'

21.12'

N 44° 38' 20" W

15'

EL. 170.0

EL. 171.0

EL. 172.0

EL. 173.0

EL. 174.0

30'-0"

EL. 172'-6"

EL. 172'-6"

PAVED SERVICE YARD
2 1/2" ASPHALT
6" GRAVEL BASE

COVERED LOADING AREA

16'-0" CLEAR

EL. 172.9

EL. 74.0

174.0

8" CONCRETE CROSSING

235'-0"

N 0° 36'50" E

ROAD

EL. 173.0

EL. 175.0

20'-0"

60'-0"

10'-0"

8" CONCRETE CROSSING

EL. 175.0

EL 174.0

EL. 176.0

LEGEND

EXISTING CONTOUR LINE

PROPOSED CONTOUR LINE

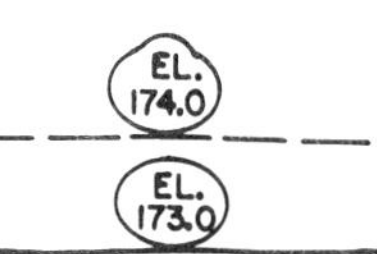

WAREHOUSE (1)

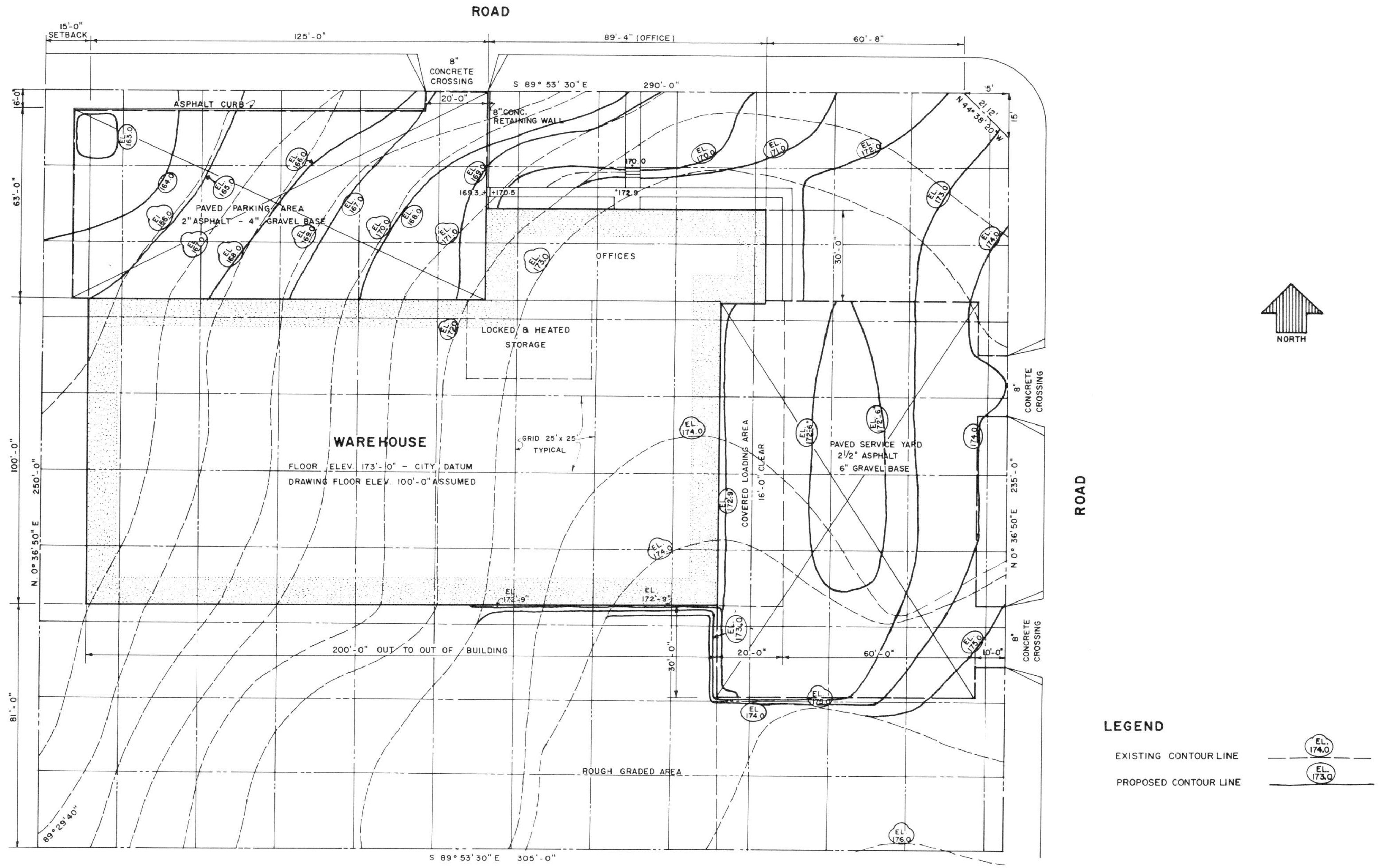

ROAD

15'-0" SETBACK

125'-0"

89'-4" (OFFICE)

60'-8"

8" CONCRETE CROSSING

S 89° 53' 30" E 290'-0"

ASPHALT CURB

20'-0"

8" CONC. RETAINING WALL

N 44° 38' 20" W

21.12'

5'

15'

16'-0"

63'-0"

PAVED PARKING AREA
2" ASPHALT - 4" GRAVEL BASE

OFFICES

LOCKED & HEATED STORAGE

WAREHOUSE

FLOOR ELEV. 173'-0" - CITY DATUM
DRAWING FLOOR ELEV. 100'-0" ASSUMED

GRID 25' x 25' TYPICAL

COVERED LOADING AREA
16'-0" CLEAR

PAVED SERVICE YARD
2 1/2" ASPHALT
6" GRAVEL BASE

30'-0"

100'-0"

250'-0"

N 0° 36' 50" E

200'-0" OUT TO OUT OF BUILDING

20'-0"

60'-0"

10'-0"

235'-0"

N 0° 36' 50" E

ROAD

81'-0"

ROUGH GRADED AREA

89° 29' 40"

S 89° 53' 30" E 305'-0"

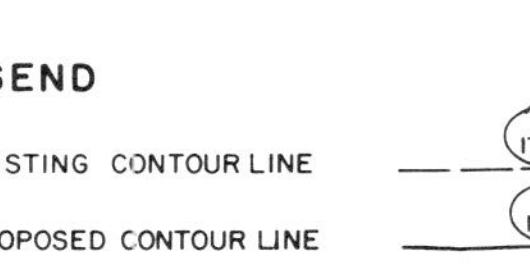

LEGEND

EXISTING CONTOUR LINE (EL. 174.0)

PROPOSED CONTOUR LINE (EL. 173.0)

SITE PLAN

WAREHOUSE

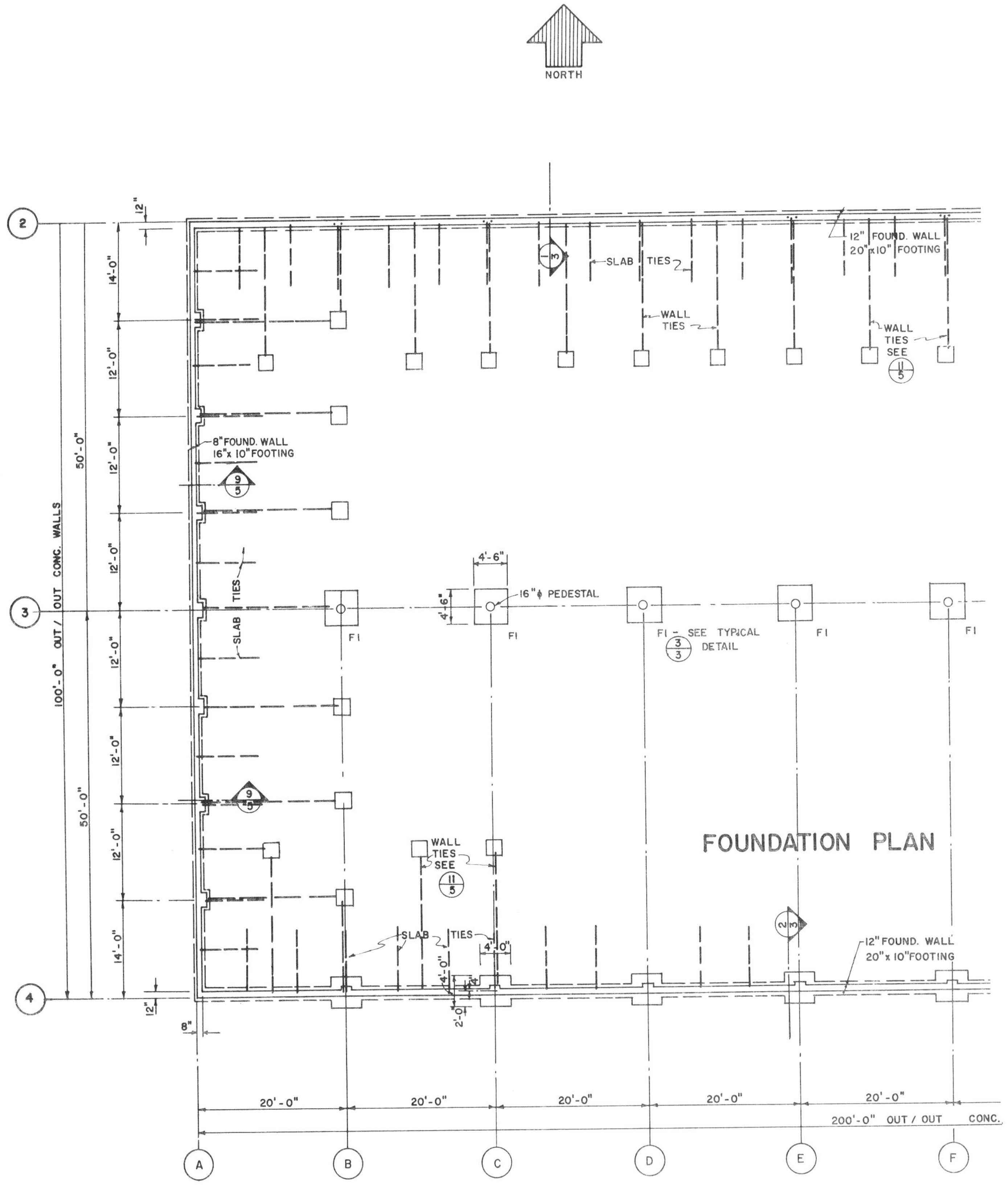

FOUNDATION PLAN

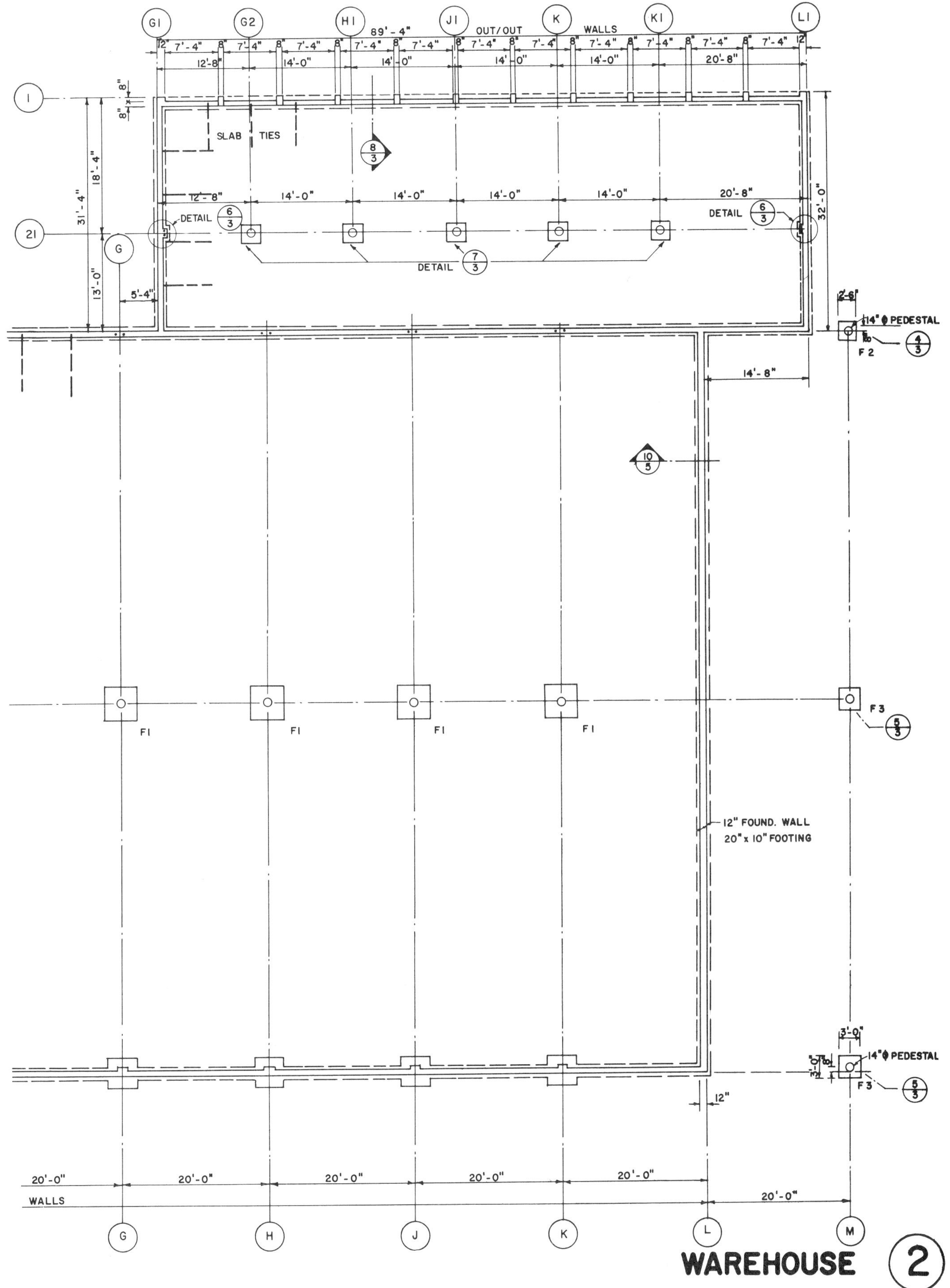

89'-4" OUT/OUT WALLS
SLAB TIES
DETAIL 6/3
DETAIL 7/3
14" ϕ PEDESTAL
F 2
F 1
F 3
12" FOUND. WALL
20" x 10" FOOTING
WALLS

WAREHOUSE 2

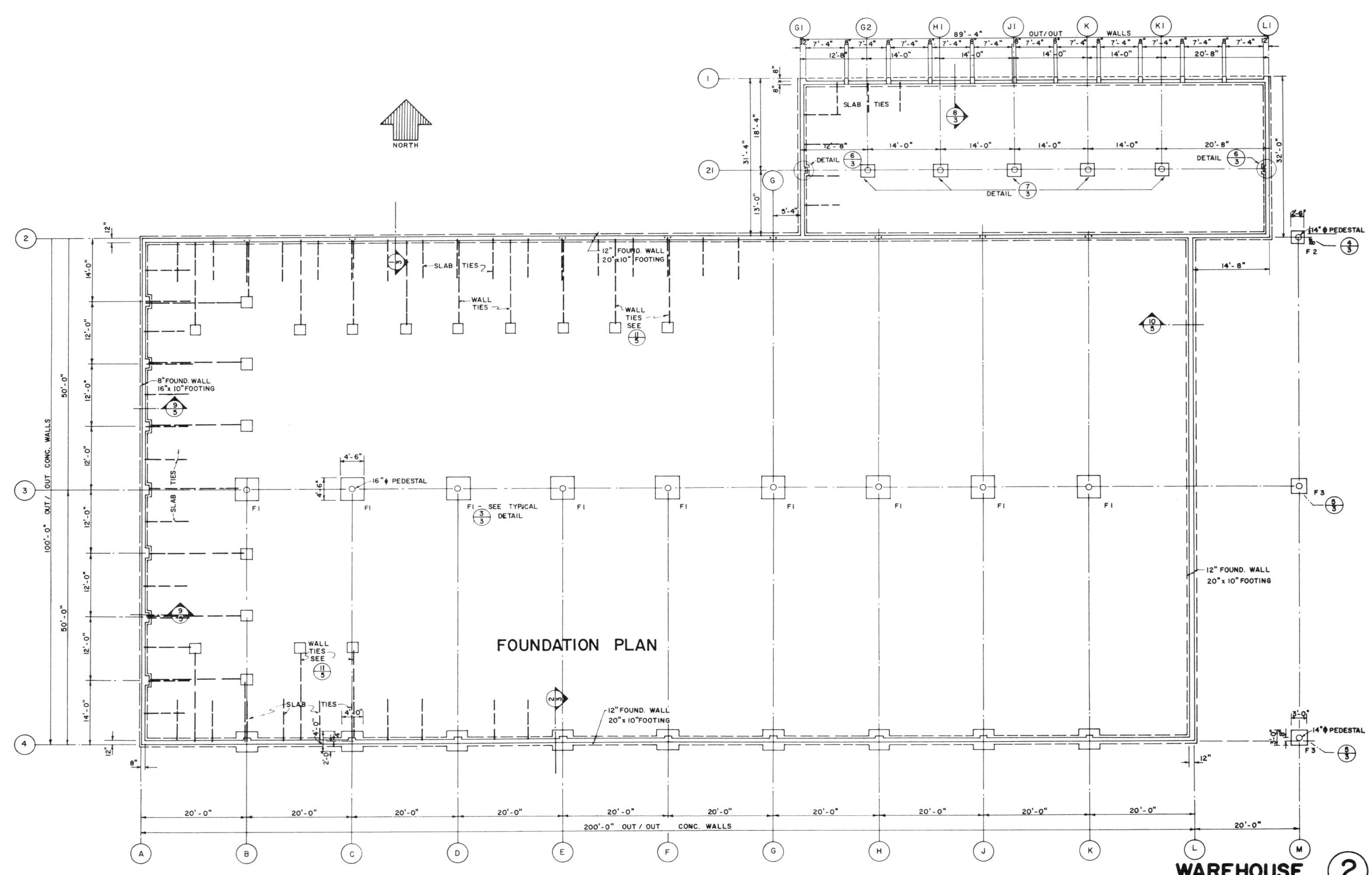

FOUNDATION PLAN
NORTH
89'-4" OUT/OUT WALLS
SLAB TIES
WALL TIES
WALL TIES SEE 11/5
DETAIL 6/3
DETAIL 7/3
12" FOUND. WALL 20"x10" FOOTING
8" FOUND. WALL 16"x10" FOOTING
16" ϕ PEDESTAL
14" ϕ PEDESTAL
F1 - SEE TYPICAL 3/3 DETAIL
F1
F2
F3
100'-0" OUT / OUT CONC. WALLS
200'-0" OUT / OUT CONC. WALLS
WAREHOUSE 2

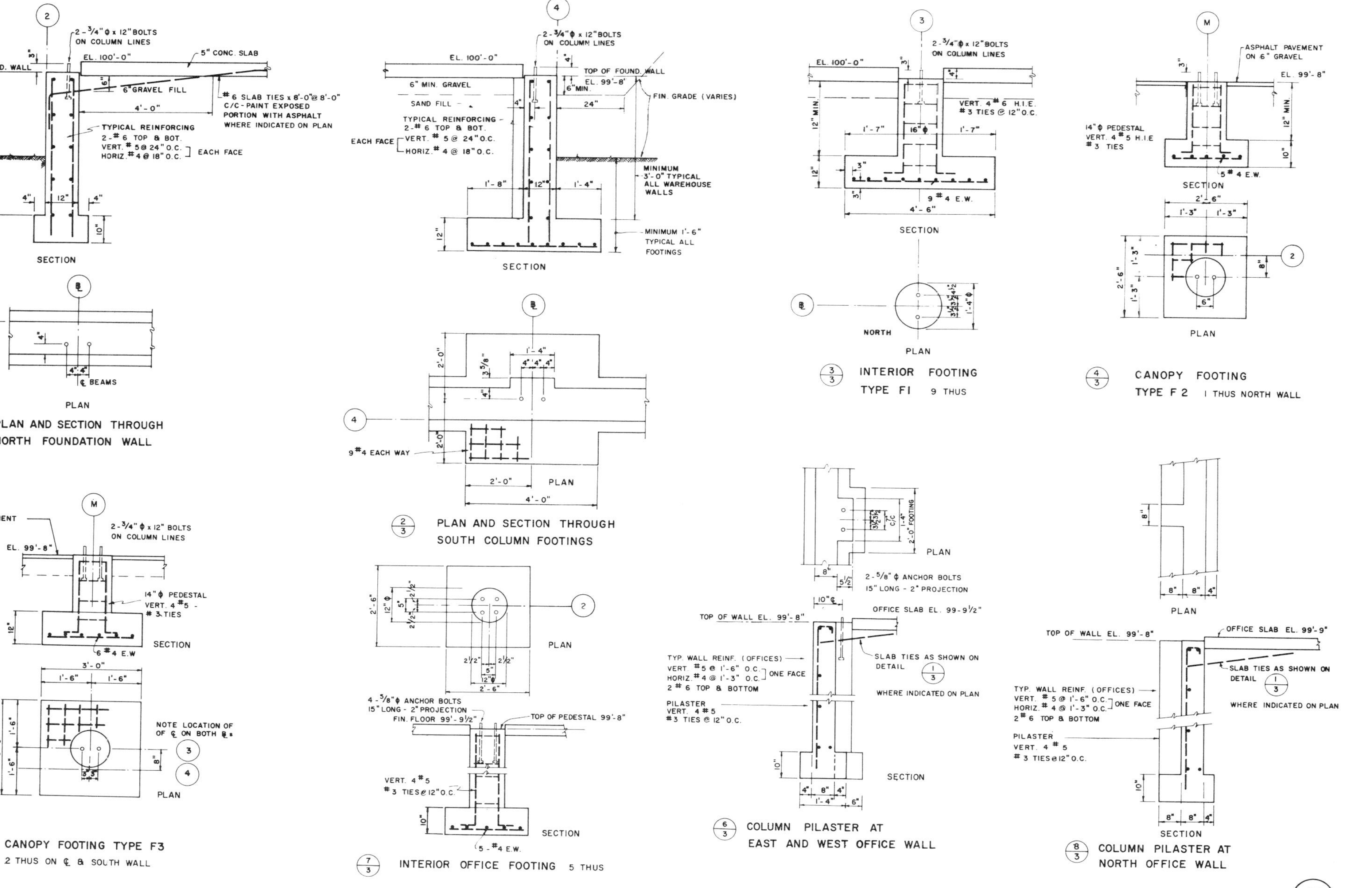

FOOTING DETAILS (SEE OTHERS ON SHEET 5)

WAREHOUSE 3

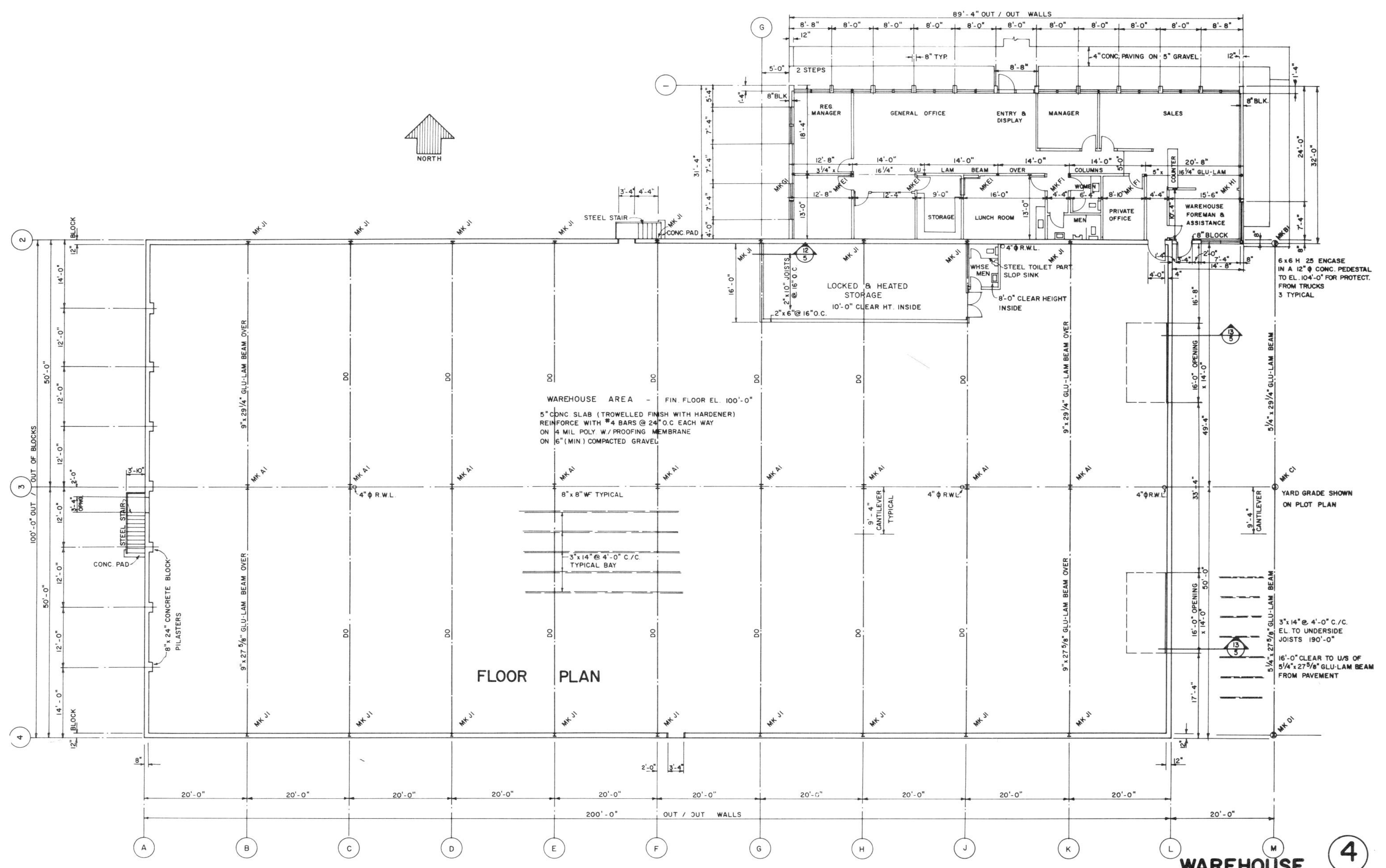
WAREHOUSE
4
FLOOR PLAN
NORTH
WAREHOUSE AREA - FIN. FLOOR EL. 100'-0"
5" CONC. SLAB (TROWELLED FINISH WITH HARDENER)
REINFORCE WITH #4 BARS @ 24" O.C EACH WAY
ON 4 MIL POLY W/PROOFING MEMBRANE
ON 6" (MIN) COMPACTED GRAVEL
LOCKED & HEATED STORAGE
10'-0" CLEAR HT. INSIDE
2"x10" JOISTS @ 16" O.C
2"x6" @ 16" O.C.
REG MANAGER
GENERAL OFFICE
ENTRY & DISPLAY
MANAGER
SALES
STORAGE
LUNCH ROOM
WOMEN
MEN
PRIVATE OFFICE
WAREHOUSE FOREMAN & ASSISTANCE
COUNTER
WHSE MEN
STEEL TOILET PART.
SLOP SINK
8'-0" CLEAR HEIGHT INSIDE
16 1/4" GLU-LAM BEAM OVER COLUMNS
4" CONC. PAVING ON 5" GRAVEL
2 STEPS
8" BLK.
8" BLOCK
STEEL STAIR
CONC. PAD
8"x24" CONCRETE BLOCK PILASTERS
9"x29 1/4" GLU-LAM BEAM OVER
9"x27 5/8" GLU-LAM BEAM OVER
5 1/4"x29 1/4" GLU-LAM BEAM
5 1/4"x27 5/8" GLU-LAM BEAM
8"x8" WF TYPICAL
3"x14" @ 4'-0" C/C TYPICAL BAY
4" ϕ R.W.L.
9'-4" CANTILEVER TYPICAL
16'-0" OPENING x 14'-0"
6x6 H 25 ENCASE IN A 12" ϕ CONC. PEDESTAL TO EL. 104'-0" FOR PROTECT. FROM TRUCKS 3 TYPICAL
YARD GRADE SHOWN ON PLOT PLAN
3"x14" @ 4'-0" C/C. EL. TO UNDERSIDE JOISTS 190'-0"
16'-0" CLEAR TO U/S OF 5 1/4"x27 5/8" GLU-LAM BEAM FROM PAVEMENT
89'-4" OUT / OUT WALLS
200'-0" OUT / OUT WALLS
100'-0" OUT / OUT OF BLOCKS

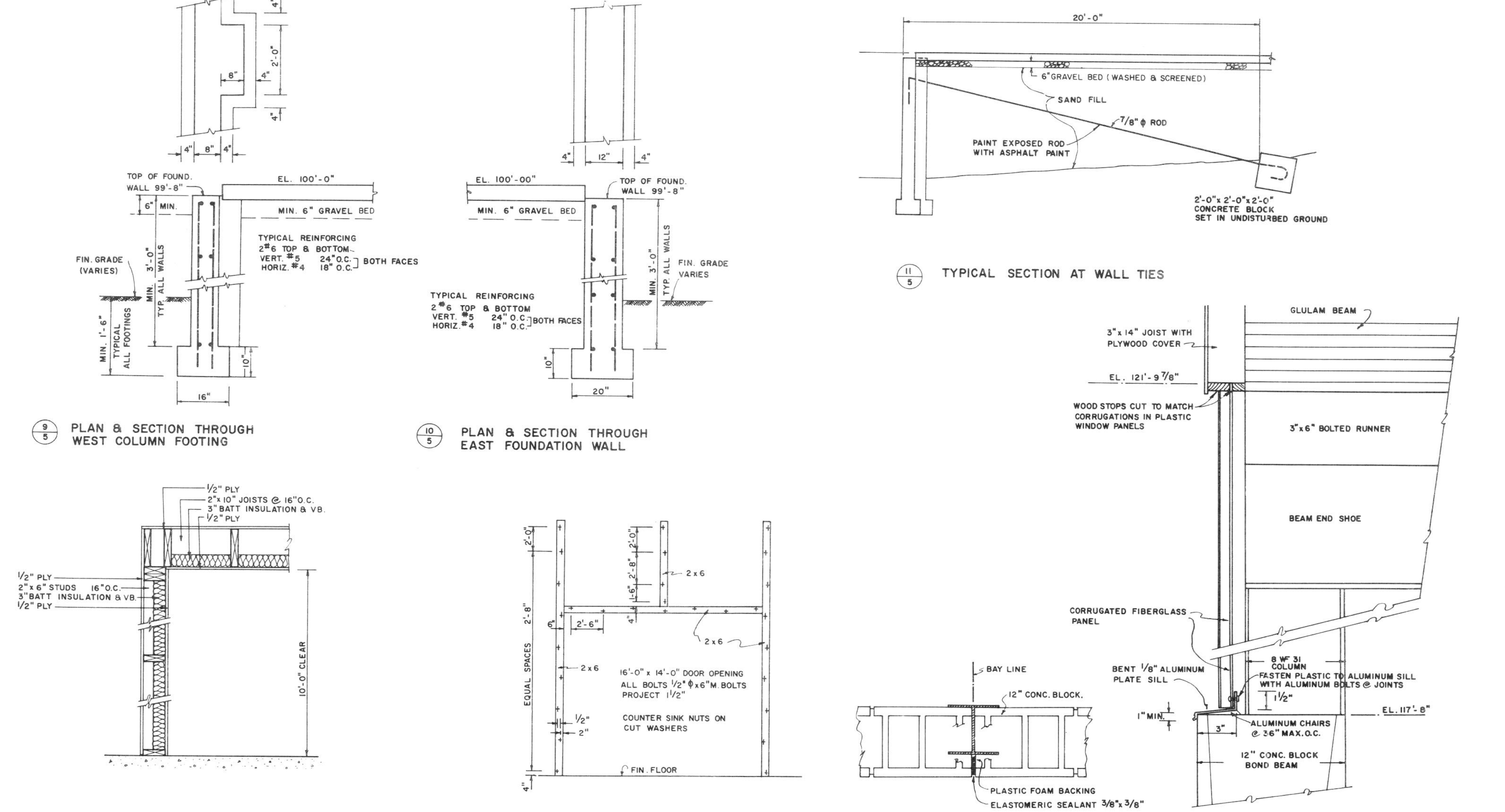

9/5 PLAN & SECTION THROUGH WEST COLUMN FOOTING

10/5 PLAN & SECTION THROUGH EAST FOUNDATION WALL

11/5 TYPICAL SECTION AT WALL TIES

12/5 SECTION THROUGH WEST WALL OF STORAGE AREA (MENS TOILET SIMILAR)

13/5 WOOD FRAMING FOR 2 - 16' x 14' HI-LIFT O.H. DOORS

15/5 CONTROL JOINT DETAIL (TYPICAL @ EACH BAY LINE NORTH & SOUTH WALLS)

14/5 WINDOW HEAD & SILL DETAIL FOR WAREHOUSE

WALL SECTIONS & DETAILS (WAREHOUSE)

WAREHOUSE 5

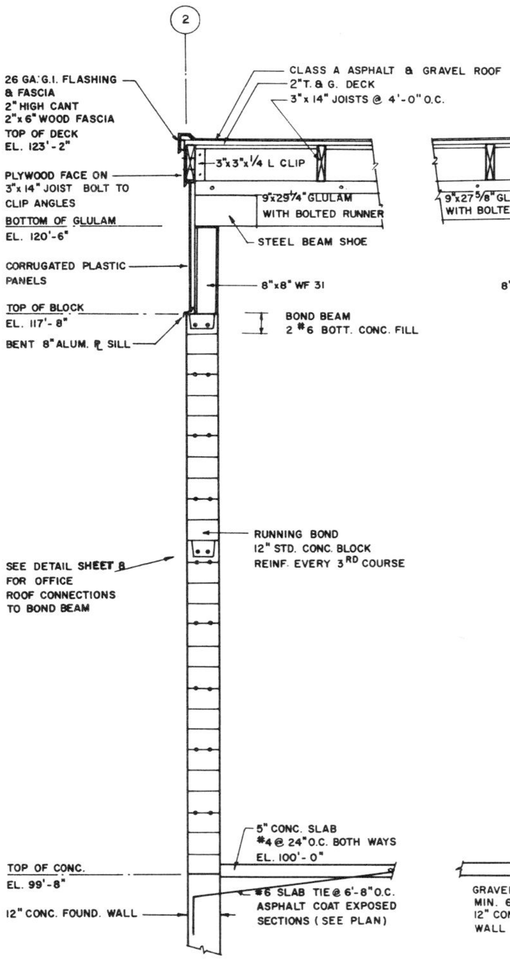

SECTION THROUGH NORTH WALL

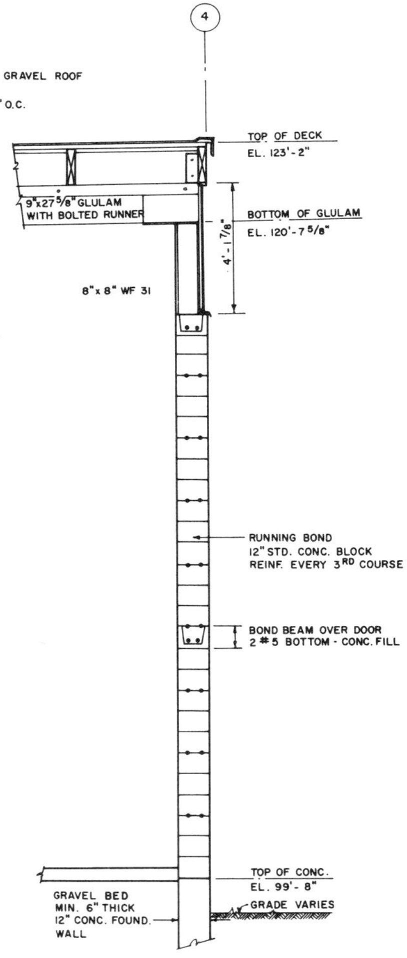

2

26 GA. G.I. FLASHING & FASCIA
2" HIGH CANT
2"x 6" WOOD FASCIA
TOP OF DECK
EL. 123'-2"

CLASS A ASPHALT & GRAVEL ROOF
2" T. & G. DECK
3"x 14" JOISTS @ 4'-0" O.C.

3"x3"x1/4 L CLIP

PLYWOOD FACE ON
3"x 14" JOIST BOLT TO
CLIP ANGLES

9"x29 1/4" GLULAM
WITH BOLTED RUNNER

BOTTOM OF GLULAM
EL. 120'-6"

STEEL BEAM SHOE

CORRUGATED PLASTIC
PANELS

8"x8" WF 31

TOP OF BLOCK
EL. 117'-8"

BOND BEAM
2 #6 BOTT. CONC. FILL

BENT 8" ALUM. ℞ SILL

RUNNING BOND
12" STD. CONC. BLOCK
REINF. EVERY 3RD COURSE

SEE DETAIL SHEET 8
FOR OFFICE
ROOF CONNECTIONS
TO BOND BEAM

5" CONC. SLAB
#4 @ 24" O.C. BOTH WAYS
EL. 100'-0"

TOP OF CONC.
EL. 99'-8"

#6 SLAB TIE @ 6'-8" O.C.
ASPHALT COAT EXPOSED
SECTIONS (SEE PLAN)

12" CONC. FOUND. WALL

SECTION THROUGH SOUTH WALL

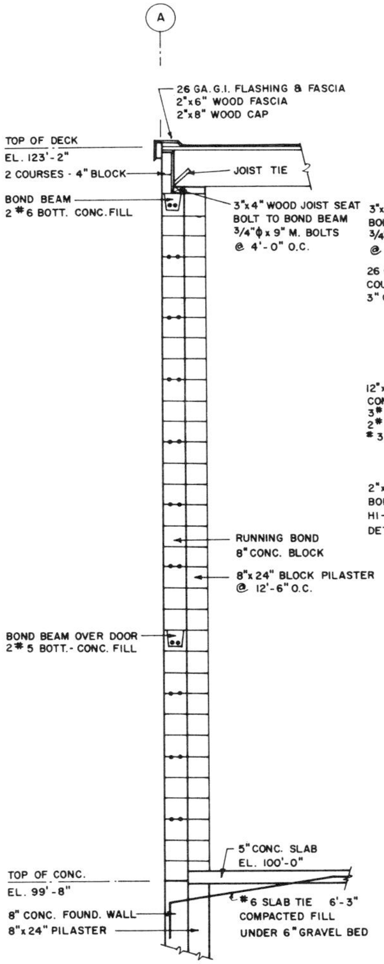

4

TOP OF DECK
EL. 123'-2"

9"x27 5/8" GLULAM
WITH BOLTED RUNNER

BOTTOM OF GLULAM
EL. 120'-7 5/8"

4'-1 7/8"

8"x 8" WF 31

RUNNING BOND
12" STD. CONC. BLOCK
REINF. EVERY 3RD COURSE

BOND BEAM OVER DOOR
2 #5 BOTTOM - CONC. FILL

TOP OF CONC.
EL. 99'-8"

GRADE VARIES

GRAVEL BED
MIN. 6" THICK
12" CONC. FOUND.
WALL

SECTION THROUGH WEST WALL

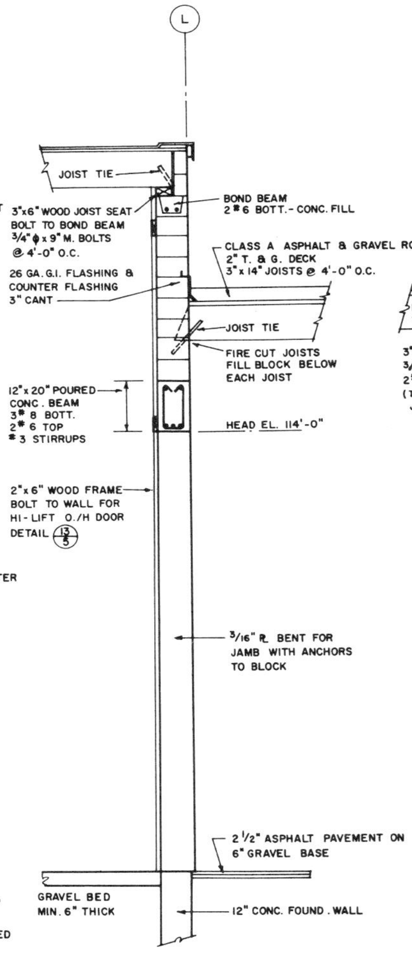

A

26 GA. G.I. FLASHING & FASCIA
2"x6" WOOD FASCIA
2"x8" WOOD CAP

TOP OF DECK
EL. 123'-2"

2 COURSES - 4" BLOCK

JOIST TIE

BOND BEAM
2 #6 BOTT. CONC. FILL

3"x 4" WOOD JOIST SEAT
BOLT TO BOND BEAM
3/4" ϕ x 9" M. BOLTS
@ 4'-0" O.C.

RUNNING BOND
8" CONC. BLOCK

8"x 24" BLOCK PILASTER
@ 12'-6" O.C.

BOND BEAM OVER DOOR
2 #5 BOTT.- CONC. FILL

5" CONC. SLAB
EL. 100'-0"

TOP OF CONC.
EL. 99'-8"

8" CONC. FOUND. WALL
8"x 24" PILASTER

#6 SLAB TIE 6'-3"
COMPACTED FILL
UNDER 6" GRAVEL BED

SECTION THROUGH EAST WALL

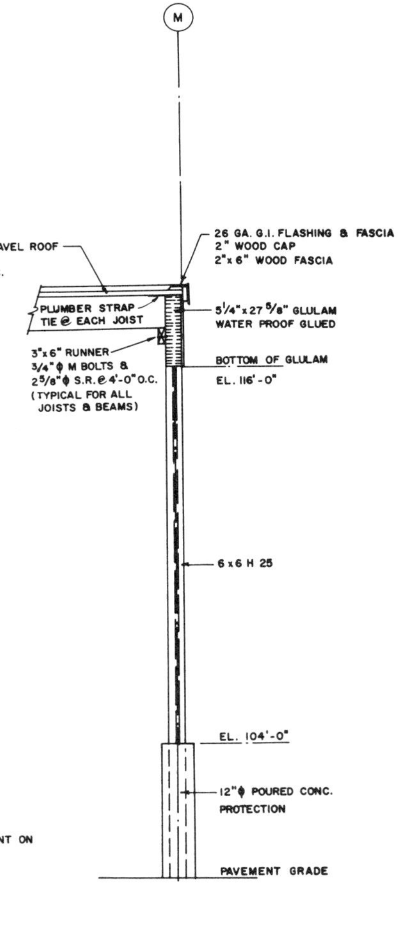

L

JOIST TIE

BOND BEAM
2 #6 BOTT.- CONC. FILL

3"x6" WOOD JOIST SEAT
BOLT TO BOND BEAM
3/4" ϕ x 9" M. BOLTS
@ 4'-0" O.C.

CLASS A ASPHALT & GRAVEL ROOF
2" T. & G. DECK
3"x 14" JOISTS @ 4'-0" O.C.

26 GA. G.I. FLASHING &
COUNTER FLASHING
3" CANT

JOIST TIE

FIRE CUT JOISTS
FILL BLOCK BELOW
EACH JOIST

12"x 20" POURED
CONC. BEAM
3 #8 BOTT.
2 #6 TOP
#3 STIRRUPS

HEAD EL. 114'-0"

2"x 6" WOOD FRAME
BOLT TO WALL FOR
HI-LIFT O./H DOOR
DETAIL 13/5

3/16" ℞ BENT FOR
JAMB WITH ANCHORS
TO BLOCK

2 1/2" ASPHALT PAVEMENT ON
6" GRAVEL BASE

GRAVEL BED
MIN. 6" THICK

12" CONC. FOUND. WALL

SECTION THROUGH CANOPY SUPPORT BEAM

M

26 GA. G.I. FLASHING & FASCIA
2" WOOD CAP
2"x 6" WOOD FASCIA

PLUMBER STRAP
TIE @ EACH JOIST

5 1/4"x 27 5/8" GLULAM
WATER PROOF GLUED

3"x6" RUNNER
3/4" ϕ M BOLTS &
2 5/8" ϕ S.R. @ 4'-0" O.C.
(TYPICAL FOR ALL
JOISTS & BEAMS)

BOTTOM OF GLULAM
EL. 116'-0"

6 x 6 H 25

EL. 104'-0"

12" ϕ POURED CONC.
PROTECTION

PAVEMENT GRADE

WAREHOUSE 

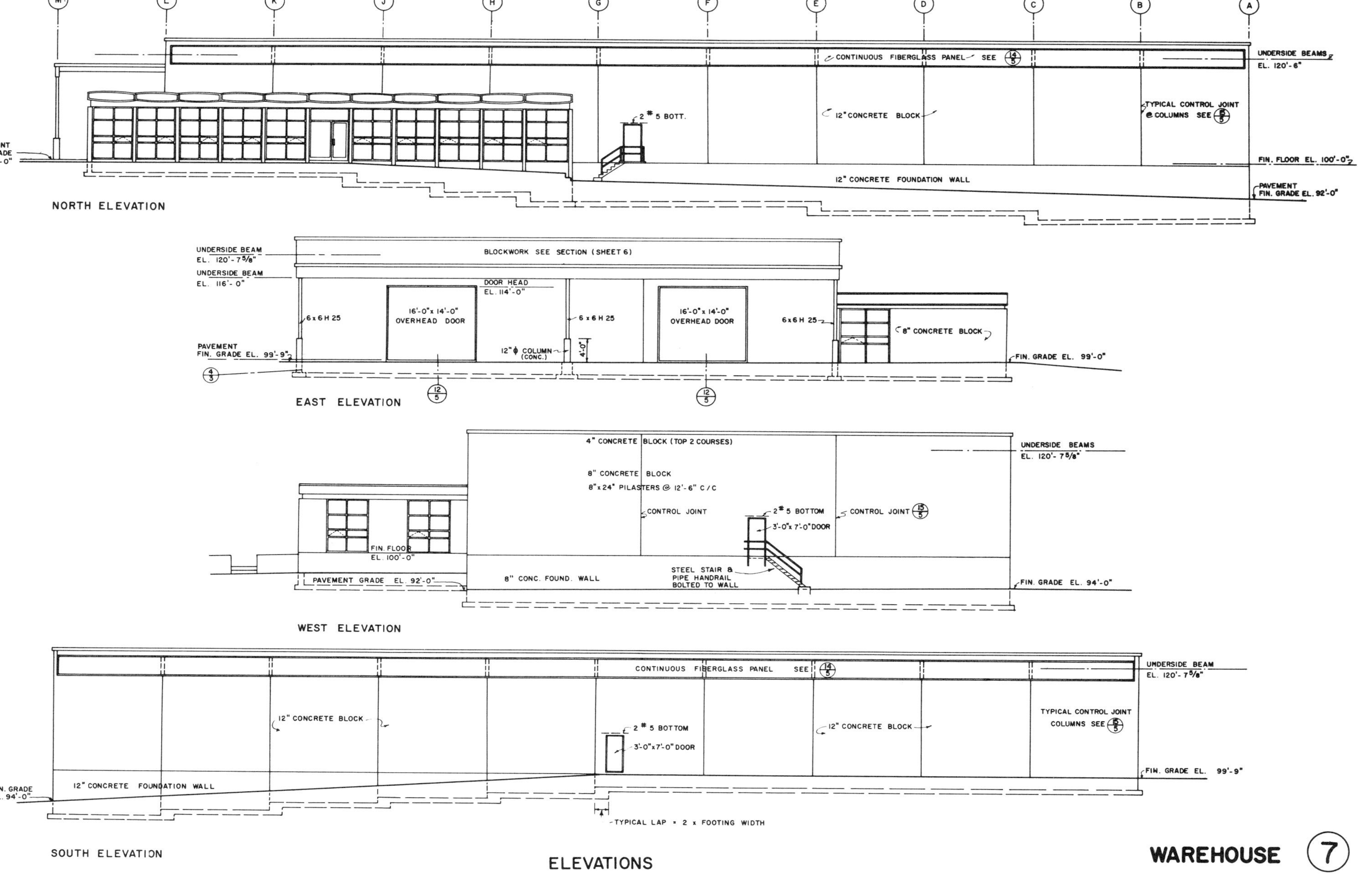

ELEVATIONS

WAREHOUSE 7

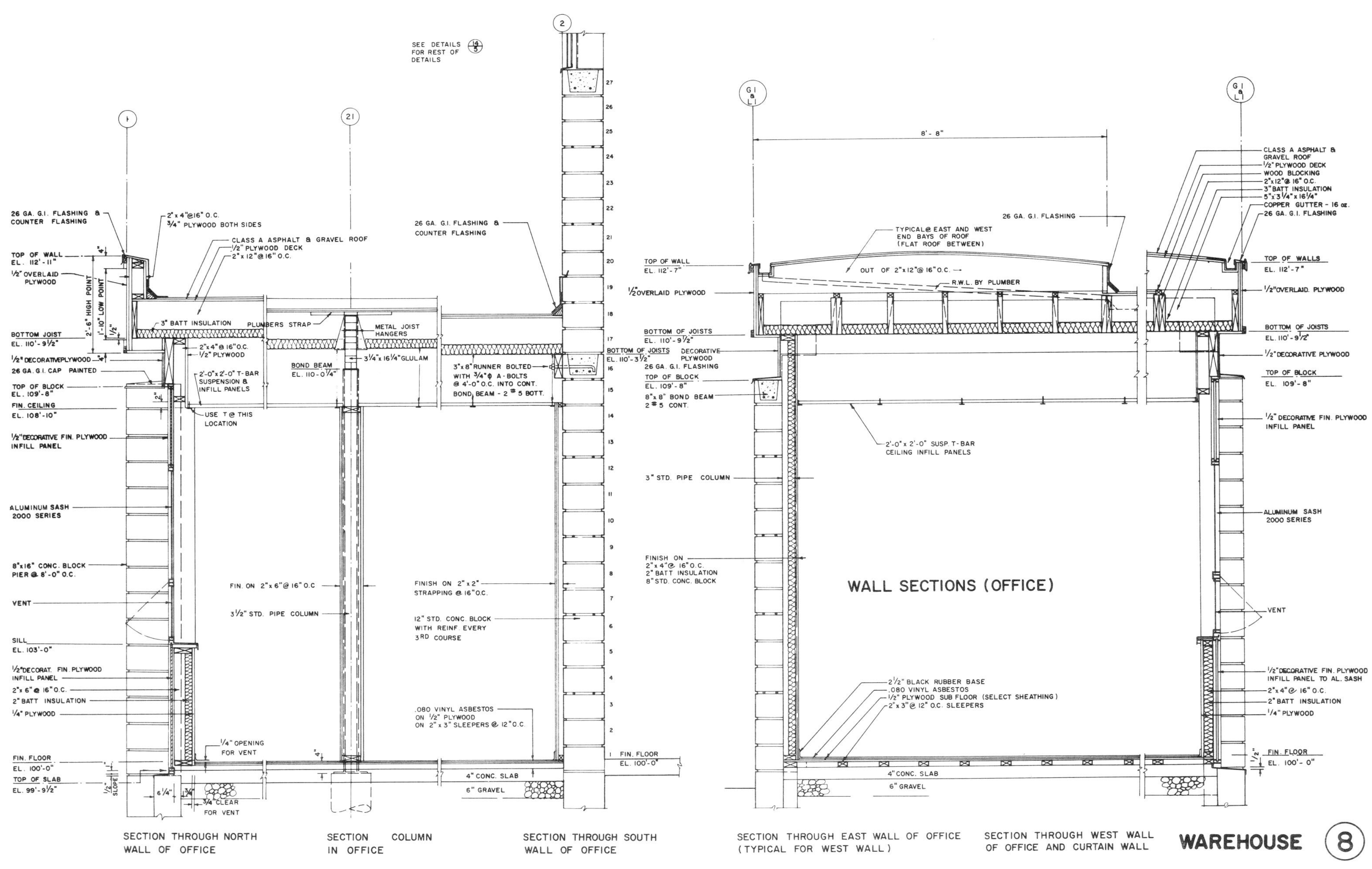
WALL SECTIONS (OFFICE)
SEE DETAILS 14/5 FOR REST OF DETAILS
26 GA. G.I. FLASHING & COUNTER FLASHING
2"x4"@16" O.C.
3/4" PLYWOOD BOTH SIDES
CLASS A ASPHALT & GRAVEL ROOF
1/2" PLYWOOD DECK
2"x12"@16" O.C.
TOP OF WALL EL. 112'-11"
1/2" OVERLAID PLYWOOD
2'-6" HIGH POINT
1'-10" LOW POINT
3" BATT INSULATION
PLUMBERS STRAP
METAL JOIST HANGERS
BOTTOM JOIST EL. 110'-9 1/2"
2"x4"@16" O.C.
1/2" PLYWOOD
1/2" DECORATIVE PLYWOOD
26 GA. G.I. CAP PAINTED
2'-0"x2'-0" T-BAR SUSPENSION & INFILL PANELS
BOND BEAM EL. 110-0 1/4"
3 1/4"x16 1/4" GLULAM
3"x8" RUNNER BOLTED WITH 3/4"ϕ A-BOLTS @ 4'-0" O.C. INTO CONT. BOND BEAM - 2 # 5 BOTT.
TOP OF BLOCK EL. 109'-8"
FIN. CEILING EL. 108'-10"
USE T @ THIS LOCATION
1/2" DECORATIVE FIN. PLYWOOD INFILL PANEL
ALUMINUM SASH 2000 SERIES
8"x16" CONC. BLOCK PIER @ 8'-0" O.C.
VENT
SILL EL. 103'-0"
1/2" DECORAT. FIN. PLYWOOD INFILL PANEL
2"x6"@16" O.C.
2" BATT INSULATION
1/4" PLYWOOD
FIN. FLOOR EL. 100'-0"
TOP OF SLAB EL. 99'-9 1/2"
SLOPE
6 1/4"
3/4" CLEAR FOR VENT
1/4" OPENING FOR VENT
FIN. ON 2"x6"@16" O.C.
3 1/2" STD. PIPE COLUMN
FINISH ON 2"x2" STRAPPING @ 16" O.C.
12" STD. CONC. BLOCK WITH REINF. EVERY 3RD COURSE
.080 VINYL ASBESTOS ON 1/2" PLYWOOD ON 2"x3" SLEEPERS @ 12" O.C.
4" CONC. SLAB
6" GRAVEL
BOTTOM OF JOISTS EL. 110'-3 1/2"
FIN. FLOOR EL. 100'-0"
8'-8"
26 GA. G.I. FLASHING
TYPICAL @ EAST AND WEST END BAYS OF ROOF (FLAT ROOF BETWEEN)
OUT OF 2"x12"@16" O.C.
R.W.L. BY PLUMBER
TOP OF WALL EL. 112'-7"
1/2" OVERLAID PLYWOOD
BOTTOM OF JOISTS EL. 110'-9 1/2"
DECORATIVE PLYWOOD
26 GA. G.I. FLASHING
TOP OF BLOCK EL. 109'-8"
8"x8" BOND BEAM 2 # 5 CONT.
2'-0"x2'-0" SUSP. T-BAR CEILING INFILL PANELS
3" STD. PIPE COLUMN
FINISH ON 2"x4"@16" O.C. 2" BATT INSULATION 8" STD. CONC. BLOCK
2 1/2" BLACK RUBBER BASE
.080 VINYL ASBESTOS
1/2" PLYWOOD SUB FLOOR (SELECT SHEATHING)
2"x3"@12" O.C. SLEEPERS
4" CONC. SLAB
6" GRAVEL
CLASS A ASPHALT & GRAVEL ROOF
1/2" PLYWOOD DECK
WOOD BLOCKING
2"x12"@16" O.C.
3" BATT INSULATION
5"x3 1/4"x16 1/4"
COPPER GUTTER - 16 oz.
26 GA. G.I. FLASHING
TOP OF WALLS EL. 112'-7"
1/2" OVERLAID PLYWOOD
BOTTOM OF JOISTS EL. 110'-9 1/2"
1/2" DECORATIVE PLYWOOD
TOP OF BLOCK EL. 109'-8"
1/2" DECORATIVE FIN. PLYWOOD INFILL PANEL
ALUMINUM SASH 2000 SERIES
VENT
1/2" DECORATIVE FIN. PLYWOOD INFILL PANEL TO AL. SASH
2"x4"@16" O.C.
2" BATT INSULATION
1/4" PLYWOOD
FIN. FLOOR EL. 100'-0"
SECTION THROUGH NORTH WALL OF OFFICE
SECTION COLUMN IN OFFICE
SECTION THROUGH SOUTH WALL OF OFFICE
SECTION THROUGH EAST WALL OF OFFICE (TYPICAL FOR WEST WALL)
SECTION THROUGH WEST WALL OF OFFICE AND CURTAIN WALL
WAREHOUSE 8

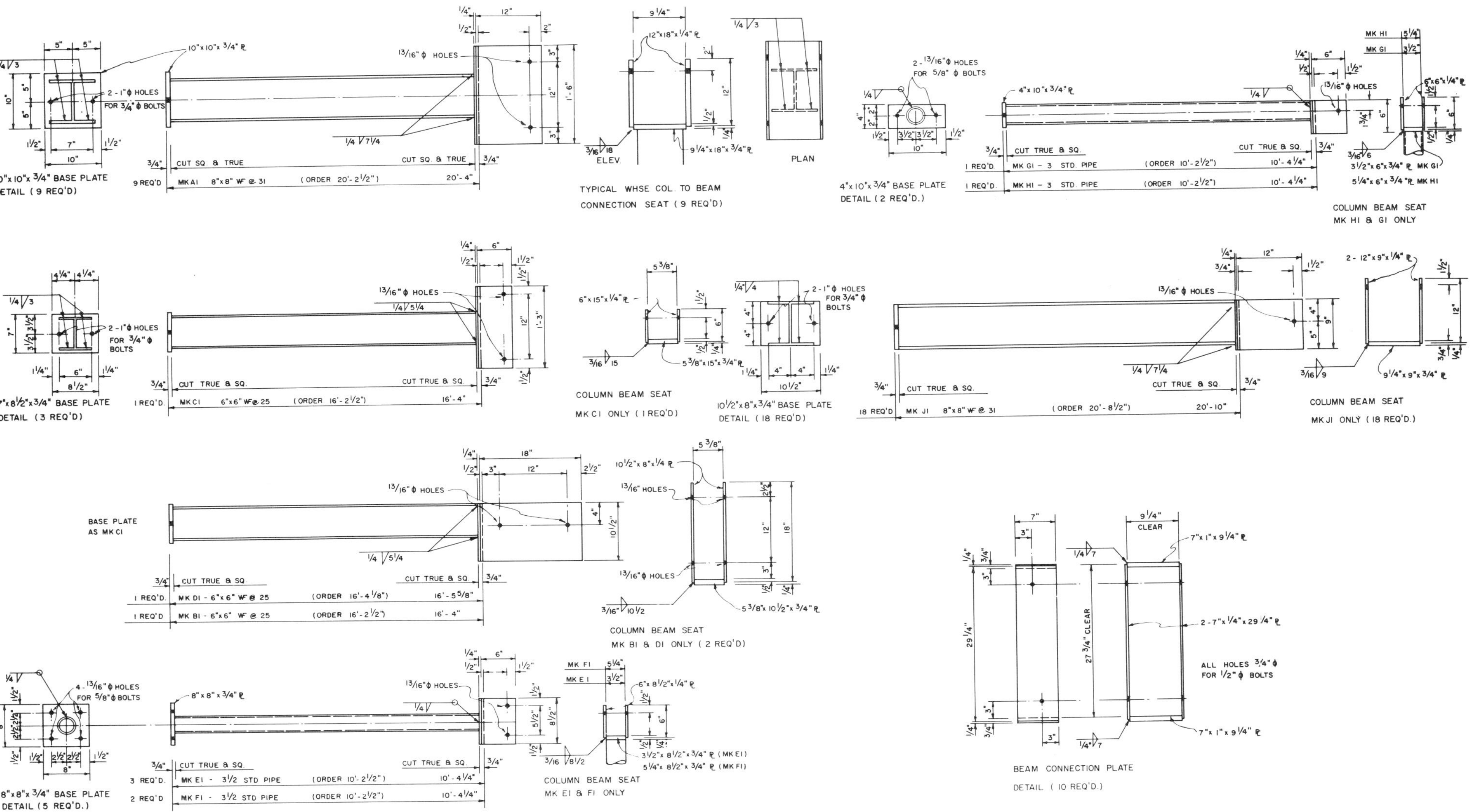

STEELWORK

WAREHOUSE 9

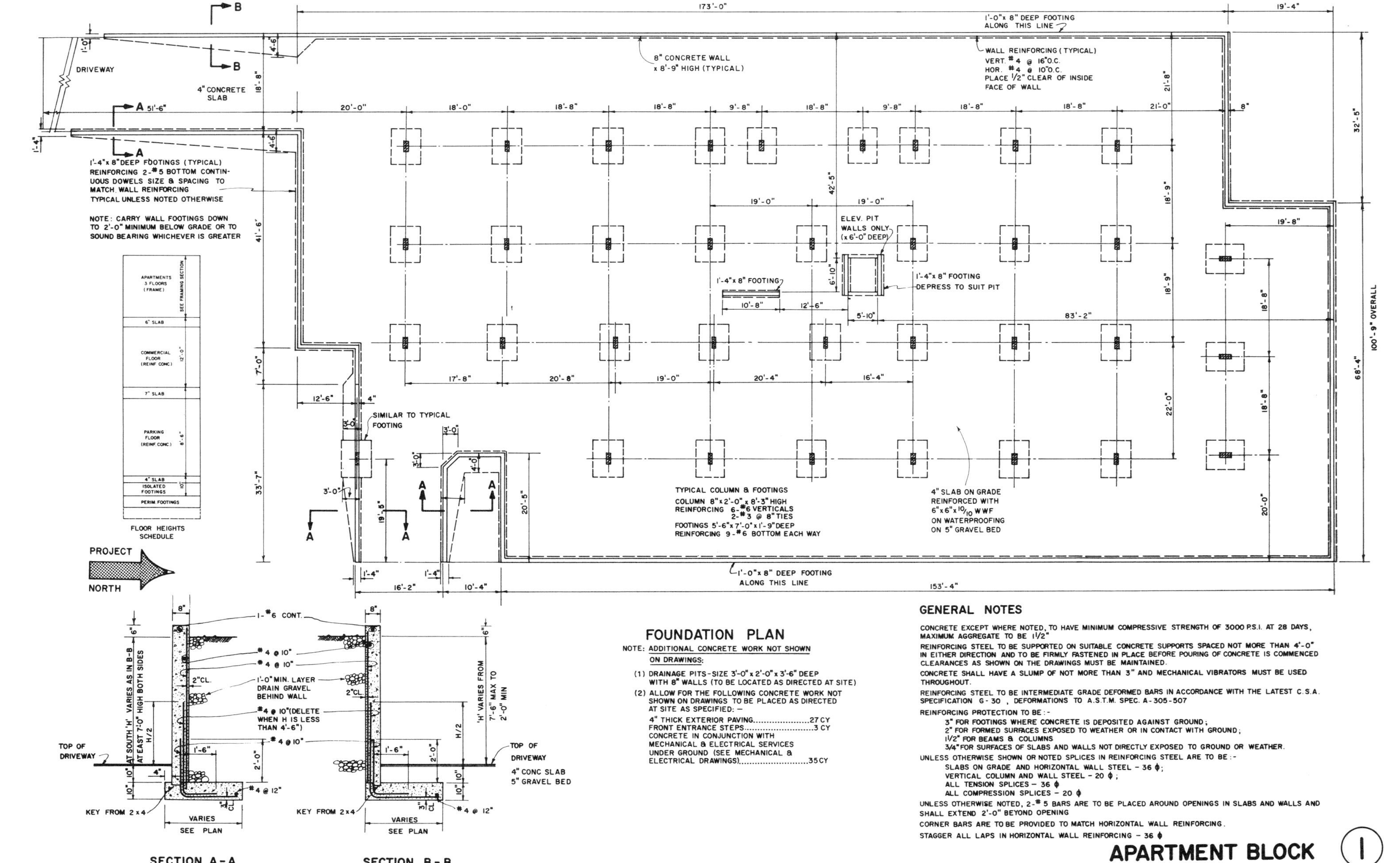

192'-4" OVERALL
173'-0"
19'-4"
1'-0"x 8" DEEP FOOTING ALONG THIS LINE
8" CONCRETE WALL x 8'-9" HIGH (TYPICAL)
WALL REINFORCING (TYPICAL)
VERT. #4 @ 16" O.C.
HOR. #4 @ 10" O.C.
PLACE 1/2" CLEAR OF INSIDE FACE OF WALL
DRIVEWAY
4" CONCRETE SLAB
1'-4"x 8" DEEP FOOTINGS (TYPICAL) REINFORCING 2-#5 BOTTOM CONTINUOUS DOWELS SIZE & SPACING TO MATCH WALL REINFORCING TYPICAL UNLESS NOTED OTHERWISE
NOTE: CARRY WALL FOOTINGS DOWN TO 2'-0" MINIMUM BELOW GRADE OR TO SOUND BEARING WHICHEVER IS GREATER
ELEV. PIT WALLS ONLY (x 6'-0" DEEP)
1'-4"x 8" FOOTING
1'-4"x 8" FOOTING DEPRESS TO SUIT PIT
SIMILAR TO TYPICAL FOOTING
FLOOR HEIGHTS SCHEDULE
PROJECT NORTH
TYPICAL COLUMN & FOOTINGS
COLUMN 8"x 2'-0" x 8'-3" HIGH
REINFORCING 6-#6 VERTICALS 2-#3 @ 8" TIES
FOOTINGS 5'-6"x 7'-0"x 1'-9" DEEP
REINFORCING 9-#6 BOTTOM EACH WAY
4" SLAB ON GRADE REINFORCED WITH 6"x 6"x 10/10 WWF ON WATERPROOFING ON 5" GRAVEL BED
1'-0"x 8" DEEP FOOTING ALONG THIS LINE
153'-4"
100'-9" OVERALL
68'-4"
32'-5"
83'-2"
42'-5"
1-#6 CONT.
#4 @ 10"
#4 @ 10"
1'-0" MIN. LAYER DRAIN GRAVEL BEHIND WALL
#4 @ 10" (DELETE WHEN H IS LESS THAN 4'-6")
#4 @ 10"
#4 @ 12"
AT SOUTH 'H' VARIES AS IN B-B
AT EAST 7'-0" HIGH BOTH SIDES
'H' VARIES FROM 7'-6" MAX TO 2'-0" MIN
TOP OF DRIVEWAY
4" CONC SLAB
5" GRAVEL BED
KEY FROM 2 x 4
VARIES
SEE PLAN
SECTION A-A
SECTION B-B
FOUNDATION PLAN
NOTE: ADDITIONAL CONCRETE WORK NOT SHOWN ON DRAWINGS:
(1) DRAINAGE PITS-SIZE 3'-0"x 2'-0"x 3'-6" DEEP WITH 8" WALLS (TO BE LOCATED AS DIRECTED AT SITE)
(2) ALLOW FOR THE FOLLOWING CONCRETE WORK NOT SHOWN ON DRAWINGS TO BE PLACED AS DIRECTED AT SITE AS SPECIFIED: –
4" THICK EXTERIOR PAVING......27 CY
FRONT ENTRANCE STEPS......3 CY
CONCRETE IN CONJUNCTION WITH MECHANICAL & ELECTRICAL SERVICES UNDER GROUND (SEE MECHANICAL & ELECTRICAL DRAWINGS)......35 CY
GENERAL NOTES
CONCRETE EXCEPT WHERE NOTED, TO HAVE MINIMUM COMPRESSIVE STRENGTH OF 3000 P.S.I. AT 28 DAYS, MAXIMUM AGGREGATE TO BE 1 1/2"
REINFORCING STEEL TO BE SUPPORTED ON SUITABLE CONCRETE SUPPORTS SPACED NOT MORE THAN 4'-0" IN EITHER DIRECTION AND TO BE FIRMLY FASTENED IN PLACE BEFORE POURING OF CONCRETE IS COMMENCED CLEARANCES AS SHOWN ON THE DRAWINGS MUST BE MAINTAINED.
CONCRETE SHALL HAVE A SLUMP OF NOT MORE THAN 3" AND MECHANICAL VIBRATORS MUST BE USED THROUGHOUT.
REINFORCING STEEL TO BE INTERMEDIATE GRADE DEFORMED BARS IN ACCORDANCE WITH THE LATEST C.S.A. SPECIFICATION G-30, DEFORMATIONS TO A.S.T.M. SPEC. A-305-507
REINFORCING PROTECTION TO BE:-
3" FOR FOOTINGS WHERE CONCRETE IS DEPOSITED AGAINST GROUND;
2" FOR FORMED SURFACES EXPOSED TO WEATHER OR IN CONTACT WITH GROUND;
1 1/2" FOR BEAMS & COLUMNS
3/4" FOR SURFACES OF SLABS AND WALLS NOT DIRECTLY EXPOSED TO GROUND OR WEATHER.
UNLESS OTHERWISE SHOWN OR NOTED SPLICES IN REINFORCING STEEL ARE TO BE:-
SLABS ON GRADE AND HORIZONTAL WALL STEEL – 36 ϕ;
VERTICAL COLUMN AND WALL STEEL – 20 ϕ;
ALL TENSION SPLICES – 36 ϕ
ALL COMPRESSION SPLICES – 20 ϕ
UNLESS OTHERWISE NOTED, 2-#5 BARS ARE TO BE PLACED AROUND OPENINGS IN SLABS AND WALLS AND SHALL EXTEND 2'-0" BEYOND OPENING
CORNER BARS ARE TO BE PROVIDED TO MATCH HORIZONTAL WALL REINFORCING.
STAGGER ALL LAPS IN HORIZONTAL WALL REINFORCING – 36 ϕ
APARTMENT BLOCK
1

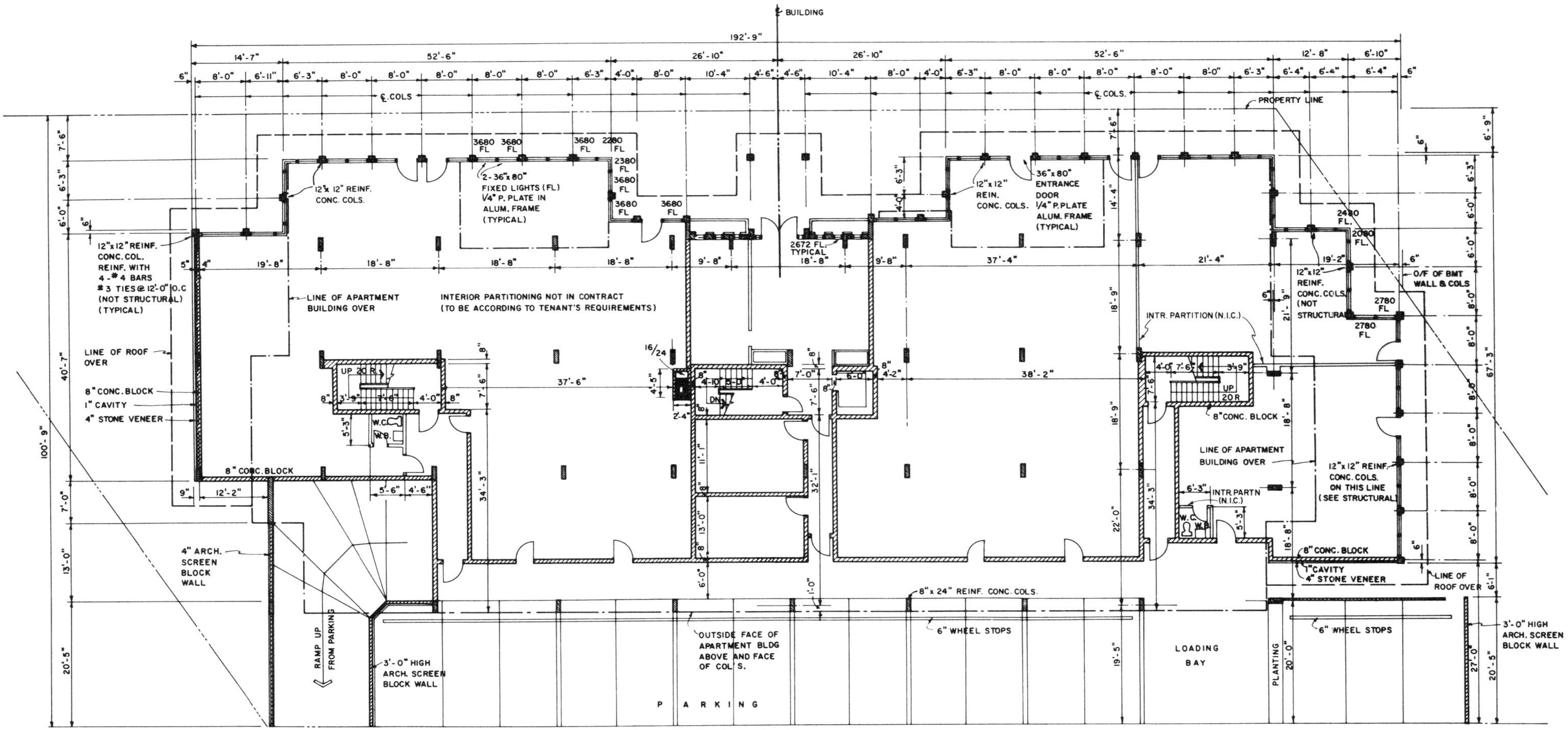

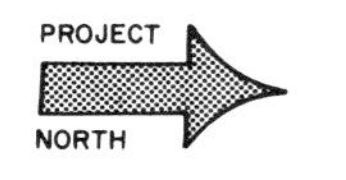

FLOOR PLAN – COMMERCIAL AREA

NOTES:

12" x 12" REINFORCED CONCRETE COLUMNS ON THIS FLOOR ARE NOT STRUCTURAL (EXCEPT 6-COLUMNS AT NORTH END SHOWN ON STRUCTURAL DRAWINGS) CHECK LOCATION OF COLUMNS WITH ARCHITECT BEFORE INSTALLING.

EXTR DOORS AT REAR OF RENTAL SPACES TO BE LOCATED TO SUIT TENANTS.

INTR PARTITIONS (OTHER THAN CONC BLOCK) ARE NOT IN CONTRACT.

APARTMENT BLOCK

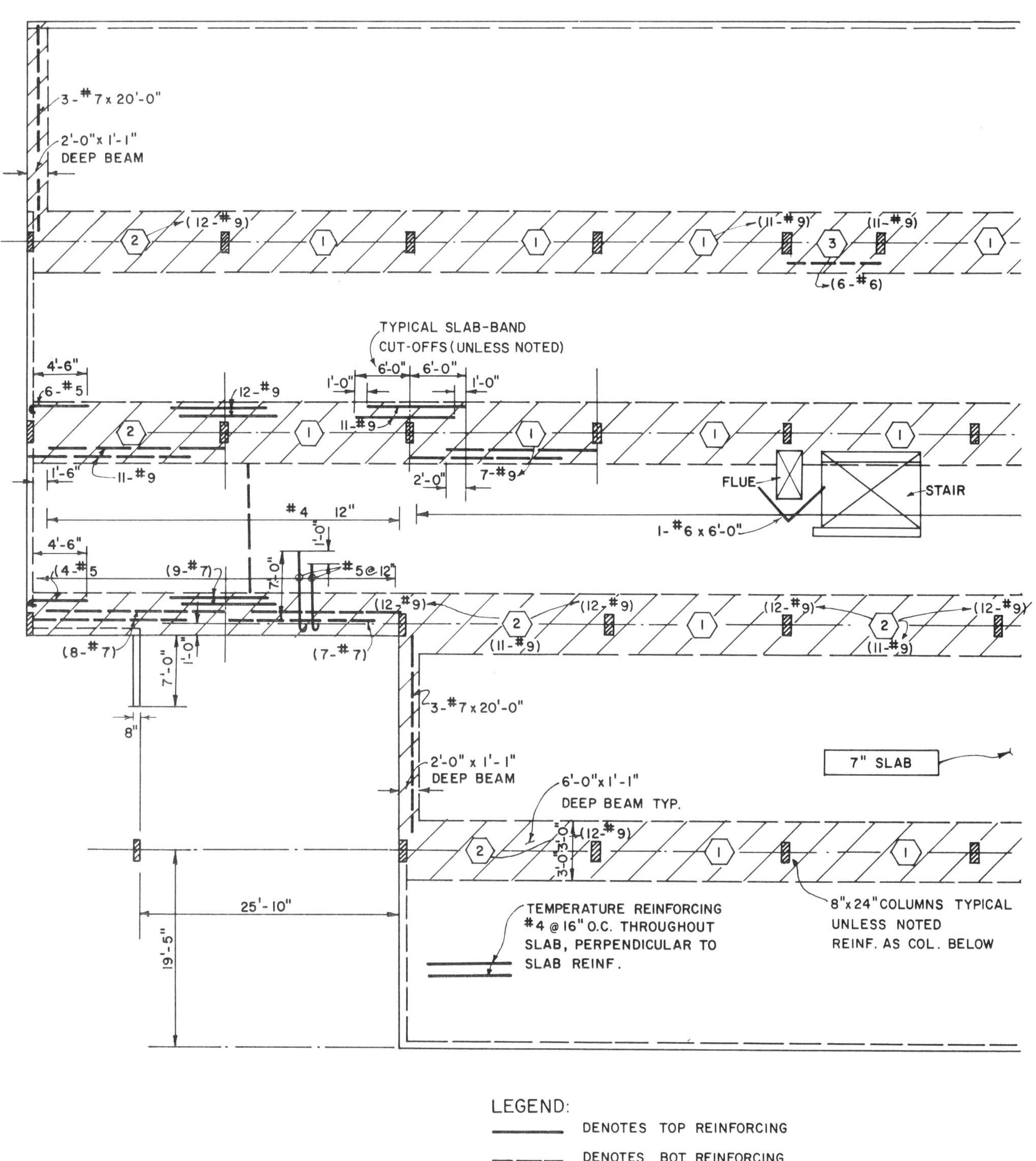

CEILING PLAN – PARKING STRUCTURE

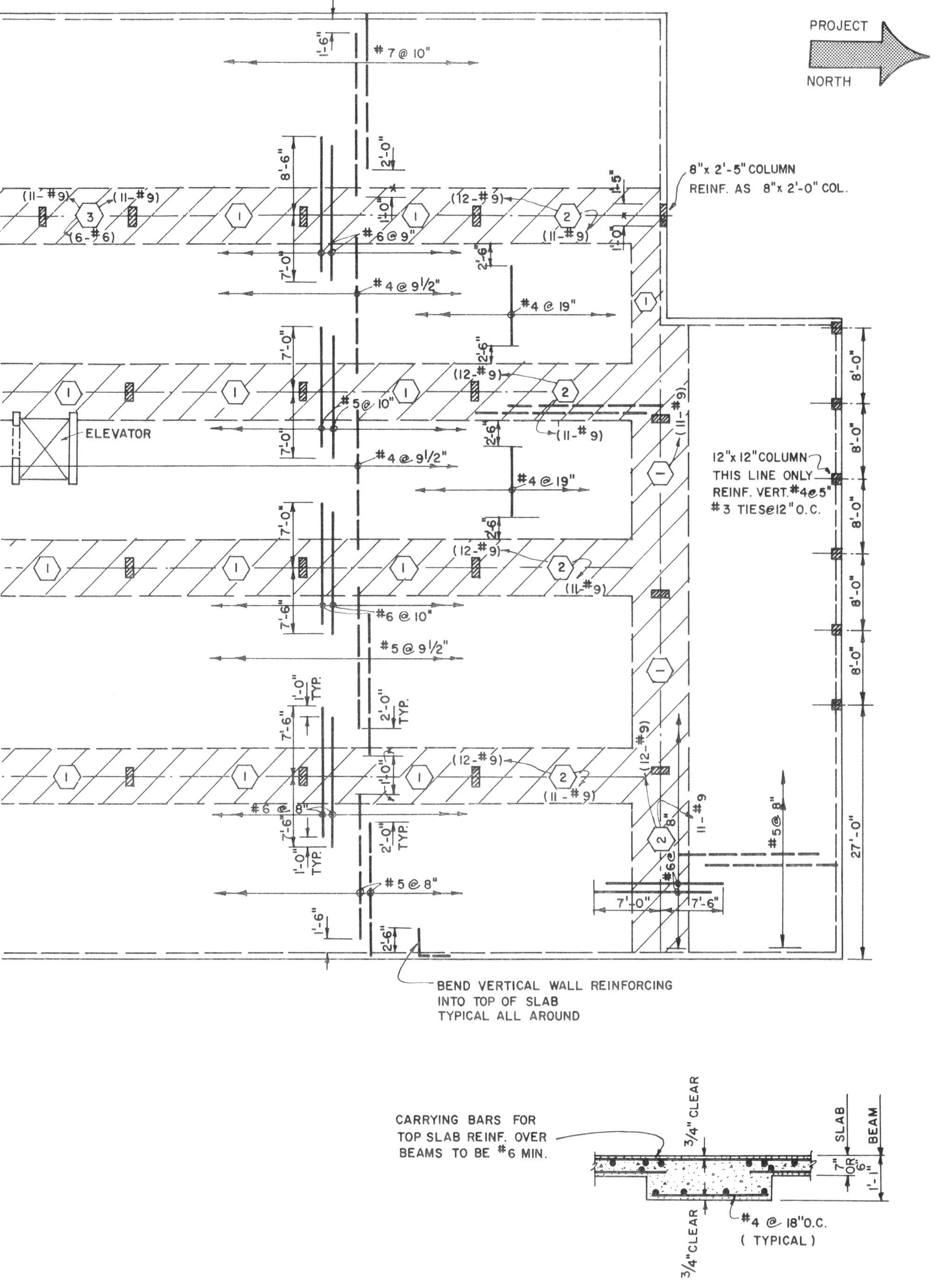

TYPICAL SLAB-BAND SECTION

APARTMENT BLOCK 3

CEILING PLAN - PARKING STRUCTURE

APARTMENT BLOCK (3)

PROJECT NORTH

8"x 2'-5" COLUMN REINF. AS 8"x 2'-0" COL.

12"x 12" COLUMN THIS LINE ONLY REINF. VERT. #4@5" #3 TIES@12" O.C.

3-#7 x 20'-0" 2'-0"x 1'-1" DEEP BEAM

TYPICAL SLAB-BAND CUT-OFFS (UNLESS NOTED)

6'-0"x 1'-1" DEEP BEAM TYP.

7" SLAB

FLUE

STAIR

ELEVATOR

1-#6 x 6'-0"

8"x 24" COLUMNS TYPICAL UNLESS NOTED REINF. AS COL. BELOW

TEMPERATURE REINFORCING #4 @ 16" O.C. THROUGHOUT SLAB, PERPENDICULAR TO SLAB REINF.

BEND VERTICAL WALL REINFORCING INTO TOP OF SLAB TYPICAL ALL AROUND

LEGEND:

DENOTES TOP REINFORCING

DENOTES BOT. REINFORCING

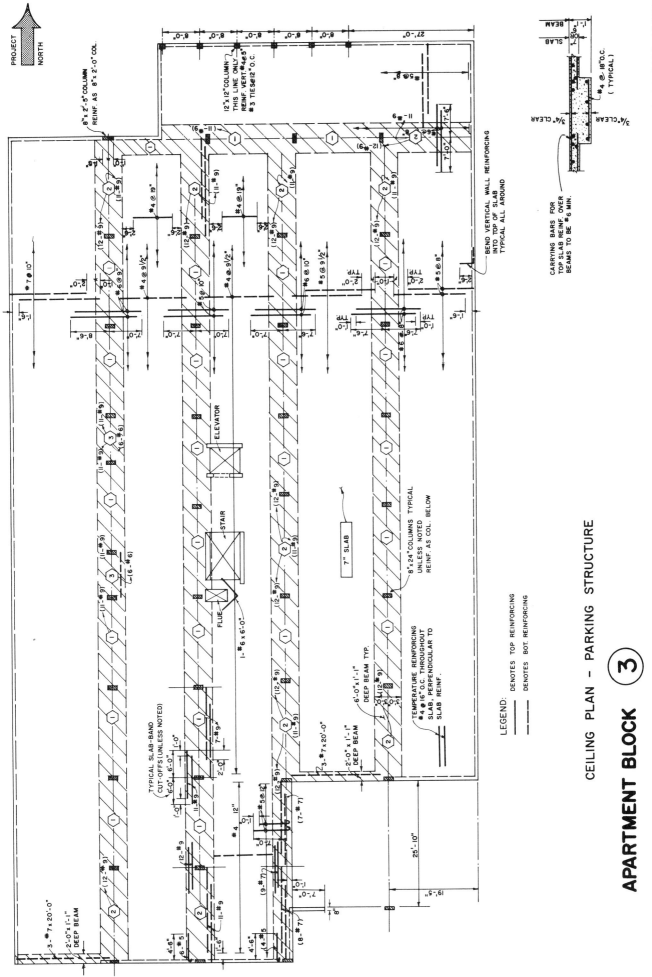

TYPICAL SLAB-BAND SECTION

CARRYING BARS FOR TOP SLAB REINF. OVER BEAMS TO BE #6 MIN.

3/4" CLEAR

3/4" CLEAR

#4 @ 18" O.C. (TYPICAL)

SLAB 7" OR 6"

BEAM 1'-1"

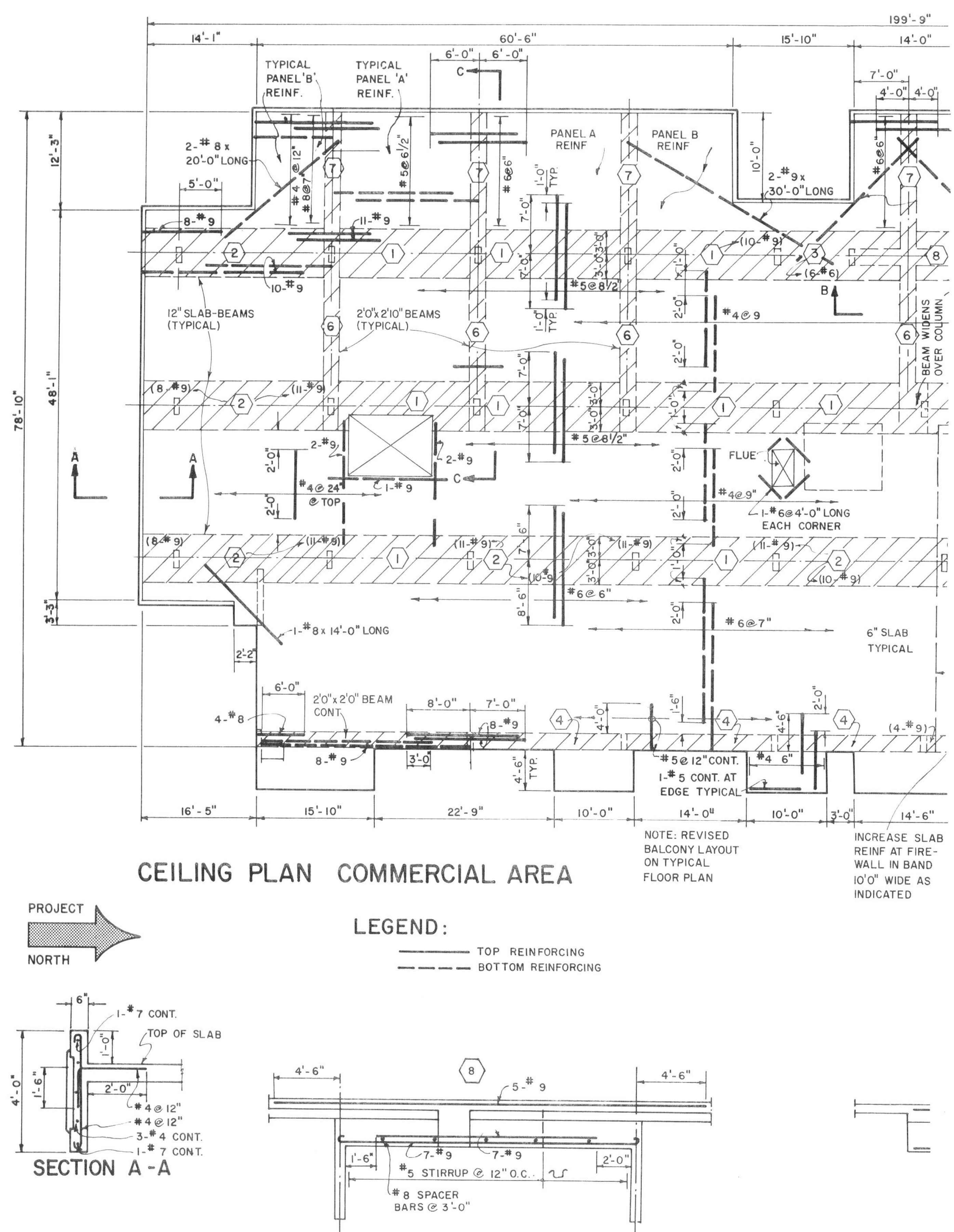

199'-9"
14'-1"
60'-6"
15'-10"
14'-0"
6'-0"
6'-0"
7'-0"
4'-0"
4'-0"
TYPICAL PANEL 'B' REINF.
TYPICAL PANEL 'A' REINF.
PANEL A REINF
PANEL B REINF
2-#8 x 20'-0" LONG
2-#9 x 30'-0" LONG
#4 @ 12"
#8 @ 7
#5 @ 6 1/2"
#6 @ 6"
8-#9
11-#9
10-#9
(10-#9)
(6-#6)
5'-0"
10'-0"
1'-0" TYP.
7'-0"
3'-0"
#5 @ 8 1/2"
#4 @ 9
12" SLAB-BEAMS (TYPICAL)
2'0" x 2'10" BEAMS (TYPICAL)
BEAM WIDENS OVER COLUMN
(8-#9)
(11-#9)
2-#9
#4 @ 24" @ TOP
1-#9
FLUE
#4 @ 9"
1-#6 @ 4'-0" LONG EACH CORNER
(10-#9)
#6 @ 6"
#6 @ 7"
1-#8 x 14'-0" LONG
6" SLAB TYPICAL
12'-3"
48'-1"
78'-10"
3'-3"
2'-2"
2'-0"
8'-6"
6'-0"
2'0" x 2'0" BEAM CONT
4-#8
8'-0"
7'-0"
8-#9
3'-0
4'-6" TYP.
4'-0"
1'-6"
4'-6"
#5 @ 12" CONT.
#4 6"
1-#5 CONT. AT EDGE TYPICAL
(4-#9)
16'-5"
15'-10"
22'-9"
10'-0"
14'-0"
10'-0"
3'-0"
14'-6"
NOTE: REVISED BALCONY LAYOUT ON TYPICAL FLOOR PLAN
INCREASE SLAB REINF AT FIRE-WALL IN BAND 10'0" WIDE AS INDICATED
CEILING PLAN COMMERCIAL AREA
PROJECT NORTH
LEGEND:
TOP REINFORCING
BOTTOM REINFORCING
6"
1-#7 CONT.
TOP OF SLAB
1'-0"
4'-0"
1'-6"
2'-0"
#4 @ 12"
#4 @ 12"
3-#4 CONT.
1-#7 CONT.
SECTION A-A
4'-6"
4'-6"
5-#9
7-#9
7-#9
1'-6"
2'-0"
#5 STIRRUP @ 12" O.C.
#8 SPACER BARS @ 3'-0"
SECTION B-B

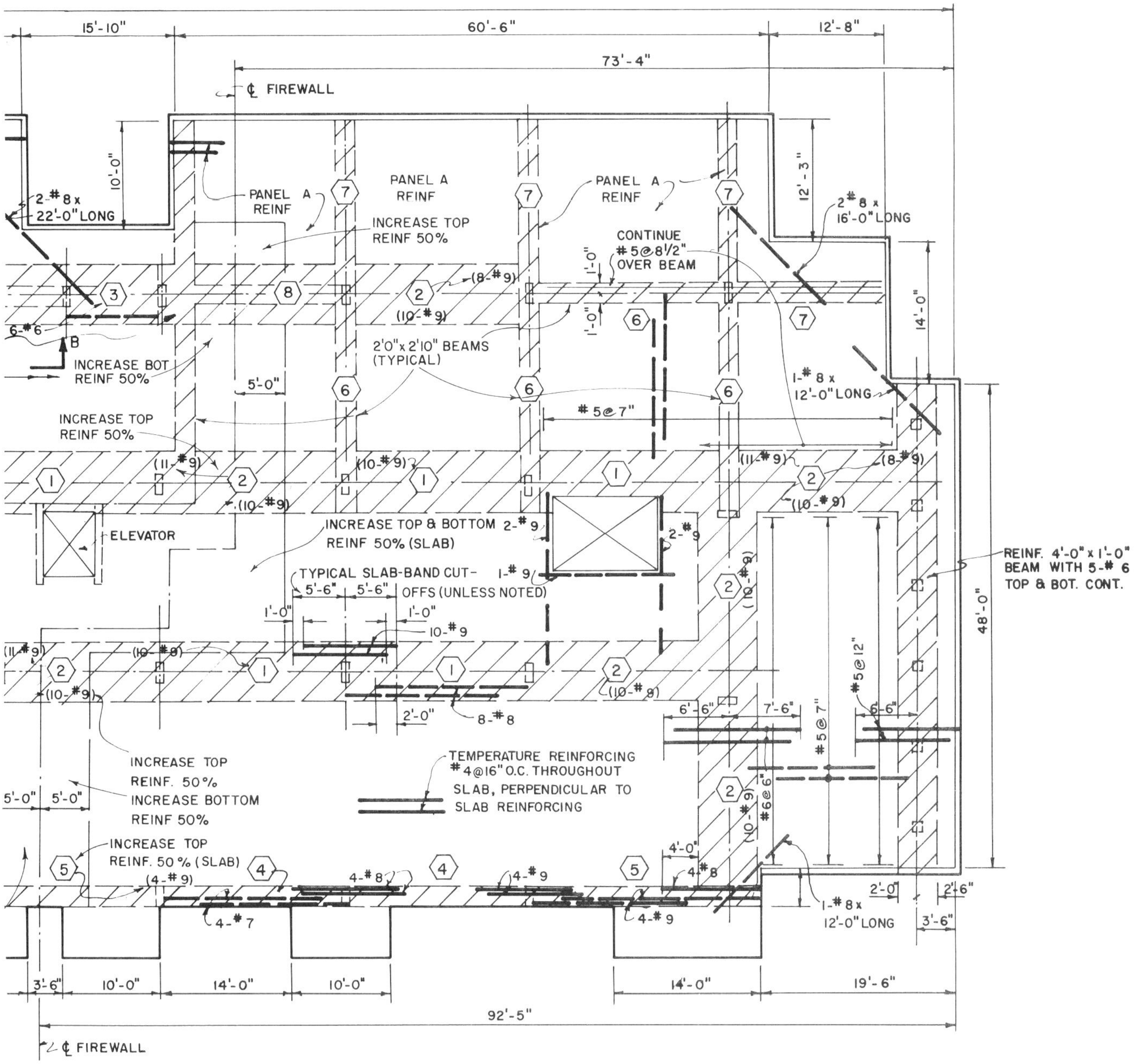

15'-10" 60'-6" 12'-8"
73'-4"
℄ FIREWALL
10'-0"
2-#8 x 22'-0" LONG
PANEL A REINF
PANEL A RFINF
PANEL A REINF
12'-3"
2-#8 x 16'-0" LONG
INCREASE TOP REINF 50%
CONTINUE #5@8½" OVER BEAM
1'-0"
(8-#9)
(10-#9)
14'-0"
6-#6
B
INCREASE BOT REINF 50%
2'0" x 2'10" BEAMS (TYPICAL)
5'-0"
1-#8 x 12'-0" LONG
#5@7"
INCREASE TOP REINF 50%
(11-#9)
(10-#9)
(8-#9)
ELEVATOR
INCREASE TOP & BOTTOM REINF 50% (SLAB)
2-#9
1-#9
REINF. 4'-0" x 1'-0" BEAM WITH 5-#6 TOP & BOT. CONT.
TYPICAL SLAB-BAND CUT-OFFS (UNLESS NOTED)
5'-6" 5'-6"
1'-0" 1'-0"
10-#9
48'-0"
#5@12"
2'-0"
8-#8
6'-6" 7'-6" 6'-6"
#5@7"
INCREASE TOP REINF. 50%
INCREASE BOTTOM REINF 50%
TEMPERATURE REINFORCING #4@16" O.C. THROUGHOUT SLAB, PERPENDICULAR TO SLAB REINFORCING
5'-0" 5'-0"
#6@6"
INCREASE TOP REINF. 50% (SLAB)
(4-#9)
4-#8
4-#9
4'-0"
4-#8
4-#7
4-#9
2'-0" 2'-6"
1-#8 x 12'-0" LONG
3'-6"
3'-6" 10'-0" 14'-0" 10'-0" 14'-0" 19'-6"
92'-5"
℄ FIREWALL

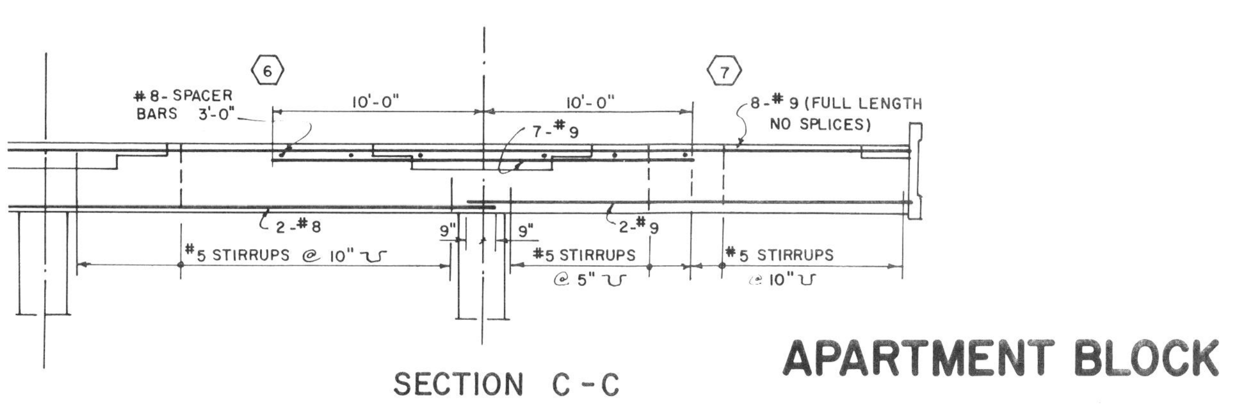

SECTION C-C

APARTMENT BLOCK 4

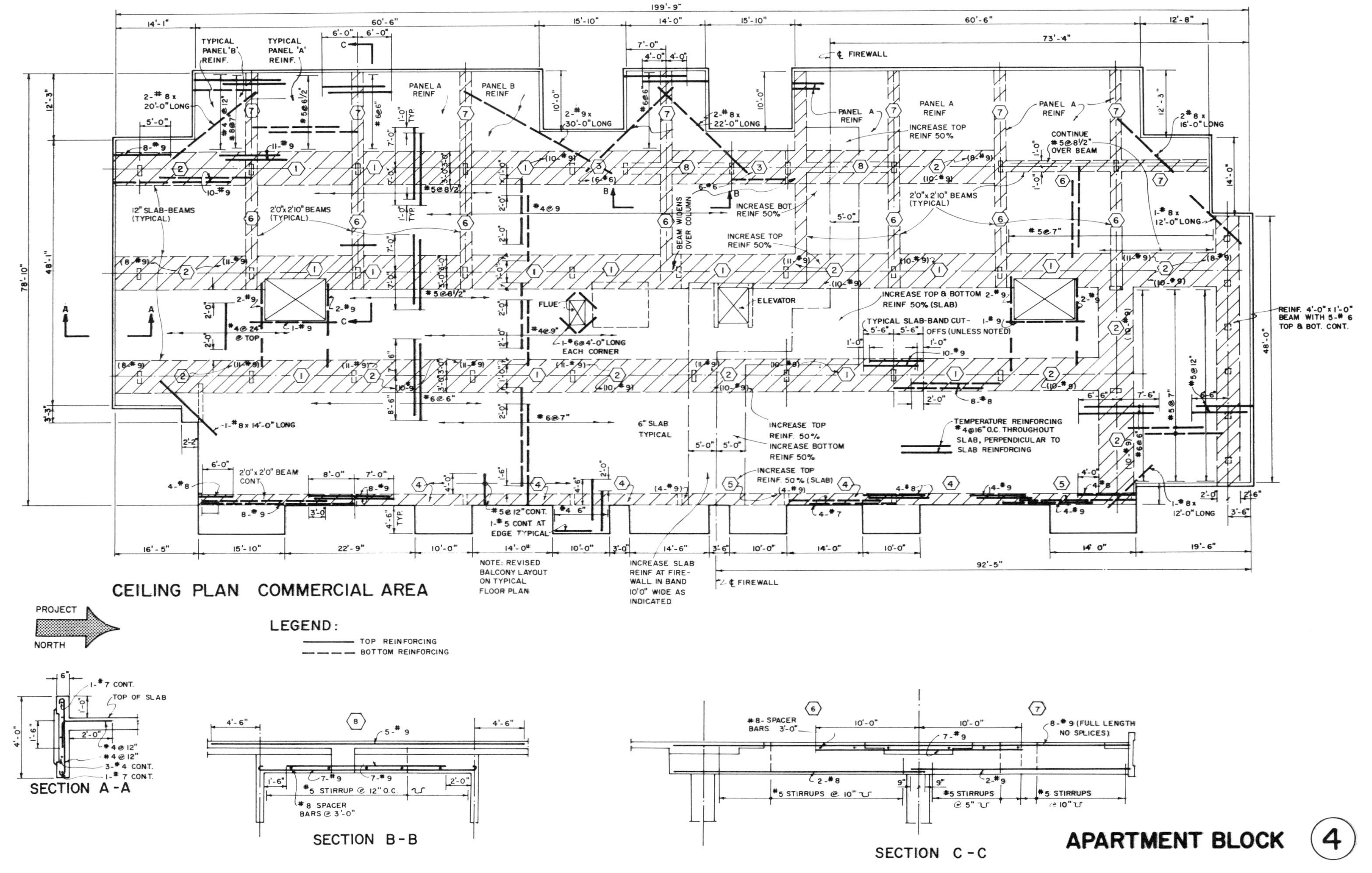

199'-9"
TYPICAL PANEL 'B' REINF.
TYPICAL PANEL 'A' REINF.
PANEL A REINF
PANEL B REINF
2-#8 x 20'-0" LONG
2-#9 x 30'-0" LONG
2-#8 x 22'-0" LONG
2-#8 x 16'-0" LONG
1-#8 x 12'-0" LONG
1-#8 x 14'-0" LONG
℄ FIREWALL
INCREASE TOP REINF 50%
CONTINUE #5@8½" OVER BEAM
12" SLAB-BEAMS (TYPICAL)
2'0" x 2'10" BEAMS (TYPICAL)
BEAM WIDENS OVER COLUMN
INCREASE BOT REINF 50%
INCREASE TOP & BOTTOM 2-#9 REINF 50% (SLAB)
TYPICAL SLAB-BAND CUT-OFFS (UNLESS NOTED)
ELEVATOR
FLUE
1-#6@4'-0" LONG EACH CORNER
REINF. 4'-0" x 1'-0" BEAM WITH 5-#6 TOP & BOT. CONT.
6" SLAB TYPICAL
INCREASE TOP REINF. 50% INCREASE BOTTOM REINF 50%
INCREASE TOP REINF. 50% (SLAB)
TEMPERATURE REINFORCING #4@16" O.C. THROUGHOUT SLAB, PERPENDICULAR TO SLAB REINFORCING
2'0" x 2'0" BEAM CONT.
#5@12" CONT. 1-#5 CONT AT EDGE TYPICAL
NOTE: REVISED BALCONY LAYOUT ON TYPICAL FLOOR PLAN
INCREASE SLAB REINF AT FIRE-WALL IN BAND 10'0" WIDE AS INDICATED
92'-5"
78'-10"
48'-1"
48'-0"
CEILING PLAN COMMERCIAL AREA
PROJECT NORTH
LEGEND:
TOP REINFORCING
BOTTOM REINFORCING
1-#7 CONT.
TOP OF SLAB
#4@12"
3-#4 CONT.
SECTION A-A
5-#9
7-#9
#5 STIRRUP @ 12" O.C.
#8 SPACER BARS @ 3'-0"
SECTION B-B
#8-SPACER BARS 3'-0"
8-#9 (FULL LENGTH NO SPLICES)
#5 STIRRUPS @ 10"
#5 STIRRUPS @ 5"
SECTION C-C
APARTMENT BLOCK 4

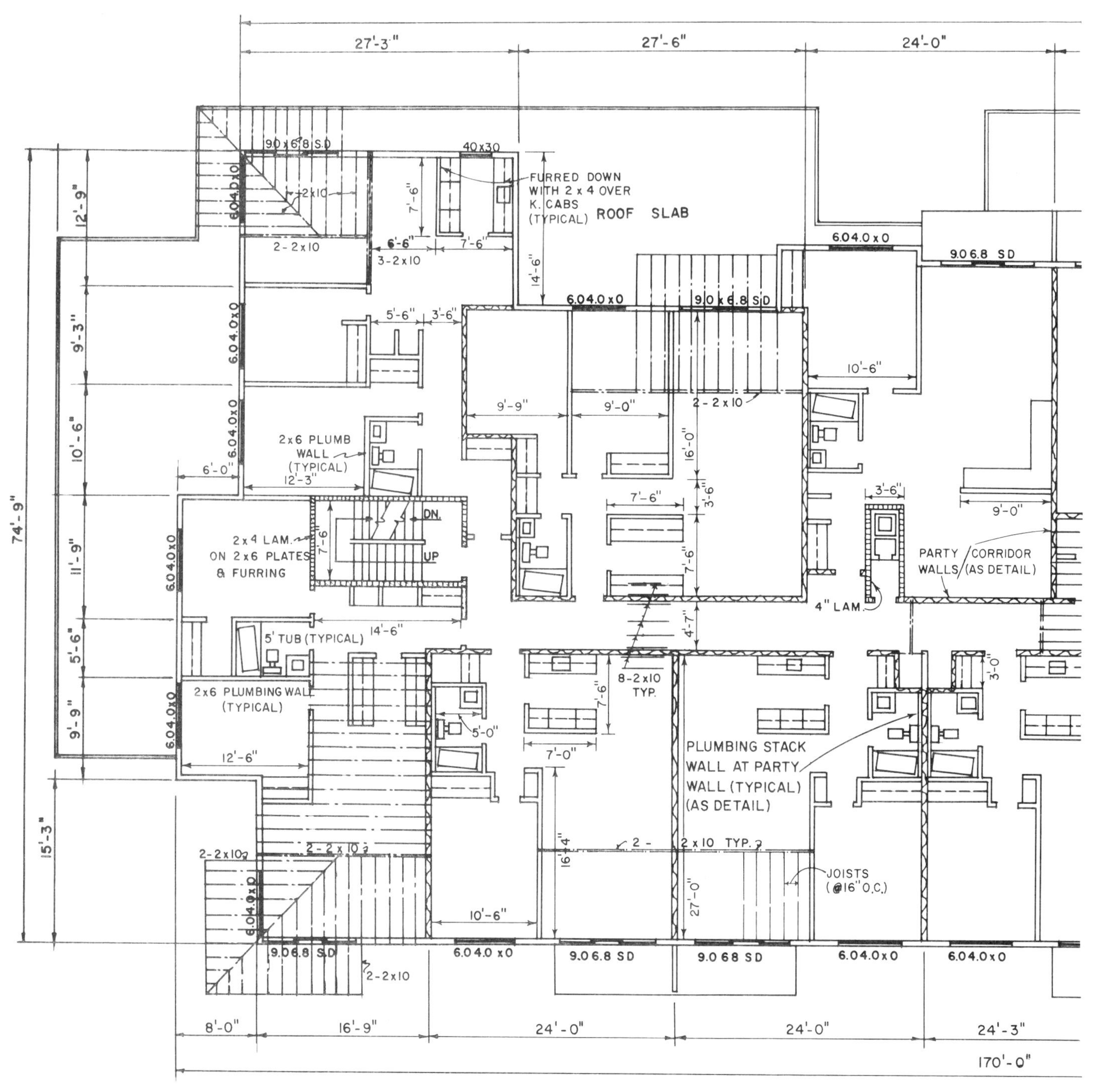

TYPICAL FLOOR PLAN

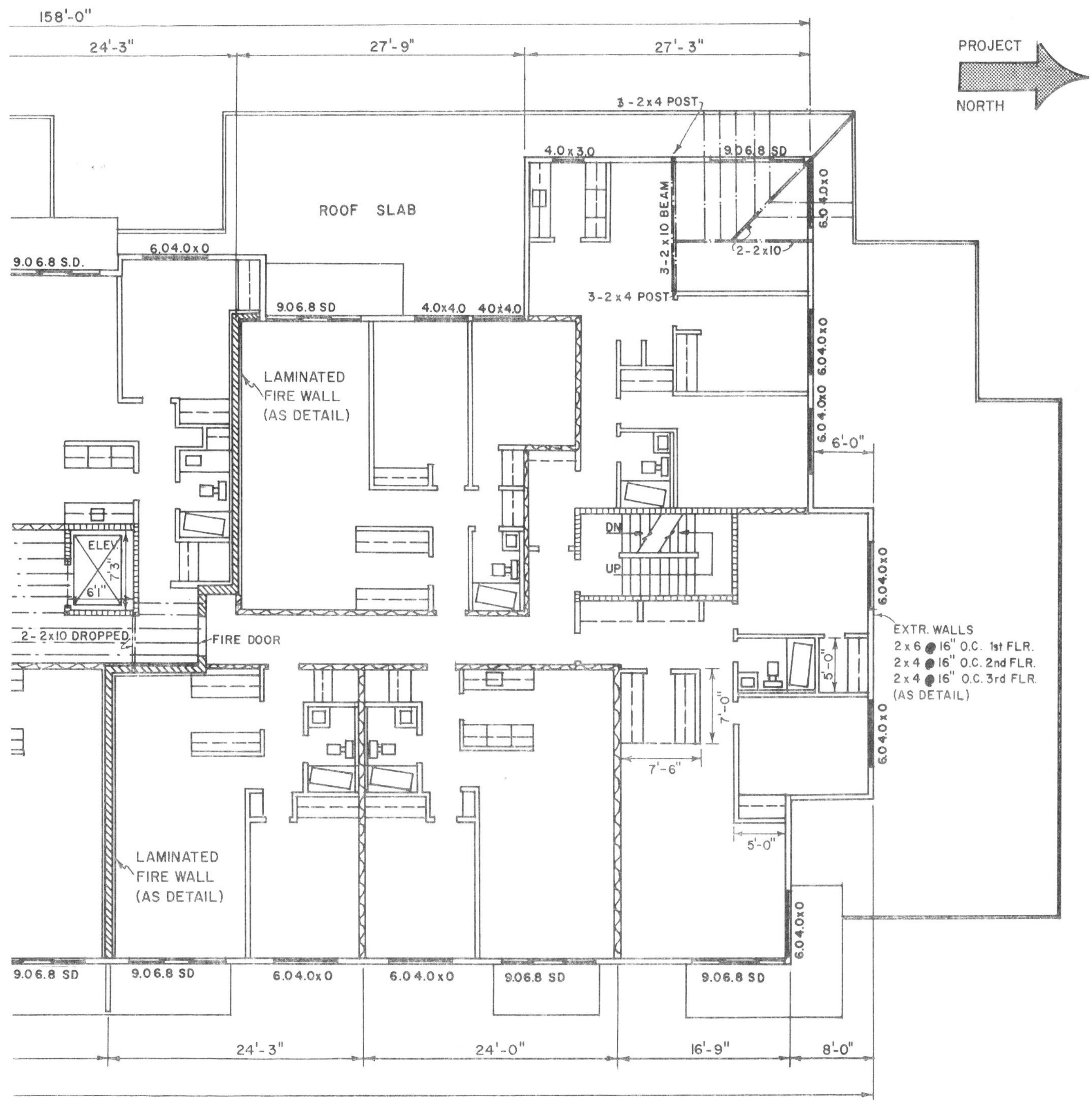

NOTES AND REVISIONS

ALL INTERNAL DIMENSIONS ARE TO FRAMING

SUITE & BALCONY LAYOUT REVISED ON EAST SIDE OF FLOOR PLAN (STRUCTURE & ELEVATIONS TO FOLLOW)

CONC BLOCK FIREWALL CHANGED TO LAMINATED WOOD AS DETAIL

CEILING JOISTS ONLY TO BATHROOMS ON 2nd AND 3rd FLOORS AND OMITTED ELSEWHERE

PROVIDE 2x4 FURRING OVER KITCHEN CPBDS. & TO TUB PANELS

PROVIDE CONT. 2x4/s AS REQUIRED TO SUPPORT WALLBOARD EDGES AT CORNERS OF WALLS & CEILINGS I.E. AT CEILING BOARD EDGES AT CORRIDOR WALLS

APARTMENT BLOCK (5)

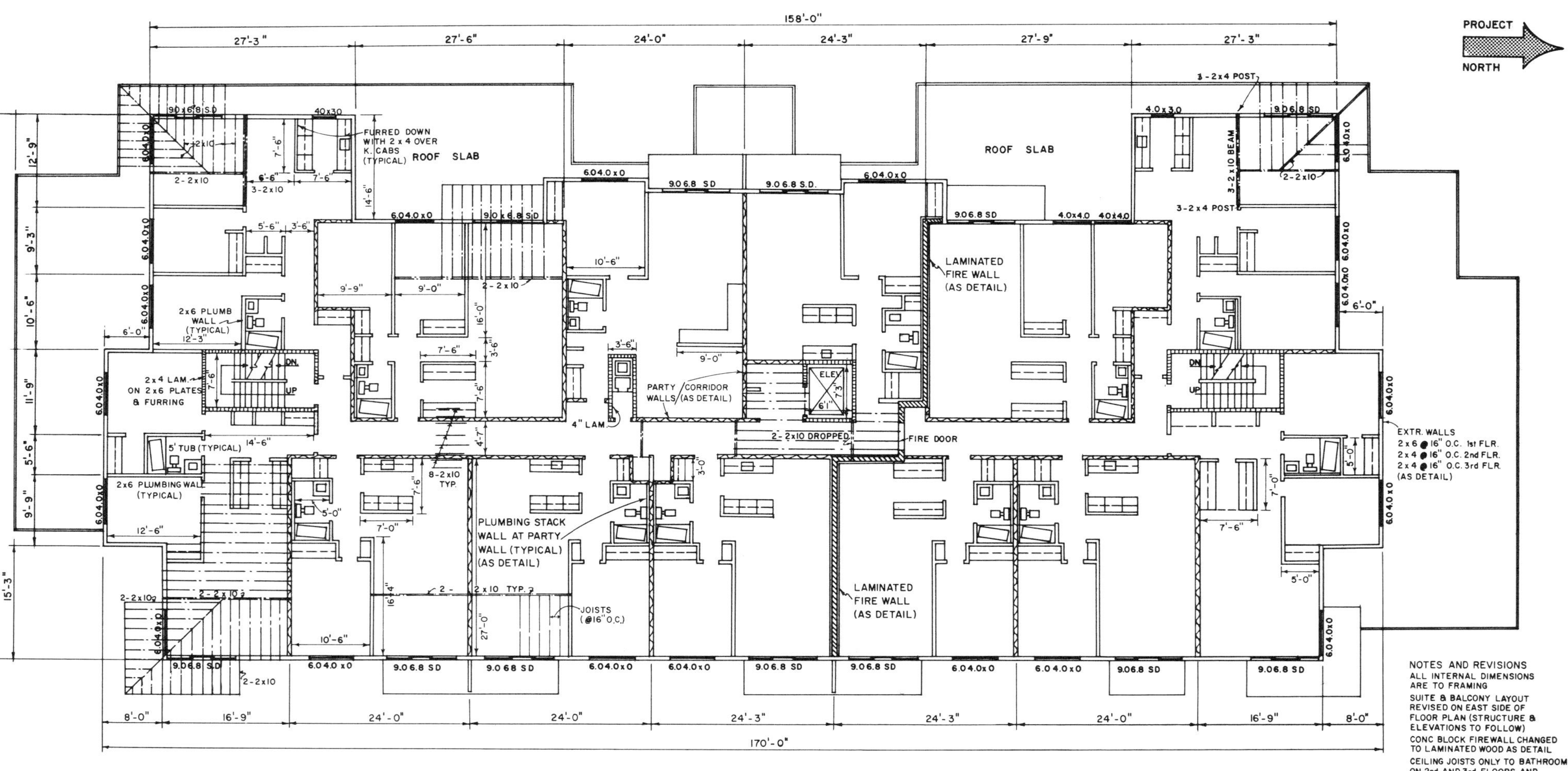

TYPICAL FLOOR PLAN

NOTES AND REVISIONS
ALL INTERNAL DIMENSIONS ARE TO FRAMING
SUITE & BALCONY LAYOUT REVISED ON EAST SIDE OF FLOOR PLAN (STRUCTURE & ELEVATIONS TO FOLLOW)
CONC BLOCK FIREWALL CHANGED TO LAMINATED WOOD AS DETAIL
CEILING JOISTS ONLY TO BATHROOMS ON 2nd AND 3rd FLOORS AND OMITTED ELSEWHERE
PROVIDE 2x4 FURRING OVER KITCHEN CPBDS. & TO TUB PANELS
PROVIDE CONT. 2x4/s AS REQUIRED TO SUPPORT WALLBOARD EDGES AT CORNERS OF WALLS & CEILINGS I.E. AT CEILING BOARD EDGES AT CORRIDOR WALLS

APARTMENT BLOCK 5

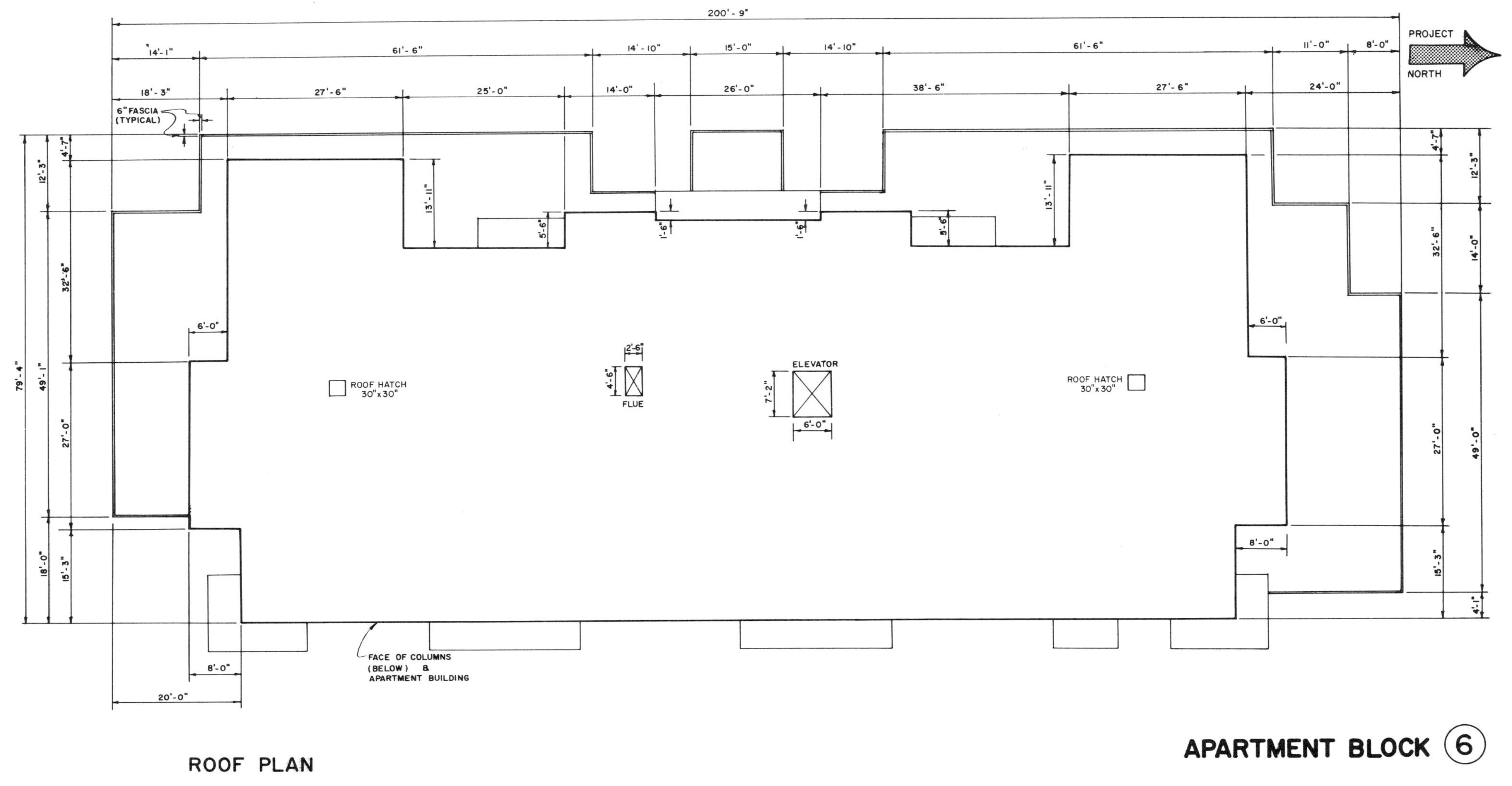
PROJECT NORTH
200'- 9"
14'-1"
61'- 6"
14'- 10"
15'-0"
14'- 10"
61'- 6"
11'-0"
8'-0"
18'- 3"
27'- 6"
25'- 0"
14'-0"
26'-0"
38'- 6"
27'- 6"
24'-0"
6" FASCIA
(TYPICAL)
79'- 4"
ROOF HATCH
30"x30"
FLUE
ELEVATOR
FACE OF COLUMNS
(BELOW) &
APARTMENT BUILDING
20'-0"
ROOF PLAN
APARTMENT BLOCK 6

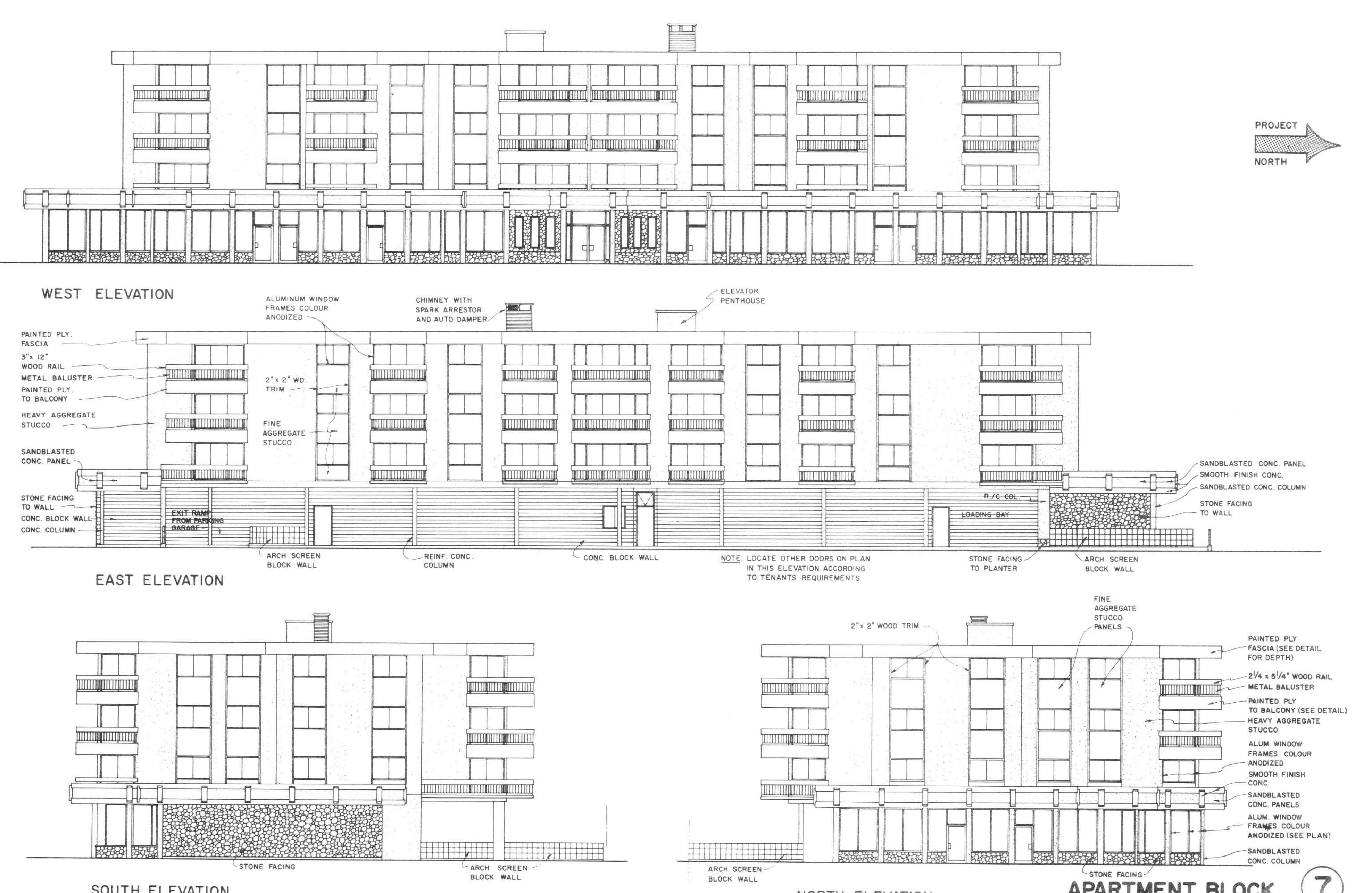
PROJECT NORTH
WEST ELEVATION
ALUMINUM WINDOW FRAMES COLOUR ANODIZED
CHIMNEY WITH SPARK ARRESTOR AND AUTO DAMPER
ELEVATOR PENTHOUSE
PAINTED PLY. FASCIA
3"x 12" WOOD RAIL
METAL BALUSTER
PAINTED PLY. TO BALCONY
HEAVY AGGREGATE STUCCO
2"x 2" WD. TRIM
FINE AGGREGATE STUCCO
SANDBLASTED CONC. PANEL
STONE FACING TO WALL
CONC. BLOCK WALL
CONC. COLUMN
EXIT RAMP FROM PARKING GARAGE
ARCH SCREEN BLOCK WALL
REINF. CONC. COLUMN
CONC. BLOCK WALL
NOTE: LOCATE OTHER DOORS ON PLAN IN THIS ELEVATION ACCORDING TO TENANTS' REQUIREMENTS
R/C COL.
LOADING BAY
STONE FACING TO PLANTER
ARCH SCREEN BLOCK WALL
SANDBLASTED CONC. PANEL
SMOOTH FINISH CONC.
SANDBLASTED CONC. COLUMN
STONE FACING TO WALL
EAST ELEVATION
STONE FACING
ARCH SCREEN BLOCK WALL
SOUTH ELEVATION
2"x 2" WOOD TRIM
FINE AGGREGATE STUCCO PANELS
PAINTED PLY FASCIA (SEE DETAIL FOR DEPTH)
2 1/4 x 5 1/4" WOOD RAIL
METAL BALUSTER
PAINTED PLY TO BALCONY (SEE DETAIL)
HEAVY AGGREGATE STUCCO
ALUM. WINDOW FRAMES COLOUR ANODIZED
SMOOTH FINISH CONC.
SANDBLASTED CONC. PANELS
ALUM. WINDOW FRAMES COLOUR ANODIZED (SEE PLAN)
SANDBLASTED CONC. COLUMN
STONE FACING
ARCH SCREEN BLOCK WALL
NORTH ELEVATION
APARTMENT BLOCK
7

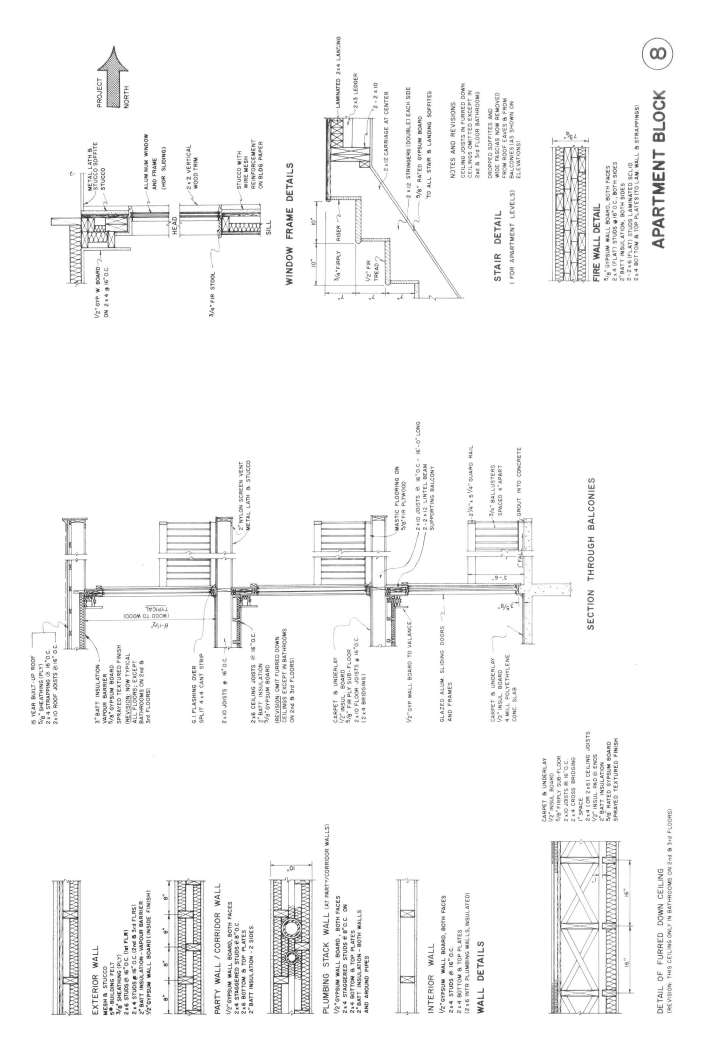

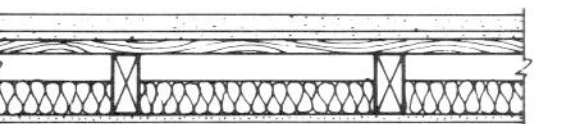

EXTERIOR WALL

MESH & STUCCO
5#-BUILDING FELT
3/8" SHEATHING (PLY)
2x6 STUDS @ 16" O.C. (1st FLR)
2x4 STUDS @ 16" O.C. (2nd & 3rd FLRS)
2" BATT INSULATION - VAPOUR BARRIER
1/2" GYPSUM WALL BOARD (INSIDE FINISH)

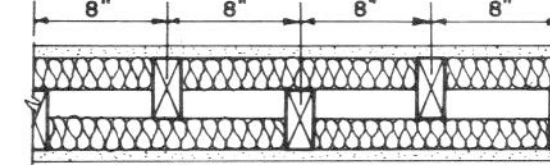

PARTY WALL / CORRIDOR WALL

1/2" GYPSUM WALL BOARD, BOTH FACES
2x4 STAGGERED STUDS @ 8" O.C.
2x6 BOTTOM & TOP PLATES
2" BATT INSULATION - 2 SIDES

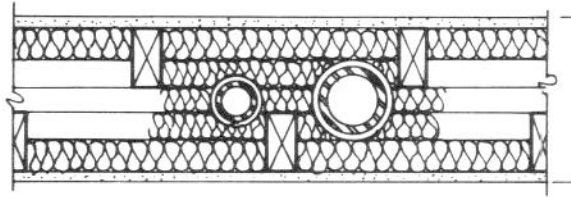

PLUMBING STACK WALL (AT PARTY/CORRIDOR WALLS)

1/2" GYPSUM WALL BOARD, BOTH FACES
2x4 STAGGERED STUDS @ 8" O.C. ON
2x4 BOTTOM & TOP PLATES
2" BATT INSULATION - BOTH WALLS
AND AROUND PIPES

INTERIOR WALL

1/2" GYPSUM WALL BOARD, BOTH FACES
2x4 STUDS @ 16" O.C.
2x4 BOTTOM & TOP PLATES
(2x6 INTR PLUMBING WALLS, INSULATED)

WALL DETAILS

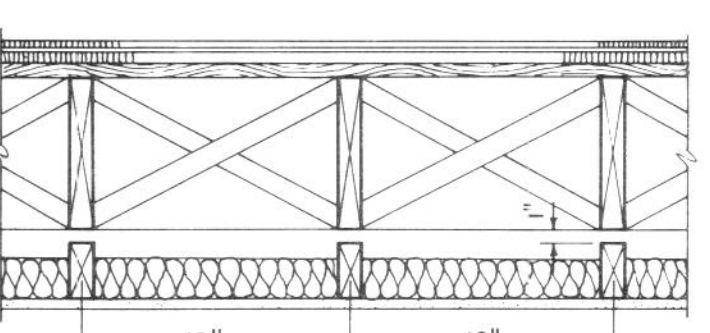

DETAIL OF FURRED DOWN CEILING
(REVISION: THIS CEILING ONLY IN BATHROOMS ON 2nd & 3rd FLOORS)

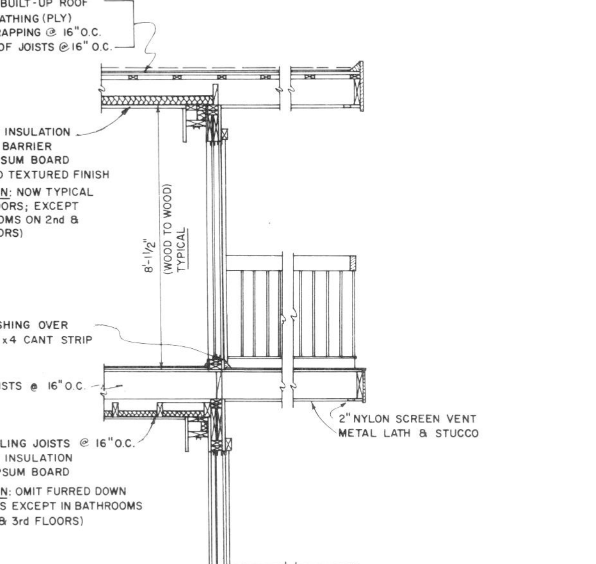
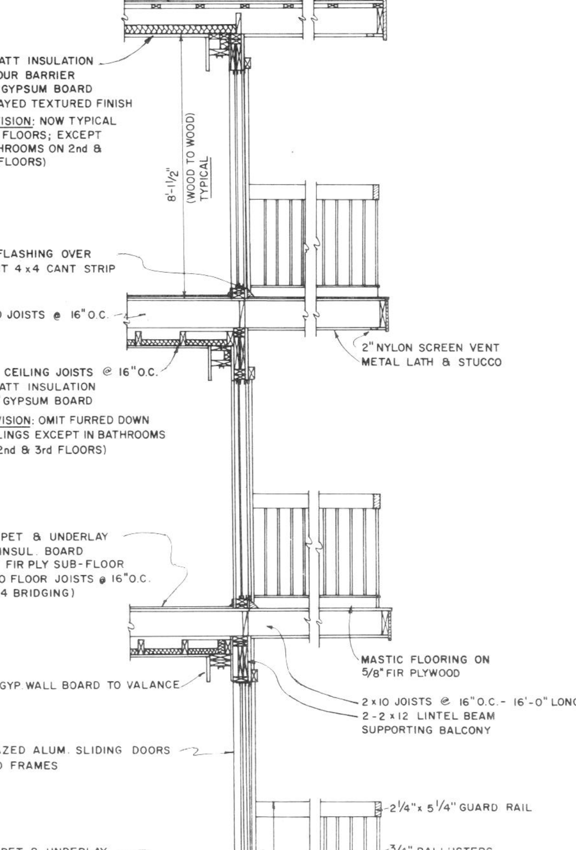

SECTION THROUGH BALCONIES

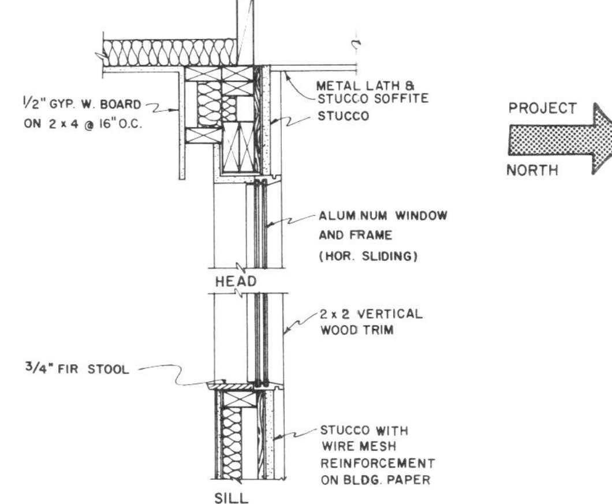

WINDOW FRAME DETAILS

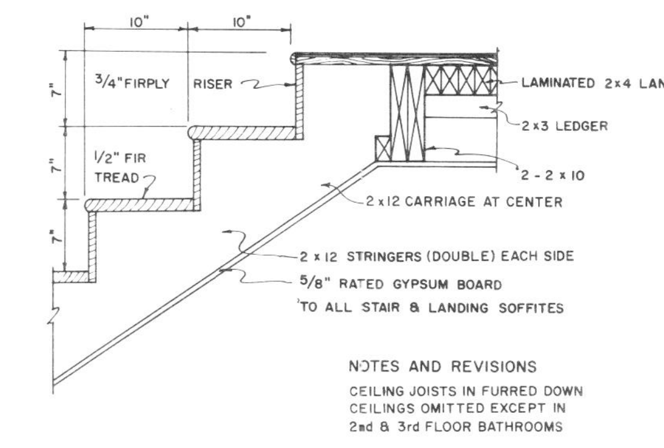

STAIR DETAIL
(FOR APARTMENT LEVELS)

NOTES AND REVISIONS

CEILING JOISTS IN FURRED DOWN CEILINGS OMITTED EXCEPT IN 2nd & 3rd FLOOR BATHROOMS

DROPPED SOFFITES AND WIDE FASCIAS NOW REMOVED FROM ROOF EAVES & FROM BALCONIES (AS SHOWN ON ELEVATIONS)

PROJECT NORTH

7 3/4"

FIRE WALL DETAIL

5/8" GYPSUM WALL BOARD, BOTH FACES
2x4 (FLAT) STUDS @ 16" O.C., BOTH SIDES
2" BATT INSULATION, BOTH SIDES
2-2x6 (FLAT) STUDS LAMINATED SOLID
2x4 BOTTOM & TOP PLATES (TO LAM. WALL & STRAPPINGS)

8